LIVING IN A
microbial world

SECOND EDITION

New to the Second Edition!
Living in a Microbial World includes free access to the Garland Science Learning System (**GSLS**) with purchase of a new book.

GSLS resources are designed to enhance learning and teaching by integrating the following tools ...

- Tutorials with assessment questions explore exciting topics and applications.
- Quizzes assess mastery of topics and provide feedback to students.
- Vocabulary review helps students with important microbiology terminology.
- Hints provided for the end-of-chapter concept questions in the book.
- Movies illuminate molecular processes and show cells in real life.
- The Instructor's Dashboard enables: selection of assignments to suit the course syllabus; customization of content; and tracking of student participation, performance, and understanding.
- Student data can be used to fine-tune lectures and tailor classroom discussion to student needs.

MASTER MICROBIOLOGY CONCEPTS WITH GARLAND SCIENCE LEARNING SYSTEM!

GS LS Garland Science LEARNING SYSTEM
powered by ROCKETMIX

INSTRUCTORS:

Start building your *Living in a Microbial World* course today ...

- Ask your bookstore to order ISBN 9780815346012 for students to receive free access to GSLS with purchase of the book.
- Go to *http://garlandscience.rocketmix.com* and register for access. Select "Instructor Access" from the main menu and submit the form.
- After registration is confirmed, you may access the course.
- The Instructor's Dashboard for each course contains an "Enrolment Link," which is a URL address unique to the course, that you provide to students.
- The link will take students to the registration portal for your course, and once they sign up, everyone will be ready to go!
- Students receive free access to your course via the redemption code on each copy of the book (ISBN 9780815346012). Students can also purchase access online when they register for the course, if they don't purchase a new book.

STUDENTS:

To get credit for your work, obtain the course "Enrolment Link" from *your instructor*. Following the instructor-provided Enrolment Link, register for the course and use the redemption code on the front of this book to access the system for free. If you don't have a new book, you may make a one-time payment with a credit card to access the system.

LIVING IN A
microbial world

SECOND EDITION

Bruce V. Hofkin

Garland Science
Taylor & Francis Group

LONDON AND NEW YORK

Garland Science
Vice President: Denise Schanck
Senior Editor: Elizabeth Owen
Editorial Assistant: Jordan Wearing
Production Editor: Deepa Divakaran
Illustrator: Matthew McClements, Blink Studio Ltd
Layout: NovaTechsetPvt Ltd
Cover Designer: Matthew McClements, Blink Studio Ltd
Copyeditor: John Murdzek
Proofreader: Sally Livitt
Indexer: Bill Johncocks

ISBNs: 9780815346012 (student edition with Garland Science Learning System)
9780815345145 (instructor copy)

Library of Congress Cataloging-in-Publication Data
Names: Hofkin, Bruce V., author.
Title: Living in a microbial world / Bruce V. Hofkin.
Description: Second edition. | New York, NY : Garland Science, Taylor & Francis Group, LLC., [2017]
Identifiers: LCCN 2016055120 | ISBN 9780815345145 (paperback)
Subjects: LCSH: Microorganisms--Textbooks. | Microbiology--Textbooks.
Classification: LCC QR41.2 .H638 2017 | DDC 579--dc23
LC record available at https://lccn.loc.gov/2016055120

Published by Garland Science, Taylor & Francis Group, LLC, an informa business,
711 Third Avenue, New York, NY 10017, USA, and 2 Park Square, Milton Park, Abingdon,
OX14 4RN, UK.

Printed in the United States of America

15 14 13 12 11 10 9 8 7 6 5 4 3 2 1

Garland Science
Taylor & Francis Group

Visit our website at http://www.garlandscience.com

SUSTAINABLE FORESTRY INITIATIVE

Certified Chain of Custody
Promoting Sustainable Forestry
www.sfiprogram.org
SFI-01681

This SFI Logo applies to the text stock only.

Preface

When the first edition of *Living in a Microbial World* was published several years ago, I felt the natural exhilaration and satisfaction that is familiar to anyone who has at last completed a complex and extensive project. Yet this feeling was tempered by the knowledge that like any life sciences textbook, it wouldn't be long before my text started to show its age. New developments in biology in general and microbiology in particular arise at a lightning pace, and writing a text that is both relevant and current is a challenging task indeed. This point was rammed home to me, within only a week or two of the first edition's publication; my Senior Editor, Mike Morales, sent me an article with a note saying that this article might prove useful for a second edition. The subject of the article instantly captured my attention: "fecal transplant." Although at the time, we were beginning to hear more and more about the importance of our microbiota, this was the first I had heard about this particular application and its potential to treat certain medical conditions. It certainly wasn't discussed in my book. Although I found the article interesting, I experienced a slight sinking feeling that my text was already dated.

Consequently, my primary task in preparing the second edition was to incorporate the new and often marvellous information that had come forward since the publication of the first edition. Every chapter was scrutinized to insure that while still providing the necessary basic information, any relevant, recent developments, especially those that change the way we think about microbes, were included. Where appropriate, examples used to illustrate key concepts were also updated, in order to provide the text with a fresh and current feel. To provide only a few examples, students are introduced in Chapter 1 to a new technique involving microorganisms in which damaged antiquities may be restored. Newly emerged pathogens such as the fungus causing white nose syndrome in bats, or the virus causing the death of baby elephants in zoos around the world are utilized in Chapters 4 and 5. In Chapter 7, students learn about quorum sensing in microorganisms, while Chapter 10 explores why coral reefs worldwide are in jeopardy and the role of symbiotic algae in this ongoing ecological crisis. In Chapter 11, readers learn more about the astonishing new findings regarding our normal microbiota, and how it might impact a remarkable range of host attributes. Chapter 14 traces the emergence and spread of Zika virus, while Chapter 15 investigates the technologies such as CRISPR/ Cas and metagenomics that are becoming increasingly important and commonplace in research laboratories.

The basic strategy of the second edition is faithful to that of the first: to present students who are new to microbiology with an understandable, scientifically meaningful, and relevant introduction to the world of microbiology. The second edition of *Living in a Microbial World* remains geared toward students taking a general microbiology or microbiology-themed course for non-science majors. The goal is still to present students with the many and often unexpected ways that microorganisms impact our daily lives, and to do so with a conversational writing style, that is not overly burdened with details. Thus the text strives to provide non-scientists with the information they need to make informed decisions about issues that affect themselves, their families, and their communities. Core topics like cell structure and function or metabolism, which have broad application to the life sciences in general, help students understand the living world in a more profound and rewarding manner.

In addition to an emphasis on how microorganisms affect us, for better or worse, in a myriad of ways, the text continues to emphasize the importance of evolutionary theory to biology in general and microbiology in particular.

The reader will find a chapter devoted to microbial evolution and evolutionary concepts are highlighted throughout the text. To streamline the text and to make it more accessible, certain traditional microbiology topics, which often receive their own chapters in other texts, have been integrated throughout other chapters. For instance, in lieu of a separate chapter on techniques, topics such as Gram and acid-fast staining are introduced in the discussion of cell wall structure.

Like the first edition, each chapter contains a number of case studies that introduce new concepts with news items, historical accounts or scenarios that are designed to engage the reader. For instance, a case added to the second edition introduces the topic of bioconversion with a case that may be of interest to those concerned with animal welfare: how an important steroid hormone used therapeutically can be produced by bacteria, rather than extracted from the urine of pregnant horses. As before, each case is followed by several questions, the answers to which should become apparent in the subsequent text section.

In these ways, I have attempted to provide readers with a meaningful and contemporary text, with which they may explore the world of microorganisms. It is my hope that after reading it, students will share my fascination with these endlessly astonishing and diverse living things.

I am excited that the second edition is able to make use of the Garland Science Learning System (GSLS) to provide additional resources that are designed to complement the book. GSLS will engage students through active learning, while instructors are able to tailor discussions to fit their class and monitor progress. There are specially written tutorials, quizzes, movies, vocabulary review, and help answering the concept questions. I encourage everyone to try it! Also new to this edition is a brief Study Skills section before the text begins, which is designed to help readers make the most of this textbook and its resources.

All textbooks are collaborative efforts, and many people have helped bring the second edition of *Living in a Microbial World* into existence. These people are recognized in the acknowledgments. Any remaining errors, however, are solely the responsibility of the author. Please help with these errors via contact at science@garland.com so that corrections can be made in subsequent printings.

Acknowledgments

No textbook is solely the work of the author whose name appears on the cover and *Living in a Microbial World*, second edition, is no exception. Microbiology educators and students from around the country reviewed drafts of each chapter, and I would like to thank them for their many helpful comments and suggestions. Specifically, I wish to thank:

Suzanne Anglehart, University of Wisconsin; Jason Arnold, Hopkinsville Community College; Linda Bruslind, Orgeon State University; Alyssa Bumbaugh, Penn State Altoona; Jean Cardinale, Alfred University; Edward Cluett, Ithaca College; David Fulford, Edinboro University of Pennsylvania; Eileen Gregory, Rollins College; Patrick Hindmarsh, Louisiana Tech University; Timothy Ladd, Millersville University; Juanita Leonhard, Illinois College; Sylvie Franke McDevitt, Skidmore College; Amy Medlock, University of Georgia; Robert Miller, Oklahoma State University; Heather Minges Wols, Columbia College Chicago; Roderick Morgan, Grand Valley State University; Karen Palin, Bates College; Carolyn Peters, Spoon River College; Greg Pryor, Francis Marion University; Rachel Robson, Morningside College; Mark Schneegurt, Wichita State University; Tammy Tobin, Susquehanna University; Jamie Welling, South Suburban College.

I am also grateful to the friends and colleagues with whom I had valuable conversations and from whom I received innumerable suggestions and food for thought as to how the text might be improved. In particular, I wish to thank Coenraad Adema, University of New Mexico; Sara Brant, University of New Mexico; Joseph Courtney, California Department of Public Health; Richard Cripps, University of New Mexico; Charlotte Kent, Centers for Disease Control and Prevention; and Eric Loker, University of New Mexico.

Many thanks also to the dedicated staff at Garland Science, who helped to make this second edition a reality. At the top of that list is Senior Editor Elizabeth Owen, whose assistance and guidance were crucial in seeing this project to fruition. Not only did Liz oversee the entire project, but her many insightful ideas and comments were instrumental in the updating and redrafting process. The editorial process, especially with regard to the artwork, was helped greatly by Editorial Assistant Jordan Wearing. Production Editor Deepa Divakaran and I worked closely together to produce the final pages. I am grateful for her professionalism, attention to detail and thoroughness, all of which contributed to the final product being of the highest standard. All new artwork for the second edition was produced by Matthew McClements of Blink Studio, who was also the artist for the first edition. As before, I am appreciative of Matt's special talent for rendering often-complex concepts into easy-to-understand and aesthetically appealing artwork. I offer my deep and sincere gratitude to all of these individuals.

And as with the first edition, I thank Leslie for being there.

INSTRUCTOR RESOURCES

Living in a Microbial World, second edition comes with a redemption code that gives students free access to original tutorials with assessments, quizzes, 24 movies designed to complement the book, vocabulary review, and hints to the end-of-chapter questions, all on the **Garland Science Learning System (GSLS)**. For students to receive a redemption code with free access to the GSLS, you must use the following ISBN when placing the book order: 9780815346012.

The Garland Science Learning System allows you to:

- Use the modules to stimulate class discussion and enhance student participation.

- Assign tutorials to track student understanding of microbiology concepts.

- Use the quizzes to gauge comprehension of textbook topics.

- Engage students' imaginations with 24 movies that bring microbiology to life.

- Customize content to match your course and suit your students' needs.

See page x to view the complete set of resources on the GSLS.

Accessing the GSLS modules is simple, just ...

1. Go to http://garlandscience.rocketmix.com and register. Select "Instructor Access" from the main menu and submit the form.

2. After registration is confirmed, you may access your course. At the top of the Instructor Dashboard you will see an "Enrolment Link," which is a URL address unique to your course that you provide students.

3. The link will take students to the registration portal for your course, and once they sign up, everyone will be ready to go!

4. The system is free to instructors. Students may access the system free of charge by using the redemption code on their textbook. If they have not purchased a new textbook with a code, they can access the GSLS by paying a fee.

OTHER INSTRUCTOR RESOURCES

Additional resources for instructors are available on the Garland Science website:
http://instructors.garlandscience.com

If you already have an approved instructor account, log in using your email and password. If you are new to Garland Science, select "Instructor Registration" and complete the form. Once you are verified as an instructor, you will be able to access the following resources.

Figures

The images from the book are available in two convenient formats, PowerPoint® and JPEG. Figures are searchable by figure number, figure name, or by keywords used in the legends.

Instructor Test Bank

Over 200 questions in a variety of formats: multiple-choice, matching, fill-in-the-blank, true/false, depth-of-understanding. Answers to all questions are supplied. Questions and answers are supplied as a Microsoft Word file and can be used in homework, examinations, and personal response systems (clickers).

Movies

The 24 movies include both animations that explain processes in detail and videos of biological processes. Each movie has a voice-over narration, and the text of the narration is located in the "Media Guide." Movies are provided in Windows and Macintosh-compatible formats for easy import into PowerPoint®. The movies are available to students in the GSLS modules.

Media Guide

This document overviews the multimedia package available for students and instructors. It also contains the text of the voice-over narration for all the movies.

STUDENT RESOURCES

Congratulations! Your textbook provides free access to the **Garland Science Learning System (GSLS),** a unique learning tool that will help you to succeed in your microbiology course.

It's easy to access the GSLS for your course:

1. Obtain the course "Enrolment Link" (a URL address unique to your course) from your instructor or the syllabus.

2. During registration, enter the redemption code from the scratch-off sticker on the front cover of the book.

3. If you haven't purchased a new book, you can access the modules by making a one-time payment with a credit card.

It is also possible to use the GSLS as an independent learner, if you don't have an instructor, or your instructor is not using the system. Information for accessing the modules independently is available at:
http://garlandscience.rocketmix.com/students

The following resources for *Living in a Microbial World* are available on the GSLS:

- Every chapter features a tutorial with assessment questions to explore exciting topics and applications.

- Quizzes assess mastery of topics and provide feedback.

- Twenty-four movies illuminate molecular processes and show cells in real life.

- Vocabulary review for each chapter helps with important microbiology terminology.

- Hints for End-of-Chapter Concept Questions.

See page x for a complete list of student resources on the GSLS.

Garland Science Learning System Contents

Chapter 1
- Tutorial: The Scientific Method (discover how to prove you're right scientifically and why a scientific theory isn't merely theoretical)
- Quiz with Feedback
- Hints for End-of-Chapter Concept Questions
- Vocabulary Review

Chapter 2
- Tutorial: Protein Structure and Denaturation (discover why a fried egg doesn't look like a raw egg and why it's important that proteins fold like socks in a clothes dryer)
- Movie: Glucose Molecule
- Quiz with Feedback
- Hints for End-of-Chapter Concept Questions
- Vocabulary Review

Chapter 3
- Tutorial: Structure and Clinical Significance of Endospores (discover why anthrax is an ideal biological weapon)
- Movie: Bacterial Flagellum Action
- Movie: Fluidity of the Lipid Bilayer
- Movie: Structure of Lipids and the Lipid Bilayer
- Movie: Phagocytosis
- Quiz with Feedback
- Hints for End-of-Chapter Concept Questions
- Vocabulary Review

Chapter 4
- Tutorial: Naming Organisms Scientifically (discover the scientist's answer to Juliet's question: What's in a name?)
- Quiz with Feedback
- Hints for End-of-Chapter Concept Questions
- Vocabulary Review

Chapter 5
- Tutorial: Viral Biosynthesis (investigate viral factories and their assembly lines)
- Movie: DNA Virus Replication
- Movie: RNA virus (+ strand) Replication
- Movie: RNA virus (– strand) Replication
- Movie: HIV Infection
- Quiz with Feedback
- Hints for End-of-Chapter Concept Questions
- Vocabulary Review

Chapter 6
- Tutorial: Koch's Postulates (discover how self-experimentation and a peptic ulcer won the Nobel Prize for Medicine)

- Quiz with Feedback
- Hints for End-of-Chapter Concept Questions
- Vocabulary Review

Chapter 7
- Tutorial: Regulation of Gene Expression (discover how bacteria turn genes on and off)
- Movie: DNA Molecule
- Movie: DNA Replication
- Movie: Translation
- Movie: Conjugation
- Quiz with Feedback
- Hints for End-of-Chapter Concept Questions
- Vocabulary Review

Chapter 8
- Tutorial: Selective and Differential Culture Media (discover how you can tell if a bacterium is a vampire)
- Movie: ATP Molecule
- Movie: Glycolysis
- Movie: ATP Synthase
- Quiz with Feedback
- Hints for End-of-Chapter Concept Questions
- Vocabulary Review

Chapter 9
- Tutorial: Natural Selection (discover why the outbreak of cholera was worse in Ecuador than in Chile)
- Quiz with Feedback
- Hints for End-of-Chapter Concept Questions
- Vocabulary Review

Chapter 10
- Tutorial: Microbes in the Environment (discover what microbes have done for you)
- Quiz with Feedback
- Hints for End-of-Chapter Concept Questions
- Vocabulary Review

Chapter 11
- Tutorial: Exotoxins and Endotoxins (investigate bacterial weapons and discover how horseshoe crabs prevent hospital infections)
- Movie: Cholera Exotoxin Mechanism
- Movie: Intracellular *Listeria* infection
- Quiz with Feedback
- Hints for End-of-Chapter Concept Questions
- Vocabulary Review

Chapter 12

- Tutorial: Adaptive Immune Response (meet the superheroes of the immune system)
- Movie: Neutrophils Moving Toward an Attractant
- Movie: Neutrophil Chasing a Bacterium
- Movie: Lymphocytes Moving to a Site of Injury
- Movie: Immunoglobulins
- Quiz with Feedback
- Hints for End-of-Chapter Concept Questions
- Vocabulary review

Chapter 13

- Tutorial: Antibiotic Targets (learn how magic bullets work)
- Quiz with Feedback
- Hints for End-of-Chapter Concept Questions
- Vocabulary Review

Chapter 14

- Tutorial: Antigenic Shift and the Flu (discover why flu is a problem year after year after year)
- Movie: Antigenic Drift
- Movie: Antigenic Shift
- Quiz with Feedback
- Hints for End-of-Chapter Concept Questions
- Vocabulary Review

Chapter 15

- Tutorial: Metagenomics (discover how you know bacteria are there if you can't see them and can't grow them)
- Quiz with Feedback
- Hints for End-of-Chapter Concept Questions
- Vocabulary Review

Chapter 16

- Tutorial: Fermentation and Food (discover why a lack of oxygen can be a good thing, especially if you like beer, wine, yogurt, cheese, or bread)
- Quiz with Feedback
- Hints for End-of-Chapter Concept Questions
- Vocabulary Review

Chapter 17

- Tutorial: Microbial Biopesticides (discover how to grow crops with added pesticides built in)
- Quiz with Feedback
- Help Answering the Concept Questions
- Vocabulary Review

Case Studies

Chapter 1
- Beauty and the bacteria 7
- Fleming revisited 10

Chapter 2
- Bacteria, salt, and cystic fibrosis 21
- Stomach pain and stomach bugs 26
- Bacterial horse helpers 29
- Bad potatoes 33
- Call in the clot busters 36

Chapter 3
- Size does matter! 44
- Plaque attack 46
- An occupational hazard 50
- The real Jurassic Park 57
- Parrot fever 63

Chapter 4
- Streams, snails, and schistosomes 70
- The heat was on! 74
- The cat's out of the bag on a protozoan parasite 79
- Bad news for bats 86

Chapter 5
- "Death" in the Rue Morgue 92
- Last laugh for the "laughing death" 105

Chapter 6
- The Spanish conquest of Mexico 110
- The assassination of President Garfield 122

Chapter 7
- DNA: form suggests function 138
- Christmas trees and transcription 142
- Protein synthesis: dead in its tracks 146
- Griffith's transforming factor 155
- End of the line for a last-line defense? 159

Chapter 8
- No fish for you! 176
- Driving under the influence (of yeast!) 178
- Canine first aid 186
- Making yogurt 193

Chapter 9
- Cells within cells 204
- Vancomycin resistance: The sequel 212
- A tale of two countries – Australia, England, rabbits, and the myxoma virus 214
- Alice in Wonderland 218

Chapter 10
- Water birds and botulism 230
- Lake Erie: A near-death experience 233
- The key to a bountiful harvest 241
- Coral reefs and Darwin's paradox 245
- A microscopic dog-eat-dog world 247

Chapter 11
- Sayonara *Salmonella* 252
- Shy mouse vs courageous mouse 255
- Laid low on the high seas 261
- Trouble in paradise 266
- A ruptured appendix 269

Chapter 12
- Influenza: exposed! 278
- Rob 279
- Laura 281
- Bonnie 295
- Oscar 299

Chapter 13
- An army marches on its stomach 308
- Hand washing 314
- A mystery illness 319
- The importance of completing prescriptions 330

Chapter 14
- Food-borne misery 340
- Tracking down a feathered felon 342
- Flu season 346
- A new bug on the block 350
- The anthrax scare 360

Chapter 15
- TB—BC (Before Columbus, that is) 365
- *E. coli's* dark side 368
- Biological blackmail 371
- The monarch butterfly: threats real and imagined 382

Chapter 16
- Flight from Egypt 390
- What's for tucker, mate? 393
- Homemade wine 396
- Vinegar's illustrious past 399
- Microbial metabolites for a feathered friend 403

Chapter 17
- Unearthing a new antibiotic 412
- Menopause, mares, and microbes 418
- The sweet smell of success 421
- Waste reduction begins at home 426

Contents

Learning Skills		**xxiii**
Chapter 1	Living in a Microbial World	1
Chapter 2	The Chemistry of Life	17
Chapter 3	The Cell: Where Life Begins	41
Chapter 4	A Field Guide to the Microorganisms	67
Chapter 5	Life's Gray Zone: Viruses and Prions	91
Chapter 6	The Microbiology of History and the History of Microbiology	109
Chapter 7	Microbial Genetics	135
Chapter 8	Metabolism and Growth	165
Chapter 9	Microbial Evolution: The Origin and Diversity of Life	197
Chapter 10	A Microbiologist's Guide to Ecology	221
Chapter 11	The Nature of Disease Part I: A Pathogen's Perspective	251
Chapter 12	The Nature of Disease Part II: Host Defense	277
Chapter 13	Control of Microbial Growth	307
Chapter 14	Epidemiology: Who, What, When, Where, and Why?	337
Chapter 15	The Future is Here: Microorganisms and Biotechnology	363
Chapter 16	Guess Who's Coming to Dinner: Microorganisms and Food	389
Chapter 17	Better Living with Microorganisms: Industrial and Applied Microbiology	411
	Glossary	433
	Figure acknowledgments	447
	Index	453

Detailed Contents

Learning Skills   xxiii

Chapter 1 Living in a Microbial World   1

All living things are composed of one or more cells   3

All living things display other observable characteristics   4

Microbiology involves the study of several distinct groups of living things   4

Viruses strain our notion of what it means to be alive   7

Microbiology is closely intertwined with the study of non-microorganisms   7

Microbiology is composed of both basic and applied components   7

Case: Beauty and the bacteria   7

Both basic and applied microbiology consist of many subdisciplines   9

A proper scientific experiment involves a series of well-defined steps   10

Case: Fleming revisited   10

If a hypothesis cannot be disproved, it may eventually become a theory   13

Looking back and looking forward   14

Garland Science Learning System   14

Concept questions   14

Chapter 2 The Chemistry of Life   17

Atoms are the basic building blocks of matter   18

Atoms are made up of smaller components called subatomic particles   18

As an atom's stability increases, its energy decreases   19

An ionic bond is formed when electrons are transferred from one atom to another   21

Case: Bacteria, salt, and cystic fibrosis   21

Covalent bonds form when atoms share electrons in their outermost shells   22

Covalent bonds can be classified as either polar or nonpolar   23

Water is essential in life processes   25

Acids and bases increase or decrease the concentration of protons in water   26

Case: Stomach pain and stomach bugs   26

Acidity is measured with the pH scale   28

Organic molecules are the building blocks of life   28

Carbohydrates function as energy storage and structural molecules   29

Case: Bacterial horse helpers   29

Lipids are relatively hydrophobic molecules and are also used for energy storage and structure   31

Proteins participate in a variety of crucial biological processes   33

Case: Bad potatoes   33

Biological reactions require enzymes functioning as catalysts   36

Case: Call in the clot busters   36

Nucleic acids direct the assembly of proteins   37

Looking back and looking forward   39

Garland Science Learning System   40

Concept questions   40

Chapter 3 The Cell: Where Life Begins   41

All cells are one of two basic types   42

Cells can be stained for enhanced observation   44

As cells get larger, their efficiency decreases   44

Case: Size does matter!   44

Prokaryotic cells have a variety of shapes   45

Many prokaryotes have extracellular structures, extending beyond the cell wall   46

Case: Plaque attack   46

Many bacteria are surrounded by an excreted layer of polysaccharides   47

Structures composed of protein may extend out from the bacterial cell   48

Most prokaryotes are protected from the exterior environment by a rigid cell wall   49

Case: An occupational hazard   50

The Gram-positive cell wall is composed of multiple layers of carbohydrates, linked by peptides   50

Gram-negative cells are surrounded by two membranes with a thin peptidoglycan layer in between   51

Acid-fast bacteria are surrounded by a thick, waxy cell wall   53

Bacteria with different cell wall types can be distinguished by specific staining techniques   53

Cell walls surround the cells of most archaea   54

Some prokaryotes lack a cell wall   54

Each cell is surrounded by a plasma membrane   55

The plasma membrane of archaea is often especially stable   56

The liquid interior of the cell forms the cell's cytoplasm   56

The prokaryotic chromosome is usually circular   56

Proteins are synthesized on the surface of ribosomes   57

Storage granules and inclusion bodies have specialized functions within cells   57

Endospores permit survival during times of environmental adversity   57

Case: The real Jurassic Park   57

Eukaryotic cells are larger and more complex than prokaryotic cells   59

The number of cells in a eukaryotic organism is variable   59

While plant and fungal cells have cell walls, animal cells do not   60

Like prokaryotes, all eukaryotic cells are surrounded by a membrane   60

The cytoplasm of eukaryotic cells contains a variety of membrane-bound organelles   61

DNA is found in the nucleus of eukaryotic cells   61

The endomembrane system functions in cellular transport   62

Case: Parrot fever   63

Chloroplasts and mitochondria function in the production and utilization of energy by the cell   64

Looking forward and looking back   65

Garland Science Learning System   65

Concept questions   66

Chapter 4 A Field Guide to the Microorganisms   67

Taxonomy is the science of biological classification   67

Taxonomy is based on a system of hierarchical groupings   68

Modern classification reflects evolutionary relatedness   68

DNA analysis provides evidence for relatedness   70

Case: Streams, snails, and schistosomes   70

Kingdoms are organized among three domains   72

Microorganisms are found in all three domains of life   73

The domain Bacteria consists of numerous lineages   73

The earliest bacteria were adapted to life on the primitive Earth   74

Case: The heat was on!   74

The tree spreads out   75

The most recently evolved lineages are found at the farthest branches of the tree   76

Although they are prokaryotic, members of domain Archaea are distinct from bacteria   76

Eukaryotic microorganisms include the protozoa and the fungi   78

Protozoa are single-celled organisms within the domain Eukarya   79

Case: The cat's out of the bag on a protozoan parasite   79

Flagella are used by the flagellates for locomotion   80

Amebae move as they extend parts of their cytoplasm   80

The beating of numerous cilia permits locomotion in the ciliates   80

Apicomplexa typically do not possess structures related to locomotion   81

Many fungi are involved in disease or ecological processes or are useful for industrial purposes   82

Fungi have one of two body plans and can be unicellular or multicellular   83

The Chytridiomycota represent a primitive phylum of fungi   84

Zygomycota often form mutualistic relationships with plants   85

Like the Zygomycota, many Basidiomycota are important plant mutualists   85

While many are decomposers, other members of the Ascomycota cause serious disease   86

Case: Bad news for bats   86

Looking back and looking forward   88

Garland Science Learning System   88

Concept questions   88

Chapter 5 Life's Gray Zone: Viruses and Prions   91

Viruses are acellular parasites, requiring a host cell in which to replicate   92

Case: "Death" in the Rue Morgue   92

Most viruses have one of a few basic structures   93

Specific viruses usually infect only certain hosts and certain cells within those hosts   94

Replication of animal viruses proceeds through a series of defined steps   95

Viruses must first attach to and enter an appropriate host cell   95

Once they have entered the cell, viruses release their protein coat prior to replication   97

The final two stages are the assembly of new virions and their release from the host cell   97

DNA viruses must replicate their DNA and use it to direct synthesis of viral proteins   98

RNA viruses rely on RNA rather than DNA as their genetic material   99

Retroviruses use their RNA to produce DNA   99

Not all viral infections cause symptoms or kill host cells   100

Viruses damage host cells in several ways   101

Like animals, plants are susceptible to many viral infections   102

Bacteriophages are viruses that infect bacteria   102

Phages can influence the number of bacteria and even the diseases that they cause   104

Prions are infectious proteins   105

Case: Last laugh for the "laughing death"   105

Looking back and looking forward   106

Garland Science Learning System   108

Concept questions   108

Chapter 6 The Microbiology of History and the History of Microbiology   109

If microbiologists wrote history...   110

Case: The Spanish conquest of Mexico   110

Disease turned the tide in the Peloponnesian War between Athens and Sparta   110

The fall of the Roman Empire was accelerated by several disease outbreaks   112

The Black Death altered European society forever   113

European explorers imported many diseases into the New World   114

Unlike the Americas, Africa was not fully colonized until the 1900s   116

Several important events of the nineteenth century owe their outcomes to microorganisms   117

Napoleon's dreams of empire were compromised by disease   117

Irish history was irrevocably changed by a disease affecting potatoes   118

French efforts to build a canal in Panama were thwarted by mosquito-borne disease   119

A bacterial metabolic process aided the British navy in World War I   120

Polish doctors spared their villages from World War II's devastation with microbial assistance   121

Some infectious disease is caused by non-microorganisms   121

The science of microbiology has a vibrant history   122

Case: The assassination of President Garfield   122

Microorganisms were first discovered in the seventeenth century   124

The science of microbiology was born in the second half of the nineteenth century   124

The germ theory of disease was advanced by the understanding of the need for hospital sanitation   126

The development of pure culture technique allowed rapid advances in microbiology   127

Koch was the first to link a specific microorganism to a particular disease   127

The germ theory led to major advances in disease control   128

With the development of the first vaccines, the field of immunology was born   129

Antimicrobial drugs are a twentieth century innovation   130

Into the modern era   131

Looking back and looking forward   132

Garland Science Learning System   133

Concept questions   133

Chapter 7 Microbial Genetics   135

DNA is an information molecule   135

The structure of DNA is the key to how it functions   136

DNA is found on chromosomes   137

Each DNA molecule can be accurately copied into two new DNA molecules   138

Case: DNA: Form suggests function   138

DNA replication begins at a site called the origin   140

A cell's characteristics are largely determined by the proteins it produces   140

In transcription, a gene's DNA is used to produce complementary RNA   142

Case: Christmas trees and transcription   142

While very similar, transcription in eukaryotes varies from that of bacteria   144

Amino acids are assembled into protein during translation   145

Case: Protein synthesis: Dead in its tracks   146

Gene activity is often carefully regulated   149

Many bacterial genes are regulated simultaneously in groups of genes called operons   150

Some operons are regulated through repression   151

Bacteria often coordinate their gene expression through chemical communication   151

Eukaryotes use various mechanisms to regulate gene expression   152

The environment influences the nature of genetically determined characteristics   152

Genetic variation can arise through diverse mechanisms   153

Sexual reproduction permits the generation of new genetic combinations   153

Horizontal gene transfer allows for genetic recombination in asexually reproducing organisms   154

Case: Griffith's transforming factor   155

Transformation relies on the ability of some bacteria to absorb DNA from the environment   156

Viruses may shuttle genes between bacteria   156

In conjugation, small portions of DNA are transferred directly from one cell to another   157

Transposons are short DNA sequences that can move from one site in the genome to another   157

Genetic recombination in prokaryotes has great significance for humans   158

Mutations are the original source of genetic variation   159

Case: End of the line for a last-line defense?   159

Mutations are caused by many factors   161

Many mutations are repaired before they can affect phenotype   162

Mutations are the raw material of evolution   163

Looking back and looking forward   163

Garland Science Learning System   164

Concept questions   164

Chapter 8 Metabolism and Growth   165

Metabolism is similar in all living things   166

Energy released from food molecules is used for other processes that require energy   166

Cells convert the energy in biological molecules into ATP   167

Energy is transferred from one molecule to another via oxidation and reduction   168

Biological molecules are typically oxidized in a series of small steps   169

Cell respiration refers to the oxidation of biological molecules   170

Carbohydrates are the primary source of energy used in cell respiration   170

Cell respiration occurs in three stages   171

In glycolysis, glucose is partially oxidized, forming two smaller molecules called pyruvate   172

In the Krebs cycle, pyruvate from glycolysis is completely oxidized and the energy released is transferred as electrons to NAD^+ and FAD   173

$NADH$ and $FADH_2$ are oxidized in electron transport, providing the energy for ATP synthesis   174

Different microorganisms may utilize different final electron acceptors   176

Case: No fish for you!   176

Protons that are pumped across a membrane in electron transport provide an energy source for ATP synthesis   177

Incomplete glucose oxidation results in fermentation   178

Case: Driving under the influence (of yeast!)   178

Molecules other than glucose can be used to generate ATP   180

Autotrophs produce their own biological molecules   180

In photosynthesis, energy in sunlight is used to produce biological molecules   181

Chemoautotrophs use chemical energy in the way that photoautotrophs use solar energy   183

Microbial metabolism influences growth rate   184

Microbial growth refers to population growth   184

Environmental factors influence microbial growth rate   186

Oxygen affects microbial growth in various ways   186

Case: Canine first aid   186

Like oxygen, temperature has a major impact on metabolism and growth   188

Microorganisms have an optimal environmental pH   190

The growth of most, but not all, microorganisms is inhibited by high salt concentration   190

Like all living things, microbes require certain nutrients to grow efficiently   191

Microbial populations pass through a sequence of phases called the growth curve   192

As the environment changes, microbial metabolism and therefore the growth curve changes in response   193

Case: Making yogurt   193

Looking back and looking forward   195

Garland Science Learning System   195

Concept questions   196

Chapter 9 Microbial Evolution: The Origin and Diversity of Life   197

Conditions on the early Earth were very different than they are today   198

The first biological molecules were formed from nonbiological precursors   198

Genetic information may have originally been encoded in RNA instead of DNA   200

The first cells required a membrane and genetic material   201

The first prokaryotes are thought to have arisen approximately 3.5 billion years ago   202

Certain important metabolic pathways evolved in a defined sequence   203

Eukaryotes evolved from prokaryotic ancestors   204

Case: Cells within cells   204

Eukaryotic cells first arose from certain archaeal ancestors   206

Multicellular life arose from colonies of unicellular eukaryotes   207

The origin of viruses is obscure   208

Prions may have arisen from abnormal host proteins   209

Evolution explains life's diversity   209

Natural selection is the driving force of evolution   210

Microorganisms are subject to the laws of natural selection   212

Case: Vancomycin resistance: The sequel   212

Natural selection can influence the virulence of disease-causing organisms   214

Case: A tale of two countries—Australia, England, rabbits, and the myxoma virus   214

Microorganisms often influence the evolution of their hosts   216

Morning sickness may have evolved to protect the developing fetus   216

Prions may have affected evolution of the brain   217

Bacteria have influenced the formation of insect species   217

Sex may have evolved in response to selection pressure imposed by pathogens   218

Case: Alice in Wonderland   218

Looking back and looking forward   219

Garland Science Learning System   220

Concept questions   220

Chapter 10 A Microbiologist's Guide to Ecology   221

Ecology is the study of how living things interact with each other and the environment   222

Energy and nutrients are passed between organisms in an ecosystem   222

Nutrients are recycled in an ecosystem, whereas energy is not   224

Microbial ecology investigates the role of microorganisms in their environment   225

Microorganisms live in microenvironments   225

Environmental conditions affect the growth rate of microorganisms   225

Many microorganisms live in biofilms   226

Soil often harbors rich microbial communities   228

Microorganisms are found deep below Earth's surface and even inside solid rock   229

Many microorganisms are adapted to life in freshwater   230

Many lakes have several distinct ecological zones   230

Case: Water birds and botulism   230

Rivers and streams have their own unique ecological conditions   233

Water pollution can lead to severe oxygen depletion   233

Case: Lake Erie: A near-death experience   233

Although similar in many ways, the marine environment is distinct from freshwater   234

The littoral zone is typically richer in nutrients than the pelagic zone   235

Microbes in the deep sea and benthic zone require specialized adaptations   237

Cloud-dwelling bacteria may be important in promoting rainfall   238

Exobiology is the search for extraterrestrial life   238

Microorganisms are primary driving forces behind biogeochemical cycles   238

Carbon moves between living things and the environment   239

"Iron fertilization" has been suggested as a way to decrease atmospheric CO_2 levels   240

Bacteria convert nitrogen into forms that plants can absorb   241

Case: The key to a bountiful harvest   241

Like other cycles, the phosphorus cycle relies on microbial activity   243

Both organisms in a mutualism benefit from the relationship   245

Case: Coral reefs and Darwin's paradox   245

In a commensal relationship one organism is benefited while the other is unaffected   246

Microorganisms in the same environment may compete for certain resources   247

Case: A microscopic dog-eat-dog world   247

Some microorganisms are predators   248

Looking back and looking forward   249

Garland Science Learning System   249

Concept questions   249

Chapter 11 The Nature of Disease Part I: A Pathogen's Perspective 251

Contact with microorganisms only rarely results in disease 252

Hosts are colonized with normal microbiota that is usually harmless 252

Case: Sayonara Salmonella 252

The microbiota may impact many other host processes 255

Case: Shy mouse vs courageous mouse 255

To cause disease, a pathogen must achieve several objectives 256

Reservoirs provide a place for pathogens to persist before and after an infection 257

Many pathogens use animal reservoirs 257

Some pathogens rely at least in part on human reservoirs 258

Some pathogens can survive indefinitely in soil, water, or other environmental reservoirs 259

The type of reservoir used by a pathogen has implications for disease control 260

Pathogens must reach a new host via one or more modes of transmission 261

Case: Laid low on the high seas 261

Contact transmission may be direct or indirect 261

Many pathogens can move through the air in an aerosol released by an infected individual 261

Intestinal pathogens most commonly rely on food- or water-borne transmission 262

Vector transmission relies on insects or other invertebrates to transmit infectious organisms 263

Some pathogens can pass from a female to her offspring 263

Pathogens gain access to the host through a portal of entry 263

Once they have entered, pathogens must adhere to the host 264

Most pathogens must increase in number before they cause disease 264

Hosts can be infected with multiple pathogens at the same time 265

Successful pathogens must at least initially evade host defenses 265

Pathology can be the result of the host's immune response 265

Bacterial pathogens cause disease in several different ways 266

Some pathogenic bacteria produce toxins 266

Case: Trouble in paradise 266

Exotoxins are bacterial proteins that interfere with cellular processes in specific ways 267

Endotoxins are components of the Gram-negative outer membrane 269

Case: A ruptured appendix 269

Some bacteria produce enzymes with pathogenic properties 270

Bacterial virulence is sometimes regulated by quorum sensing 270

Viruses cause disease by interfering with the normal activities of the cells they infect 271

Different eukaryotic parasites affect their hosts in diverse ways 272

Pathogens leave the host through a portal of exit 273

Symptoms of disease often assist the pathogen in its transmission to a new host 274

Looking back and looking forward 274

Garland Science Learning System 274

Concept questions 275

Chapter 12 The Nature of Disease Part II: Host Defense 277

When infection cannot be blocked, first innate immunity and then adaptive immunity is activated 277

Case: Influenza: exposed! 278

Plan A: Barriers to entry may prevent infection before it occurs 279

Case: Rob 279

Barriers to entry do not always work 281

Case: Laura 281

If barriers to entry fail, the pathogen is confronted by the cells and molecules of innate immunity 282

Inflammation and fever are key components of innate immunity 283

Special chemical messengers help combat viral infections 283

When innate mechanisms fail to eliminate an infection, the adaptive immune response is activated 284

Antigen-presenting cells activate those cells responsible for adaptive immunity 285

Antigen-presenting cells migrate to lymphatic organs to activate adaptive immunity 286

Antigen-presenting cells activate helper T cells to initiate an adaptive response 287

Some activated helper T cells activate cytotoxic T cells to initiate a cell-mediated response 289

Helper T cells also activate B cells to initiate a humoral immune response 290

In a humoral response, several different classes of antibodies may be produced 292

A successful adaptive response culminates in the elimination of the pathogen 293

An adaptive response is not always successful   294

Case: Bonnie   295

Subsequent exposure to the same pathogen results in a stronger and faster adaptive response   295

Vaccines induce immunological memory without causing disease   297

HIV is an ongoing public health emergency   299

Case: Oscar   299

HIV causes immune system suppression   300

Host versus pathogen: a summary   302

Looking back and looking forward   303

Garland Science Learning System   304

Concept questions   304

Chapter 13 Control of Microbial Growth   307

Physical control of microorganisms involves manipulation of environmental factors   308

Temperature is an effective means to control microbial growth   308

Case: An army marches on its stomach   308

Liquids that cannot tolerate elevated temperatures may be filtered   311

Radiation can be used to kill microorganisms in some situations   311

Certain foods are preserved by drying   312

Chemical methods can control microorganisms on living and nonliving material   313

Various chemicals have antiseptic and disinfectant properties   313

Soaps and detergents remove microorganisms from surfaces   314

Case: Hand washing   314

Heavy metals are toxic to cells   315

Antimicrobial chemotherapy has led to major improvements in health   315

An ideal antimicrobial drug inhibits microorganisms without harming the host   316

Antibiotics are antimicrobial compounds that were initially produced by microorganisms   317

Many factors influence the selection of the most appropriate antibiotic in a given situation   317

Some drugs work faster than others   318

Antibiotics are either bactericidal or bacteriostatic   318

Drugs differ in their spectra of activity   319

Case: A mystery illness   319

Drug selection may be based on where in the body an infection has occurred   320

A drug's possible side effects are always an important consideration   321

Drug resistance is a serious and growing problem   321

Antibiotics work by interfering with specific bacterial structures or enzymes   321

Many antibiotics inhibit cell wall synthesis   322

Antiribosomal drugs inhibit bacterial protein synthesis   323

Some antibiotics target bacterial DNA   324

Selective toxicity, while possible, is harder to achieve against eukaryotic pathogens   324

Antiviral drugs must interfere with a particular step in the viral replicative cycle   326

Acyclovir acts by interfering with viral DNA replication   326

Other viral infections are now amenable to antiviral drug therapy   327

Newer, innovative antiviral drugs are on the horizon   329

Antibiotic misuse has led to the problem of drug resistance   330

Case: The importance of completing prescriptions   330

Failure to complete a course of antibiotics contributes to resistance   331

Other forms of antibiotic misuse contribute to resistance   332

Antibiotic resistance can result in superinfections   333

New strategies provide options for circumventing antibiotic resistance   334

Looking back and looking forward   335

Garland Science Learning System   336

Concept questions   336

Chapter 14 Epidemiology: Who, What, When, Where, and Why?   337

The first modern epidemiological study identified cholera as a water-borne disease   338

Florence Nightingale found that improved hygiene reduced the likelihood of typhus   339

An epidemic is a sudden increase in the number of cases of a specific disease   340

A single contaminated site can give rise to a common source epidemic   340

Case: Food-borne misery   340

Host-to-host epidemics spread from infected to uninfected individuals   342

Epidemics can occur for biological, environmental, and/or social reasons   342

Case: Tracking down a feathered felon   342

Epidemics become more likely when fewer people in a population are resistant   343

Influenza epidemics are a regular occurrence   346

Case: Flu season   346

Influenza is an RNA virus with a segmented genome   346

Influenza virus undergoes regular genetic change   347

Genetic drift and shift set the stage for influenza epidemics and pandemics   348

Epidemiological investigations may reveal the causes of new disease outbreaks   349

Case: A new bug on the block   350

A case definition helps health authorities determine if unusual cases are related   350

Time, place, and personal characteristics of a new disease provide clues to the disease's identity   351

Case–control studies can pinpoint a common risk factor among affected individuals   352

Emerging and re-emerging diseases are new or changing diseases that are increasing in importance   353

Zika virus illustrates several important points about disease emergence   354

Emerging diseases can be categorized as one of four basic types   358

Bioterrorism is an ongoing threat   360

Case: The anthrax scare   360

Looking back and looking forward   361

Garland Science Learning System   361

Concept questions   362

Chapter 15 The Future Is Here: Microorganisms and Biotechnology   363

The genetic revolution began through an understanding of microbial genetic exchange mechanisms   363

Sequencing techniques can be used to reveal the sequence of nucleotides in a DNA sample   364

Case: TB—BC (before Columbus, that is)   365

Large amounts of specific DNA sequences can be obtained with the polymerase chain reaction   365

Newer forms of PCR have been developed for specific purposes   366

Gel electrophoresis can be used to determine if a PCR was successful   366

Amplified DNA can be sequenced   367

The entire genomes of many organisms have been sequenced   368

Case: E. Coli's dark side   368

Genetic material taken directly from environmental samples can be sequenced   370

Foreign genes can be introduced into the genomes of different organisms   371

Case: Biological blackmail   371

Bacterial restriction enzymes are useful for cutting DNA at specific sites   372

Genes of interest can be cloned by inserting them into bacteria   373

Other types of cloning vectors may be used in certain situations   376

Powerful new tools for altering genes are becoming available   377

DNA technology has provided a better understanding of genes and how they function   378

DNA technology has numerous medical applications   379

Vaccines may now be created using recombinant DNA technology   380

Genetic conditions in humans may one day be corrected with gene-editing techniques   381

Genetically modified crops are now a reality   382

Case: The monarch butterfly: threats real and imagined   382

Biotechnology has become routine in many industrial processes   384

Biotechnology raises important ethical and safety concerns   385

Looking back and looking forward   386

Garland Science Learning System   387

Concept questions   387

Chapter 16 Guess Who's Coming to Dinner: Microorganisms and Food   389

The beginnings of a beautiful friendship   390

Case: Flight from Egypt   390

Microbial activity can help to preserve the quality of some foods   390

Fermented dairy products and grains have been used for thousands of years   391

Some fungi and bacteria are consumed directly as food   393

Case: What's for tucker, mate?   393

In the absence of oxygen, some microorganisms undergo fermentation, releasing specific waste products   395

Flour from grains provide a sugar source in bread-making   395

Wine is produced from fermented fruit juice   396

Case: Homemade wine   396

Beer making relies on the fermentation of grains   397

Various other plant materials can be used to produce liquors   398

A long list of other plant products are fermented to produce popular food items   399

Microbes produce various products used to season or add flavor to food   399

Case: Vinegar's illustrious past   399

Fermented milk is the basis of making cheese and yogurt   401

Certain meat products require fermentation   403

Bacteria help to prepare coffee beans for roasting   403

Microbes produce many vitamin supplements   403

Case: Microbial metabolites for a feathered friend   403

Amino acids are commercially produced with microbial assistance   404

Certain food items may contribute to a healthy microbiota   404

Time to eat: a few microbial recipes   405

Looking back and looking forward   409

Garland Science Learning System   409

Concept questions   410

Chapter 17 Better Living with Microorganisms: Industrial and Applied Microbiology   411

Microorganisms have numerous commercial applications   412

Case: Unearthing a new antibiotic   412

Many microorganisms produce metabolites with commercial potential   413

Microorganisms producing promising metabolites usually must undergo strain improvement   414

Potentially valuable microbes must grow well in an industrial setting and must not pose undue risks to humans or the environment   415

A defined series of steps are followed to move production from the laboratory to the factory   415

Antibiotics are microbial metabolites that are produced commercially   417

Medically valuable steroid hormones can be produced by microorganisms   418

Case: Menopause, mares, and microbes   418

Microbial metabolites have many other, nonmedical uses   419

Some microorganisms themselves are used as materials   420

New strategies are required to combat environmental pollution   421

Case: The sweet smell of success   421

Microorganisms are used to digest harmful chemicals through the process of bioremediation   422

Both the environmental context and the microbe being used determine how bioremediation is conducted   423

Bioremediation can prove valuable in many different settings   424

Bacteria may help solve the solution of plastic waste   425

Fungi can be effective bioremediation agents   425

Microorganisms can help reduce solid waste and improve soil quality through composting   426

Case: Waste reduction begins at home   426

Plastics may be replaced by biodegradable, microbially produced alternatives   427

Microorganisms may help meet the demand for alternative energy   428

Looking back and looking forward   431

Garland Science Learning System   431

Concept questions   431

Glossary   433

Figure acknowledgments   447

Index   453

Learning Skills

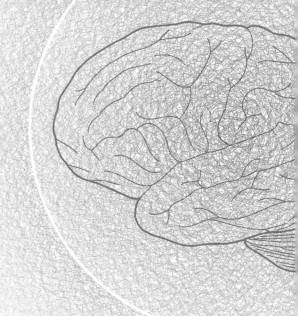

Learning is a skill, so before you start to learn about microbiology, here are some tips about how to use your brain effectively …

You have to be motivated to learn

The first step is to turn your brain on by wanting to learn and believing you can. In order to want to learn, you have to be motivated and able to answer the question "Why am I learning all this?". Motivation can be internal (driven by you) or external (driven by someone or something else), with internal motivation being by far the most powerful. Internal motivation comes from curiosity and the desire for self-improvement; you feel good when you learn something new, so learning itself is the goal. External motivations, such as passing an exam, tend to concentrate on negative thoughts, such as not passing the exam, and relying on them won't use your brain effectively. However, external motivations are a good starting point and can develop into an internal motivation to learn: subjects you find difficult or boring are simply those in which you have yet to develop interest. Once you develop an interest in a subject you'll want to learn more about it.

Memorizing is only a part of learning

Statements such as "What do I have to know?" or "What's the least I've got to study?" or "What's the minimum to pass the course?" indicate a predisposition to memorize rather than understand. Memorizing is seductive, because you can recognize and recall many facts in a short time, but they will quickly be forgotten, making memorization an inefficient use of your brain. Do not make the mistake of thinking that memorization is the key aspect of mastering a new subject. Nothing could be farther from the truth, and relying on memorization alone is a sure recipe for failure. A certain amount of memorization is necessary to provide you with the basic vocabulary in a new subject. It is valuable in the way that memorizing new words helps when your goal is to learn a new language. But memorizing vocabulary lists is not the same as actually speaking a language. Memorization simply gives you the vocabulary you need to delve into the important concepts and ideas that are the heart of any scientific field of inquiry.

Understanding is a crucial step on the road to learning

To truly learn something, you first need to understand it. Understanding comes from making connections to your existing knowledge. It is more difficult to connect unfamiliar ideas, and you probably think that the ideas in this book are unfamiliar and alien to you, but they're not. If you've had a cold, eaten yogurt, learned about the Spanish conquistadors, read Edgar Allan Poe, or noticed the "Now wash your hands" sign, you're familiar with microbiology. This book is full of examples like this and they'll help you understand and learn about microbiology.

You need time and space to learn

Once you understand an idea, you are ready to learn. Learning involves using the material and developing the skills to apply it to new situations. In other words, you need to practice with the material, just as you would practice to learn to play an instrument, a new sport, or to solve math problems. When you practice, you can take advantage of the most powerful learning tool there is; the opportunity to make mistakes in a situation where those mistakes are not costly, and then to learn from your mistakes. Over time, your performance improves, just as it always improves if you conscientiously practice a new skill. In schoolwork, this practice should take place when you study. If you make mistakes then, they will have no long-term negative impact, and you can strengthen weak spots in your knowledge, before you're asked about them in an exam or other assignment. So, how does one practice microbiology? One easy way is that after you have read about a topic, and think you understand it, try explaining it to another person. Can you do it? If you have nobody to explain it to, simply explain it out loud to an empty room or even your cat asleep on a chair. Unless the topic is very simple, you will make errors the first few times you try to explain it. In that case, simply open your book, re-examine the topic, and try again, until you can explain the topic well. Every time you correct your errors, your understanding and knowledge grow deeper.

Learning takes time, so make a good amount of time available for learning. You won't learn about any subject if you leave it until the night before the exam. Make time every week and make sure you've had enough sleep and enough to eat before you start. You also need space in your head so remove as many distractions as possible, find somewhere quiet, and turn social media off for a while.

Break it down into small chunks

Don't try and learn too much at one go: large tasks are dispiriting and demotivating because they appear insurmountable. This book is broken down into chapters, and the chapters are broken down into sections. Don't try and read the whole book in one go, don't try and learn a whole chapter in one go, take it a section at a time and you'll be amazed how far you get. Remember to have short breaks every 20 minutes or so to refresh your brain.

A wide range of resources are available to you, many of which are provided with this book. Use them all!

Use the headings (they're in blue) to get an overview of the chapter

Make notes in the book, it has wide margins for this purpose

Use the glossary and vocabulary review to explain any unfamiliar terminology

Use the case studies and tutorials to find familiar and interesting situations

Write explanations in your own words

Use the figures to follow processes

Draw your own figures

Use the videos and animations to help with visualization

Use the questions and answers to test your learning

Don't fall into the trap of thinking the answers to questions, instead, write them down and be sure any correct answers aren't lucky guesses. Be honest with yourself. The Garland Science Learning System (GSLS) resources include a quiz for each chapter, which gives you the correct answer as you go along and also lets you know why the correct answer is correct. Write down your answers to the concept questions at the end of each chapter and then look up the hints on the GSLS. Once you've answered all the questions, try and explain what you've learned to someone else.

IN SUMMARY

- **Want to learn**
- **Believe you can learn**
- **Aim for understanding not memorization**
- **Practice with the material and learn from your mistakes**
- **Make time and space to learn**
- **Use all the resources available**
- **Test yourself as you go along**
- **Enjoy it!**

LIVING IN A MICROBIAL WORLD

We live in a microbial world. Ask a friend to name a living thing and he or she will likely pick a dog, a daisy, or some other familiar animal or plant. The vast majority of living things on Earth, however, are **microorganisms**—bacteria, viruses, fungi, and protozoa, too small to see with the unaided eye (**Figure 1.1a**). The numerical advantage that microorganisms have over larger living things is truly mind boggling. A single teaspoon of soil can hold millions or billions of bacteria and fungi. A drop of seawater teems with microscopic life. And larger animals and plants provide shelter for literally trillions of smaller living things. At this moment, more bacteria than all the humans that have ever lived are quietly going about their business in your intestine.

Microorganisms, also frequently called **microbes**, live everywhere (**Figures 1.1b** and **1.1c**). They have been found frozen in ice and inside solid rock. They live in the deepest parts of the ocean and in boiling hot springs. They are found in habitats that are as acidic as battery acid or in briny water so salty that no fish could survive.

Many people lump all microorganisms together as germs—tiny organisms that are of little importance to us unless they make us sick. Nothing could be further from the truth. This book will not only demonstrate how diverse these fascinating organisms are but will also explore the many and often surprising ways they affect our lives. Indeed, it is difficult to open a newspaper or turn on the radio and not read or hear something about the way microorganisms impact humans.

It was recently announced, for instance, that researchers had genetically modified bacteria in a way that allows the diagnosis of liver cancer. In a nutshell, the bacteria, which normally live peacefully in the intestine, were altered in a way that allowed them to pass through the gut wall and enter the liver, where they naturally gravitate toward any cancer cells that are present. If the microbes find and colonize a liver tumor, they have been engineered to produce a chemical that is visible as a color change in the urine, when the chemical is excreted. Another surprising recent report is that each of us is surrounded by a unique microbial cloud of specific organisms that emanate from our bodies. Although we all harbor many species in common, the precise makeup of our microbial cloud is as individual as a fingerprint. The researchers at the University of Oregon who made the discovery even suggest that these clouds may have a forensic application; they could be used to detect, for example, whether someone had passed through a room. And if you visit Amsterdam, be sure you visit the Micropia Zoo, the world's first microbial zoo. Much of the zoo, which opened in 2014, looks like a laboratory, complete with rows of microscopes connected to giant television screens. Visitors can also look through a window at a genuine

All case studies have a few questions at the end, the answers to which will become apparent as you read the sections following the case.

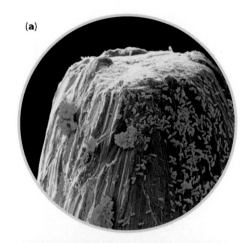

Figure 1.1 The microbial world around us. (a) Bacteria growing on the end of a pin. (b) Microorganisms live almost everywhere. When surfaces in the environment (a doorknob, a body part, or a kitchen sink, for example) are swabbed with a moist sterile swab, microorganisms that adhere to the swab can be transferred to a nutrient agar plate that supports microbial growth. After approximately 24 hours at a favorable temperature, all of these samples of environmental material show heavy microbial growth. Each visible spot on the plate represents millions of microbial cells, known as a colony. (c) Even environments such as this hot spring, too extreme to support plant or animal life, are conducive to the survival of certain microorganisms.

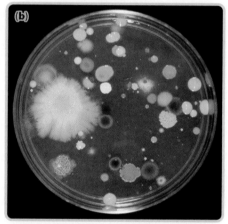

laboratory where different kinds of microbes are being grown in Petri dishes and test tubes.

Certainly there is no shortage of news stories to remind us that some microorganisms cause disease. In December 2013, the worst and most widespread epidemic of Ebola virus in history erupted in West Africa, riveting the world's attention. The virus kills an estimated 70% of those who become infected, and so far, there are well over 11,000 confirmed deaths. The Zika virus epidemic in the Americas, which started in Brazil in early 2015, is still a serious cause for concern and we continue to hear regularly about the ongoing AIDS crisis. And it's not just humans who are affected. North American bats, for example, are currently under siege from white nose syndrome, a disease caused by the fungus *Pseudogymnoascus destructans*. This organism, originally from Europe, was inadvertently introduced into New York in 2006. Since then it has spread to at least 25 US states and five Canadian provinces, killing over 6 million bats so far. Eleven bat species, including three endangered species have been affected. But microorganisms are not always the bad guys when it comes to our health. Consider the 2015 discovery of GB Virus C. Over a billion people are believed to be infected with this virus, but unlike viruses like measles or influenza, which cause disease, the GB Virus C actually appears to slow down the progression of certain other viral diseases such as AIDS.

The study of any living thing starts with the basics, and microorganisms are no exception. We will therefore begin by investigating the basic chemistry of microbes, their structure, and how they grow. We will learn how they pass genetic information on to the next generation and how they acquire new characteristics over time. We will take a close look at how they impact, for better or worse, our environment, our health, our diet, and even our history. Along the way we will learn the following, just to provide a few examples:

- What the designation H5N1 in reference to bird flu actually signifies

- Why biological dead zones, devoid of most sea life, are an important problem in the world's oceans

- How microorganisms contributed to the fall of Rome, the conquest of the Americas, and the development of a middle class in England

- Why hydrogen peroxide is a good thing to keep in your medicine cabinet

- How disease-causing microbes develop resistance to antibiotics and how such resistance can be reversed

- What gives Swiss cheese its particular flavor

- Why stone-washed denim has nothing to do with stones

The study of microbes provides answers to all of these questions. The reasons for studying microorganisms, however, go well beyond such examples, no matter how interesting they might be. When we learn about microorganisms, we are learning about life itself. As you proceed through this text, you will find that basic life processes are the same for all living things. Most of the major discoveries made by biologists about the nature of life and how it works were first made in microorganisms; biologists then found that the same discoveries applied to other forms of life. Biologists have been attracted to microorganisms because, in many respects, they are simpler, they are easier to work with, and they yield experimental results faster than animals or plants. Furthermore, many of the most important issues in the world today—issues that on the surface may seem to have little to do with biology, such as global warming, rising energy costs, and international terrorism—involve microorganisms to greater or lesser degrees. Consequently, for the newcomer to biology, there is no better place to start than the world of microorganisms.

We will learn why life itself depends on microorganisms. Although we may not see them and although some of them cause more than their share of misery and disease, we cannot live without them. Microorganisms were here millions of years before us, and they will no doubt be here long after we are gone. Before we begin to tell their tale, however, let's consider a few key characteristics that all living things, microbial or not, have in common.

All living things are composed of one or more cells

One of the most fundamental principles in all of biology is the **cell theory**—the notion that all living things are composed of **cells**. Cells are highly organized structures that are always surrounded by a membrane. Cells are the basic unit of life, and all living things are composed of one or more cells (**Figure 1.2**). The interior of cells is a watery mixture of many molecules and structures that carry out various activities. Microorganisms are most commonly **unicellular**; they consist of only a single cell. Some microbes, on the other hand, as well as all animals and plants, are **multicellular**, meaning that they are composed of many cells.

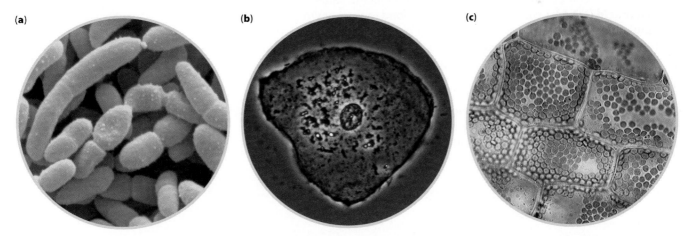

(a)   **(b)**   **(c)**

Figure 1.2 Cell types. All cells are surrounded by a cell membrane. (a) *Acetobacter*, an example of bacterial cells. These bacteria are commonly used to make vinegar. Because each individual cell constitutes an individual bacterium, most bacteria are considered to be unicellular. Like almost all bacterial cells, each cell is small and surrounded by a rigid cell wall. The genetic material of bacterial cells is not enclosed within a membrane. Cells that lack a membrane for this purpose are called prokaryotic cells. (b) An animal cell from the inside of the mouth. Note the densely stained nucleus in the cell's interior. The nucleus, which contains the cell's genetic material, is surrounded by a nuclear membrane. Cells with such a membrane are called eukaryotic cells. (c) Plant cells. In addition to a nucleus, plant cells also often contain chloroplasts (seen as green membrane-bound structures), the site of photosynthesis. Note the more regular shape of the plant cells, compared with animal cells. This is due to the presence of a cell wall, exterior to the cell membrane. Such a wall is absent in animal cells. Individual plants and animals are both composed of many cells and are consequently called multicellular organisms.

Cells fall into two general types, called prokaryotic and eukaryotic. Most microorganisms have simpler and usually smaller cells known as **prokaryotic cells** (see Figure 1.2). This name comes from the Greek for "before the nucleus," and it reminds us that in these cells the genetic material is not surrounded by a nuclear membrane. This is in contrast to the larger and more complex **eukaryotic cells** (from the Greek for "true nucleus"), in which a membrane surrounding the genetic material forms the nucleus. Eukaryotic cells also have a variety of other small structures called **organelles** (little organs), which carry out specific functions required by the cell. Eukaryotic cells are thus more compartmentalized than prokaryotic cells, with many specific activities taking place in discrete locations. In prokaryotic cells, although these same activities occur, the precise location within the cell where they occur is less defined. Eukaryotic cells are found in animals and plants, as well as some microorganisms.

All living things display other observable characteristics

What exactly is life? Perhaps surprisingly, such a simple question does not have an easy answer. One unifying characteristic of living things is that they are composed of cells (**Figure 1.3a**). These cells, prokaryotic and eukaryotic alike, share a number of defining characteristics, and the presence of these characteristics usually allows us to separate the living and nonliving worlds.

First of all, living cells typically have an internal environment that is different from their surroundings. It may be more or less salty or different in terms of acidity. Moreover, cells have mechanisms to maintain these differences, a process called **homeostasis** (**Figure 1.3b**). Just to provide one familiar example of homeostasis, your body temperature is kept relatively constant, about 37°C, regardless of the environmental temperature. In humans, critical cellular processes function best at this temperature. Should temperature-regulating homeostasis fail, a cell's ability to carry out these processes will decline, and if homeostasis cannot be reestablished, the cell will ultimately die.

Furthermore, all living things have the ability to reproduce (**Figure 1.3c**), which requires a blueprint, usually encoded in a molecule called **deoxyribonucleic acid (DNA)**. Living things can also respond to their environment; they can alter their behavior as environmental conditions change. Certain environmental cues may cause cells to move toward or away from a stimulus. Cells respond to chemical signals in their environment, and often these signals are released by other cells (**Figure 1.3d**). This represents a form of cell-to-cell communication, and such communication is also a hallmark of living systems. Another important characteristic of living things is the ability to assimilate and use energy, a process known as **metabolism** (**Figure 1.3e**). Finally, all life changes over time. Via the process of **evolution**, the characteristics of living things may change over many generations (**Figure 1.3f**).

Microbiology involves the study of several distinct groups of living things

Microbiology is distinguished from other subdisciplines within biology in that it deals primarily with microscopic organisms. It actually covers quite a bit of ground because microorganisms are such a diverse lot—far more diverse than animals or plants.

Bacteria are small, most commonly single-celled, prokaryotic organisms (**Table 1.1**). The other major group of prokaryotes is the **archaea**. Although archaea look much the same as bacteria, there are many structural differences between these two groups, and they are only distantly related.

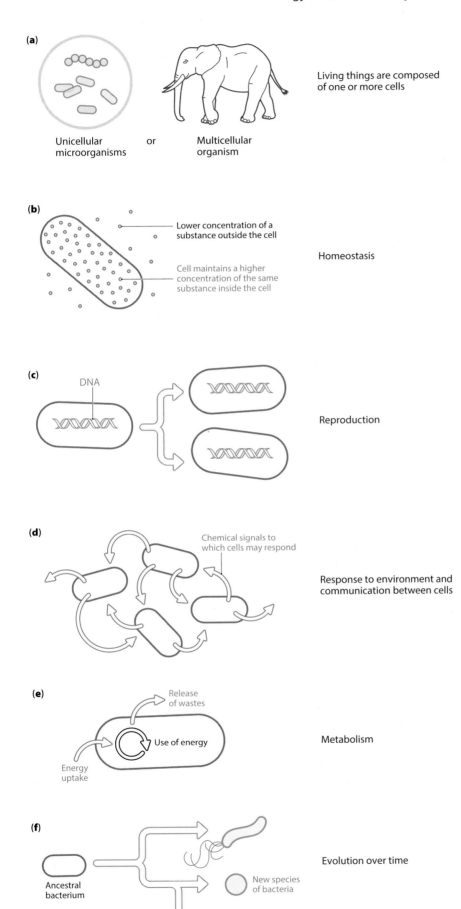

(a)

Unicellular
microorganisms

or

Multicellular
organism

Living things are composed
of one or more cells

(b)

Lower concentration of a
substance outside the cell

Cell maintains a higher
concentration of the same
substance inside the cell

Homeostasis

(c)

DNA

Reproduction

(d)

Chemical signals to
which cells may respond

Response to environment and
communication between cells

(e)

Release
of wastes

Use of energy

Energy
uptake

Metabolism

(f)

Ancestral
bacterium

New species
of bacteria

Evolution over time

**Figure 1.3 Some characteristics of
life.** (a) Living things are composed of one or
more cells. (b) For some environmental factors,
homeostasis allows cells to maintain different
conditions within the cell than those of the
surrounding environment. (c) All living things
are able to reproduce. Genetic information is
transferred between parent and offspring in the
form of a molecule called DNA. (d) Cells respond
to chemical signals that are released by other
cells or that are present in the environment.
(e) They also require energy to carry out various
cellular functions. (f) They undergo evolutionary
change over time.

Table 1.1 **Some characteristics of major types of microorganisms. Bacteria and archaea have prokaryotic cells, while protozoa, fungi, and algae have eukaryotic cells. Neither viruses nor prions are composed of cells and are therefore termed acellular.**

MICROORGANISM	CELL TYPE	CELL WALL	PHOTOSYNTHESIS
Bacteria	Prokaryotic	Yes, almost all	Some
Archaea	Prokaryotic	Some	Some
Protozoa	Eukaryotic	In some life cycle stages	No
Fungi	Eukaryotic	Yes	No
Algae	Eukaryotic	Yes	Yes
Viruses	Acellular (not composed of cells)	No	No
Prions	Acellular (protein only)	No	No

Other microorganisms have eukaryotic cells (see Table 1.1). Eukaryotic microorganisms include many fungi and the protozoa (**Figure 1.4**). Algae likewise have eukaryotic cells, and the smaller microscopic algae are considered to be microorganisms as well. Animals and plants are also composed of more complex eukaryotic cells, meaning that, in spite of the size difference, the cells of microorganisms such as baker's yeast (a fungus) and malaria parasites (protozoa) have more in common with human or other animal cells than they do with prokaryotic bacteria or archaea.

The **viruses** are particularly interesting and unusual. Viruses are the ultimate parasite; they are utterly unable to reproduce unless they gain access to the interior of an appropriate cell. Once inside, however, they essentially hijack the cell and use the cell's machinery to carry out their own replication. The cell, forced to divert energy and other resources to viral replication, may eventually die. Once they have successfully replicated, new viral particles leave the cell and go on to infect other cells to repeat the replicative cycle. Because they rely so heavily on the cells that they parasitize, viruses are unusually simple. They consist of little more than genetic material surrounded by a protein coat, and they have few of the structures found in other microorganisms. Many of humanity's worst scourges are various viruses, such as influenza and HIV

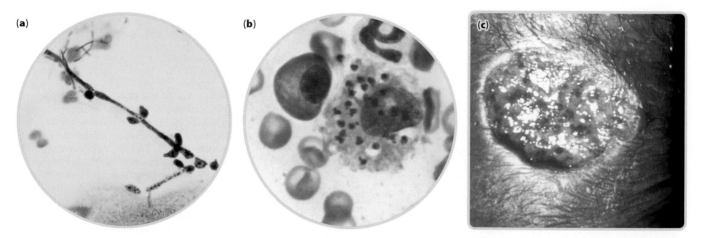

(a) **(b)** **(c)**

Figure 1.4 **Eukaryotic microorganisms.** (a) *Candida albicans*, the fungal pathogen responsible for vaginal yeast infections. (b) *Leishmania donovani* (small purple bodies) inside larger mammalian white blood cells. This protozoan parasite is transmitted through the bite of infected sand flies. (c) A skin lesion in a patient suffering from a *Leishmania* infection.

(Figure 1.5). Finally, there are the **prions**, infectious proteins, which make even viruses look complex by comparison (Figure 1.6). Prions are responsible for a set of odd diseases that includes mad cow disease.

Viruses strain our notion of what it means to be alive

Not all living things display all the characteristics of life, and microorganisms especially test the boundaries separating the living from the nonliving world. Sometimes, for instance, the capacity for independent movement is considered a characteristic of life, but many microorganisms are incapable of such movement. Viruses are particularly problematic; they seem to straddle the border between life and nonlife. Viruses are not composed of cells. On their own, they cannot reproduce, and they carry out little if any independent metabolism. Yet viruses are able to attach to and gain entry into the cells of other organisms. Once inside, they begin to replicate and certainly act like living things. So, are they alive? Such a question is perhaps more philosophical than biological, but it highlights the fact that the division between life and nonlife is not as sharp as one might think.

Microbiology is closely intertwined with the study of non-microorganisms

The scope of microbiology does not end with the microorganisms themselves. In many areas of microbiology, we cannot neglect the biology of multicellular eukaryotes such as ourselves. When we discuss how certain microorganisms cause disease, for example, it is important to understand not only what infectious microbes do to us but also how we defend against them. Consequently, we must also consider the immune system and how it helps us fend off a never-ending microbial onslaught.

Because microbiology is such a large and complex field, most microbiologists specialize within one of many specific fields of expertise. Before we begin our study of the organisms themselves, let us briefly consider some of the diverse areas of study within the science of microbiology. To do so, we will begin with a case study.

Microbiology is composed of both basic and applied components

CASE: BEAUTY AND THE BACTERIA

Many of the greatest artistic masterpieces have suffered greatly from the passage of time. Modern museums are designed to protect these priceless artifacts, but in many cases, the damage is already done; centuries of storage or display under less-than-optimal conditions, such as an old cathedral, where artworks may be exposed to soot from burning candles, moisture, light, and air pollutants take their toll, resulting in the accumulation of dirt and the degradation of the original paint. Here to help is *Pseudomonas stutzeri*, a common bacterium that digests many of the common pollutants found in artwork. Italian scientists have found that small amounts of these bacteria applied to damaged frescos can remove up to 80% of the accumulated contaminants in as little as 12 hours. To cite one example, *The Conversion of St Efisio*, painted in the late fourteenth century by the Italian artist Spinello Aretino, was on display for hundreds of years on the interior

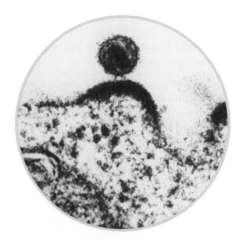

Figure 1.5 Human immunodeficiency virus (HIV). Newly formed viral particles can be seen leaving the much larger infected human cell. Unlike other microorganisms, viruses are not composed of cells, and they cannot replicate outside an appropriate cell of another organism. Consequently, they are called obligate parasites (parasitism is an absolute requirement). Once inside an appropriate cell, viruses commandeer the host-cell machinery to facilitate their own replication. Depending on the virus, such cellular hijacking may or may not kill the cell.

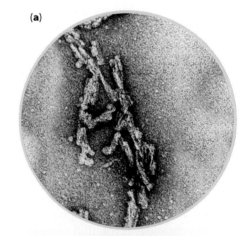

Figure 1.6 Prions. (a) Prion proteins. Prions are unique infectious agents, consisting of protein only. (b) Prions cause neurological disorders. This sheep suffers from a prion disease known as scrapie. Infected sheep behave erratically, often scraping themselves against hard objects, causing the loss of their wool coat.

wall of the famed Camposanto Cemetery in Pisa and was badly damaged. By applying *P. stutzeri* to the fresco, the original beauty of the artwork has been largely restored.

1. **What type of microbiologists would concern themselves with studies of this kind?**
2. **How does *Pseudomonas stutzeri* dissolve away accumulated material from the artwork?**
3. **Are there other types of microorganisms that damage, rather than help protect, artwork?**

Microbiology, like biology in general, has both **basic** and **applied** components. A biologist studying topics in basic biology is interested in a more complete understanding of life processes. Learning exactly how specific cells utilize energy, or how different organisms are related to each other, are both examples of basic biological questions.

Much of our understanding of basic biological processes comes from the study of microorganisms (**Figure 1.7a**). With their simpler cellular structure, microorganisms provide the microbiologist with a useful starting point for teasing apart cellular processes. They are also easier to maintain in the laboratory than plants or animals, and they are relatively easy to manipulate experimentally. Because they reproduce so quickly, experimental results can be obtained far more rapidly with them than with slower growing and more slowly reproducing multicellular organisms. And finally, as we have already noted and as we will highlight throughout this text, many fundamental biological processes are very similar whether we are talking about a single-celled bacterium or a large animal composed of trillions of cells. Consequently, lessons learned from microorganisms are often useful in understanding the basic biology of animals, plants, and other organisms.

Certain bacteria and other living things that have been closely studied for these and similar reasons are referred to as **model organisms**. The bacterium *Escherichia coli*, for example, is sometimes referred to as the microbial lab rat (rats are an animal model organism), because of the intensity with which it has been studied. Many seminal biological discoveries, such as the discovery of DNA and how it functions, were first elucidated by the use of this bacterial species. *E. coli* continues to reveal secrets with applications to living things in general. For instance, in Chapter 15 we will see how many breakthroughs of the biotechnology revolution came to us courtesy of this microorganism.

Figure 1.7 Microbiologists at work.
(a) Basic microbiology seeks to gain a more complete understanding of biological processes, focusing on microorganisms. Here, a microbiologist uses a device called a pipetter to transfer material from one tube to another. (b) Using knowledge gained through basic research, the applied microbiologist attempts to solve practical problems. Here, an environmental microbiologist investigates the effect of microorganisms on a lake's water quality.

The basic biologist is not trying to solve a practical problem. That's the job of the applied biologist. Applied biologists use the knowledge gained in basic biology to solve real-world problems. For example, a basic researcher might determine the exact way that a certain virus enters human cells. An applied scientist may then use this information to design a new antiviral drug that would thwart the entry of the virus. The division between basic and applied science, however, is not a sharp one, and it is not uncommon for a scientist to work on both basic and applied aspects of a particular problem.

Both basic and applied microbiology consist of many subdisciplines

Within the broad area of basic microbiology, a scientist might wish to study any number of specific topics. A microbial geneticist, for example, may choose to study how the genetic material of a particular microorganism is turned on or shut off in response to changing environmental conditions. A microbial ecologist might opt to investigate how a community of microorganisms living in the soil influences the availability of nutrients for plants. A microbiologist interested in adaptations to extreme environments may wish to know how certain microorganisms withstand levels of radiation that would kill an animal.

Applied microbiology likewise encompasses many diverse areas of interest. Medical microbiologists try to better understand the role of microorganisms in infectious disease, with an eye toward prevention or treatment. Public health microbiologists help educate the public about how to best avoid such disease, whereas epidemiologists track the spread of disease in the community and attempt to predict future outbreaks.

Food microbiologists, on the other hand, utilize microorganisms to produce a large portion of what we eat, from cheese to bread to soy sauce. Agricultural microbiologists often utilize knowledge gained through the study of microbial ecology to enhance crop growth or reduce the likelihood that valuable food plants will suffer from disease (**Figure 1.7b**). Industrial microbiologists investigate the manner in which bacteria and fungi can be used to produce a wide variety of industrial and commercial products.

Increasingly, microbiologists, along with other biologists, have entered the new biological frontier of genetic engineering and recombinant DNA technology. Using a variety of techniques and molecular tools provided by microbes, these scientists have been able to introduce foreign DNA into new organisms, creating cells with novel capabilities. As we will discover, because of this powerful new technology, we now have bacteria that produce insulin for human diabetics and potential cures for genetic diseases, as well as a whole host of other potentially exciting, sometimes scary applications.

These are just a few of the many avenues of investigation open to the microbiologist. We will discover others as we proceed through the text. And what about the scientists who are using bacteria to restore irreplaceable artwork (**Figure 1.8a**)? Those who work to protect paintings and other art objects from

Figure 1.8 Microorganisms and art; friends and foes. (a) A portion of the fresco *The Conversion of St Efisio* by the Italian artist Spinello Aretino (1350–1410). The upper photo has yet to be restored. The lower photo is following treatment with *Pseudomonas stutzeri*. (b) *Starry Night* by Vincent van Gogh (1853–1890). Faint areas of brownish and greenish discoloration are due to the breakdown of crocoite, a mineral used by many artists in the nineteenth century to formulate certain colors. Bacteria are responsible for crocoite digestion and consequently the discoloration. (c) The dark discoloration of these Mayan ruins in Yucatan, Mexico, is likewise caused by various microorganisms.

damage of any sort are generally called restoration scientists. This new field, utilizing microbial assistance, as yet has no formal name. Perhaps it should be called restoration microbiology or artifact microbiology. Regardless, this is an example of applied microbiology. Like other applied microbiologists, these scientists are grounded in basic microbiology—in this instance, the precise manner in which certain bacteria break down the art contaminants.

Pseudomonas stutzeri do not offer their services out of the goodness of their hearts. The bacteria are simply digesting materials that they can then use themselves to meet their nutritional or metabolic needs. They do so by secreting proteins called enzymes that catalyse biological reactions—in this case, the digestion of the molecules accumulating on the artwork. We will take a close look at enzymes and how they function in Chapter 2.

Other microbes are not so accommodating. Many bacteria and fungi are capable of colonizing artwork, and digesting components of the paint (**Figure 1.8b**). Others cause discoloration to stone relics as they grow on or even beneath their surfaces (**Figure 1.8c**). It is not always clear how to prevent such damage without further damaging the artwork itself. It appears that restoration microbiologists will have plenty to keep them busy for years to come.

Microbiology, like all fields of biology, is a **science**. By this we mean that microbiology is a process of learning about nature by observation and experiment. Scientists gain knowledge about the natural world in a precise manner known as the **scientific method**. Before we end this introductory chapter and begin our study of microorganisms in earnest, let us consider exactly how scientific knowledge is acquired.

A proper scientific experiment involves a series of well-defined steps

CASE: FLEMING REVISITED

The British scientist Alexander Fleming discovered the first antibiotic in 1929. He noticed that when certain molds contaminated his bacterial culture plates, the bacteria were unable to grow close to the mold. Fleming reasoned that the mold was secreting a substance that killed the bacteria. This proved to be the case, and the mold was identified as *Penicillium notatum*, now called *Penicillium chrysogenum*. Fleming named the secreted antibiotic penicillin (**Figure 1.9a**).

Two microbiology students, Eleonora and John, learn of Fleming's work in their class, and later, when they must perform a small, independent research project as part of their laboratory grade, they decide to repeat Fleming's work with a different yet closely related mold, *Penicillium roqueforti*. This mold is commonly used in the production of blue cheese. The mold digests milk sugars, and the waste products that are released contribute to the taste of the cheese. The mold also gives blue cheese its characteristic blue streaks (**Figure 1.9b**). Does it also, like *P. chrysogenum*, have antibacterial properties?

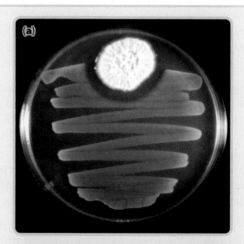

Figure 1.9 *Penicillium* molds. (a) Penicillin secreted by the mold *Penicillium chrysogenum* prevents the bacteria from growing near the mold colony. (b) *Penicillium roqueforti*, seen as blue streaks, is used to make blue cheese.

To find out, Eleonora and John buy a block of blue cheese and transfer mold cells onto nutrient agar plates, which will support the growth of a wide range of bacteria as well as the *P. roqueforti* mold. They inoculate 12 plates with lawns of the bacterium *Staphylococcus epidermidis*. Lawns are prepared by swabbing a plate with a uniform coating of bacteria. The bacteria then grow over the entire surface of the plate (**Figure 1.10a**). Eleonora and John, using an inoculating loop, next streak *P. roqueforti* in a zigzag pattern over six of their plates (**Figure 1.10b**). The other six plates are streaked with an uninoculated loop without the mold. All 12 plates are then incubated at 35°C for 24 hours. The two students then examine their plates, comparing those with and without *P. roqueforti* for evidence of antimicrobial activity.

1. **What hypothesis are Eleonora and John testing?**
2. **Which of their nutrient agar plates are serving as controls? Which are the experimental plates? Why are the control plates critical?**
3. **What is the experimental variable in this experiment? What are the important control variables?**
4. **Why is it important to use six plates with and six plates without the mold, rather than just one of each?**

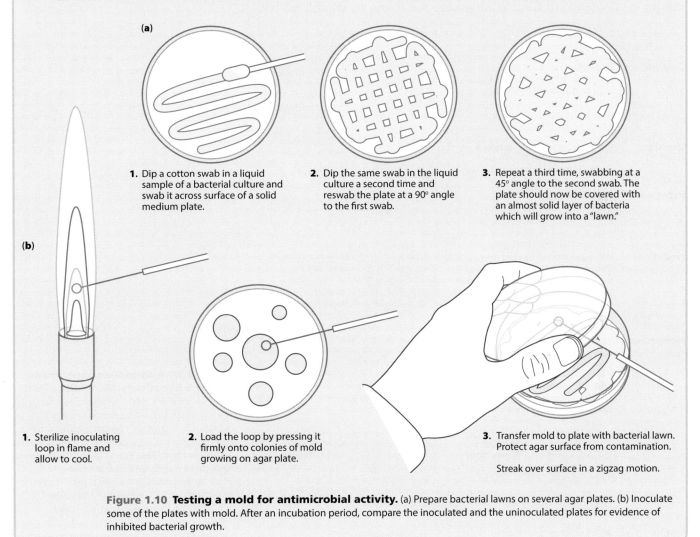

(a)

1. Dip a cotton swab in a liquid sample of a bacterial culture and swab it across surface of a solid medium plate.

2. Dip the same swab in the liquid culture a second time and reswab the plate at a 90° angle to the first swab.

3. Repeat a third time, swabbing at a 45° angle to the second swab. The plate should now be covered with an almost solid layer of bacteria which will grow into a "lawn."

(b)

1. Sterilize inoculating loop in flame and allow to cool.

2. Load the loop by pressing it firmly onto colonies of mold growing on agar plate.

3. Transfer mold to plate with bacterial lawn. Protect agar surface from contamination.

Streak over surface in a zigzag motion.

Figure 1.10 Testing a mold for antimicrobial activity. (a) Prepare bacterial lawns on several agar plates. (b) Inoculate some of the plates with mold. After an incubation period, compare the inoculated and the uninoculated plates for evidence of inhibited bacterial growth.

The scientific method and the process of scientific inquiry begin with an observation. Fleming noticed that *Staphylococcus* was unable to grow close to a mold contaminant. Observations such as these lead to questions. How did the mold inhibit bacterial growth?

To answer such questions, the scientist formulates an appropriate **hypothesis**. A hypothesis is a tentative explanation for a specific question. To be valid, a hypothesis must be testable. In other words, it must be possible to collect evidence that either supports or refutes the hypothesis. To gather such evidence, one must be able to make predictions based on the hypothesis. Experiments or observations of future events are then used to determine whether the prediction was realized. If so, the hypothesis is supported. If the prediction proves to be inaccurate, the hypothesis is rejected as false. In the case of the mold Eleonora and John are testing, their hypothesis is that *P. roqueforti* produces compounds that can inhibit bacterial growth. They therefore predict that *Staphylococcus* will be unable to grow close to the mold.

The next step is to test the hypothesis with a valid experiment. Such an experiment will compare an **experimental group** with a **control group** (**Figure 1.11**). The experimental group is the group in which a crucial factor will be manipulated. In the control group, that same factor will be left

Observation: *Penicillium notatum* releases a chemical that inhibits the growth of *Staphylococcus* bacteria.

Zone of inhibition, no bacterial growth

Penicillium notatum

Question: Does the *Penicillium roqueforti* used to make blue cheese also inhibit bacterial growth in this way?

Blue cheese

Hypothesis: *P. roqueforti* inhibits bacterial growth in the same manner as *P. notatum*.

Prediction: If *P. roqueforti* is streaked over a lawn of *Staphylococcus,* a zone of inhibition will be observed around the mold.

Zone of inhibition observed

Lawn of *Staphylococcus*

Experiment: 12 *Staphylococcus* lawns prepared. On six experimental plates, *P. roqueforti* is streaked. On six control plates, streaking is performed with an uninoculated loop

Experimental variable: Presence or absence of *P. roqueforti*
Control variables: Nutrients in agar, incubation temperature, length of incubation, etc.

Streak of *P. roqueforti*

Control streak without mold

Experimental plate

Control plate

Results: (**a**) No difference between experimental and control plates?

(**b**) Difference between experimental and control plates?

No zones of inhibition

Zone of inhibition

No zone of inhibition

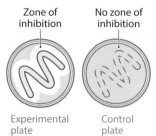

Experimental plate

Control plate

Experimental plate

Control plate

Conclusion: **Hypothesis rejected**   **Hypothesis accepted**

Figure 1.11 A controlled experiment to test a hypothesis. After making an initial observation that leads to a specific question, the scientist formulates a proper hypothesis. A proper hypothesis is a tentative answer to the question that has been posed and it must predict future events. If these events do not occur, the hypothesis is rejected and a new hypothesis must be generated. If the events predicted by the hypothesis occur, the hypothesis is supported. In this example, the hypothesis that *P. roqueforti* can inhibit bacterial growth by secreting antibacterial compounds is being tested experimentally. The experimenter prepares bacterial lawns and streaks *P. roqueforti* on half of these lawns. These represent the experimental group. On the remaining bacterial lawns, no *P. roqueforti* is added. These represent the control group, which will be compared with the experimental group to determine whether the added *P. roqueforti* has had any effect on bacterial growth. The only difference between the two groups of plates is the presence or absence of *P. roqueforti*. This difference is the experimental variable. All other factors (for example, temperature or provided nutrients) must be the same in the two groups. These are the control variables.

unchanged, as a means of comparison. If the manipulation of this factor results in a specific change as predicted by the hypothesis, and if that change does not occur in the unmanipulated control group, the hypothesis is supported.

For example, in Eleonora and John's experiment, the experimental group consists of those bacterial lawns onto which the mold was streaked. No mold was placed on the control plates. Consequently, if their hypothesis that *P. roqueforti* is able to produce antibiotic compounds is correct, bacterial growth on the experimental and control plates should be different. If the hypothesis is incorrect, bacterial growth on the two types of plates will be similar, and the hypothesis can be rejected.

The experimental and control groups in any experiment should be identical for all factors except one. In our students' experiment, all plates contained the same nutrients and they were incubated at identical temperatures. These represent **controlled variables**. The only difference was the **experimental variable**—in this case, the presence or absence of mold. After 24 hours, the two students compared bacterial growth on their experimental plates streaked with mold with their control plates lacking it. Because they carefully controlled all variables except one, any differences in bacterial growth on the two groups of plates could be attributed to the mold rather than some other factor. Note that Eleonora and John even streaked the control plates with an empty inoculating loop. This added controlled variable ensures that if they do see reduced bacterial growth on experimental plates, it must be due to the mold rather than the physical act of streaking.

Following completion of the experiment, the experimental and control plates are compared. Without the control group there would be no way to interpret the effect of the experimental variable. A significant inhibition of growth on the experimental plates indicates that the hypothesis may be tentatively accepted. No difference in bacterial growth on the two sets of plates indicates that the hypothesis should be rejected. When hypotheses are rejected in this way, a scientist has to develop a new hypothesis to explain the initial observation. A new experiment with a different experimental variable is then conducted. Keep in mind that in most situations there may be numerous plausible hypotheses. The task of the scientist, in this case, is to rule out as many of these competing hypotheses as possible. Ideally, only a single hypothesis remains to explain the initial observation.

Recall that the students in our case inoculated several experimental and control plates, rather than just one of each. It is almost always true that the larger the number of treatments, the more meaningful the results. Replication reduces the amount of variance due to unaccounted-for events. For example, the students might have accidentally prepared one of their bacterial lawns improperly. If they put too few organisms on an experimental plate, they may have incorrectly assumed that *P. roqueforti* was unable to inhibit the growth of *S. epidermidis*. In truth, these results might merely have reflected improper inoculation. By repeating the experiment, the likelihood of such erroneous interpretation can be reduced.

If a hypothesis cannot be disproved, it may eventually become a theory

If a hypothesis is rigorously and repeatedly tested and is never rejected, it may in time become part of a **theory**. A theory, such as the theory of gravity, is an important principle supported by a large body of experimental evidence. In this text we have already mentioned the cell theory, and we will encounter other biological theories, such as the theory of biological evolution and the germ theory of disease. A theory can never be proven beyond all doubt, but a scientific theory is a powerful concept; a large body of evidence has accumulated to support it. Until proven otherwise, it remains the best explanation available to account for a given phenomenon. In other words, a theory has withstood the test of time. In everyday speech, however, a theory may refer to merely an idea or a guess that has not been tested or lacks any supporting evidence. Because the word has two very different meanings, it is often a source of great confusion among the general public. Indeed, it is common to hear nonscientists erroneously dismiss a fundamental scientific principle as "only a theory," when in fact it has the weight of scientific evidence behind it.

Looking back and looking forward

In this chapter we have laid the groundwork for what is to come. In these first few pages we have learned a little about who the microorganisms are, and something about where they are found and what characteristics they possess. We have also come to know what a microbiologist is, the sorts of things such a scientist might study, and how scientific inquiry is conducted.

Having gained some insight into exactly what we will be studying, it is time to begin. In Chapter 2 we introduce some of the basic chemical principles that are necessary for an understanding of microbiology. As we proceed through the fundamentals of biological chemistry, we will see that even when discussing atoms, the basic units of all matter, the implications for microorganisms and the manner in which they interact with us are ever-present and often unexpected.

Garland Science Learning System

- http://garlandscience.rocketmix.com/students/

- Discover how (scientifically) to prove you're right and why a scientific theory isn't merely theoretical

- Test your knowledge of this chapter by taking the quiz

- Familiarize yourself with the terminology used in this chapter by using the vocabulary review

- Get help with the answers to the Concept questions

Concept questions

These questions are designed to help you start thinking like a microbiologist. The answers are not always simply found in the text. Instead, you will need to take the concepts about which you have learned and apply them to new situations. Some of the questions may not even have just a single correct answer. Help is provided as part of the GSLS resources, which can be accessed through http://garlandscience.rocketmix.com/students/.

1. It is stated in this chapter that the question of whether or not viruses are alive might be more of a philosophical question than a biological one. Explain what you think is meant by this.

2. Nonscientists often refer to all microorganisms as "germs." Is it reasonable to lump all microorganisms together with this catch-all term? Why or why not?

3. Open a newspaper or magazine, or go to an online news site and find a story that in some way discusses bacteria, a virus, or another microorganism. Are the scientists whom the story describes engaged in basic or applied microbiology?

4. A drug manufacturer develops a new antiviral drug, which it hopes will be effective in preventing colds. First, a group of 1000 volunteers is assembled. All are comparable in terms of age and general health. Of these volunteers, 500 (group A) are given the drug and told to take it as prescribed throughout the cold season. The remaining 500 individuals (group B) are given a placebo (a sugar pill that has no antiviral activity) to take. None of the volunteers know whether they are getting the drug or the placebo. After 5 months, all 1000 volunteers are asked whether or not they developed colds.

   a. What is the hypothesis being tested in this experiment?

   b. Which group is the experimental group? Which group is the control group?

   c. What is the experimental variable? What might some of the controlled variables be?

   d. What results might allow us to tentatively accept the hypothesis?

   e. What results would cause us to reject the hypothesis?

5. In a class you read an article that describes the prevailing theory as to why the dinosaurs went extinct. Data are provided to show that dinosaur eradication was caused by the impact of a gigantic meteorite that threw enormous quantities of dust into the atmosphere. This caused both a decline in Earth's temperature and a loss of photosynthetic vegetation, resulting in dinosaur extinction. Animals that were both better able to survive the cooler temperatures and less dependent on a huge diet of plant matter, such as primitive mammals, for example, survived. You tell a friend about this, and his response is, "That's just a theory. I don't believe it." In what way is your friend misunderstanding you?

THE CHEMISTRY OF LIFE

2

Shamokin Creek drains approximately 350 km^2 in east-central Pennsylvania. It used to be prime fishing habitat, with large populations of pike, sunfish, and other aquatic species. But the Shamokin Creek basin is also rich in coal, which was mined extensively from about 1840–1950. In the late 1990s, surveys were conducted to determine the ecological health of Shamokin Creek, as well as other streams in the area. Certain stretches of the creek were found to be extremely acidic and utterly devoid of all fish (**Figure 2.1**). Other indicators of ecological health, such as populations of aquatic plants and invertebrates, were also absent or greatly decreased. Restoration efforts began in 2000. Although the goal is a return to environmental health by 2020, Shamokin Creek remains one of the most polluted waterways in Pennsylvania.

And Shamokin Creek is unfortunately not unique. Mines can be death-traps as far as fish living in nearby streams are concerned. The culprits are bacteria called *Acidithiobacillus ferrooxidans*, which utilize chemical compounds in mine waste, releasing their own acidic waste product. Yet as depressing as this story is, it does remind us how vital chemistry is when it comes to understanding the microbial world and how micro-organisms affect us.

Figure 2.1 Damage to a stream resulting from mining.
This stream has been badly acidified through the activity of bacteria. Mining activity releases sulfur compounds into the environment, and certain bacteria thrive on these compounds. Acids are released as a waste product of bacterial metabolism, and as bacterial numbers explode, nearby streams become increasingly acidic.

All case studies have a few questions at the end, the answers to which will become apparent as you read the sections following the case.

Indeed, all matter is composed of chemical compounds and the rules governing how these compounds form or interact are identical, whether that matter is living or nonliving. That means that if we really want to understand biological processes, some introductory chemistry is essential. This may be particularly true in microbiology. Most of the organisms we will discuss in this text owe their impact to the chemical changes they cause in their environments. When we consider the role microorganisms play in composting, how bacteria in your mouth cause cavities, why certain bacteria are harmless while others can cause serious health problems, or how specific microorganisms are used to produce sourdough bread or sauerkraut, we will really be considering aspects of microbial chemistry.

A comprehensive study of chemistry is well beyond the scope of this text. Rather, our goal here in Chapter 2 is to review those basic chemical principles necessary for us to fully appreciate how microorganisms interact with the world around them. We will begin by looking at atoms, the basic structural units of matter, and then investigate how atoms combine with each other to form molecules. One familiar molecule is water, and we will pay special attention to this remarkable substance, upon which all life depends. Finally, we will discuss the important biological molecules that are found in all living things, including microbes, and the roles they play in the chemistry of life.

Atoms are the basic building blocks of matter

All matter in the universe is composed of one or more **elements**. An element is a substance that cannot be broken down into other substances by chemical reaction. There are 94 naturally occurring elements known. Many, such as sulfur and calcium, are quite familiar to us. Others, including lanthanum and vanadium, are more obscure. The basic unit of an element is the **atom**. In other words, an atom is the smallest unit that can be identified as a specific element. We can speak, for instance, about an atom of magnesium, but if that atom is broken into its smaller component parts, it is no longer recognizable as magnesium.

Only about 25 elements are commonly found in microorganisms and other living things. The four most important are hydrogen, carbon, nitrogen, and oxygen, which make up approximately 96% of an organism's mass. The remaining 4% is composed of small amounts of other elements such as potassium and iron.

Atoms are made up of smaller components called subatomic particles

Atoms themselves are composed of still smaller subatomic particles. It is the number and kind of these particles in an atom that determine what kind of element it is and the chemical properties it possesses. Three of these subatomic particles, **electrons**, **neutrons**, and **protons**, are of particular importance. The smallest atoms, those of the element hydrogen, have only one proton, no neutrons, and one electron. An atom of uranium, on the other hand, is relatively large and is composed of 92 protons, 146 neutrons, and 92 electrons. Carbon typically has six protons, six neutrons, and six electrons. Nitrogen has seven each of these subatomic particles, while oxygen has eight.

The core of an atom, called the **nucleus**, is composed of protons and neutrons (**Figure 2.2a**). All atoms have a characteristic **atomic number**, which

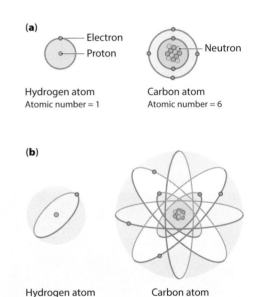

(a)

Electron
Proton

Hydrogen atom
Atomic number = 1

Neutron

Carbon atom
Atomic number = 6

(b)

Hydrogen atom   Carbon atom

Figure 2.2 Hydrogen and carbon contain different numbers of subatomic particles. (a) A hydrogen atom, with a single positively charged proton in its nucleus, has an atomic number of 1. Its atomic mass is also equal to 1. Carbon, with a nucleus consisting of six protons and six uncharged neutrons, has an atomic number of 6 and an atomic mass of 12. A complete atom consists of the tightly packed nucleus, composed of protons and neutrons, surrounded by orbiting, negatively charged electrons. The schematic representation is highly simplified. (b) Electrons are located in specific electron shells surrounding the nucleus, and at any one time electrons may be in any of many possible locations.

corresponds to the number of protons in the nucleus. A proton carries a positive charge, while neutrons are electrically neutral. Electrons have negative charges and typically move around the nucleus at a specified distance in what is referred to as the electron's shell (**Figure 2.2b**). The protons and neutrons in the nucleus make up almost 100% of an atom's mass, and a given atom's **atomic mass** is simply the sum total of its neutrons and protons. Thus, the atomic masses of hydrogen, carbon, nitrogen, and oxygen are typically 1, 12, 14, and 16, respectively (**Table 2.1**; also see Figure 2.2).

Because electrons have a negative charge, they are attracted to the positively charged protons in the nucleus. However, because electrons are in motion around the nucleus and therefore have energy, they are not pulled into the nucleus. Electrons are found in characteristic **electron shells** at specific distances from the nucleus. Those electrons closest to the nucleus are held by the nucleus most tightly and have less energy than those electrons farther away from the nucleus. Each shell can hold only a specific number of electrons. In larger atoms with more electrons, additional shells are needed, and electrons in more distant shells are under less nuclear attraction. These more distant electrons consequently have more energy.

The first electron shell holds a maximum of two electrons. Hydrogen, with only one electron, has its single electron in this first shell (see Figure 2.2a). Helium, with two electrons, fills up this first shell, while lithium, which has three electrons, has a full innermost shell and a single electron in the second shell. The second shell holds a maximum of eight electrons. Neon, with a total of 10 electrons, has a full first and second shell. Sodium (11 electrons) and magnesium (12 electrons) begin the process of filling up the third shell. The largest known atoms require up to eight shells to hold all their electrons.

As an atom's stability increases, its energy decreases

Energy always tends toward its lowest state. Consider the water in a river as it reaches a waterfall. As the water falls down, it loses energy; it cannot flow back up the waterfall on its own. It would have to be pumped back up through an input of mechanical energy. By moving down the waterfall, the water has moved from a higher to a lower energy state. Since it is easier for water to lose energy than it is to gain it, the lower energy state is less likely to change; it is therefore more stable.

The same is true of atoms. They are more likely to interact with other atoms when they have more energy and are therefore less chemically stable. Atoms reach their lowest possible energy level, and thus their greatest chemical stability, when their outermost electron shell is full. When this shell is *not* full, an atom is **reactive**, meaning it will tend to gain or lose electrons (**Figure 2.3**). An atom tends to interact with other atoms and remain reactive until it either

ELEMENT	PROTONS	NEUTRONS	ELECTRONS	ATOMIC NUMBER	ATOMIC WEIGHT
Hydrogen (H)	1	0	1	1	1
Carbon (C)	6	6	6	6	12
Nitrogen (N)	7	7	7	7	14
Oxygen (O)	8	8	8	8	16

Table 2.1 Atomic characteristics of elements common in living things.

		Energy level (electron shell)			
Atomic number	Element	I	II	III	IV
1	Hydrogen	O			
2	Helium	OO			
6	Carbon	OO	OOOO		
7	Nitrogen	OO	OOOOO		
8	Oxygen	OO	OOOOOO		
10	Neon	OO	OOOOOOOO		
11	Sodium	OO	OOOOOOOO	O	
12	Magnesium	OO	OOOOOOOO	OO	
15	Phosphorus	OO	OOOOOOOO	OOOOO	
16	Sulfur	OO	OOOOOOOO	OOOOOO	
17	Chlorine	OO	OOOOOOOO	OOOOOOO	
18	Argon	OO	OOOOOOOO	OOOOOOOO	
19	Potassium	OO	OOOOOOOO	OOOOOOOO	O
20	Calcium	OO	OOOOOOOO	OOOOOOOO	OO

Figure 2.3 Shell configurations of some atoms. The number of electrons in an atom's outermost shell determines its chemical reactivity. The red circles indicate electrons in unfilled outermost electron shells. Atoms with such electrons are chemically reactive. Electrons in filled shells are indicated by blue circles. Atoms with no reactive outer electrons, such as helium, neon, and argon, are unreactive. Notice that the four elements most common in living organisms—hydrogen, carbon, nitrogen, and oxygen—are all chemically reactive.

gains enough electrons to fill the outermost shell or loses enough outer electrons to lose its outer shell entirely. Either way, an atom becomes stable and will no longer interact with other atoms.

Sodium, for instance, with 11 electrons, typically has a full inner shell (two electrons), a full second shell (eight electrons), and one electron in the third shell. In this uncombined state, sodium is electrically neutral because the 11 negative electrons are perfectly balanced by the 11 positively charged protons in the nucleus (**Figure 2.4**). Sodium, however, is reactive in this form, and because of its lack of stability, it tends to lose its outermost electron. If this occurs, sodium becomes stable because its 10 electrons fill both the first and second shells. But it is no longer electrically neutral. It has lost a negatively charged electron, so there are now 11 protons but only 10 electrons, giving sodium a charge of +1. Charged yet stable atoms are called **ions**. Calcium, in its reactive state, has two electrons in its outermost shell. To achieve stability, a calcium atom loses both of these electrons. Because it now has two more protons than it has electrons, the charge on a stable calcium ion is +2. Fluorine, on the other hand, has seven outer electrons in a shell that can hold eight. Fluorine can become stable by gaining a single electron, and in the process it becomes an ion with a charge of –1. Note that some atoms such as helium and neon are stable without gaining or losing electrons, because their outermost shells are already filled (see Figures 2.3 and 2.4). Such atoms are already at maximum stability and do not form ions. Because they are so stable, they do not interact with other atoms.

Perhaps the most important property of an atom is how it interacts or combines with other atoms. When two or more atoms combine, the linkage between them is called a **chemical bond**. An atom's electrons are of critical importance in determining the types of bonds an atom can form and with which other atoms it can form them. Not all bonds form in exactly the same way and we will next inspect the principal types of bonds that atoms may form with each other.

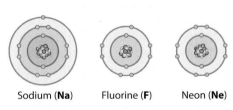

Sodium (**Na**)   Fluorine (**F**)   Neon (**Ne**)

Figure 2.4 Atoms of three elements.
A sodium atom becomes stable by losing one electron from its outermost shell. Once this negatively charged electron is lost, the sodium atom becomes a *positively* charged ion. A fluorine atom becomes stable by gaining an electron to complete its outermost shell. By gaining an electron, the fluorine atom becomes a *negatively* charged ion. A neon atom with a complete outermost shell neither gains nor loses electrons and is unreactive.

An ionic bond is formed when electrons are transferred from one atom to another

CASE: BACTERIA, SALT, AND CYSTIC FIBROSIS

Cystic fibrosis (CF) is an inherited disease affecting about one in every 3000 people of Northern European descent. In normal individuals, chlorine atoms, which exist as negatively charged chloride ions, can easily cross the membrane of lung cells because these cells have special chlorine-transporting proteins on their surfaces. Various mutations can affect these transport proteins and in those with CF, these transporters do not function properly. Chlorine therefore cannot enter cells efficiently. Prevented from entering lung cells, the ions readily bind to sodium ions, which carry a positive charge. The combined chlorine and sodium form the salt sodium chloride. High salt levels help to explain why bacteria proliferate in the lungs of CF sufferers. Normally, cells lining the lungs produce natural protein antibiotics that protect the lungs from bacterial colonization. However, the high salt concentration in the lungs of CF patients inactivates many of these proteins, allowing the bacteria to proliferate (**Figure 2.5**). Current research is underway to synthesize similar antibacterial proteins that are salt-resistant and that may be used therapeutically in CF patients to help control bacterial infections.

1. **Why do chlorine atoms form negatively charged ions, while sodium atoms form positively charged ions?**
2. **What exactly is a salt? Why do these charged atoms combine to form salt in the manner described?**
3. **How does the salt interfere with the antibacterial protein in the lungs of cystic fibrosis patients and thus increase the likelihood of bacterial infections?**

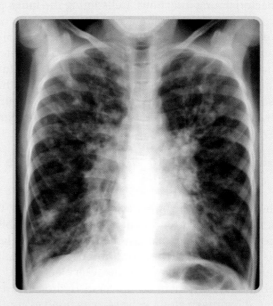

Figure 2.5 Chest X-ray of a cystic fibrosis patient. Light-colored areas in the lungs indicate areas of bacterial infection. Cystic fibrosis patients are often plagued by salty, thick mucus in the lining of the lungs and associated air passages that is difficult to cough up.

If atoms become stable simply by gaining or losing electrons, why don't they just go ahead and gain or lose them? Why should we ever encounter unstable, uncharged atoms in nature? An atom will not ordinarily lose electrons unless another atom that tends to gain electrons is nearby. The atom that requires an electron to achieve stability literally takes the electron from the atom that tends to lose electrons, forming both a positive and a negative ion in the process. In other words, atoms typically do not spontaneously gain or lose electrons. They do so because they are close to other atoms that attract their electrons more strongly or weakly than they do. The degree to which an atom attracts electrons is described as that atom's **electron affinity**. Atoms that attract electrons strongly are said to have high electron affinity. Those that attract electrons weakly have low affinity.

When one or more electrons are transferred in this manner, the positive and negative ions attract each other by virtue of their opposite charges. Such an attraction, called an **ionic bond**, holds the two ions formed in just this way. Sodium (chemical symbol Na) has one outermost electron, which it can lose to chlorine (chemical symbol Cl) with seven outer electrons. Sodium thus forms a +1 ion (Na^+), while chlorine forms a –1 ion (Cl^-). The opposing charges hold the two atoms together. The ionic compound formed, sodium chloride (Na^+Cl^-), is shown in **Figure 2.6**. Note that even though each atom in the compound is charged, the negative and positive charges balance each other exactly, making the **salt** electrically neutral. In the case of cystic fibrosis (CF), since the chlorine ions cannot enter the cells, they combine with the sodium in this way.

Sodium chloride is the chemical name for common table salt, one crystal of which contains many sodium and chlorine ions packed together. Sodium chloride is just one type of salt. Magnesium fluoride (MgF_2) is another salt, containing one magnesium and two fluorine atoms ionically bonded.

Now that we know why sodium chloride forms in the lungs of CF patients, we can better understand part of the reason why such patients often suffer from bacterial lung infections. Cells lining the lungs produce proteins that act as natural antibiotics, protecting the lungs from bacterial colonization. Many of these compounds only work efficiently in relatively low-salt environments. In the lungs of CF patients, salt tends to interfere with and inactivate these proteins, opening the door to infection. Interestingly, the faulty chlorine transporters that result in salt formation in the lungs of CF patients are also found in other parts of the body. This explains why the saliva and perspiration of people with this disease are also unusually salty.

Covalent bonds form when atoms share electrons in their outermost shells

Ionic bonds are not the only way atoms can interact. Consider two reactive atoms, both of which tend to achieve chemical stability by gaining rather than losing electrons. In this case, neither atom is likely to completely donate an electron to the other. The two atoms may obtain a full outer shell, however, and thus become stable, by sharing outer electrons. Consider hydrogen, the most abundant element in the universe. Hydrogen rarely exists as individual H atoms. A hydrogen atom, however, can bond to another hydrogen atom to form hydrogen gas (H_2).

Recall that each hydrogen atom has only a single electron in its first electron shell, which holds a maximum of two electrons. When two hydrogen atoms come in contact, neither can take an electron from the other, because their ability to attract electrons is exactly the same (that is, they have the same affinity). Instead, the two hydrogen atoms share their electrons, providing each atom with the two outer electrons required to achieve stability. When two atoms share electrons in their outer shells in this way, they have formed a **covalent bond** (Figure 2.7).

Hydrogen can bond with other atoms as well. An important example is water, the molecular formula of which is H_2O. A reactive oxygen atom has six outer electrons in a shell that holds a maximum of eight. Oxygen thus needs two additional electrons to become stable, and it can obtain them by sharing two electrons—one with each of two hydrogens (Figure 2.7b).

Hydrogen also commonly bonds to carbon, forming an important class of molecules called **hydrocarbons**. The simplest hydrocarbon is methane, which is composed of one carbon atom and four hydrogens (CH_4). Carbon, with four outer electrons, can form four covalent bonds by sharing each of these electrons with hydrogen (Figure 2.7c). Methane is an important greenhouse gas that contributes to global warming. Because much of the methane in our atmosphere is a microbial waste product, we will discuss this gas more fully in Chapter 10, where we investigate the role that microbes play in environmental processes.

Two atoms may form more than one covalent bond with each other in certain cases. For instance, atmospheric oxygen generally is in the form of O_2. Each oxygen atom has six outer electrons in its second shell. Both atoms require an additional two electrons to become stable. Consequently, they will share a total of four electrons, resulting in two covalent bonds. In carbon dioxide (CO_2),

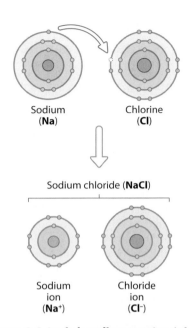

Figure 2.6 Ionic bonding. Ionic bonds form between negative and positive ions, forming salts. The red arrow in this figure indicates the transfer of electrons from one atom to another. The atom losing one or more electrons becomes a positively charged ion, while the atom gaining one or more electrons becomes a negatively charged ion. Ions with opposing charges attract each other, forming an ionic bond. In order to achieve its stability, sodium transfers its outermost electron to one chorine atom, which becomes stable by gaining the electron. Sodium forms a +1 ion, while chlorine forms a −1 ion; the two are attracted and form the salt sodium chloride.

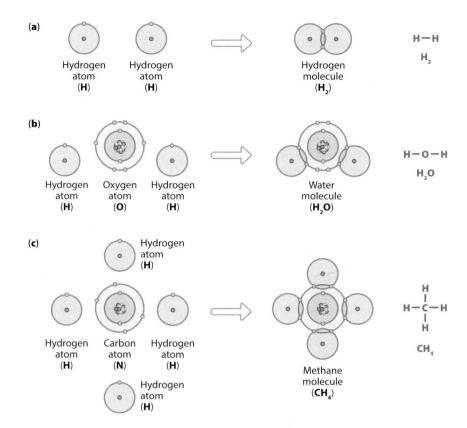

(a)

Hydrogen atom (**H**)   Hydrogen atom (**H**)   Hydrogen molecule (**H₂**)   H—H
H_2

(b)

Hydrogen atom (**H**)   Oxygen atom (**O**)   Hydrogen atom (**H**)   Water molecule (**H₂O**)   H—O—H
H_2O

(c)

Hydrogen atom (**H**)

Hydrogen atom (**H**)   Carbon atom (**N**)   Hydrogen atom (**H**)   Methane molecule (**CH₄**)

Hydrogen atom (**H**)

H
|
H—C—H
|
H

CH_4

Figure 2.7 Covalent bonding. Covalent bonds form between reactive atoms that share electrons. Note that in all examples the shared electrons (in blue) are in the outermost shells of the atoms involved. Unshared electrons are in gray. The nucleus is depicted in the center of each atom. In the simplified stick models next to each atomic model, each line represents a shared electron pair. (a) Each hydrogen atom needs one additional electron to achieve stability. To gain the necessary electron, the two atoms share their electrons to form a molecule of stable hydrogen gas (H_2). (b) Oxygen has a full inner shell, but in its second shell, which can hold eight electrons, there are only six. Oxygen can become stable by sharing two outer electrons with two atoms of hydrogen, forming a molecule of water (H_2O). (c) Carbon, with four outer electrons, can form four covalent bonds with hydrogen, forming methane (CH_4).

recall that each carbon atom can form four covalent bonds, while each oxygen can form two. The single carbon thus forms two bonds with each oxygen.

Covalent bonds can be classified as either polar or nonpolar

You are no doubt familiar with the word "polar." The Earth with its north and south poles and a magnet with positively and negatively charged ends are common examples of polar objects, because they are different at their two ends. Likewise, some covalent bonds are termed **polar covalent bonds**, because the combination of participating atoms carries a slight positive charge at one end and a slight negative charge at the other. Other covalent bonds lack these opposing charges at the opposite ends. In other words, in terms of charge, the two ends are both neutral. Such bonds are called **nonpolar covalent bonds**.

Why do some covalently bound molecules have slight opposing charges at opposite ends while others do not? To explain, consider the examples of hydrogen gas (H_2), in which the covalent bond is nonpolar, and ammonia (NH_3), in which the single nitrogen atom is bound to three hydrogens by three polar covalent bonds.

When two hydrogen atoms are covalently bonded, each atom has an identical affinity for the shared electrons. In other words, the electrons are attracted equally to the nucleus of each hydrogen atom, and they are shared equally. This is not the case, however, when nitrogen binds hydrogen to form NH_3. Remember that protons in the nucleus attract the negatively charged electrons. Nitrogen has seven protons in its nucleus, whereas hydrogen has only one. Therefore, when nitrogen forms covalent bonds with hydrogen, the electrons are actually more attracted to the nitrogen than they are to the hydrogen. The difference in attraction is slight. Even though hydrogen has only a single proton to attract its shared electrons, those electrons are in hydrogen's first shell, and thus closer to the nucleus than they are in nitrogen; nitrogen's edge in number of positively charged protons is to some degree offset by the closeness of hydrogen's nucleus to its electrons. Nevertheless, the difference in attraction is large enough to ensure that the shared electrons are not equally shared. The nitrogen tends to pull the electrons away from hydrogen and toward itself. Because the electrons are negatively charged and tend to cluster closer to nitrogen, the nitrogen atom gains a very slight, partial negative charge. At the same time, the three hydrogens to which nitrogen is bound obtain partial positive charges, because the negatively charged electrons are pulled away from them and toward the nitrogen.

This gives ammonia a polarity, meaning that there is a difference in electrical charge at each of its ends. The covalent bonds holding nitrogen to each hydrogen are therefore polar covalent bonds (**Figure 2.8a**). When there is no charge difference at either end of a molecule, as is the case with H_2 gas, the result is a nonpolar covalent bond (**Figure 2.8b**). Oxygen gas (O_2) and the carbon–hydrogen bonds found in hydrocarbons such as methane are also nonpolar. This may sound like a fairly trivial point, but in reality the consequences of this are enormous. As we will see, molecules that are composed of mostly polar covalent bonds and those that are composed of mostly nonpolar covalent bonds behave very differently.

Water is an important example of a polar molecule (see Figure 2.8a). The partial positive charges on the hydrogen atoms of one molecule will be attracted to the partial negative charge on the oxygen of a different water molecule. This means that all water molecules are attracted to each other because of opposing partial charges. Such interactions between separate molecules are termed **hydrogen bonds** (**Figure 2.9**). Each hydrogen bond

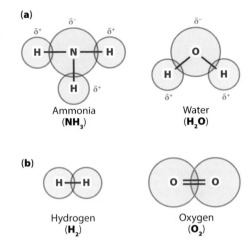

Figure 2.8 Polar and nonpolar covalent bonds. (a) Ammonia and water molecules are held together by polar covalent bonds. Because nitrogen in ammonia or oxygen in water attracts electrons more strongly than hydrogen, nitrogen and oxygen both obtain partial negative charges, represented by δ−. The hydrogens in these molecules, with less attraction (lower affinity) for the electrons, each obtain a partial positive (δ+) charge. (b) In hydrogen and oxygen molecules, electrons are shared equally. Because there are no partial charges, the covalent bonds holding the atoms together are nonpolar covalent bonds.

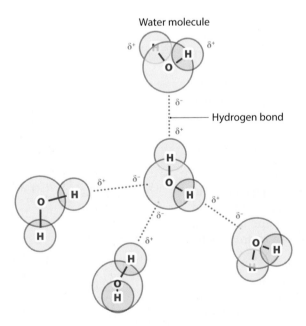

Figure 2.9 Hydrogen bonding between water molecules. Partial charges on separate water molecules attract each other, forming hydrogen bonds. Note that hydrogen bonds are bonds between separate molecules, while a covalent bond is a bond within a single molecule.

is very weak, but collectively, many hydrogen bonds can exert considerable attraction. For instance, when water is boiled, individual water molecules will not break off and leave solution as vapor until enough heat is applied to disrupt the hydrogen bonds holding the water molecules together.

It is not just water molecules that attract each other. Ammonia, as we have seen, has a partial negative charge on nitrogen and a partial positive charge on each hydrogen. Ammonia will form hydrogen bonds not only with water but also with other ammonia molecules. Nonpolar compounds, on the other hand, which lack partial charges, cannot form hydrogen bonds with water and other polar compounds.

Water is essential in life processes

Water is fundamental to all life. As we will discuss in Chapter 9, life itself evolved in water and all biochemical reactions take place in an aqueous or water environment. Environments that are totally lacking in water are devoid of living things. If water is present, it is likely that life will be present as well. For instance, researchers have found microorganisms in pockets of liquid water over 10 meters deep in frozen Antarctic lakes, in boiling hot oceanic volcanic vents, and even in water vapor found in clouds.

The ability or inability of a compound to interact with water is a crucial characteristic of that compound. We have already seen that polar compounds can interact with water. Nonpolar molecules cannot; lacking partial charges, they are unable to form hydrogen bonds with water. A simple example is ordinary table sugar, also called sucrose ($C_{12}H_{22}O_{11}$). Sucrose is largely polar in nature. When you place a teaspoon of sugar in water, the individual sugar molecules form hydrogen bonds with the water and disperse through the solution; the sugar dissolves. Alternatively, when you drop a piece of fat into water, nothing happens. Fat molecules are mostly made of hydrocarbons, which are nonpolar and consequently cannot dissolve in water.

It is not only polar compounds that will dissolve in water; anything with electrical charges will be attracted to water's partial positive and negative charges. For instance, when you place an ionically bonded compound such as table salt (NaCl) into water, positively charged sodium ions are attracted to the partial negative charges on the oxygen in water molecules. Similarly, negatively charged chlorine ions are attracted by the partial positive charges on the hydrogens. Ultimately these attractive forces pull apart the ionic compound, separating it into individual ions. Hence, the salt also dissolves in water. Consider the odd fact that even in saltwater there is little or no actual salt; it is largely dissolved into individual ions. It is only when water is removed, for instance by evaporation, that the positive and negative ions reassociate to form salt.

Polar covalent and ionic compounds that interact well with water in this fashion are termed **hydrophilic**, from the Greek term for "water loving." Nonpolar covalent compounds that cannot dissolve in water are referred to as **hydrophobic** or "water fearing." While molecules such as salt or oil are purely hydrophilic or hydrophobic, respectively, many molecules have both hydrophilic and hydrophobic regions and will consequently interact with water to greater or lesser degrees.

Acids and bases increase or decrease the concentration of protons in water

CASE: STOMACH PAIN AND STOMACH BUGS

A peptic ulcer is no picnic. Although some individuals have only mild symptoms, others suffering these breaks in the stomach lining experience nausea, vomiting, and loss of appetite. More serious symptoms include bleeding into the stomach. For decades, conventional medical wisdom held that these ulcers were caused by stress, consumption of spicy foods, or an over-secretion of stomach acid. Not anymore. Two Australian physicians, Drs Barry Marshall and Robin Warren, have conclusively shown that most peptic ulcers are caused by the bacterium *Helicobacter pylori*. When they first proposed the idea, they were ridiculed by the medical community, in part because it was believed that the stomach, with a pH of 2–4 (and as low at 1–2 after a meal), was too acidic to allow bacteria to grow. Ultimately, however, it was conclusively demonstrated that these bacteria can thrive in the stomach and that they are the primary culprits behind peptic ulcers. Such ulcers are now routinely treated with antibiotics.

1. **What exactly is acidity? What does a pH of 2–4 really mean?**
2. **How do *Helicobacter pylori* survive and grow in such an acidic environment?**
3. **How can the growth of these bacteria result in a break in the stomach lining?**

Before we can answer these questions, we need to learn a little bit about acids and bases. Not only will this discussion allow us to better understand how bacteria are involved in ulcers, but also, as we saw in this chapter's introduction, how mining and microorganisms conspire to harm freshwater ecosystems, as well as many other aspects of microbiology.

Recall that all atoms in water carry partial charges. Occasionally, the combined strength of attraction between oxygen and hydrogen on different water molecules actually pulls a water molecule apart. Oxygen atoms with partial negative charges can sometimes pull a hydrogen atom's proton off its molecule, leaving the shared electron pair behind with the remaining portion of the water molecule. This remnant of water picks up a negative charge, because it now has both of the previously shared electrons, and it is termed a hydroxide or OH^- ion. By the same token, the proton that was pulled off into solution is a positive H^+ ion. In pure water, every time an OH^- is formed, an H^+ is also generated. Thus, the concentration of these two ions stays exactly equal. The reaction is considered reversible, because an OH^- and an H^+ might easily attract each other and recombine to re-form H_2O.

Pure water, however, is exceedingly rare. Water usually exists as a **solution**. A solution is composed of a liquid and whatever is dissolved in it. Remember that for something to dissolve in water, it must be hydrophilic in nature. Many hydrophilic substances, when placed in water, form numerous hydrogen bonds with water molecules and might on occasion lose a proton into the solution. In this case, the concentration of H^+ is now higher than the concentration of OH^-.

Anything that raises the concentration of H^+ in this manner is termed an **acid**. As an example, consider hydrochloric acid (HCl). Hydrochloric acid is an ionic compound. Hydrogen loses an electron to chlorine, and the attraction between the positive hydrogen and negative chlorine forms the ionic

(a)

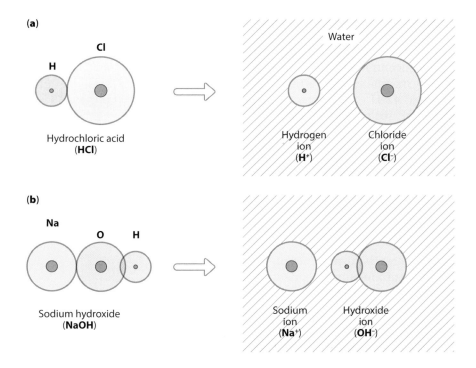

Figure 2.10 Concentration of H$^+$ ions in acidic and basic solutions. (a) If hydrochloric acid is added to water, it will dissociate into H$^+$ and Cl$^-$ ions, raising the concentration of H$^+$ ions in solution. (b) When a base such as NaOH is placed in water, it will dissociate into Na$^+$ and OH$^-$ ions. Because OH$^-$ will readily combine with free H$^+$ ions to form water, the concentration of H$^+$ in solution decreases.

bond. When placed in water, however, the positively charged hydrogen is readily pulled off into solution by the oxygen atoms in water molecules, which each carry a partial negative charge. When the acid dissociates in this manner, negative chlorine ions are left in solution (**Figure 2.10a**).

In contrast to an acid, a **base** is any compound that lowers the concentration of H$^+$ ions in solution. For example, sodium hydroxide (NaOH) is a compound consisting of a positive sodium ion (Na$^+$) ionically bonded to a negative hydroxide ion (OH$^-$). When placed into water, sodium hydroxide dissociates into Na$^+$ and OH$^-$. The negatively charged hydroxide ions readily bind to any H$^+$ ions that may be present in solution and thus lower the acidity by essentially acting as proton sponges (**Figure 2.10b**).

Acids and bases are damaging to living tissue. Hydrogen ions are simply naked protons, and these positively charged particles are extremely reactive. They will essentially latch onto anything with a negative charge, and in doing so, they disrupt the structure of molecules to which they attach. Bases are also highly reactive (bleach, for example). Because of their high affinity for hydrogen ions, they tend to remove protons from various molecules, damaging their structure.

So how does *H. pylori* manage to survive in the stomach (**Figure 2.11a**)? The bacteria employ several strategies that permit them to thrive under exceedingly acid conditions. For one thing, they tend to live in the lower parts of the stomach and the upper reaches of the small intestine, where the acidity is not quite as great. Second, the proteins we eat are mainly digested in the stomach. A compound called urea is a waste product of this digestion. The bacteria are able to break down this urea to CO$_2$ and ammonia. Ammonia is highly basic, so it neutralizes the acid in the vicinity of the bacteria. In other words, the bacteria create small islands of neutrality within the acid stomach that are more compatible with survival. The growth of the bacteria causes damage to the lining of the stomach, which leads to inflammation called gastritis. This inflammation results in the peptic ulcer and can ultimately perforate the stomach lining (**Figure 2.11b**).

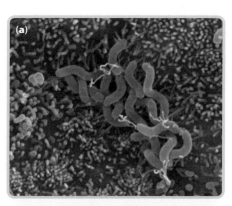

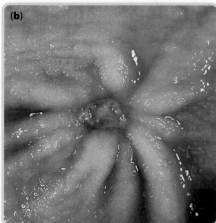

Figure 2.11 *Helicobacter pylori* and peptic ulcers. (a) Spiral-shaped *Helicobacter pylori* growing on the lining of the stomach. (b) The small, oval whitish area in the center of the image is a peptic ulcer, caused by the growth of *H. pylori*. Although up to half of all humans harbor these bacteria in their stomachs, only about 20% of those infected develop peptic ulcers.

Acidity is measured with the pH scale

To measure the acidity of any solution, we rely on the **pH scale** (**Figure 2.12**). This scale provides a way to exactly quantify the concentration of H^+ ions. The scale runs from 1 to 14, with a pH of 7 indicating that the solution is neutral because the concentration of H^+ is equal to that of OH^-. As H^+ ions are released into solution by an acid, the value on the pH scale drops below 7, whereas the pH value is greater than 7 when the concentration of H^+ is less than that of OH^-. The pH scale is logarithmic, meaning that a solution of pH 6 is actually 10 times more acidic than a solution of pH 7. The stomach, for example, with a pH of roughly 2 is 10 times the acidity of vinegar with a pH of 3 and 100 times more acidic than tomatoes, with a pH of 4.

Organic molecules are the building blocks of life

Living things can be distinguished from inanimate matter by the presence of **organic molecules**: molecules largely composed of carbon atoms bonded to one another and to hydrogen. Other types of atoms may also be bonded to carbon. With four outer electrons, carbon is able to form four stable covalent bonds. Carbon atoms therefore can link together to form complex chain and ring structures (**Figure 2.13**).

Carbon is unique in its ability to form such structures. The covalent bonds formed by carbon are very strong because the shared electron pairs are only in the second electron shell and therefore close to the positively charged nucleus. Because of the relative strength of these bonds, carbon compounds are quite stable in the environment and are thus ideal for serving as life's building blocks. As we will see in this section, all important biological molecules are based on modifications of hydrocarbon compounds.

Silicon, with four outermost electrons, can form four covalent bonds. You might conclude, then, that silicon compounds could also serve as the basis for biological molecules. Silicon, however, has its outermost electrons in the third electron shell. Because these electrons are further from the nucleus

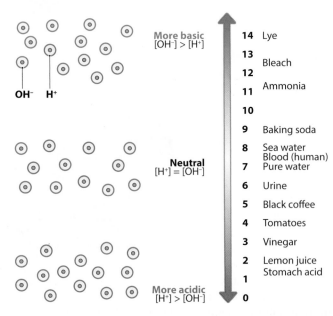

Figure 2.12 The pH scale, showing the pH of several common substances. A pH of 7 indicates that the concentrations of H^+ and OH^- are equal. A pH greater than 7 indicates that the concentration of OH^- exceeds that of H^+. A pH less than 7 means that the concentration of H^+ is greater than that of OH^-.

and therefore at higher energy, any covalent bonds formed by silicon will be inherently less stable than those formed by carbon. Because they are less stable, long chains of silicon do not persist. In spite of this fact, science fiction writers and filmmakers sometimes describe their aliens as silicon-based life forms, unlike the carbon-based life forms found here on Earth. But such beings must remain in the realm of fantasy; the laws of chemistry forbid their actual existence.

Organic molecules differ in other ways besides their number of carbon and hydrogen atoms. In many cases, one or more hydrogen atoms of a hydrocarbon molecule are replaced by a different atom or group of atoms. For example, the hydrocarbon ethane (C_2H_6) can be converted into ethyl alcohol (C_2H_5OH), also known as ethanol, by replacing one of the hydrogen atoms with an OH group (**Figure 2.14**). An OH group is known as a **hydroxyl group**, and it is very polar. As a result, ethane is nonpolar and hence hydrophobic, whereas ethanol is polar and hence hydrophilic.

Four types of modified hydrocarbons are of particular importance in microorganisms and other living things: **carbohydrates**, **lipids**, **proteins**, and **nucleic acids**. Due to the manner in which they are modified, these four classes of biological molecules have certain characteristic properties and are used by living things for specific purposes. In the remainder of this chapter we will consider each of these four types of biological molecules.

Carbohydrates function as energy storage and structural molecules

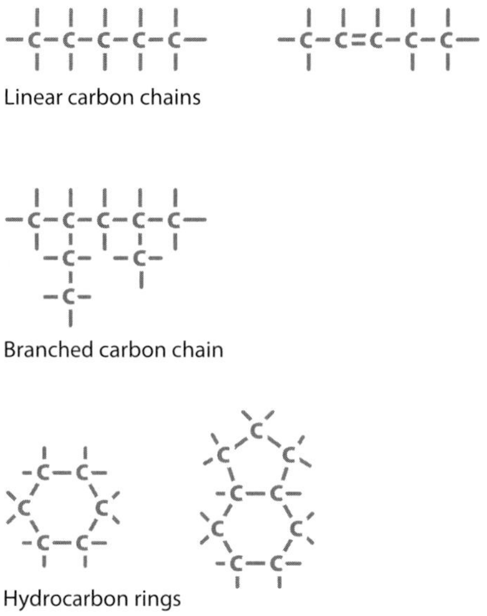

Figure 2.13 Some hydrocarbon structures. Each carbon forms four covalent bonds, while each hydrogen forms one. Carbon has unique structural flexibility because of its ability to form long chains and other structures. Carbon can form single or double bonds with other carbons and may form linear chains, branched chains, or simple or complex ring structures. Each dash indicates a covalent bond between carbon and hydrogen.

CASE: BACTERIAL HORSE HELPERS

Lucky is a 16-year-old Arabian stallion who lives on a grass-filled pasture. Much of his diet consists of the rich Kentucky bluegrass regularly available to him. When he suddenly loses his appetite, his owner becomes alarmed, especially after noticing that Lucky's stools are looser and darker than normal. He calls the veterinarian, who suspects bacterial diarrhea. The veterinarian prescribes a course of ampicillin for Lucky. The veterinarian also provides the owner with a tube of Bene-Bac® with instructions that Lucky should be given a daily dose while on antibiotics. Although the bacteria that cause equine diarrhea are often resistant to ampicillin, Lucky responds well to the treatment, and recovers fully without complications. Bene-Bac® is a veterinary preparation commonly used in situations such as this. It consists of a paste containing live cultures of various bacterial species, considered to be beneficial bacteria for horses and other herbivores.

1. **What is meant by beneficial bacteria?**
2. **Why would the veterinarian recommend Bene-Bac® for Lucky, specifically while he is being treated for bacterial diarrhea?**

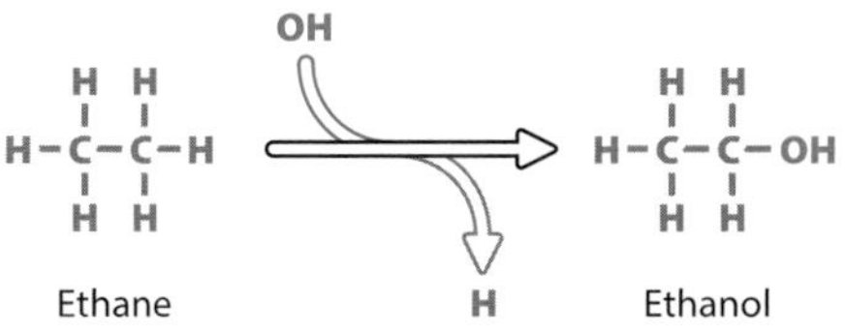

Sugars, starches, and cellulose are familiar examples of carbohydrates. This group of molecules is principally composed of carbon, hydrogen, and oxygen. Carbohydrates are important structural and energy-storage compounds in microorganisms and other living things. An important characteristic of sugars is their ability to dissolve in water. When you put table sugar into a cup of coffee, the sugar dissolves because the many polar regions on the sugar molecules form hydrogen bonds with the water in the coffee. This causes the individual sugar molecules to disperse throughout the solution.

The size of carbohydrate molecules spans the gamut from small to very large. The simplest carbohydrates are composed of a single sugar molecule and are

Figure 2.14 Substitution in hydrocarbon molecules. Ethane is composed entirely of nonpolar covalent bonds. It is therefore hydrophobic and unable to dissolve in water. If, however, a hydrogen in ethane is replaced by a hydroxyl (OH) group, the result is ethanol, an example of a class of compounds called alcohols. Because the bond between O and H in the hydroxyl group is a polar covalent bond, each of these two atoms obtains a partial charge, allowing it to form hydrogen bonds with water. The molecule is now hydrophilic and can dissolve in water.

termed **monosaccharides**. A given monosaccharide may contain from three to six carbons. Many important sugars, such as glucose and fructose, are six-carbon sugars. Both of these monosaccharides have the chemical formula $C_6H_{12}O_6$, but the manner in which these atoms are assembled is slightly different in each one. Molecules such as these, which have identical chemical formulas but different structures, are known as **isomers** (**Figure 2.15a**).

Two monosaccharides can be covalently linked together to form a **disaccharide**. The reaction that forms the bond is called a **condensation reaction** (**Figure 2.15b**). In a condensation reaction, a molecule of water is removed as a bond is formed between the two sugars. Two glucose molecules can be joined in this way to form maltose (malt sugar), while a glucose and a fructose can form the disaccharide sucrose (ordinary table sugar). Lactose (milk sugar) is the disaccharide formed by joining glucose with a different six-carbon sugar called galactose.

Condensation reactions are reversible. In other words, a disaccharide can be split back into monosaccharides by reinserting a molecule of water. Such a reaction is termed a **hydrolysis reaction**.

To either remove or replace water to either form or break apart a disaccharide, a specific biological catalyst is required. Such catalysts are called **enzymes**. Enzymes are chiefly proteins (there are a few exceptions), and each enzyme is highly specific, participating in only one or a few reactions. The enzyme that acts on lactose, for instance, is called lactase and it participates only in this reaction. Microbes able to synthesize this enzyme are able to break lactose into monosaccharides, which can then be used for energy. Other microorganisms lacking this enzyme cannot utilize lactose. When dairy products (like milk) that contain lactose become spoiled, it is generally because lactose-digesting bacteria have broken down the disaccharide. The bacterial waste product, lactic acid, gives the milk its sour taste.

Many carbohydrates consist of hundreds or thousands of monosaccharides, bonded together to form **polysaccharides**. Starch, for instance, is a polysaccharide consisting of many glucose molecules. Starch is the primary molecule used by plants for energy storage and is present in many foods that are derived from plants. When an animal eats food containing starch, the carbohydrate is digested via hydrolysis. The fact that animals, including humans, can digest starch indicates that we have the specific enzymes necessary to do so. One such enzyme found in human saliva cleaves the large starch molecules into many maltose disaccharides. The maltose sugars can then be digested to glucose in the intestine by the enzyme maltase. These individual

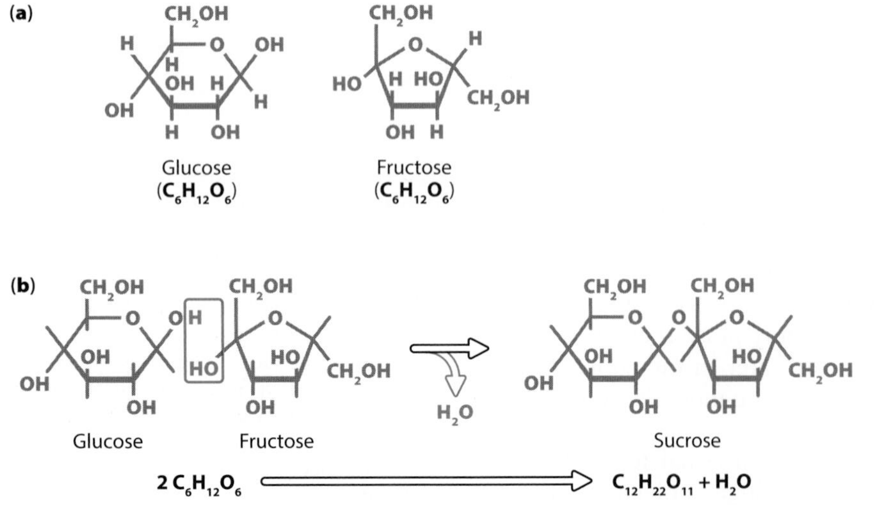

(a)

Glucose
($C_6H_{12}O_6$)

Fructose
($C_6H_{12}O_6$)

(b)

Glucose   Fructose   Sucrose

$$2\ C_6H_{12}O_6 \Longleftrightarrow C_{12}H_{22}O_{11} + H_2O$$

Figure 2.15 Monosaccharides and disaccharides. (a) Glucose and fructose both have the same chemical formula ($C_6H_{12}O_6$) but different structures. They are therefore isomers of each other. (b) Two monosaccharides can be linked via condensation to form a disaccharide. In a condensation reaction an OH is removed from one sugar and an H is removed from the other sugar. The OH and H join together, forming a molecule of water.

glucose sugars are used to meet immediate energy needs. Alternatively, an animal might store extra glucose in the form of glycogen as a reserve to meet future energy demands. Glycogen and starch are very similar in structure—both molecules consist of covalently linked glucose sugars. Starch and glycogen differ only in the exact position of the bonds holding the individual sugars together (**Figure 2.16**).

Cellulose is another polysaccharide composed of glucose subunits. All plant cells are surrounded by a rigid cellulose wall, which provides the cell with structural support. Animals, therefore, encounter cellulose whenever they eat plants. Animals, however, lack the enzyme cellulase, so they are unable to digest the bonds between the glucose sugars in cellulose. A certain amount of such indigestible material or dietary fiber in our diet has been equated with a healthy gastrointestinal tract.

Which brings us back to Lucky. How can horses and other herbivores survive on a diet of grass, which is mostly cellulose (**Figure 2.17**)? In the case of humans, even though cellulose is composed of glucose, and thus a potentially rich source of energy, we cannot use it, because we lack the necessary enzymes. Many bacteria, however, produce cellulase, allowing them to use the glucose in cellulose. Herbivores like Lucky have innumerable bacteria in their intestine that do much of their digestion for them. When a horse eats grass, these beneficial intestinal bacteria break down the cellulose in the grass. There is far more glucose released than the bacteria can use, and the excess is available to the horse. But in Lucky's case, in addition to these normal bacteria, he was also infected with organisms causing his diarrhea. When the veterinarian gave him an antibiotic to destroy these pathogens, many of his normal bacteria were also killed. Consequently, he could temporarily become somewhat less able to digest grass. The bacteria found in the Bene-Bac® paste, however, help to replace these organisms. Once Lucky was removed from the antibiotics, additional intestinal bacteria were acquired from the environment as he grazed normally.

Lipids are relatively hydrophobic molecules and are also used for energy storage and structure

Lipids are familiar to us as fats, oils, and cholesterol. Like carbohydrates, they are used by microorganisms and other living things for structure and energy storage. If you compare a lipid molecule and a carbohydrate molecule, however, you will note several important differences. First, both molecules are primarily composed of hydrogen, carbon, and oxygen, but the lipid contains less oxygen and has more carbon–hydrogen bonds than the

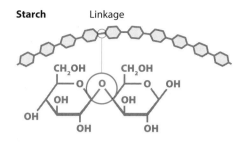

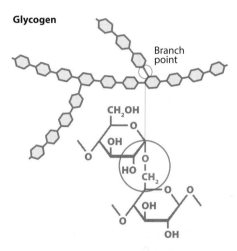

Figure 2.16 Representative polysaccharides. Both starch and glycogen are long polysaccharides composed entirely of glucose. The only difference is the extensive branching typical of glycogen.

Figure 2.17 Herbivores feeding. Herbivores like sheep are able to survive on a cellulose-rich diet, thanks to the billions of cellulose-digesting bacteria living in their intestines.

carbohydrate. As we will see in Chapter 8, carbon–hydrogen bonds are the primary type of bonds that cells use as an energy source. The energy in the shared electrons in such bonds is harvested and ultimately utilized by cells for all energy-requiring processes, such as movement, transport of materials in and out of the cell, and construction of molecules. Because a lipid has more carbon–hydrogen bonds, it has more potential energy per unit weight than carbohydrates do. Calories are a measure of such potential energy. Ice cream has more calories than an equally sized portion of bread, reflecting the high lipid content of the ice cream relative to the bread and the higher proportion of energy-yielding carbon–hydrogen bonds.

A second important difference is that lipids such as fats and oils do not dissolve in water (**Figure 2.18**). Because they lack hydrophilic regions, lipids are generally very hydrophobic. With few or no atoms capable of hydrogen-bonding with water, lipids will not disperse through solution but tend to aggregate together and interact with the water as little as possible.

A typical fat molecule provides an example of lipid structure (**Figure 2.19a**). Each fat molecule is composed of one molecule called glycerol and three fatty acid chains. Each fatty acid is linked to glycerol via a condensation reaction between the glycerol and the three fatty acids. Because there are three reactions, three molecules of water are produced in fat synthesis. To digest a fat, rather than producing water, water is added via a hydrolysis reaction. The water introduced into these reactions reconstitutes the glycerol and fatty acids. Just as each reaction in the formation or breakdown of a carbohydrate requires a specific enzyme, so does lipid synthesis or digestion. Such enzymes are collectively called **lipases**.

The hydrocarbon chains in a fatty acid may be of variable length. Stearic acid, found in beef fat, consists of 18 carbons (**Figure 2.19b**). Butyric acid, found in butterfat, has only four carbons. Furthermore, the carbons in the fatty acid chain may be bound to other carbons via single or double covalent bonds. Because carbon can only form a total of four covalent bonds, it can form fewer bonds with hydrogen if it is double-bonded to another carbon. In **saturated fats**, carbons are linked exclusively by single covalent bonds. Carbons are thus bound to the maximum number of hydrogens; they are saturated with as much hydrogen as they can possibly bind. If there are double bonds between the carbons in a fatty acid, the fat is termed **unsaturated** because it holds fewer than the maximum number of hydrogens (Figure 2.19b).

Saturated and/or longer fatty acids tend to make a fat more solid, because individual fat molecules can be packed together more tightly. Shorter hydrocarbon chains and/or double bonds make a fat more liquidlike due to the looser packing of the individual fat molecules. The only difference between a fat and an oil is that the fat is solid at room temperature while the oil is liquid.

Consider the natural style peanut butter you might buy at the market. The unsaturated fatty acids in peanut oil give it a liquid consistency that must be stirred in with the nonlipid portion prior to use. Because some people find this inconvenient, manufacturers often bubble hydrogen gas through the peanut oil. This converts some of the unsaturated fatty acids to saturated ones, thereby making the product more solid at room temperature. These saturated oils are associated with an increased risk of heart disease, but at least you don't have to stir your peanut butter before spreading it on an English muffin.

Figure 2.18 Oil and water do not mix. In Oahu's Pearl Harbor, diesel fuel continues to leak from the USS *Arizona*, more than half a century after the ship sank. Diesel fuel, like other hydrocarbon fuels, is unable to interact with water owing to its hydrophobic nature. Consequently, it forms an oil slick on the water's surface. New technologies using oil-digesting microbes are currently being considered as a way to combat oil spills. This topic will be more fully explored in Chapter 17.

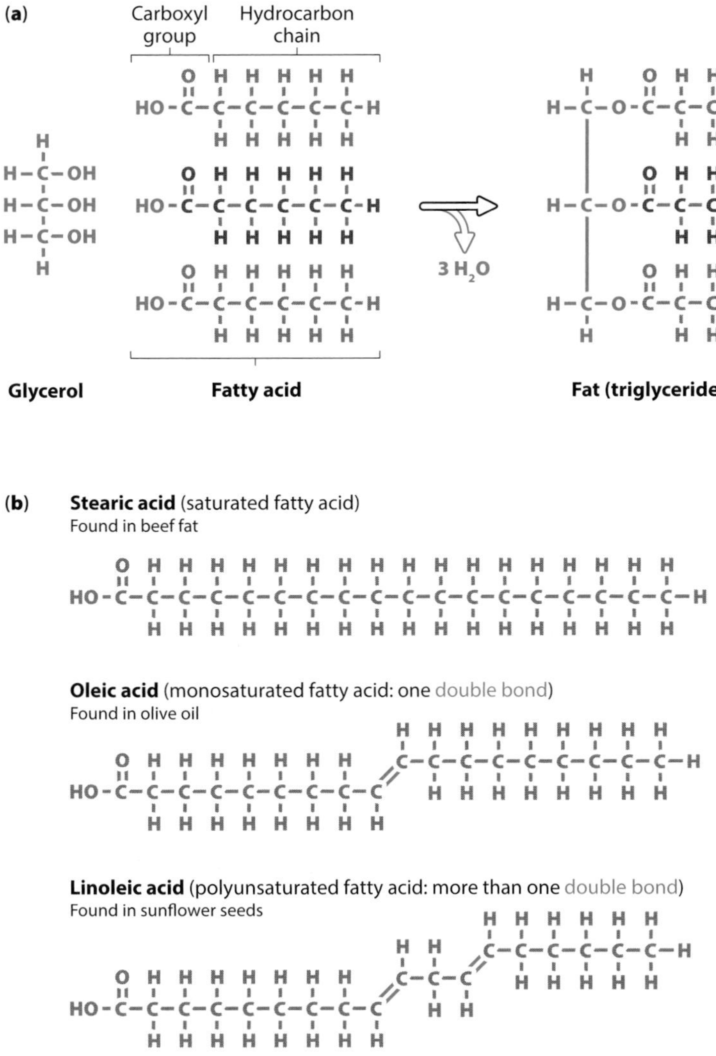

(a) Glatter | Fatty acid | Fat (triglyceride) molecule

(b) **Stearic acid** (saturated fatty acid)
Found in beef fat

Oleic acid (monosaturated fatty acid: one double bond)
Found in olive oil

Linoleic acid (polyunsaturated fatty acid: more than one double bond)
Found in sunflower seeds

Figure 2.19 Fats and fatty acids.
(a) Formation of a fat. Three fatty acids are linked to one molecule of glycerol via three condensation reactions. (b) Representative fatty acids. Note that although all three happen to have equal numbers of carbon atoms (18), they differ in the number of hydrogens. Saturated fatty acids, such as stearic acid, have no carbon–carbon double bonds and therefore are bound to a maximum number of hydrogens; they are saturated with hydrogen. Oleic acid, with one carbon–carbon double bond, is termed monounsaturated. Other fatty acids, such as linoleic acid, have more than one carbon–carbon double bond and are called polyunsaturated fatty acids. Each double bond reduces the potential number of carbon–hydrogen bonds. Note the twists in the structure of the unsaturated fatty acids, caused by the C=C double bonds.

Proteins participate in a variety of crucial biological processes

CASE: BAD POTATOES

On April 19, 2015, about 80 people attended a church potluck in Lancaster, Ohio. Within a few days, several individuals who were at the potluck sought medical treatment for symptoms that included difficulty breathing, nausea, dry mouth, abdominal pain, and double vision. These patients were diagnosed as suffering from botulism. Ultimately there were 21 confirmed and 10 additional suspected cases of this disease. One patient, a 54-year-old woman, died. Several others required prolonged hospitalization and mechanical ventilation to assist with breathing. Investigations traced the outbreak to home-canned potatoes, which had been used to make potato salad for the event. Those who did not consume the potato salad were unaffected.

1. What is the precise cause of botulism?
2. Is there anything specific about potatoes that make them a likely source of botulism? What, if any, is the link to home canning?
3. What other foods have similar risks and why?

Botulism is a rare disease, but when it strikes it can be spectacularly deadly. The culprit is the bacterium *Clostridium botulinum* (**Figure 2.20**), but the microorganism does not cause the disease directly. Rather, in appropriate environments, *C. botulinum* produces a protein that is highly toxic to the nerves that operate muscles. Since the ability to breathe depends on the action of the muscular diaphragm, exposure to the protein toxin generally results in the sorts of respiratory problems seen in some of the patients. Botulism toxin is perhaps the most lethal toxin known, but why the connection with certain foods? To answer that, we first need to understand something about a third class of biological molecules—the proteins.

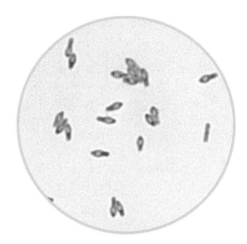

Figure 2.20 *Clostridium botulinum,* **the agent of botulism.** When exposed to oxygen, *C. botulinum* bacteria persist as metabolically inactive spores (small green structures). However, when such spores contaminate oxygen-free environments such as an improperly canned food item, they germinate into active bacterial cells (elongated red structures containing the spores). These cells then synthesize the protein toxin that causes botulism.

Proteins are large molecules that are involved in diverse functions, ranging from structure to host defense. They are involved in transport into and out of cells and intracellular communication, and they are essential as enzymes. And, as we saw in our case, they can sometimes cause life-threatening medical problems.

The basic building blocks of proteins are **amino acids**. There are 20 different amino acids, and while they all have similar structures—all contain an NH_2 group and a COOH group—they differ in terms of their **R group** (**Figure 2.21**). It is the R group that gives each amino acid its unique properties. Those with polar R groups are hydrophilic, meaning they can form hydrogen bonds with each other or with other hydrophilic compounds. Those with positively or negatively charged R groups are likewise hydrophilic and will attract molecules or ions with opposite charges. Nonpolar R groups render an amino acid hydrophobic.

The number of amino acids in a protein is highly variable. A small protein might have a few dozen, while a larger protein may have hundreds. Amino acids are linked together in a manner that should by now be familiar. A molecule of water is removed enzymatically, resulting in the formation of a covalent bond. An OH, removed from one amino acid, joins with an H, removed from an adjacent amino acid, to form the water molecule (**Figure 2.22**) (again, a condensation reaction). The new covalent bond is termed a **peptide bond**, and a short chain of amino acids (not yet a complete protein) is called a **peptide**.

A protein, however, is not merely a long, straight chain of amino acids. Rather, the amino acids fold up into a complex, and intricately folded, three-dimensional structure. The precise shape is crucial to a protein's function. Many proteins interact with other molecules much like pieces of a jigsaw puzzle. If the pieces don't fit, they cannot come together.

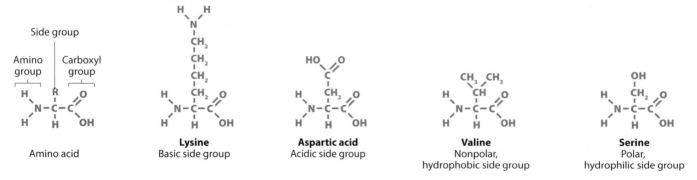

Figure 2.21 Amino acids, the subunits of proteins. Representative amino acids are shown. All 20 amino acids can be characterized as having basic, acidic, nonpolar, or polar side groups (also called R groups).

H
H
O
N-C-C
H
R
OH
H
R
O
N-C-C
H
H
OH
H₂O
Peptide
bond

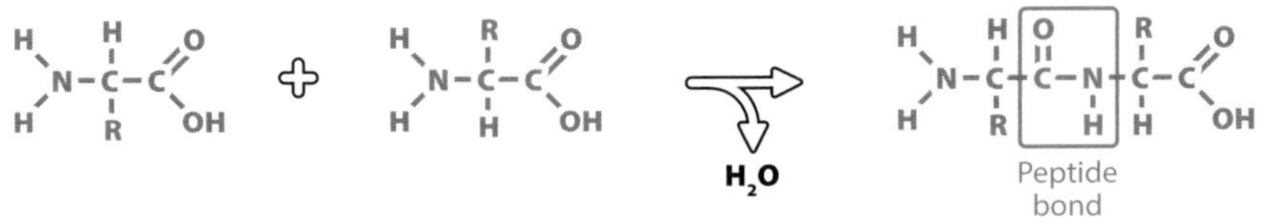

Figure 2.22 Peptide bonding. A peptide bond, highlighted in blue, forms by condensation between two amino acids.

The precise sequence of amino acids in a protein is the starting point for determining the specific shape of that protein. As we have discussed, amino acids differ in their properties and some are more likely to form bonds, such as hydrogen bonds, than others. The order of amino acids in a protein, therefore, affects overall protein structure. Different types of bonds may be formed between different amino acids in the protein as they are brought into contact with each other (**Figure 2.23**). A protein that is in its precise, three-dimensional shape is said to be in its **native** conformation.

Anything that affects these numerous interactions may cause the protein to change shape and any alteration of shape is likely to reduce protein function. If the three-dimensional shape of a protein is completely destroyed, the protein is rendered nonfunctional. A protein that has unfolded and lost its three-dimensional shape is said to be **denatured**. Many environmental factors, including high temperature and pH extremes, can disrupt weak attractive forces and denature a protein (**Figure 2.24**). A familiar example is the process by which milk spoils. As we discussed earlier in this chapter, when lactose-digesting bacteria gain access to milk, they release an acid waste as they digest the milk sugar. Besides giving the milk a sour taste, the acid also denatures proteins in the milk, converting them from the liquid to the solid curdled state.

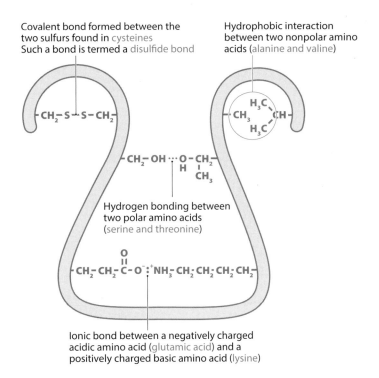

Covalent bond formed between the two sulfurs found in cysteines
Such a bond is termed a disulfide bond
Hydrophobic interaction between two nonpolar amino acids (alanine and valine)
CH2-S-S-CH2
CH3
H3C
CH
H3C
CH2-OH···O-CH2
H
CH3
Hydrogen bonding between two polar amino acids (serine and threonine)
O
CH2-CH2-C-O⁻ : ⁺NH3-CH2-CH2-CH2-CH2
Ionic bond between a negatively charged acidic amino acid (glutamic acid) and a positively charged basic amino acid (lysine)

Figure 2.23 Protein three-dimensional structure. Many interactions between amino acids take place to give a protein molecule its intricately folded shape. This diagram shows several of them.

Like other proteins, the botulism toxin requires a precise three-dimensional shape to adversely affect anyone unlucky enough to consume it. Only when folded up properly can it latch on to nerve cells and prevent them from releasing the chemicals that cause muscle contraction. The potato salad implicated in the Ohio outbreak was tainted with protein toxin that was likewise in its native shape. Had the potatoes been well heated after they were removed from their cans, it is likely that the toxin would have been denatured by the heat and consequently unable to cause disease.

The *C. botulinum* bacteria that produce the toxin cannot grow in the presence of oxygen. Therefore, most cases of botulism are associated with improperly canned food. The bacteria are present in the environment as nongrowing spores (see Figure 2.20), but if they get into a vacuum-packed can that was improperly sterilized, they begin to grow in the can and secrete toxin in the process. Spores on fresh food exposed to oxygen pose no risk because they are unable to grow. Interestingly, not all canned foods are equally risky, even without heating. You would have less to fear from highly acidic foods because low pH will also denature the toxin. Canned sardines packed in vinegar, for instance, would be less worrisome than sardines packed in oil. Canned tomatoes, because of their low pH, are an unlikely source of botulism contamination. But other canned foods, such as mushrooms or potatoes, which are at neutral pH and may be served cold, are occasionally implicated in botulism.

Biological reactions require enzymes functioning as catalysts

CASE: CALL IN THE CLOT BUSTERS

Dr Creek is a cardiologist in a large urban hospital. While on call late one night, he is summoned to the emergency room (ER) to treat an apparent heart attack victim. Upon arriving in the ER, he confirms that the patient has indeed suffered a severe heart attack and he quickly administers supportive measures. Among the treatments he orders is the immediate administration of intravenous (IV) streptokinase over the next hour. Streptokinase is actually a bacterial enzyme that is often used to treat heart attack victims. Dr Creek notes with satisfaction that blood flow to the heart tissue significantly increases, raising hopes for the patient's safe recovery.

1. **What exactly is an enzyme?**
2. **How does streptokinase increase blood flow to the patient's heart tissue? Can any enzyme do this, or is there something special about streptokinase?**

We have already referred to enzymes and the crucial role they play in biological reactions, such as the condensation and hydrolysis reactions that form or digest other biological molecules. Almost all biological reactions require a specific enzyme if they are to take place and almost all enzymes are proteins. Enzymes are so important that much of the difference that we observe between living things reflects the fact that different organisms have different enzymes and are therefore capable of different biological reactions.

We have recently discussed, for instance, the intestinal bacteria that digest cellulose. Not all bacteria, however, can digest this sugar. Those that cannot lack the necessary enzymes. Fortunately for our heart attack victim and for cardiologists like Dr Creek, certain species of *Streptococcus* bacteria produce the bacterial enzyme streptokinase. Streptokinase is a protein-digesting

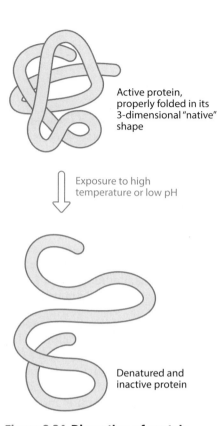

Active protein, properly folded in its 3-dimensional "native" shape

Exposure to high temperature or low pH

Denatured and inactive protein

Figure 2.24 Disruption of protein shape by environmental factors. Proteins can be denatured by conditions such as high temperature or low pH. Because protein function depends on shape, alterations in a protein's three-dimensional structure tend to reduce or eliminate protein activity.

enzyme known as a **protease** (the names of most enzymes end with "ase" and often provide clues as to what the enzyme acts upon). Most enzymes are very specific, and a protease will not necessarily digest any protein. In the case of streptokinase, the targeted protein is fibrin, an important component of blood clots. The enzyme interacts with fibrin, facilitating its digestion.

When *Streptococcus* bacteria invade host tissue, they use streptokinase to help free themselves when they are trapped inside a clot. These bacteria thus use streptokinase to help them spread through host tissues. Cardiologists take advantage of this bacterial enzyme to increase the survival rate of heart attack patients. Heart attacks are often caused when clots form in the arteries that supply the heart with blood. Without adequate blood supply, heart tissue quickly begins to die. But the immediate IV administration of streptokinase can dissolve these clots, restoring normal blood flow and preventing excessive damage to heart tissue.

When we say that enzymes facilitate specific biological reactions, this is not to say that a given reaction is impossible if the proper enzyme were missing. Enzymes, however, greatly speed up the rate at which reactions take place— often billions of times. For all practical purposes, life itself depends on enzymes. Fibrin might eventually break down without streptokinase, but neither *Streptococcus* bacteria nor heart attack patients can wait days or weeks until it does. To efficiently digest fibrin, the enzyme is required.

Because enzymes are highly specific, each one participates in only one or a very small number of reactions. A cellulose-digesting enzyme does not digest other carbohydrates. Streptokinase digests fibrin but not other molecules. This means that cells need a large number of different enzymes to carry out all of their necessary biochemical reactions.

Enzymes are necessary because biological molecules, as a rule, are fairly stable. They have to be, or they would spontaneously decompose and life itself would become impossible. Charcoal (a hydrocarbon), for example, placed in your barbecue does not spontaneously burst into flames. Even though burning the charcoal releases energy, a small amount of energy must first be added to get charcoal burning. This energy input might be in the form of a lighted match. Yet anyone who has ever lit a barbecue knows that one match rarely does the trick; you would have to use many matches to ordinarily light the coals. Or you could use a catalyst, like lighter fluid, to lower the number of matches you need to ignite the charcoal. With the catalyst, the amount of energy needed to get the reaction started, known as the **activation energy**, is greatly reduced. Enzymes also act as catalysts by lowering the activation energy needed for a reaction to occur (**Figure 2.25**).

Enzymes accomplish this by temporarily binding to the molecule to be altered (the **substrate**) at a special binding site on the enzyme called the **active site** (**Figure 2.26**). When the substrate is bound by the enzyme's active site, it is physically altered such that it begins to take on the shape of the final **product** of the reaction. In other words, the fit of the enzyme in the active site is not a perfect fit. When it is bound, the bonds in the substrate are stressed and can then be more readily altered into the product. In the case of streptokinase, these bonds are the peptide bonds in the fibrin protein. When these bonds are stressed, it becomes much easier for the required digestion to occur.

Nucleic acids direct the assembly of proteins

Living things typically produce thousands of different proteins, and in a very real way, an organism's structure and function are a consequence of the

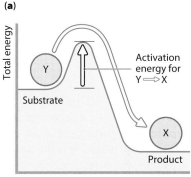

(a)

Uncatalyzed reaction pathway

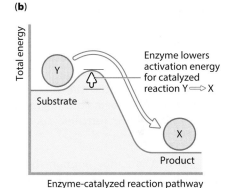

(b)

Enzyme-catalyzed reaction pathway

Figure 2.25 Effect of enzymes on activation energy. (a) Compound Y (a substrate) is in a relatively stable state, and energy is required to convert it to compound X (a product), even though X is at a lower overall energy level than Y. This conversion will not take place unless compound Y can acquire enough activation energy from its surroundings to undergo the reaction that converts it into compound X. This energy may be provided by increasing the temperature. The precise amount of energy needed to allow the reaction to proceed is called the activation energy. The reverse reaction (X → Y) will require a much larger input of energy to occur and consequently will take place only rarely. (b) Energy barriers for specific reactions can be lowered by catalysts. Enzymes are particularly effective catalysts because they greatly reduce the activation energy for the specific reactions they catalyze.

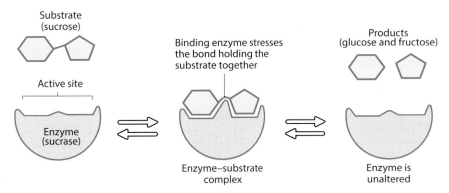

Figure 2.26 Enzyme action. In this example, the enzyme sucrase catalyzes the digestion of the disaccharide sucrose. The reaction is reversible; sucrase can also catalyze the synthesis of sucrose from the monosaccharides glucose and fructose. Enzyme specificity is a function of the active site, which can bind only specific substrates. Once the substrate is bound, an enzyme–substrate complex is formed. A change in the enzyme's shape strains the bonds in the substrate, reducing the energy necessary to break or rearrange them. The products are then released, leaving the enzyme unchanged and free to combine with new substrate molecules.

proteins it makes. Proteins also determine what carbohydrates and lipids are made or digested by an organism, because the synthesis or degradation of these molecules requires specific enzymes. Different organisms, however, make different proteins. It is the task of the fourth and final group of biological molecules, the nucleic acids, to specify exactly which proteins an organism can produce.

Deoxyribonucleic acid (DNA) is the genetic material in microorganisms and other living things. DNA is composed of individual building blocks called **nucleotides** (**Figure 2.27a**). Each nucleotide has three principal parts: a five-carbon sugar (deoxyribose), a phosphate (PO_4^{-3}) group, and a nitrogen-containing molecule with basic properties, known as a nitrogenous base. A nucleotide many have any of four different nitrogenous bases (**Figure 2.27b**). Cytosine and thymine are single-ring bases called **pyrimidines**, whereas the **purines**, adenine and guanine, have a double-ring structure.

A strand of DNA may contain thousands or millions of nucleotides, all covalently bound between the phosphate on one nucleotide and the sugar on the next. The entire DNA molecule consists of two such strands wound around each other like two spiral staircases in what is known as a **double helix** (**Figure 2.28**). Note how the two helices orient toward each other, with backbones of sugar and phosphate on the outside and the nitrogenous bases projecting in toward the interior of the helix. The bases on opposite sides of the helix are able to hydrogen-bond with each other, holding the two helices together. This base pairing is highly specific; a purine bonds only with a pyrimidine. Furthermore, adenine can only properly bond to thymine, while guanine pairs only with cytosine.

The sequence of nitrogenous bases on a strand of DNA specifies the exact amino acids that will be incorporated into a protein. The bases therefore represent a code, utilized for protein production.

Much of an organism's DNA is organized into **genes**. A gene is a sequence of DNA nucleotides that codes for a single protein or part of a protein. A single bacterium may have several thousand genes, indicating that it can produce several thousand proteins.

(a)

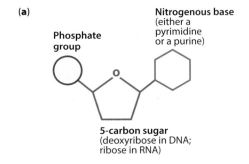

Phosphate group

Nitrogenous base (either a pyrimidine or a purine)

5-carbon sugar (deoxyribose in DNA; ribose in RNA)

(b)

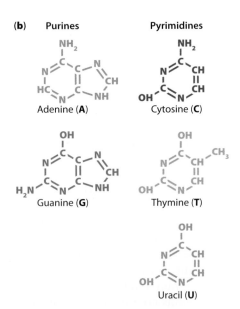

Purines

Adenine (**A**)

Guanine (**G**)

Pyrimidines

Cytosine (**C**)

Thymine (**T**)

Uracil (**U**)

Figure 2.27 Nucleotide and base structure. (a) A single DNA nucleotide is composed of a molecule of deoxyribose, a phosphate, and a nitrogenous base. In RNA, the sugar is ribose instead of deoxyribose. (b) Nitrogenous bases. Thymine is found only in DNA; in RNA, thymine is replaced by uracil.

Although DNA is composed of genes and contains the exact instructions for protein synthesis, DNA does not actually link the amino acids together to form the protein. Before a protein can actually be synthesized, the information encoded in a gene must be transferred to a second type of nucleic acid called **ribonucleic acid (RNA)**. RNA acts as a messenger, carrying DNA's genetic message to the actual site where amino acids are assembled into proteins (**Figure 2.29**). This flow of genetic information, from DNA to RNA to protein, is a fundamental process common to living things, microbial or otherwise. A convenient way to think about the different roles of DNA and RNA is to think of DNA as a cookbook, containing recipes for various proteins. RNA acts as the cook, using the recipes to construct the specified proteins. We will carefully consider this important concept in Chapter 7.

The structure of RNA is very similar to that of DNA, with three major differences. First, unlike DNA, RNA is usually a single-stranded molecule. Second, the sugar found in an RNA nucleotide is ribose rather than deoxyribose. Finally, RNA uses the nitrogenous base uracil in place of thymine (see Figure 2.27b).

Looking back and looking forward

The fundamental chemical principles governing the behavior of matter are identical, whether we are discussing the living or nonliving world. Understanding these principles therefore pays dividends when we try to understand how living things interact with each other and with their environments. A solid grounding in this chemistry is especially important when considering microorganisms, because so much of their impact is due to chemical changes they cause in their environments.

With some of these basic chemical concepts in hand, and in particular the chemistry of key biological molecules, we are now prepared to move on to an investigation of how biological molecules are assembled into cells—the basic unit of life. In Chapter 3, we will take a close look at cell structure and function. In doing so, we will come to understand, among other things, why some cells divide faster than others, how an antibiotic like penicillin kills some but not all bacteria, and how some cells can persist for hundreds or even perhaps millions of years, as inert spores in ancient clay pottery or in the stomachs of amber-encased insects.

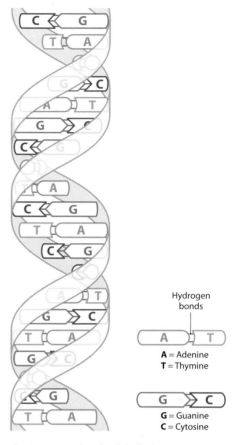

Figure 2.28 The double helix. Note the specific hydrogen bonding of adenine with thymine and guanine with cytosine.

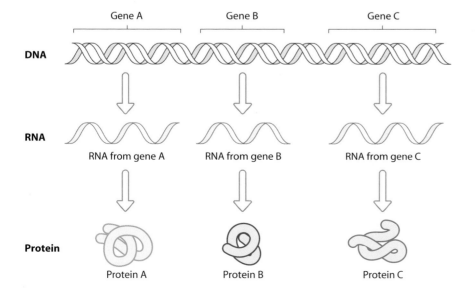

Figure 2.29 The flow of genetic information. Genetic information flows from DNA to RNA and from RNA to protein. Collectively this flow of information is known as gene expression.

Garland Science Learning System

- http://garlandscience.rocketmix.com/students/

- Discover why a fried egg doesn't look like a raw egg and why it's important that proteins fold like socks in a clothes dryer

- Test your knowledge of this chapter by taking the quiz

- Familiarize yourself with the terminology used in this chapter by using the vocabulary review

- Get help with the answers to the Concept questions

Concept questions

These questions are designed to help you start thinking like a microbiologist. The answers are not always simply found in the text. Instead, you will need to take the concepts about which you have learned and apply them to new situations. Some of the questions may not even have just a single correct answer. Help is provided as part of the GSLS resources, which can be accessed through http://garlandscience.rocketmix.com/students/.

1. Is an ionic bond more like a polar covalent bond or a nonpolar covalent bond?

2. Referring back to the case on *Helicobacter pylori* and how it survives in the stomach, the actual equation describing the digestion of urea is

$$CO(NH_2)_2 + H_2O \rightarrow 2NH_3 + CO_2$$
Urea   Ammonia Carbon dioxide

Thinking back to our description of how biological molecules are synthesized or digested, what do you think is the role of water here? What sort of reaction would you call this?

3. When a large ship sinks in the ocean, it will dissolve if given enough time. Why is this? Would the same thing happen if the ship were submerged in a nonpolar solution?

4. When H_2SO_4 is placed into water, it converts into H^+ and HSO_4^-. Why does this happen only when the molecule is placed in water? On the basis of this information, what type of molecule is H_2SO_4?

5. Imagine a room with 100 seats in it. One hundred ten people enter the room. In situation A, 100 people race to occupy the seats, pushing out of the way the 10 others, who don't seem to care much about sitting down. These 10 people don't get a seat. In situation B, 110 people once again enter the room, but 10 people, for some reason, run screaming in fear out of the room, leaving the seats for the 100 other people. If you consider the seats as water molecules and the 10 people who don't get seats as hydrophobic molecules, which situation, A or B, is a more accurate representation of the way they would respond to water?

6. Hair is basically composed of the protein keratin. Imagine a woman gets her naturally very straight hair done in a perm. The hair stylist applies certain chemicals to the hair and then sets the hair in the newly desired shape. What do you think the function of the chemicals is? On her way home, the woman stops at a store to pick up some shampoo. Assuming she wants to maintain her perm for as long as possible, does it matter if she gets a pH-balanced shampoo or not?

7. Very few microorganisms are able to digest lignin, a very large and complex carbohydrate found in plant material. What do you think distinguishes those microorganisms that are able to digest lignin from those that cannot?

8. We opened this chapter with a discussion of *Acidithiobacillus ferrooxidans* and how it affects streams near mines. This bacterium is unusual because it can live at pH as low as 3. Many other bacteria do best at about neutral pH (pH = 7). How many times more acidic is a pH of 3 than a pH of 7?

9. Suppose a particular DNA molecule contains 30% guanine nucleotides and 20% adenine nucleotides. How much thymine and cytosine does it contain?

10. Lifeguards often keep a bottle of vinegar in their lifeguard station. If a swimmer is stung by a jellyfish, the lifeguard will pour vinegar on the skin to reduce the stinging. Knowing that jellyfish toxin is a protein, and that vinegar is acidic, explain why the vinegar is effective.

THE CELL: WHERE LIFE BEGINS

Robert Hooke might be one of the most prolific inventors you've never heard of. Among his many creations were the balance wheel in mechanical watches, cross hairs in optical instruments, and the universal joint. The Englishman also made key scientific contributions. He developed some of the first hypotheses regarding the origin of fossils and he is often credited with the discovery of Jupiter's giant red spot. And when he peered through a crude microscope at a slice of cork from tree bark in 1665, he was intrigued to see that the cork appeared to be divided up into small individual components, invisible to the naked eye. Because each of these units resembled a small room or chamber, he called them cells.

Over time it became clear that what was true for trees was also true for all other living things. One of the foundations of modern biology, the **cell theory**, states that all living things, from the smallest bacterium to the largest redwood tree, are composed of one or more cells. As we will see in Chapter 5, viruses are acellular, and could therefore be an exception, depending on one's definition of life. Yet even viruses must enter cells in order to reproduce, and for those organisms that clearly show all the features of living things, the cell theory has remained valid for well over 150 years.

The cell theory also states that the cell is the fundamental unit of life. What exactly does this mean? In Chapter 1 we reviewed some of the features that distinguish living things from inanimate objects. Living things respond to their environment and they utilize energy. They reproduce and pass genetic information on to progeny. All life has evolved from other living things, and a living organism is highly organized compared with a nonliving object. Cells possess all of these features. Biological molecules, including proteins or DNA, discussed in Chapter 2, do not. Although such molecules and other subcellular components are organized into living cells, they themselves are not living.

Consequently, if we really want to understand living things, microbial or otherwise, we must investigate those entities in which life begins: cells. Here in Chapter 3, we will take a close look at cell structure and how this structure relates to function. By understanding basic cell structure, we learn a great deal about how a particular organism grows, reproduces, and otherwise goes about the business of living. We will discuss the features that all cells share, as well as those that account for the enormous diversity of cell types seen in nature. As we will see, microorganisms with different cell structures present us with very different problems when it comes to understanding and successfully coping with them. Thus, in addition to a consideration of basic cell biology, we will learn why, when you have seen one cell, you most certainly have not seen them all.

All case studies have a few questions at the end, the answers to which will become apparent as you read the sections following the case.

All cells are one of two basic types

Most of the organisms with which you are most familiar, including animals, plants, and fungi, are composed of **eukaryotic cells** (**Figure 3.1a**). Alternatively, many of the microorganisms we will consider in this text have **prokaryotic cells** (**Figure 3.1b**). When you visit the doctor because of strep throat, or if you suffer through a case of traveler's diarrhea on an overseas trip, you are experiencing problems caused by prokaryotes. On the other hand, if you use vinegar on your salad, or enjoy a cup of yogurt for lunch, you can thank the prokaryotes that produced them. Much of the oxygen we breathe is produced by prokaryotes and when animals and plants die, the nutrients in these organisms are returned to the environment, in large part due to prokaryotic activity. Bacterial cells are prokaryotic, as are the cells of another large group of microscopic, single-celled organisms—the archaea. Before we discuss the differences between prokaryotic and eukaryotic cells, however, let's consider some of the things they have in common.

Both prokaryotic and eukaryotic cells are composed of the same basic biological molecules—the carbohydrates, lipids, proteins, and nucleic acids described in Chapter 2. In both cell types, genetic information is encoded in DNA, and the production of proteins, while not identical, is very similar. Prokaryotes and eukaryotes both use the same sorts of nutrients to meet their nutritional and energy needs. Both harvest energy in similar ways.

All cells are surrounded by a membrane, composed largely of lipids and protein (see Figure 3.1). In other words, cells are membrane-bound. This **plasma membrane** (also commonly called the **cell membrane**) separates the inside of the cell from the outside world. The fluid-filled interior of the cell is called the **cytoplasm**. The cytoplasm contains a variety of small, subcellular structures that allow the cell to carry out its various functions. The nature of these structures varies considerably according to cell type. Frequently there are additional structural features exterior to the cell membrane. These may include a rigid cell wall or other structures involved in cell movement or adherence. The membrane also has special structures to allow substances to move from outside to inside the cell and vice versa.

There are also important differences that distinguish the two cell types. In addition to the cytoplasmic structures referred to above, eukaryotic cells generally have a variety of membrane-bound **organelles** (little organs) located in the cytoplasm (see Chapter 1, p. 4). As we will see, each organelle is associated with specific activities. The most prominent of these structures is the **nucleus**, which surrounds the cell's genetic material. Membrane-bound organelles, including the nucleus, are absent in prokaryotic cells. The names "eukaryotic" and "prokaryotic" are derived from the Greek terms for "true nucleus" and "before the nucleus," respectively. Prokaryotic cells do have genetic material, but it is present in the cytoplasm rather than surrounded by its own membrane. They also contain a handful of other internal structures without membranes, such as the ribosomes, utilized to link amino acids into proteins. Figure 3.1 illustrates the principal prokaryotic and eukaryotic cell features that will be described in this chapter.

The first difference between prokaryotic and eukaryotic cells that you might notice is their size disparity. Most prokaryotic cells are only a few micrometers long, whereas eukaryotic cells are often more than 10 times larger (**Figure 3.2**). As we will shortly see, there are some fascinating exceptions to this rule.

(a) Typical animal cell (eukaryote)

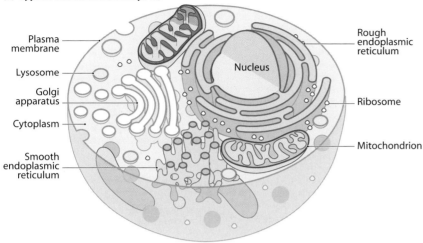

Plasma membrane
Lysosome
Golgi apparatus
Cytoplasm
Smooth endoplasmic reticulum
Rough endoplasmic reticulum
Nucleus
Ribosome
Mitochondrion

Typical plant cell (eukaryote)

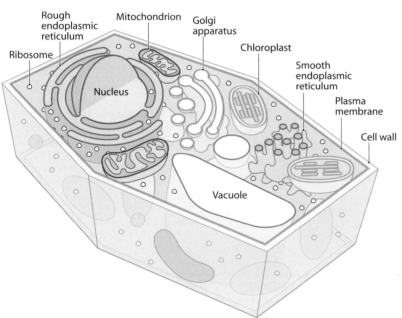

Rough endoplasmic reticulum
Ribosome
Mitochondrion
Golgi apparatus
Chloroplast
Smooth endoplasmic reticulum
Plasma membrane
Cell wall
Nucleus
Vacuole

(b) Typical bacterial cell (prokaryote)

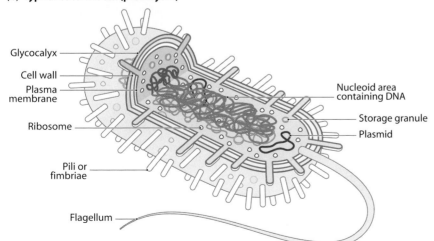

Glycocalyx
Cell wall
Plasma membrane
Ribosome
Pili or fimbriae
Flagellum
Nucleoid area containing DNA
Storage granule
Plasmid

Figure 3.1 Eukaryotic and prokaryotic cells. (a) Animal and plant cells are examples of eukaryotic cells. They differ from prokaryotes in that their DNA is always surrounded by a nuclear membrane ("eukaryote" comes from the Greek for "true nucleus"). Eukaryotic cells are typically more complex than prokaryotic cells, with a number of subcellular membrane-bound structures called organelles, which carry out various cell functions. Not all eukaryotic cells have all organelles. (b) A prokaryotic cell. Both bacteria and archaea have prokaryotic cells. The term "prokaryote" comes from the Greek for "before the nucleus," and prokaryotic cells are characterized by the lack of a nuclear membrane surrounding the DNA. Certain features of this cell, such as the pili and the flagella, are not present in all prokaryotes.

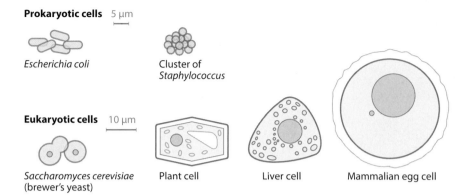

Prokaryotic cells 5 μm

Escherichia coli

Cluster of
Staphylococcus

Eukaryotic cells 10 μm

Saccharomyces cerevisiae
(brewer's yeast)

Plant cell

Liver cell

Mammalian egg cell

Figure 3.2 Sizes of various cells. Typical prokaryotic cells may be between 2 and 8 μm in length [1 micrometer (μm) = 1/1,000,000 or 1×10^{-6} meter]. The size of eukaryotic cells varies greatly but on average is about 10 times that of prokaryotic cells. To provide a sense of comparison, your thumb is approximately 20,000 μm across.

Cells can be stained for enhanced observation

Most cells are microscopic, so to see them requires the use of a microscope. Even with a microscope, cells would be difficult to see if they were not stained prior to observation. Staining the cell increases the contrast between the cell and its background, permitting easy viewing. We generally stain the cells we wish to view with a basic chemical such as methylene blue. Recall from Chapter 2 that bases accept protons and therefore acquire positive charges when in solution. Plasma membranes, on the other hand, carry negative charges. The positively charged stain adheres to the negatively charged plasma membrane, providing the necessary color and contrast for the cell to stand out from its background (**Figure 3.3**).

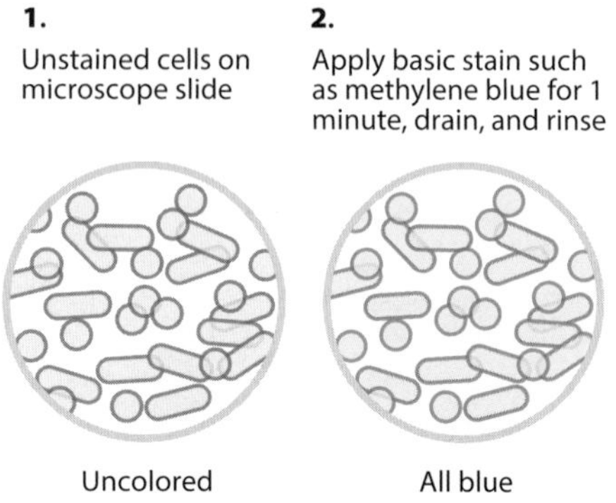

1.
Unstained cells on microscope slide

2.
Apply basic stain such as methylene blue for 1 minute, drain, and rinse

Uncolored

All blue

Figure 3.3 The simple stain. Cells are stained prior to examination under the microscope to increase contrast with the illuminated background. The basic stains, carrying positive charges, adhere to the negatively charged plasma membrane of the cells.

As cells get larger, their efficiency decreases

CASE: SIZE DOES MATTER!

Not all bacteria are so tiny. When *Epulopiscium fishelsoni* was discovered off the Australian coast in 1988, it was hailed as the largest prokaryotic organism known. With a length of up to 0.57 mm, and a width of 0.06 mm, they were easily visible to the naked eye and had a volume more than a million times greater than the cells of typical bacteria. But in 1997 a new heavyweight champion was found. *Thiomargarita namibiensis* (**Figure 3.4**), discovered in deep water bottom sediments off the coast of southwest Africa, has up to three times the volume of *E. fishelsoni*. Strains of this behemoth have since been found off the Mexican coast as well. Besides their gigantic proportions, *T. namibiensis* is unusual in other ways. For instance, the cytoplasm of *T. namibiensis* cells contains liquid-filled vacuoles, which take up much of the available space.

1. **How would the metabolism of *T. namibiensis* compare with that of ordinary-sized bacteria?**
2. **Would you expect these cells to reproduce more quickly or more slowly than other prokaryotes?**
3. **What is the relationship between the large size of these bacteria and the unusual vacuoles in the cytoplasm?**

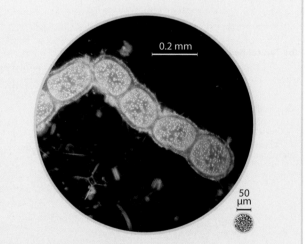

0.2 mm

50 μm

Figure 3.4 *Thiomargarita namibiensis*. This species, with cells ranging from 0.1 to 0.3 mm long, currently holds the record for world's largest bacteria. In comparison, *E. coli* and other normal-sized bacteria (seen in the small image to the lower right) are an average of 2 μm, approximately 0.7% the size of *T. namibiensis*. The bacteria are seen here as a chain of spherical cells.

Newcomers to microbiology often mistakenly assume that since eukaryotic cells are usually so much larger and more complex than prokaryotic cells, they must somehow be better at carrying out their functions. They are not. In many ways prokaryotic cells, because of their simplicity, are actually more efficient. Small size in particular gives prokaryotes an edge when it comes to metabolic efficiency—the ability to use energy effectively.

All cells must absorb nutrients from their environment and excrete waste products. As cell size increases, the need for nutrients and waste excretion goes up. All traffic into and out of the cell must cross the plasma membrane (see Figure 3.1). Having a large amount of membrane relative to cell volume ensures an efficient transport of materials into and out of the cell. But as cells increase in size, even though the total surface area goes up, the ratio of surface area to volume decreases (**Figure 3.5**). Small cells therefore have more membrane surface area per unit of volume and are thus more efficient at moving materials across the membrane. This superior transport into and out of the cell translates into a more efficient metabolism, which in turn means the cells can grow and reproduce more quickly. This is a principal reason why bacteria reproduce rapidly compared with eukaryotes. Typical bacteria like *Escherichia coli* might divide every 30 minutes or so under ideal conditions. The time between divisions for a eukaryotic cell is more frequently measured in hours or days. Like eukaryotes, *T. namibiensis* discussed in our case, has a relatively low surface-to-volume ratio. This results in a relatively inefficient metabolism, and we would expect this species to reproduce slowly compared with more typical bacteria. The large, fluid-filled vacuoles in the cytoplasm, however, take up over 80% of the cell's volume. The vacuoles, therefore, reduce the actual amount of cytoplasm, effectively raising the surface-to-volume ratio for this species and negating to some extent the disadvantages associated with large size.

Prokaryotic cells have a variety of shapes

The shape of prokaryotic cells is distinctive, and while some have highly unusual forms, most cells fall into one of a few easily recognizable groups: rod-shaped **bacilli** (singular, bacillus) (**Figure 3.6a**), spherical **cocci** (singular, coccus) (**Figure 3.6b**), or spiral-shaped **spirilla** (singular, spirillum) (**Figure 3.6c**). Vibrio species have comma-shaped cells (**Figure 3.6d**). Cocci may form pairs of cells (diplococci; singular, diplococcus) or chains of streptococci (singular, streptococcus).

When bacteria divide, they often separate into individual cells, but in some species the newly formed cells remain joined together and form distinctive

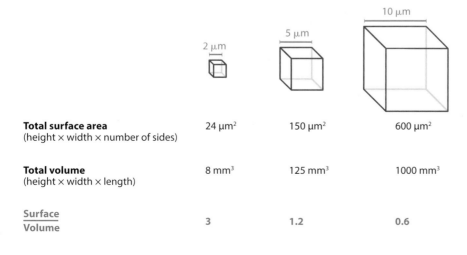

	2 μm	5 μm	10 μm
Total surface area (height × width × number of sides)	24 μm²	150 μm²	600 μm²
Total volume (height × width × length)	8 mm³	125 mm³	1000 mm³
Surface / Volume	3	1.2	0.6

Figure 3.5 Surface-to-volume ratio and metabolic efficiency. Consider the cubes depicted to represent cells of three different sizes. Although the total surface area increases as size increases, the volume increases more rapidly, meaning the surface-to-volume ratio decreases as size increases. The result is that as cells get larger, their membrane area cannot keep up with the increased demand for membrane transport. Metabolic efficiency therefore declines.

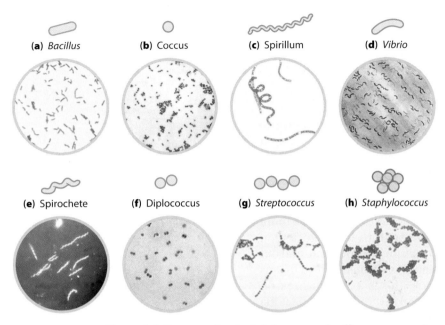

(a) *Bacillus*   **(b)** Coccus   **(c)** Spirillum   **(d)** *Vibrio*

(e) Spirochete   **(f)** Diplococcus   **(g)** *Streptococcus*   **(h)** *Staphylococcus*

Figure 3.6 Common bacterial shapes and cell arrangements.
(a) *Bacillus*. (b) Coccus or spherical bacterium. Each round sphere in the photograph is an individual bacterial cell. (c) Spirillum. (d) *Vibrio*. (e) Spirochete. (f) Diplococcus. (g) *Streptococcus*. (h) *Staphylococcus*.

patterns. Cocci that divide and remain attached in long chains are called **streptococci** (**Figure 3.6g**). Those that divide into grapelike clusters are called **staphylococci** (**Figure 3.6h**). The prefixes "strepto-" and "staphylo-" come from the Greek for "chain" and "grape," respectively. Bacilli generally form individual rods, but some divide into long chains reminiscent of delicatessen sausage links. Such cells are called **streptobacilli**.

Most prokaryotic cells have structures exterior to the cell wall that function in protection, adherence, and movement. We will begin our exploration of prokaryotic cells by examining these extracellular structures. We will then move inward to the cell wall and the plasma membrane. Finally, we will examine those structures found in the cell's cytoplasm.

Many prokaryotes have extracellular structures, extending beyond the cell wall

CASE: PLAQUE ATTACK

David and Jennifer are a happily married couple, both in their late thirties. They have one son, Kyle, age 11. Like all married couples, they have their differences. David is extremely conscientious (Jennifer would say obsessive) regarding dental care. He brushes his teeth without fail after each meal and regularly flosses at least twice a day. Jennifer finds this behavior annoying, and when David is out of town on business, as he frequently is, she tends to be rather lax regarding her own and Kyle's dental hygiene. When David takes his son to the dentist for a routine checkup, he is shocked when the dentist reports that not only does Kyle have several cavities, but his teeth are also heavily covered with plaque (**Figure 3.7**). Two additional visits to the office will be required for cleaning and

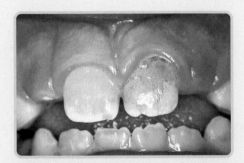

Figure 3.7 Dental plaque. Bacteria adhere to the enamel of teeth with secreted extracellular polysaccharides that make up the glycocalyx. The combination of polysaccharides and the bacteria that adhere to them are known as dental plaque.

fillings. The dentist explains that cavities are caused by bacterial infections and that to reduce the likelihood of additional problems, Kyle must follow the traditional advice—namely, brush regularly and avoid sugary snacks. David is upset to hear about the cavities and asks the dentist the following questions:

1. **What exactly is dental plaque?**
2. **What is the role of bacteria in plaque and cavity formation? What characteristics do these bacteria possess?**
3. **Why are sugar-rich foods associated with cavities?**

Beyond the cell wall, many prokaryotes come equipped with a variety of structures or appendages, which are involved in tasks such as adherence to surfaces, protection, and movement. Here we will review some of the most important of these extracellular structures.

Many bacteria are surrounded by an excreted layer of polysaccharides

Bacterial cells are often surrounded by a gel-like layer, composed of polysaccharides, called the **glycocalyx**. When the glycocalyx forms a thick, regular shell-like structure around the cell, it is called a **capsule** (**Figure 3.8**). The capsule may help the cell to adhere to various surfaces, or it may provide protection against the host's immune system. For instance, *Streptococcus pneumoniae* can cause lower respiratory tract infections. This species also is responsible for many of the middle ear infections commonly seen in young children. Not all strains of *S. pneumoniae*, however, can cause disease. Those bacteria without capsules typically pose no threat to health; it is only encapsulated strains that are worrisome. To understand why, consider that as a first line of defense against invading microorganisms, white blood cells of the immune system try to engulf the invading organisms, quite literally swallowing them whole through a process called phagocytosis. Bacterial capsules, however, impede efficient phagocytosis, meaning encapsulated cells are better able to survive this initial immune onslaught and go on to reproduce and initiate disease. Cells lacking capsules are quickly engulfed and destroyed by phagocytic cells. The capsule is a good example of a **virulence factor**—a characteristic that increases the disease-causing capacity (or virulence) of the microorganism.

Other bacterial cells secrete an irregular and diffuse glycocalyx called a **slime layer**. Like a capsule, its primary function is to aid in adherence or to protect the bacteria. *Streptococcus mutans*, for instance, secretes a slime layer that allows the bacterial cells to adhere to teeth. Other bacteria can also adhere to this sticky, sugary material, and the combination of bacteria and glycocalyx on teeth is commonly referred to as dental plaque. It is unlikely that David, the assiduous brusher and flosser in our case, has much plaque on his teeth; regular routine dental care removes this bacterial coating. Kyle, on the other hand, and perhaps Jennifer as well, by not flossing regularly, allow plaque buildup in tooth crevices. Sugar, or more specifically the sucrose found in sweet food products, adds to the problem, but perhaps not in the way you think. The sucrose itself does not directly cause plaque or cavities. *Streptococcus mutans* uses the sucrose in food to synthesize its glycocalyx. This problem is especially bad when the sucrose comes in the form of candy or other sticky substances. When a caramel or gum drop sticks to the teeth, we essentially have the bacterial equivalent of room service. Cavities are formed when the metabolizing bacteria release an acid waste product that damages the tooth enamel. Perhaps Kyle and Jennifer should at the very least start chewing sugarless gum. Such gum reduces the likelihood of cavities because the extra saliva generated while chewing helps to wash away bacterial acids and thus reduces the time during which such acids can damage tooth enamel.

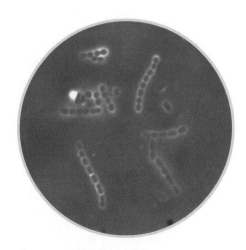

Figure 3.8 The bacterial capsule. The capsules produced by these streptococci are seen as the clear "halo" surrounding each cell.

Structures composed of protein may extend out from the bacterial cell

Pili (singular, pilus) are short protein fibers that are arranged in a helical manner, forming a cylinder with a hollow core. These structures, also called **fimbriae**, extend out beyond the cell wall (**Figure 3.9**) and allow the cell to adhere to specific surfaces. Often they function as a crucial virulence factor, because if bacteria are unable to adhere in certain locations, they are unlikely to cause disease. *Escherichia coli* provides an excellent example. Generally speaking, *E. coli* are harmless bacteria that occur by the billions in the human intestine. However, certain strains of *E. coli* have a variety of potent virulence factors, among which are especially efficient pili. These pili permit the cells to bind tightly to epithelial cells in the intestine. Other virulence factors may contribute to the watery diarrhea that many of us know as traveler's diarrhea.

Sex pili, although made of the same protein as fimbriae, are longer and have an entirely different function. A bacterial cell known as the **donor** can grow a sex pilus, which connects it with another cell known as the **recipient**, in a process called **conjugation**. Once the cells are brought together by the sex pilus, small segments of DNA may be transferred from the donor to the recipient. The recipient is therefore genetically altered and may acquire characteristics present in the donor. Because this process is so important in bacterial genetics, and because the genes transferred via conjugation often code for important virulence factors, we will discuss conjugation more thoroughly in Chapter 7.

Similar to pili, **flagella** are thin protein tubes that extend out from the cell surface. They are much longer than pili, however, and they are used for locomotion rather than adherence. Prokaryotic flagella are able to spin, much like a boat's propeller, moving the cell through its environment. Each flagellum terminates in the cell's cytoplasm, where it is rotated by a motor protein in an energy-dependent manner. Depending on the species, a bacterium may have one, a few, or many flagella, and they may be present in any of several arrangements (**Figure 3.10**).

Not all bacteria and archaea have flagella, and those that do not are incapable of independent movement. Prokaryotes that can move independently are termed **motile**; they use their flagella to move toward a favorable stimulus (exhibiting **positive taxis**) or away from unfavorable conditions

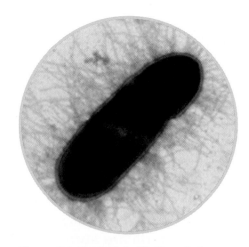

Figure 3.9 Bacterial pili. Also called fimbriae, these structures, composed of protein, allow the cell to tightly adhere to surfaces. In some cases pili act as a virulence factor, because if disease-causing organisms cannot adhere to the surface of host tissues, they are far less likely to cause disease.

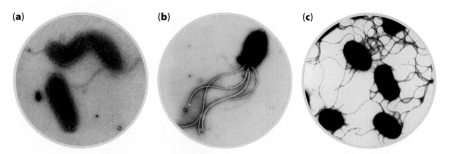

(a)   **(b)**   **(c)**

Figure 3.10 Bacterial flagella. Flagellated bacteria may have a variable number of flagella, in any one of several arrangements. (a) Bacteria with a single flagellum are termed monotrichous (single-haired). (b) A lophotrichous (having a tuft of hair) bacterium, with several flagella at one end. (c) These bacteria have flagella over their entire cell surface and are termed peritrichous (hairy all around).

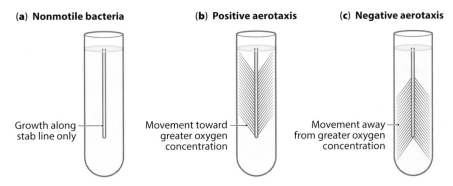

(a) Nonmotile bacteria

Growth along stab line only

(b) Positive aerotaxis

Movement toward greater oxygen concentration

(c) Negative aerotaxis

Movement away from greater oxygen concentration

Figure 3.11 Experimental demonstration of bacterial motility. Motile bacteria have flagella, while nonmotile bacteria do not. In this example, bacteria have been inoculated into a semisolid deep—a medium that permits bacterial growth and, because of its semisolid nature, permits motile bacteria to move. Bacteria are collected on an inoculating needle and stabbed directly into the medium. (a) Nonmotile bacteria can grow only along the stab line. Motile bacteria may show taxis. (b) A motile species that requires oxygen for growth will move toward the surface of the medium where oxygen is abundant. This is an example of positive aerotaxis. (c) A motile bacterial species for which oxygen is toxic will move away from the surface, displaying negative aerotaxis.

(exhibiting **negative taxis**). For example, bacteria that are moving toward a favorable temperature are said to be exhibiting positive **thermotaxis**. A different species with different temperature requirements might move away from this same temperature, displaying negative thermotaxis. Movement toward or away from certain chemicals is likewise called positive or negative **chemotaxis**, respectively. Some bacteria require oxygen for survival, while for other species, oxygen is a toxin. These respective bacteria, if motile, might therefore demonstrate positive or negative **aerotaxis** (**Figure 3.11**).

Most prokaryotes are protected from the exterior environment by a rigid cell wall

Prokaryotic cells face many hazards in their environment. The plasma membrane in particular can burst if the cell absorbs excess water. Such bursting of the cell membrane is called **lysis**. Furthermore, many chemicals commonly found in a cell's environment might easily damage the plasma membrane. Bile, for instance, produced by the liver, aids in the breakdown of dietary fat in the intestine. Yet bacteria live in the intestine by the trillions, in spite of having a plasma membrane composed largely of lipid. Why are their membranes not also digested? With very few exceptions, prokaryotes have a protective **cell wall** exterior to the plasma membrane. The cell wall thus serves as a barrier that protects the more vulnerable membrane beneath it from environmental perils.

Most bacteria have one of three basic cell wall types: Gram-positive, Gram-negative, or acid-fast. Although each type of cell wall allows bacteria to persist in the face of environmental hazards, they differ in important ways. To some extent, cell wall type influences where the cells are found and the rate at which they grow. In certain cases, as we will see, the kinds of diseases caused by bacteria are a function of their cell walls, and treatment will often vary depending on cell wall type. A sound knowledge of cell wall structure and function is therefore crucial to our understanding of how these microorganisms live and how they affect us.

CASE: AN OCCUPATIONAL HAZARD

Dirk, a 24-year-old carpenter, accidentally cut his left hand while working with a circular saw. During the next week the injury site became red, swollen, and painful. A few days later, he began to experience chills and fever. He then sought medical attention. His doctor, on the basis of the nature of the injury and Dirk's symptoms, suspected a *Staphylococcus* infection. The doctor sent a sample of pus from the wound to the laboratory for analysis. In the laboratory, the sample was Gram-stained and showed Gram-positive clusters of cocci, suggestive of *Staphylococcus*. The bacteria were grown in culture and confirmed as *Staphylococcus aureus*. A blood sample taken from Dirk was also positive for this species. The doctor prescribed a course of oxacillin, and within a few days Dirk had made a complete recovery.

1. **What about Dirk's case made the doctor suspect *Staphylococcus*?**
2. **What is a Gram stain and why was it used?**
3. **Why was oxacillin an appropriate drug in this case? Would it be recommended for an infection with Gram-negative bacteria as well?**

The Gram-positive cell wall is composed of multiple layers of carbohydrates, linked by peptides

The major component of the **Gram-positive** cell wall is a complex molecule composed of sugars and amino acids called **peptidoglycan** (**Figure 3.12**). Multiple layers of this molecule surround the cell, protecting it from damage and helping it maintain its shape. While this molecule is found in the cell walls of both Gram-positive and Gram-negative bacteria, it is more abundant in Gram-positive cells.

Notice in **Figure 3.13** that the polysaccharide portion of peptidoglycan is composed of two alternating sugars, *N*-acetylglucosamine (NAG) and *N*-acetylmuramic acid (NAM). Attached to each NAM is a short chain of four amino acids. The sugar backbone of peptidoglycan is identical in all bacterial species. The amino acids in the short peptide chain may vary. The amino acid in the third position on one peptide is covalently linked to an amino acid in position four on the peptide on a neighboring layer of peptidoglycan. These linkages are via **pentaglycine bridges**, each composed of five glycine amino acids (see Figure 3.13). Pentaglycine bridges hold the individual layers of peptidoglycan together and give the cell wall its strength and its ability to both maintain cell shape and prevent cell rupture. Additionally, recall from Chapter 2 that sugars are hydrophilic in nature. Molecules like bile that might damage the hydrophobic lipids of the plasma membrane are themselves hydrophobic. Because bile is unable to interact with and cross the hydrophilic cell wall, it is prevented from reaching the vulnerable cell membrane.

The thick, hydrophilic Gram-positive cell wall may consist of up to 30 or 40 layers of cross-linked peptidoglycan. Those species with a greater number of layers are able to withstand increasingly harsh environmental conditions. For example, human skin is fairly inhospitable for many bacteria. Not only is the skin too dry for most species, but salts deposited on the skin during perspiration can draw water out of bacterial cells. Those bacteria that do survive on the skin tend to be Gram-positive species with especially thick cell walls. *Staphylococcus* in particular is an important part of the skin flora. This explains why Dirk's doctor suspected a staphylococcal infection early on. Species such as *Staphylococcus aureus* may cause no particular problem on intact skin, but if the skin is broken, these bacteria can gain access to

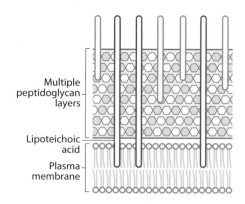

Figure 3.12 The Gram-positive cell wall. The cell wall is exterior to the plasma membrane and consists of many layers of peptidoglycan. The number of peptidoglycan layers is variable. Molecules called lipoteichoic acid link the cell wall to the plasma membrane.

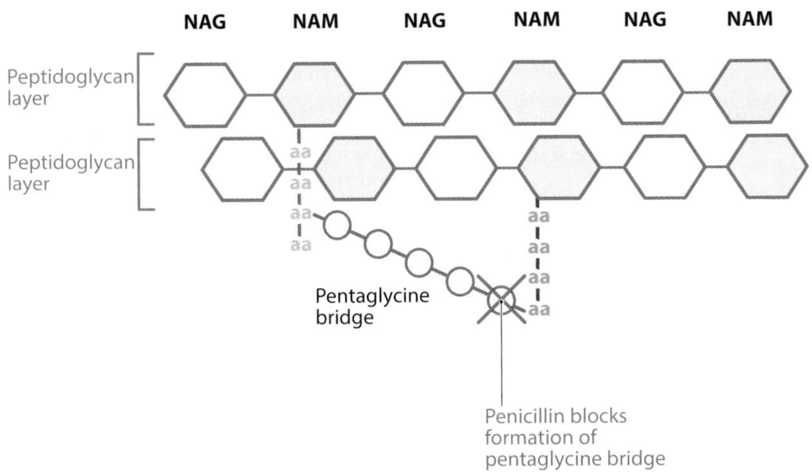

Figure 3.13 Pentaglycine bridges. A single layer of peptidoglycan consists of a polysaccharide made of alternating *N*-actylglucosamine (NAG) and *N*-acetylmuramic acid (NAM) sugars. A short chain of four amino acids (aa) is attached to each NAM (only two are shown for clarity). Separate peptioglycan layers are linked by pentaglycine bridges (open circles), each composed of five glycine amino acids. Penicillin prevents the formation of pentaglycine bridges, weakening the cell wall of Gram-positive bacteria.

subdermal tissue and the circulatory system, causing an infection, as they did in Dirk.

Peptidoglycan is a unique bacterial molecule; it is found in no other living thing. As such, it makes an ideal target for antibacterial drugs, known as **antibiotics**. When antibiotics are developed to combat bacterial infections, it is crucial that the drug does not damage or kill human or other host cells in the process. Because peptidoglycan is lacking from all eukaryotic cells, the risk of unwanted side effects of drugs that target it is greatly reduced. The penicillins are a large and important group of antibiotics that inhibit peptidoglycan synthesis. Specifically, these drugs prevent the formation of pentaglycine bridges (see Figure 3.13). Without these linkages between the peptidoglycan layers, the cell wall is much less rigid and the plasma membrane is prone to lysis. Oxacillin, an important member of the penicillins, works in this manner, which explains why it was effective in Dirk's case. These antibiotics do not damage peptidoglycan that is already formed; they only inhibit the synthesis of new peptidoglycan. They are consequently most effective against actively dividing bacterial cells that are in the process of producing new peptidoglycan.

Gram-negative cells are surrounded by two membranes with a thin peptidoglycan layer in between

Gram-negative bacteria protect their plasma membrane in a different way. Like Gram-positive cells, they have a peptidoglycan cell wall exterior to the plasma membrane, but in Gram-negative cells, this wall is a thin structure, only one or two layers thick. Beyond this wall is an **outer membrane** (**Figure 3.14**). In some ways the outer membrane is similar to the plasma membrane, which we will describe shortly. However, the outer membrane contains a molecule unique to Gram-negative bacteria known as a bacterial **lipopolysaccharide** or **LPS**. As the name implies, LPS is composed of both lipid and sugars. The lipid portion is called **lipid A**. Sugars, called **O polysaccharides** form the exterior-most portion of the outer membrane, and like all carbohydrates, they are hydrophilic. Gram-negative cells are thus surrounded by a hydrophilic covering that prevents hydrophobic compounds like bile from reaching and digesting either the outer membrane or the plasma membrane.

Although most substances cannot easily cross the outer membrane, proteins called **porins** permit the entry of certain materials (see Figure 3.14). Porins essentially form hydrophilic tunnels through the hydrophobic membrane,

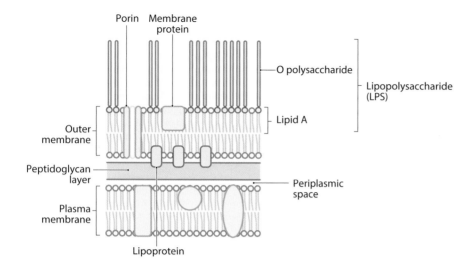

Figure 3.14 **The Gram-negative cell wall and outer membrane.** In Gram-negative bacteria there is only a thin layer of peptidoglycan (in purple), sandwiched between the plasma membrane and an outer membrane. Like the plasma membrane, the outer membrane is composed principally of lipid and proteins and includes an important component called lipopolysaccharide (LPS). LPS consists of carbohydrates called O polysaccharides, which are anchored to the outer membrane by lipid A. Hydrophilic protein channels called porins permit the passage of small hydrophilic molecules and ions across the outer membrane. Lipoprotein attaches the outer membrane to the peptidoglycan layer.

permitting small hydrophilic molecules and ions to enter. Amino acids, monosaccharides, and iron ions are examples of substances that can move through porins.

The LPS in the outer membrane is extremely important from a medical point of view. When Gram-negative bacteria die, their outer membranes break down and LPS is released into the environment. Lipid A, in particular, is recognized by the host immune system, and this recognition initiates a number of immunological events, including fever and inflammation.

Such immune response is normal and beneficial when it occurs in a controlled manner, but when the host is suddenly exposed to a large amount of LPS, the overreaction of the immune system can result in shock or even death. This is why a ruptured appendix or a perforated bowel is such a medical emergency. The intestine is home to trillions of Gram-negative bacteria. Within the confines of the bowel, these organisms generally pose no medical risk. But if the intestinal lining is breached, as it is when the appendix ruptures, bacteria flood into the normally sterile blood, causing a serious condition known as bacterial **sepsis**. As these bacteria die in large numbers, high levels of LPS are released into the bloodstream, increasing the likelihood of shock. Because of its toxic properties, LPS is also referred to as **endotoxin**.

Treating a Gram-negative sepsis is tricky. Many antibiotics, including the oxacillin used on Dirk, cannot cross the outer membrane and are thus useless in such a situation. Other antibiotics can cross this barrier, but even then, drugs must be selected with care. Many of our most powerful antibiotics, including the penicillins, act by killing bacteria and are therefore known as **bactericidal** agents. However, if bactericidal drugs were used to treat a Gram-negative sepsis, the cure might be worse than the disease, since endotoxin is released when Gram-negative bacteria die. High levels of endotoxin released by large numbers of dying bacteria could induce shock. **Bacteriostatic** drugs are therefore generally used. These antibiotics prevent bacteria from reproducing; they do not directly kill bacterial cells. By keeping bacterial numbers in check, such drugs give the immune system an opportunity to contain the infection slowly. The steady, relatively low levels of endotoxin released in such instances, while associated with certain symptoms such as fever, are unlikely to endanger the patient. Because treatment of Gram-positive and Gram-negative infections is often so different, clinicians must be able to quickly determine which type of infection they must treat.

Acid-fast bacteria are surrounded by a thick, waxy cell wall

Although most bacteria have Gram-positive or Gram-negative cell walls, a few employ a third solution to the problem of plasma membrane protection. These **acid-fast** bacteria include in their ranks some of the most important **pathogens** (disease-causing organisms) afflicting humans and animals, including *Mycobacterium tuberculosis*, the causative agent of tuberculosis. The closely related *Mycobacterium leprae*, which causes leprosy, is also acid-fast. The acid-fast cell wall contains large amounts of waxes, interlaced with peptidoglycan. Waxes are essentially derivatives of very-long-chain fatty acids that give these cells a highly impervious wall. Acid-fast cells are thus very hardy and difficult to destroy. They are resistant to many toxic chemicals as well as the phagocytic activity of immune system cells.

All this protection, however, comes with a cost. Nutrients and oxygen also enter the cells very slowly, and thus the growth of acid-fast bacteria is slow. While typical Gram-positive or Gram-negative cells may divide in less than an hour under ideal conditions, *M. tuberculosis* might require 24 hours. Antibiotics used to treat tuberculosis are also absorbed slowly, which explains why the course of treatment is so long. A typical antibiotic prescription for Gram-positive or Gram-negative bacteria might be for 5–10 days, whereas an infection with an acid-fast species might require treatment lasting 6 months or longer.

Bacteria with different cell wall types can be distinguished by specific staining techniques

Dirk's doctor ordered a **Gram stain** as part of the laboratory analysis. This is an easy way to quickly determine if bacteria are Gram-positive or Gram-negative. The simple stain discussed earlier in this chapter (see Figure 3.3) stains all cells in a similar manner. Although it provides information about cell size and shape, it cannot distinguish Gram-positive from Gram-negative cells. The Gram stain, however, takes advantage of differences in cell wall structure, allowing us to differentiate between the two cell types. The first step is to stain all cells with the basic stain crystal violet. At this stage, then, all cells will appear violet in color. When iodine is next applied, the crystal violet bound to the membrane forms large crystals that cannot escape through the cell wall. The cells are subsequently exposed to an alcohol, which dissolves the outer membrane of Gram-negative cells; the Gram-positive cell wall is unaffected. The crystal violet can then be washed out of Gram-negative cells with distilled water, leaving them colorless; the Gram-positive cells remain violet. Because colorless cells are difficult to see, a second basic stain, safranin, is applied that turns the Gram-negative cells pink. The Gram stain procedure is diagrammed in **Figure 3.15a**. Because most bacterial infections are caused by either Gram-positive or Gram-negative species, and because these two types of infections are often treated differently, a Gram stain is often an appropriate way to quickly determine which of these two types of bacteria is causing a specific problem.

To identify acid-fast cells, an **acid-fast stain** is used (**Figure 3.15b**). Cells are initially stained with carbol-fuchsin, a reddish basic stain. Staining is performed over heat to allow the stain to penetrate the thick waxy acid-fast wall. Cells are then exposed to an acid wash. Gram-positive and Gram-negative cell walls are damaged by the acid, resulting in their decolorization. Acid-fast cells, on the other hand, are not damaged by the acid and they retain their red color. A blue counterstain, methylene blue, is used to visualize

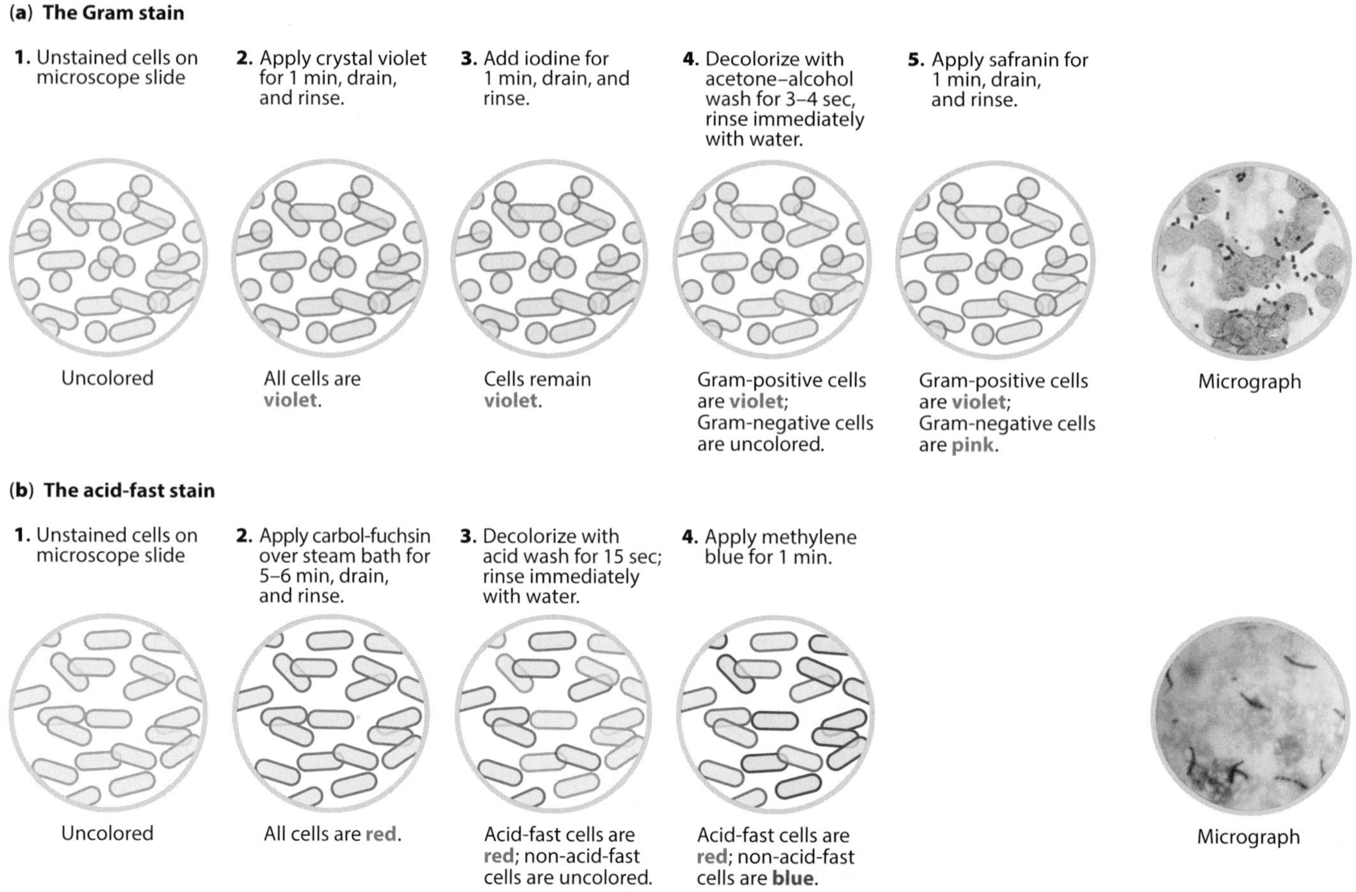

(a) The Gram stain

1. Unstained cells on microscope slide   **2.** Apply crystal violet for 1 min, drain, and rinse.   **3.** Add iodine for 1 min, drain, and rinse.   **4.** Decolorize with acetone–alcohol wash for 3–4 sec, rinse immediately with water.   **5.** Apply safranin for 1 min, drain, and rinse.

Uncolored   All cells are **violet**.   Cells remain **violet**.   Gram-positive cells are **violet**; Gram-negative cells are uncolored.   Gram-positive cells are **violet**; Gram-negative cells are **pink**.   Micrograph

(b) The acid-fast stain

1. Unstained cells on microscope slide   **2.** Apply carbol-fuchsin over steam bath for 5–6 min, drain, and rinse.   **3.** Decolorize with acid wash for 15 sec; rinse immediately with water.   **4.** Apply methylene blue for 1 min.

Uncolored   All cells are **red**.   Acid-fast cells are **red**; non-acid-fast cells are uncolored.   Acid-fast cells are **red**; non-acid-fast cells are **blue**.   Micrograph

Figure 3.15 Common staining procedures to identify bacterial cell wall type. (a) The Gram stain takes advantage of the differences in Gram-positive and Gram-negative bacteria to distinguish between these two important groups. (b) The acid-fast stain differentiates between acid-fast and non-acid-fast bacteria. Because these staining procedures stain cells with different structures differently, they are both examples of differential stains.

non-acid-fast cells. If a patient is displaying symptoms consistent with an infection caused by acid-fast bacteria, an acid-fast stain is a fast and efficient way to confirm the diagnosis.

Cell walls surround the cells of most archaea

Like bacteria, most archaea also have cell walls. These walls are variable in structure, but unlike bacterial cell walls, peptidoglycan is never present. In many cases, the cell wall is composed of a single thick layer of polysaccharides. The precise polysaccharide involved varies from species to species. Alternatively, many species have a rigid array of proteins that envelop the cell. This protein layer acts as a protective barrier that prevents potentially damaging compounds from reaching the vulnerable plasma membrane.

Some prokaryotes lack a cell wall

A few bacteria and archaea lack a cell wall entirely. *Mycoplasma* species, for instance, are bacteria that have rigid lipid molecules called **sterols** in their plasma membranes to help stabilize their structure. Nevertheless, the lack of a sturdy cell wall means that these bacteria have an extremely variable shape. One species in particular, *M. pneumoniae*, causes a relatively mild form of lower respiratory tract infection called atypical or walking pneumonia. Penicillin, which inhibits cell wall formation, has no effect on these organisms. Infections such as this must be treated with other antibiotics, which target structures other than the cell wall.

Each cell is surrounded by a plasma membrane

The plasma membrane lies just below the cell wall. This thin, flexible structure separates the cell's cytoplasm from the external environment. Its primary responsibilities are to contain the cytoplasm and to regulate what enters and leaves the cell.

Plasma membrane structure is similar in bacteria and eukaryotes. In both cell types, the plasma membrane contains a large proportion of lipids called **phospholipids** (**Figure 3.16a**). A phospholipid has a similar structure to a fat (see Chapter 2, p. 33) with one major exception: while a fat consists of three fatty acids covalently bonded to a molecule of glycerol, a phospholipid has only two fatty acid chains. A **phosphate group** (PO_4^{-3}) is bonded to the glycerol in place of the third fatty acid. This gives a phospholipid a very important property; because phosphate groups carries negative charges, they are hydrophilic. The fatty acid chains, on the other hand, are hydrophobic. The hydrophilic region is attracted to water, while the hydrophobic regions aggregate with other hydrophobic molecules, away from water. In cell membranes, this results in the formation of a lipid **bilayer**, consisting of two layers of phospholipids. The phosphate groups project out toward the extracellular environment and in toward the cytoplasm, where they can hydrogen-bond with water. The hydrophobic hydrocarbon chains of the fatty acids are sandwiched between the two layers of phosphates, away from water (**Figure 3.16b**).

The other principal components of the plasma membrane are proteins. Membrane proteins typically make up about half, but sometimes as much as 70%, of a membrane's mass. Many of these are **carrier proteins** that span the membrane from the **extracellular** side (the side of the membrane outside the cell) to the cytoplasmic side (the side on the cell's interior) and function in the transport of materials across the plasma membrane. Carrier proteins and other membrane proteins that extend across the membrane from the extracellular side to the cytoplasmic side are collectively called

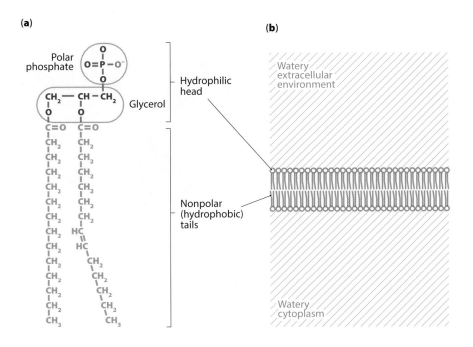

Figure 3.16 The plasma membrane.
Phospholipids are the major lipid component of bacterial plasma membranes. (a) Diagram and chemical composition of a phospholipid. The phosphate group, because of its negative charge, is hydrophilic. The fatty acid chains, composed of carbon and hydrogen only, are nonpolar and thus hydrophobic in nature. (b) In a watery environment, phospholipids form a bilayer. Hydrophilic phosphate groups interact with water on both the exterior (extracellular) and interior (cytoplasmic) sides of the membrane. Hydrophobic fatty acids are found in the membrane's interior, away from water.

transmembrane proteins. Other proteins, called **peripheral proteins**, are bound to the membrane surface, where they carry out diverse functions.

The phospholipids in the membrane are not bound to each other. The bilayer that they form merely results from the way in which the phosphates in phospholipids are attracted to water and the fatty acid tails are not. If it were not for water, the phospholipid bilayer would not form. The plasma membrane is therefore not a rigid structure. Rather, it has the consistency of oil. Because of its fluidity, it is in a constant state of flux, and the membrane phospholipids are continually moving around and changing position within the membrane. Neither are the proteins bound to the phospholipids. Their behavior in the phospholipid bilayer is akin to that of icebergs floating in a lipid sea. Because the membrane is dynamic and the constituent molecules are able to move, the membrane is said to conform to the **fluid mosaic model** of membrane structure (**Figure 3.17**).

Because of its vital role, it seems reasonable that the cell membrane would make a tempting target for antibiotics. Certainly many compounds can disrupt the plasma membrane and kill bacterial cells, but the plasma membrane is rarely the target of antibiotic activity. Since the basic structure of the plasma membrane is nearly identical in bacterial and eukaryotic cells, any antibiotic that disrupts bacterial membranes is likely to damage our membranes as well.

The plasma membrane of archaea is often especially stable

The basic design of the archaeal plasma membrane is similar to that of bacteria and eukaryotes, consisting of two hydrophilic surfaces and a hydrophobic core. Likewise, there are numerous proteins, both transmembrane and peripheral, associated with the phospholipids. But, as we will discuss further in Chapter 4, many archaea are adapted to live at very high temperatures. Certain features of the archaeal membrane make the membrane more stable, and this enhanced stability is thought to reflect their adaptation to temperature extremes. The hydrocarbon chains are somewhat different from those in bacteria and eukaryotes, as is the precise manner in which these chains bond to glycerol. Furthermore, in those archaea adapted to the highest temperatures, membrane stability is enhanced even further, with covalent bonds between the hydrocarbons in the two phospholipid layers. This effectively turns the phospholipid bilayer into an especially stable monolayer. This unusual feature is never observed in bacteria or eukaryotes.

The liquid interior of the cell forms the cell's cytoplasm

The cytoplasm is the liquid portion inside the cell surrounded by the plasma membrane. It is composed primarily of water and is the site of most of the cell's metabolic reactions. Although the prokaryotic cytoplasm does not contain the membrane-bound organelles found in eukaryotic cells, a small number of important subcellular structures are present.

The prokaryotic chromosome is usually circular

Prokaryotes typically have a single, circular loop of double-stranded DNA called the **chromosome**. Unlike the chromosomes of eukaryotic cells, which are surrounded by a nuclear membrane, prokaryotic chromosomes are free in the cytoplasm. The area in the cytoplasm containing the chromosome is called the **nucleoid area** (**Figure 3.18**). Many prokaryotic cells, in

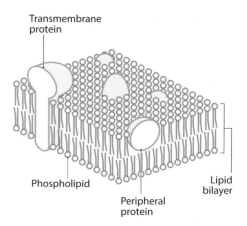

Figure 3.17 The fluid mosaic model of membrane structure. The plasma membrane, composed mainly of proteins embedded in a phospholipid bilayer, has the consistency of oil. Because of its fluid nature, membrane components are in a constant state of motion within the membrane. Proteins that extend all the way through the phospholipid bilayer are called transmembrane proteins. Proteins that attach to the membrane's surface are called peripheral proteins.

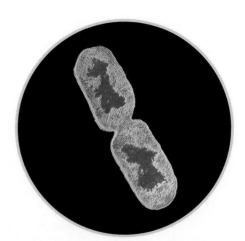

Figure 3.18 The bacterial nucleoid area. In this photo, taken with a transmission electron microscope and subsequently colored, the nucleoid area, containing the chromosome, appears red. Cell division of one original cell into two new daughter cells is almost complete. The grainy appearance of the cytoplasm is due to the presence of numerous ribosomes.

addition to their chromosome, have one or more much smaller, circular DNA molecules called **plasmids**. Although plasmids generally carry a very small number of genes, these genes often code for virulence factors, which can be passed to another cell via conjugation. This topic will be explored in Chapter 7.

Proteins are synthesized on the surface of ribosomes

Ribosomes are composed of RNA and protein. Ribosomes are the site of protein synthesis, and it is on their surface that amino acids are linked together into proteins. The bacterial cytoplasm contains up to 20,000 ribosomes, which give the cell a grainy appearance when viewed with an electron microscope (see Figure 3.18).

Although ribosomes are found in both prokaryotic and eukaryotic cells, prokaryotic ribosomes are smaller and have a somewhat different structure than eukaryotic ribosomes. Many common antibiotics, such as erythromycin, bind to and interfere with the smaller bacterial ribosomes, arresting protein synthesis. Because these drugs do not bind to larger eukaryotic ribosomes, they generally can be used safely without risk to the host.

Storage granules and inclusion bodies have specialized functions within cells

As the name implies, **storage granules** serve as repositories for various nutrients. Granules composed of carbohydrate or lipid, for instance, often serve as reservoirs for carbon, nitrogen, or phosphate. Inclusion bodies do not store nutrients but may have other specialized functions. Some aquatic bacteria, for example, have gas-filled inclusion bodies in their cytoplasm. These protein-coated structures allow the cell to float near the surface, providing optimum conditions for their growth.

Endospores permit survival during times of environmental adversity

CASE: THE **REAL** JURASSIC PARK

We've all seen the movies in which extinct creatures are brought back to life. And although some scientists think it is at least theoretically possible to retrieve extinct organisms under certain conditions, we have yet to witness an actual resurrected dinosaur or dodo. That may not be precisely true, however, when it comes to bacteria. When a bee fossilized in amber was found in the Dominican Republic, the ancient insect was dated to being about 25 million years old. When the amber was cracked open and material from the bee's gut was extracted, bacterial endospores were found. Amazingly the spores germinated when placed on an appropriate growth medium, and developed into living, bacterial cells. An analysis of their DNA indicated that they were a species of *Bacillus* that is no longer found today. In other words, the researchers had resurrected a formerly extinct species!

1. **What are endospores and how do they survive so long?**
2. **Do all bacteria produce endospores?**

Endospores are unique bacterial structures that form in response to nutrient depletion or other unfavorable environmental conditions. As endospores, the bacteria can persist in the face of a harsh environment, remaining able to resume growth should conditions improve in the future.

As the environment deteriorates, an endospore-forming bacterium responds by first replicating its chromosome (**Figure 3.19**). The newly replicated chromosome is then walled off from the rest of the cytoplasm by a portion of the plasma membrane. This membrane barrier, which divides the cell into two unequal compartments, is called the **septum**. The smaller of these two compartments is next engulfed by the larger compartment, forming a double membrane around the new chromosome. Layers of peptidoglycan are produced between the two membrane layers, and the entire structure is next surrounded by a thick coat of protein. When endospore formation is complete, the cell lyses, releasing the spore into the environment.

Endospores are metabolically inactive. Because of their thick coat, they can remain dormant in the environment for years or decades, impervious to the effects of heat, desiccation, radiation, and chemical agents. If and when conditions for growth become favorable, an endospore can germinate. As conditions improve, the permeability of the coat increases, and as water enters, the endospore swells, eventually rupturing the coat. The developing cell then emerges, resuming metabolic activity and reproduction. Endospores thus represent a unique adaptation, permitting cells to persist over long periods of severe conditions.

How long can endospores survive? This remains a difficult question to answer, but certainly most can survive at least a year, and many remain viable for much longer. Endospores have been recovered from compacted clay in Roman ruins dated from about 90 AD. When placed in a warm oxygen-rich environment, some of these endospores germinated and were identified as *Thermoactinomyces vulgaris*. Even more remarkable was the discovery described in our case. It boggles the mind to think that any living thing that was alive when the Alps were forming and Antarctica was gaining its ice cap, has been returned to the modern world. A close inspection of these Lazarus-like bacteria found them to be similar to *Bacillus sphericus*, which is still found in the digestive system of bees in the Dominican Republic today. But it was different enough that researches considered it to be a different species, perhaps the species from which *B. sphericus* evolved.

Not all bacteria can form endospores. Most common endospore formers are species of *Clostridium* and *Bacillus*. Recall the case in Chapter 2, where a large number of individuals became dangerously ill after ingesting potato salad made from canned potatoes that were contaminated with *Clostridium botulinum* endospores. Other important endospore formers include *C. tetani*, which causes tetanus, and *C. perfringens*, the causative agent of gas gangrene. Endospores of these organisms are found in the environment, but germination and growth can occur only in the absence of oxygen. When the endospores of *C. tetani*, for instance, enter the body through a puncture wound, the favorable, oxygen-free conditions permit germination. As the bacteria begin to grow, they release a powerful toxin that results in tetanus, also known as lockjaw.

Bacillus anthracis, the causative agent of anthrax, is a Gram-positive rod that is also an endospore former. Its endospores can survive in the soil for decades. Anthrax is mainly a disease of animals, particularly livestock that may become exposed to endospores while grazing. Unlike the *Clostridium* species described above, *B. anthracis* requires oxygen for growth. Humans

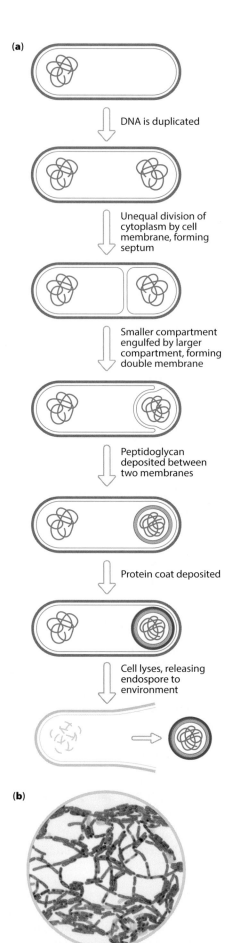

(a)

DNA is duplicated

Unequal division of cytoplasm by cell membrane, forming septum

Smaller compartment engulfed by larger compartment, forming double membrane

Peptidoglycan deposited between two membranes

Protein coat deposited

Cell lyses, releasing endospore to environment

(b)

Figure 3.19 Bacterial endospores. (a) The process of endospore formation. The released endospore is protected by a double membrane, with peptidoglycan (in purple) between the two membranes, as well as an exterior protein coat (in blue). (b) Endospores (green) in the cytoplasm of bacilli (red).

occasionally become infected when viable endospores come in contact with broken skin (they may contract cutaneous anthrax) or are respired into the lungs (where they cause pulmonary anthrax). The *Bacillus* endospores germinate in these moist, oxygen-rich environments, and the emerging cells resume reproduction. Cutaneous anthrax, although it causes skin ulceration and other symptoms, is unlikely to be lethal. Pulmonary anthrax, on the other hand, results in a rapidly progressing, potentially fatal form of pneumonia, for which immediate treatment is required.

Because it can be so deadly and because it forms endospores, *B. anthracis* is worrisome as a potential bioterrorism tool. Most nonendospore-forming bacteria, no matter how deadly, would be unable to persist in the environment long enough to infect many people. We will revisit this topic in Chapter 14.

Eukaryotic cells are larger and more complex than prokaryotic cells

Let's now turn our attention to eukaryotic cells. All eukaryotes have cells that are structurally similar. Typically they are larger than prokaryotic cells and are far more complex internally. Like prokaryotic cells, they are all bound by a plasma membrane. Their most distinctive feature is that they contain many organelles, each surrounded by a membrane, such as the nuclear membrane enveloping the genetic material. Eukaryotic cells may or may not have a cell wall or extracellular appendages (see Figure 3.1).

Different organelles perform different functions, and they allow the compartmentalization of eukaryotic cells; that is, they allow complex functions to take place in specific regions of the cell. For instance, powerful enzymes used to digest nutrients are restricted to specific organelles, ensuring that these enzymes do not damage the cell itself. Other organelles are involved with biomolecule synthesis, energy use, and intracellular transport.

The number of cells in a eukaryotic organism is variable

Most prokaryotes are **unicellular** (single-celled) organisms (see Chapter 1, p. 3). Some eukaryotes are also unicellular, but many others are **multicellular** (many-celled); larger eukaryotic organisms consist of billions or trillions of cells. All eukaryotic cells, whether they are unicellular organisms or part of a larger multicellular plant or animal, share many important structural features. Eukaryotic cells are highly diverse, however, making it difficult to describe a typical example. Protozoa, for instance, are eukaryotes and are generally unicellular; each individual cell must constitute a fully independent, self-contained entity, carrying out all critical life functions. For instance, it must be able to find and ingest food and generate energy without assistance from other cells. In a multicellular animal or plant, alternatively, all cells work together as a single organism. Consequently, different cells specialize in different functions. Fat cells, for example, store energy, while muscle cells allow movement and nerve cells conduct electrical impulses. In humans, there are about 200 cell types, each type specialized for specific functions. An individual liver or lung cell is incapable of independent living because of its dependence on other cells in the body. Each cell type has only those organelles and structures required to carry out its particular role.

In multicellular eukaryotes, groups of cells of one or more types often work together for particular functions. Such groupings are called **tissues**. Examples in animals include nervous, muscle, and epithelial tissue (**Figure 3.20a**). Nervous tissue consists not only of the nerve cells themselves (the

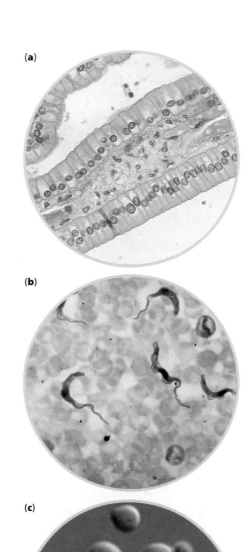

(a)

(b)

(c)

Figure 3.20 Representative eukaryotic cells. (a) Epithelial tissue in a multicellular animal. Epithelial tissue is found lining body surfaces. Note the darkly stained nuclei. (b) *Trypanosoma brucei* (in dark purple), the causative agent of African sleeping sickness, is a unicellular protozoan parasite. Red blood cells are seen in pale red. (c) *Saccharomyces cerevisiae*, referred to commonly as brewer's yeast, is a unicellular fungal microorganism.

neurons), but also a variety of other cell types that surround, protect, and nourish the neurons. Groups of tissues that work together to carry out specific tasks are called **organs**. Examples in animals include the liver, brain, and skin.

Many eukaryotes are of interest to microbiologists. More children are killed each year by the protozoan *Plasmodium,* the causative agent of malaria, than any other single microorganism, prokaryotic or eukaryotic. Countless people are infected with other protozoans, including amoebas and the trypanosomes that cause sleeping sickness (**Figure 3.20b**). Fungal microorganisms, such as *Saccharomyces cerevisiae* (**Figure 3.20c**), permit bread to rise; others give flavor to cheese and recycle nutrients in the environment. They also spoil our food and cause athlete's foot.

We will revisit these organisms in more detail in Chapter 4. Here we will focus on important features of eukaryotic cells, especially those aspects that distinguish them from prokaryotic cells.

While plant and fungal cells have cell walls, animal cells do not

As previously discussed, the vast majority of prokaryotic cells are surrounded by a rigid cell wall. Among eukaryotes, cell walls are the rule in plants and fungi, where they help to maintain cell shape, as they do in prokaryotes. Plant cells are surrounded by a cell wall composed of cellulose, a polysaccharide consisting of a long chain of glucose molecules. Fungal cell walls may also be composed of cellulose, but many are constructed out of chitin, a nitrogen-containing polysaccharide. The presence of a cell wall in large, multicellular eukaryotes is an adaptation to a stationary mode of life. Because neither plants nor fungi move, their cells profit from the protection and rigidity provided by the cell wall. Cell walls are lacking in both protozoa and animals. All animals are motile, at least during part of their lives. A rigid cell wall surrounding their cells would be incompatible with locomotion.

Like prokaryotes, all eukaryotic cells are surrounded by a membrane

The eukaryotic cell membrane is structurally similar to that described for bacteria (see Figure 3.17). Both are composed of phospholipid bilayers associated with proteins. As in bacteria, the plasma membrane separates the cell cytoplasm from the extracellular environment and regulates transport into and out of the cell. Many proteins in the eukaryotic cell membrane are covalently linked to polysaccharides that extend into the extracellular environment (**Figure 3.21**). These **glycoproteins** play important roles in

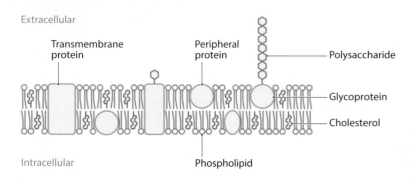

Figure 3.21 The animal cell membrane. The structure of the cell membrane in eukaryotes is very similar to that described for prokaryotes. In this diagram of the animal cell membrane, note the presence of membrane glycoproteins and cholesterol, which provide added flexibility and strength to the membrane. Cholesterol is characteristic of animal cell membranes only. It is absent from the cell membranes of other eukaryotes.

cell-to-cell communication. Such communication is especially important in multicellular organisms, where activities in different cells must be coordinated.

Most eukaryotic plasma membranes contain lipids called sterols in addition to the phospholipids. Sterols are rarely found in prokaryotic plasma membranes. These molecules give the membrane added durability and flexibility. In animal cells, cholesterol serves this function, while ergosterol is found in the membranes of fungal cells. Because it is sufficiently different in structure from cholesterol, ergosterol is a frequent target for antifungal drugs.

Many eukaryotic cells have either flagella or **cilia**, which permit cell movement or generate a water current over the cell surface (**Figure 3.22**). Cilia are shorter than flagella but otherwise have a similar structure; both are composed of numerous rod-shaped proteins called microtubules. They are able to beat back and forth very rapidly, as many as 40 times a second. Cilia may help move a cell through its environment, or in multicellular organisms, they may move material over the cell surface. In the female mammal reproductive tract, for instance, the gentle sweeping of ciliated cells moves eggs from the ovary through the fallopian tube to the uterus.

Figure 3.22 Eukaryotic cilia. Cilia may provide motility in single-celled organisms. In multicellular organisms they help to move fluid over the surface of the ciliated cell.

The cytoplasm of eukaryotic cells contains a variety of membrane-bound organelles

The fluid-filled cytoplasm contains many organelles, as well as various cytoplasmic proteins and other structures such as ribosomes. A system of protein filaments in the cytoplasm called the **cytoskeleton** acts as a scaffolding for the cell (**Figure 3.23**). These long, slender protein filaments assemble into a dense mesh immediately beneath the plasma membrane. They also form an extensive network throughout the cytoplasm that anchors organelles at specified locations, and they function as a sort of microscopic rail system, along which intracellular transport occurs.

The cytoskeleton was once believed to be a unique feature of eukaryotic cells. In recent years, however, it has become evident that prokaryotes also have cytoskeletons, composed of proteins generally similar to those of eukaryotes. The functions carried out by the prokaryotic cytoskeleton, such as maintenance of shape and cell division, are also largely similar those of its eukaryotic counterpart.

We will conclude our tour of the eukaryotic cell by briefly describing the principal organelles found in the cytoplasm that distinguish eukaryotic from prokaryotic cells. Additional information about these important structures will be provided in subsequent chapters, when we discuss their specific roles in microbial processes or in microbe–host interaction. Remember that many eukaryotic cells have very specific functions, especially in multicellular organisms. Consequently, cells differ in terms of both the types and numbers of organelles present.

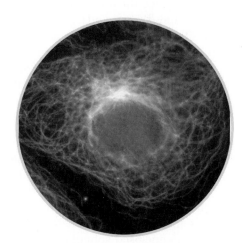

DNA is found in the nucleus of eukaryotic cells

The word "eukaryotic" comes from the Greek for "true nucleus." The nucleus is the largest organelle in the cell and it is bounded by a double-layered membrane called the **nuclear envelope**, which surrounds the DNA. The envelope has many pores in its surface to facilitate transport between the

Figure 3.23 The cytoskeleton. The cytoskeleton appears as a dense meshwork of long protein fibers. The fluorescent-microscopic technique used to take this image reveals only the cytoskeleton and leaves other structures obscured. Ordinarily the cytoskeletal elements would extend throughout the eukaryotic cell cytoplasm.

cytoplasm and the interior of the nucleus. Among its various functions, the nuclear membrane is thought to help the cell regulate the rate at which the information stored in DNA is used to make proteins.

The endomembrane system functions in cellular transport

We have learned that cells are usually busy places, with lots of transport of substances in and out. This transport occurs in several ways. Certain small, hydrophobic molecules can simply move across the plasma membrane. Specialized transport proteins in the membrane move other relatively small hydrophilic substances into and out of the cell as needed. Prokaryotes, however, cannot import or export large molecules. Consider the peptidoglycan found in bacterial cell walls. Bacteria cannot secrete preassembled peptidoglycan. Rather, they secrete small subunits of this material, which then assemble outside the cell. Likewise, bacteria cannot import large molecules. Such molecules must be digested outside the cell. Smaller subunits can then be brought in via transport proteins. Eukaryotes, on the other hand, can move many large molecules or particles directly into or out of the cell. To do this, they rely on a group of organelles collectively known as the **endomembrane system**. Cells that are actively involved in such import or export appear to have cytoplasm that is crammed with membrane, much of which is part of this cellular transport system.

To explain how this intricate system works, we will first consider how a large protein is exported out of the cell. We will then describe the opposite problem: how cells transport large substances in, and what then happens to them once they are inside the cell.

As we have discussed, amino acids are linked into proteins on ribosomes. If a protein is destined to be exported from the cell, soon after protein synthesis begins, the ribosome attaches to the complex system of double membranes that extend throughout the cytoplasm called the **endoplasmic reticulum (ER)**. The presence of ribosomes on part of the ER surface gives it a knobby appearance, and it is called **rough ER** (**Figure 3.24a**). Other regions of the ER that lack ribosomes are called **smooth ER**. Both types of ER are seen in most eukaryotic cells. In multicellular organisms, cells specialized for the production and secretion of specific proteins can be recognized by their especially extensive network of rough ER. Immune system cells called plasma cells, specialized for the production and secretion of antibodies into the blood, are an example (**Figure 3.24b**).

As a protein is being produced on the surface of rough ER, the protein is threaded through the ER membrane into the interior space, or **lumen**, of the ER (**Figure 3.25**). Once in the lumen, the protein is transported through the ER, toward the membrane. Along the way, the protein assumes its three-dimensional shape and may be modified by the addition of sugars. When it reaches the furthest portion of the ER, the protein buds off in a small, membrane-enclosed sphere called a **transport vesicle**. The vesicle moves along the cytoskeleton until it reaches a different organelle called the **Golgi apparatus**.

The primary function of the Golgi apparatus is to receive materials from the ER and direct them to their ultimate destination. This organelle is composed of a series of flattened membrane-bound sacs (see Figure 3.25). When a transport vesicle containing a protein buds off the ER, it fuses with the Golgi apparatus, releasing its protein into the series of membrane sacs. As the

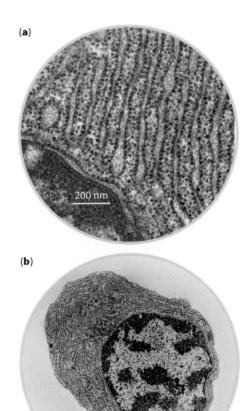

(a)

200 nm

(b)

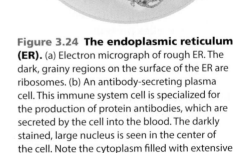

Figure 3.24 The endoplasmic reticulum (ER). (a) Electron micrograph of rough ER. The dark, grainy regions on the surface of the ER are ribosomes. (b) An antibody-secreting plasma cell. This immune system cell is specialized for the production of protein antibodies, which are secreted by the cell into the blood. The darkly stained, large nucleus is seen in the center of the cell. Note the cytoplasm filled with extensive rough ER.

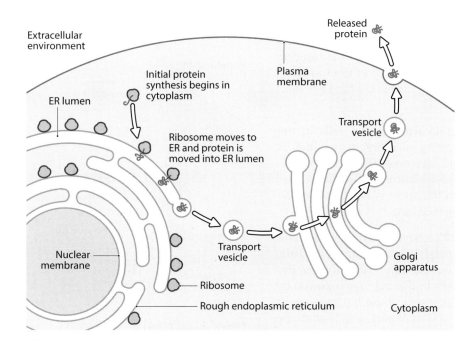

Figure 3.25 The path of exported proteins in eukaryotic cells. Protein synthesis begins on a ribosome (in dark gray) that is free in the cytoplasm. If the protein (in red) is to be used within the cell, the ribosome remains free in the cytoplasm, but if the protein is to be exported, the ribosome attaches to the ER. As the protein is being made, it is threaded into the ER lumen. Once inside the lumen, it is transported through the ER. Along the way the protein folds up into its specific three-dimensional shape. Eventually the protein buds off the ER in a transport vesicle, which moves along the cytoskeleton to the Golgi apparatus. The transport vesicle fuses with the membranes of the Golgi apparatus and moves through the Golgi's series of membrane-bound sacs. The protein, which may be modified within the Golgi apparatus, buds off in another transport vesicle and moves to the plasma membrane. When the transport vesicle fuses with the plasma membrane, the protein is released. This process is known as exocytosis.

protein is transported through this stack of membranes, it is further modified and routed to its final destination. This destination may be the plasma membrane, where the protein will serve as a membrane protein, or it may be another organelle within the cell. Alternatively, the protein may be transported to the cell surface in another membrane-bound vesicle that will fuse with the plasma membrane and release the protein to the extracellular environment. This process, known as **exocytosis**, is unique to eukaryotes; it does not occur in bacteria. All vesicle transport from the ER to the Golgi apparatus and from the Golgi apparatus to the final destination is along cytoskeletal elements. Other molecules are brought into the cell when they first bind to the cell surface and are next surrounded by portions of the plasma membrane. The membrane forms a vesicle around the cell, which is then brought into the cytoplasm by **endocytosis**.

CASE: PARROT FEVER

Thinking that their two sons should have a pet, Joe and Charlotte buy the boys a pair of cockatiels. The birds initially seem healthy, but within a few weeks they appear inactive and listless. Concerned about the birds, Joe takes them to the veterinarian, who diagnoses a case of psittacosis, otherwise known as parrot fever. The veterinarian recommends giving the birds food that has been impregnated with the antibiotic tetracycline. She also tells Joe to exercise great care while cleaning the cage for the next few weeks. Curious about the disease, Joe reads up on psittacosis. He finds that it is caused by the bacterium *Chlamydia psittaci*, which is most common in parrots and related birds. Pet stores that sell birds are a regular source of *C. psittaci* infection. Joe also learns that these bacteria are unusual in that during an infection, they are engulfed but not easily destroyed by phagocytic immune system cells.

1. **How do the bacteria resist destruction by phagocytic cells?**
2. **Why did the veterinarian caution Joe about cleaning the cockatiels' cage?**

When a large molecule is brought into a cell by endocytosis, the transport vesicle containing the imported material is usually routed along the cytoskeleton to another organelle called the **lysosome** (**Figure 3.26**). The lysosomes contain powerful digestive enzymes. The transport vesicle and the lysosome fuse, releasing the vesicle's cargo into the lysosome interior. The lysosome's enzymes then digest whatever was carried in the transport vesicle. In many cases, the digested products are nutrients, which are then available to the cell.

Lysosomes are especially important in certain white blood cells of the immune system. When such cells contact invading microorganisms, such as a bacterium or a virus, they first adhere to and then engulf them by extending a portion of the plasma membrane around the intruder. The membrane fuses around the microbe and seals it off in a transport vesicle. The vesicle then fuses with the lysosome and the microbe is digested.

This process, called **phagocytosis** (**Figure 3.27**), is a type of endocytosis, and it is crucial to a successful immune response. However, some microorganisms have devised insidious strategies to avoid destruction in lysosomes. For instance, the *C. psittaci* infecting the cockatiels in the case are ingested by phagocytic cells called macrophages. Once ingested, these bacteria are at least temporarily able to prevent fusion of the transport vesicle and the lysosome. If they reproduce sufficiently in the vesicle, they may kill the macrophage, and the dying cell releases chemicals that can stimulate inflammation.

It is not clear how the bacteria prevent lysosome fusion, but it is probably due to the secretion of a compound that alters the membrane of the transport vesicle in a way that blocks fusion. Tetracycline works well against *C. psittaci*, but the treatment can last as long as a month or more. During that time, infected birds may still be shedding bacteria in their feces. If a human breathes in dust from these droppings, he or she can develop pneumonia. Hence the veterinarian's warning about cleaning the birds' cage.

Chloroplasts and mitochondria function in the production and utilization of energy by the cell

Chloroplasts are responsible for the green color that is so characteristic of most plants and many algae. Within chloroplasts, carbon dioxide and water are converted into sugar, using the energy found in sunlight. This process is known as **photosynthesis**. Even in plants, chloroplasts are not found in all cells but only in leaves and stems, where photosynthesis occurs. **Mitochondria**, on the other hand, are found in all but a few eukaryotic cells. These organelles are involved in the release of energy from the chemical bonds in biological molecules. We will return to the topic of energy use and production in Chapter 8.

The structure of chloroplasts and mitochondria is similar (**Figure 3.28**). In both organelles a double membrane surrounds a space known as the **stroma** in chloroplasts and the **matrix** in mitochondria. Within the stroma are stacks of membranous vesicles called **thylakoids**. The various reactions of photosynthesis and energy release take place either on the surface of these membranes or in the stroma or matrix.

Some bacteria are also capable of photosynthesis, and many carry out processes associated with the mitochondrion as well. Because they do not have organelles, these processes take place in the cytoplasm or in association with the plasma membrane.

Chloroplasts and mitochondria are especially interesting because in many ways these organelles resemble bacterial cells themselves. They are

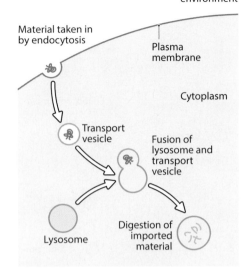

Figure 3.26 Endocytosis. Eukaryotes can bring large molecules into the cell via endocytosis. The adherence of the foreign material to the plasma membrane causes the membrane to engulf the material, eventually enclosing it in a transport vesicle. The vesicle, which moves along the cytoskeleton, may then fuse with a lysosome. Enzymes within the lysosome digest the imported material.

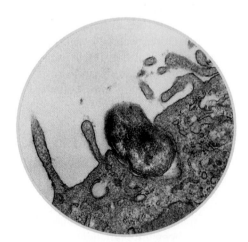

Figure 3.27 A macrophage phagocytosing a bacterial cell. Macrophages are white blood cells that play an important role in defense against microorganisms. Macrophages are able to engulf bacteria and other foreign material in a type of endocytosis called phagocytosis. The bacterium is the small, oval-shaped body in the center of the image. The macrophage is extending its membrane around the bacterium as it begins to ingest it. Normally, the ingested bacteria are killed by lysosomal enzymes. *Chlamydia psittaci* and certain other bacteria are able to at least temporarily survive inside macrophages because they can block the fusion of the transport vesicle to the lysosome.

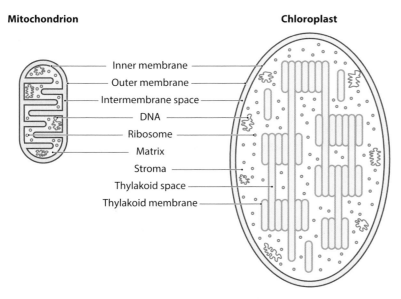

Figure 3.28 Mitochondria and chloroplasts. The general structures of a mitochondrion and a chloroplast are shown. Both of these organelles are surrounded by an outer and an inner membrane. They also both contain their own DNA molecules and make some of their own proteins.

approximately the same size and shape as some bacteria and they contain their own DNA. The DNA is organized in loops as it is in prokaryotes. Additionally, chloroplasts and mitochondria produce some of their own proteins, and these proteins are made on ribosomes that match those of bacteria in terms of size and mass. These similarities gave rise to the proposition that originally chloroplasts and mitochondria were bacteria that entered into a symbiotic relationship with other cells over a billion years ago. This concept, known as **endosymbiosis**, the union of a bacterial cell and larger cells into a mutually beneficial relationship, will be examined more closely in Chapter 9.

Looking forward and looking back

Having investigated the basic unit of life—the cell—we can now appreciate their fundamental importance to all of biology. While having many features in common, cells also display remarkable diversity. Although all cells may be categorized as either prokaryotic or eukaryotic, both of these large groupings can be further divided into numerous subcategories, each with its own specific characteristics. Microbes fall into many of these subcategories. An understanding of both the similarities and differences that we observe among cells gives us insight into various cellular processes, including growth rate, metabolic efficiency, and, in the case of microorganisms, how we treat diseases.

Chapter 4 takes a closer look at precisely how scientists form these categories and how they organize living things into a meaningful system of classification. After discussing the basic principles of classification, we will introduce the various important groups of prokaryotic and eukaryotic microorganisms. We will then see how in some ways they are similar, how in other ways they differ dramatically, and how they are related to each other. Think of Chapter 4, then, as our microbial field guide, giving us the information we need to identify and comprehend the unseen members of the microbial world.

Garland Science Learning System

- http://garlandscience.rocketmix.com/students/

- Discover why anthrax is an ideal biological weapon

- See a bacterial flagellum at work, and phagocytosis actually happen

- Test your knowledge of this chapter by taking the quiz

- Familiarize yourself with the terminology used in this chapter by using the vocabulary review

- Get help with the answers to the Concept questions

Concept questions

These questions are designed to help you start thinking like a microbiologist. The answers are not always simply found in the text. Instead, you will need to take the concepts about which you have learned and apply them to new situations. Some of the questions may not even have just a single correct answer. Help is provided as part of the GSLS resources, which can be accessed through http://garlandscience.rocketmix.com/students/.

1. Give a succinct, one-sentence definition of a cell.

2. Microbiologist and author Howard J. Rogers noted a number of years ago, "The era in which workers tended to look at bacteria as very small bags of enzymes has long passed." What do you think he meant by that?

3. You look at a cell under the microscope at high magnification. How might you tell if the cell were prokaryotic or eukaryotic?

4. The first acid-fast stain was developed by the German microbiologist Paul Erhlich in the late 1800s. After he developed it, he speculated that it might not be possible to use acid solutions to disinfect surfaces contaminated with acid-fast bacteria. Based on your understanding of the acid fast stain, why was Erhlich's speculation logical?

5. Sometimes, when doing a Gram stain, erroneous results are obtained because of errors in the Gram staining process. One mistake might be that the cells are exposed to alcohol for too short a time. If this mistake is made, would you more likely expect Gram-positive cells to be incorrectly identified as Gram-negative or Gram-negative cells to be incorrectly identified as Gram-positive?

6. Consider two new antibiotics, A and B. Antibiotic A disrupts cell membranes, whereas antibiotic B interferes with the synthesis of peptidoglycan. Which of these drugs do you think is safer to use to treat an infection in an animal and why?

7. A bacterium shows strong negative taxis to acidic conditions. It most likely

   a. is an acid-fast bacterium.

   b. has pili.

   c. has flagella.

   d. is a coccus of some type.

8. Which of the following doctors has made a mistake?

   a. Dr Adams prescribes penicillin for a *Streptococcus* (Gram-positive) infection.

   b. Dr Baker orders a drug for a Gram-negative infection of the blood. The drug causes rapid killing of the bacterial cells.

   c. Dr Curtis uses a bactericidal drug to treat a *Staphylococcus* (Gram-positive) infection.

   d. Dr Duran treats a tuberculosis patient with a drug that inhibits cell wall synthesis.

9. You visit a small island, which just days earlier experienced a volcanic explosion. Nothing appears to be alive on the island, but you collect soil samples and place them on culture media in the laboratory. Within 24 hours you find actively growing bacterial colonies on several of your plates. The bacteria that you find are most likely to

   a. have sex pili.

   b. have a glycocalyx.

   c. have capsules.

   d. be Gram-negative.

   e. have endospores.

10. Enterotoxic *E. coli* can cause severe food poisoning, unlike other strains of *E. coli*, which are generally part of the normal bacterial flora. The enterotoxic strains have various virulence factors that allow them to cause disease. These include fimbriae and toxins that inhibit protein production in eukaryotic cells. If an anti-pilin vaccine could be developed against enterotoxic *E. coli*, how might it prevent infection by this organism?

A FIELD GUIDE TO THE MICROORGANISMS

4

Lake Vostok in Antarctica is one of the coldest places on Earth. In fact, the lowest temperature ever recorded, –89°C (–128°F), was measured there in 1983. The lake is buried under about 4000 m (13,100 ft) of ice and hasn't seen the light of day since the Antarctic ice cap formed 15 million years ago (**Figure 4.1**). Lake Vostok therefore represents a sort of time capsule of prehistoric water that might tell us about ancient conditions on our planet.

Scientists have speculated about what if anything might be alive in the lake. If there are living things there, they represent life forms that were sealed off from other organisms millions of years ago. In 2013 researchers announced that they had bored all the way into the still-liquid water and recovered powerful evidence of life, both prokaryotic and eukaryotic. Others were sceptical; these indications of microbes could just be contamination by surface-dwelling organisms.

So, do microbes dwell in Lake Vostok or not? We may know soon. In 2015, Russian scientists announced that they had drilled an uncontaminated hole into the lake, hitting water at 3769 m below the surface ice. Once water samples are obtained and analysed, we may finally know if the lake truly contains prehistoric life.

But if ancient microbes are recovered, it should not be such a surprise. After all, microorganisms are already found in every conceivable environment, from superheated thermal ponds, where the water would scald your skin, to deep inside of rock formations. They live in desert soils, on your skin, and under your toenails. There are microorganisms that thrive on toxic chemicals and others that digest petroleum. It is safe to say that microorganisms exploit every possible habitat on the planet. They are so diverse that we can only guess at how many different types of microorganisms there actually are.

Here in Chapter 4 we will formally introduce the principal groups of prokaryotic and eukaryotic microorganisms to be considered throughout this text. But before we start, we need to get organized. Because all living things, including microbes, are classified in specific ways, we will begin by investigating the principles of modern biological classification and how it reflects the relatedness of all living things.

Figure 4.1 Satellite image of Lake Vostok. The largest of Antarctica's 400 known subglacial lakes, Lake Vostok covers an area of 12,500 km^2 (4830 mi^2). It lies beneath roughly 4000 m of solid ice. Could microorganisms, cut off from the rest of the world 15 million years ago when the ice cap formed, still be living in the lake?

Taxonomy is the science of biological classification

Taxonomy attempts to group organisms together on the basis of similarity. Accordingly, an accurate taxonomy reflects the relationship between organisms. Taxonomy also gives scientists a common language with which to communicate about the living world.

All case studies have a few questions at the end, the answers to which will become apparent as you read the sections following the case.

If you have ever spent time in another English-speaking country, you have no doubt occasionally been struck by the different names that are used for animals and plants in different places. Even within North America, crayfish might also be called "crawfish" or "crawdads," depending on where you are. These interchangeable terms may not be too confusing, but would you know what Australians were discussing if they mentioned "yabbies"? Such problems only worsen when you visit non-English-speaking countries. Among biologists, such problems can be more serious than minor misunderstandings. When two scientists are sharing information about an organism, there must be no doubt that they are discussing the same thing. In microbiology specifically, the consequences of such miscommunication can literally be life-threatening. For instance, antibiotics that work well against one species of *Staphylococcus* bacteria may have little or no effect against others. Two doctors discussing treatment of a *Staphylococcus* infection must be certain they are talking about the same microorganism or an inappropriate medication might be prescribed.

Taxonomy is based on a system of hierarchical groupings

Although attempts to classify animals and plants date back to Aristotle in the fourth century BC, the modern era of taxonomy began with the Swedish naturalist Carolus Linnaeus in the mid-1700s. Linnaeus developed a classification scheme based on a hierarchical sequence of groupings. In this system, the smallest, most fundamental group is the **species** (plural, species). All members of a particular species group are essentially the same type of organism. Groups of species that appeared to be similar were placed in the same **genus** (plural, genera). Thus, while the domestic dog is a unique species, *Canis familiaris*, it is placed in the genus *Canis* along with other canines, such as the wolf (*Canis lupus*) and the coyote (*Canis latrans*). Notice that each species is assigned two names in a manner called **binomial nomenclature**, first introduced by Linnaeus. The first name indicates the organism's genus, while the second identifies it as a particular species within that genus. Both the genus and the species names are Latinized and are either underlined (when hand-written) or printed in italics. The genus name is always capitalized, while the species name is always written in lowercase. Although the genus name must be written out in full the first time it is used, it may subsequently be abbreviated.

Moving up the hierarchy, those genera considered to be similar are placed in a common **family**, and related families compose an **order**. For instance, the fox is considered to be different enough from dogs to merit its own genus, *Vulpes*. Both genera, *Canis* and *Vulpes*, are in the same family of doglike animals known as Canidae, and all Canidae are in the order Carnivora along with other similar families, such as Felidae (the cats) and Ursidae (the bears). Carnivores are just one order in the **class** Mammalia (the mammals). Others include the order Primates (monkeys, apes, and humans) and the order Cetacea (whales and dolphins). Mammals, along with the classes Aves (birds) and Osteichthyes (bony fish), as well as other animals having a backbone or a backbonelike structure at least some time in their lives, are grouped together in the **phylum** (plural, phyla) Chordata, and all animals including Chordata are in the **kingdom** Animalia (**Figure 4.2**). Animals are one of several kingdoms that will be discussed in this chapter.

Modern classification reflects evolutionary relatedness

Although the basic idea of hierarchal classification and binomial nomenclature has remained the same since first proposed by Linnaeus, the way we

	Bacterium	Corn	Housefly	Domestic cat
Kingdom	None designated for bacteria	Plantae	Animalia	Animalia
Phylum	Proteobacteria	Anthophyta	Arthropoda	Chordata
Class	Gamma-proteobacteria	Monocotyledonae	Insecta	Mammalia
Order	Enterobacteriales	Commelinales	Diptera	Carnivora
Family	Enterobacteriaceae	Poaceae	Muscidae	Felidae
Genus	*Escherichia*	*Zea*	*Musca*	*Felis*
Species	*E. coli*	*Z. mays*	*M. domestica*	*F. catus*
	E. coli	**Z. mays**	**M. domestica**	**F. catus**

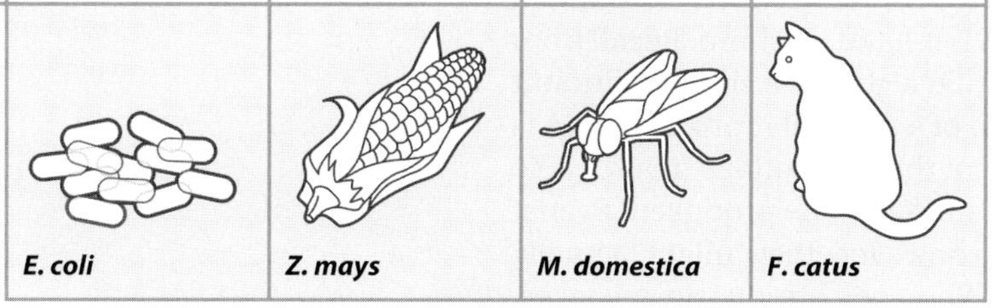

Figure 4.2 Taxonomy's hierarchy. A hierarchical classification of a bacterium, a plant, and two animals. In such a hierarchy, similar species are placed in the same genus, and similar genera are placed in the same family. Similar families are placed in the same order, and so on. Both the housefly and the domestic cat, for instance, are different enough to be placed in separate phyla, but they share certain characteristics that place them both in the kingdom Animalia.

think about and use taxonomy has changed as our understanding of living things has increased. Prior to the development of evolutionary theory, to be described in Chapter 9, organisms were grouped on the basis of similar visible features. No attempt to group organisms according to evolutionary relatedness was made, because these relationships were not recognized. However, most modern taxonomic schemes try to organize living things according to their evolutionary ancestry. Taxonomy based on a shared evolutionary history is known as **phylogenetics**. In such classification, species within a genus share a common evolutionary ancestor more recently than do species in separate genera. Two genera within the same family are more closely related than genera in different families, and so on, up the taxonomic hierarchy. Evolutionary relationships are often depicted as a tree (**Figure 4.3**). In such a tree of life, the trunk represents the ancestral organisms of all

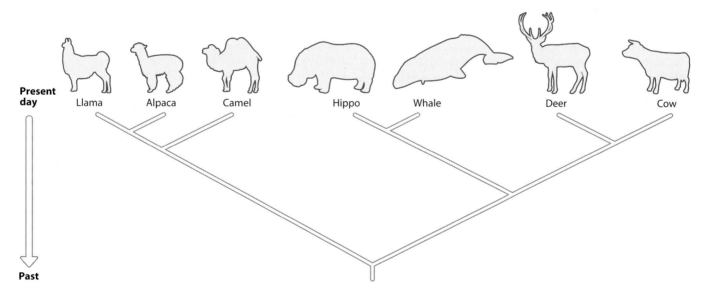

Figure 4.3 Phylogenetic tree showing the relationships between several representative mammals. Llamas share a more recent common ancestor with alpacas than they do with other animals in this tree. Thus, llamas and alpacas are more closely related to each other than they are to other animals. Llamas and alpacas, along with the next closest group, the camels, form the family Camelidae. Others in this tree share a common ancestor more recently with each other than any of them do with members of Camelidae.

species in the tree. Branches indicate where groups of organisms separated from each other in the course of their evolution. In a complete taxonomic tree, the small terminal branches would represent individual species. By examining a well-constructed phylogenetic tree, the evolutionary history of organisms can be traced back, and the relatedness of different organisms can be inferred.

DNA analysis provides evidence for relatedness

Originally, similar anatomical and biochemical features were taken as an indication of a similar evolutionary past. Since the 1980s, however, taxonomists have increasingly relied on genetic similarities to determine phylogenies. By comparing certain genes between organisms, scientists now have a far more accurate means of determining who is related to whom. Recall from our discussion of nucleic acids in Chapter 2 that a gene is a specific amount of DNA providing the code for the production of a particular protein. DNA is composed of nucleotides, and the sequence of the nucleotides in a specific gene dictates the order of amino acids in the protein to be produced. Even if two organisms produce the same protein, however, they might actually make it in slightly different forms. These differences are due to mistakes, known as **mutations**, which occur when DNA is reproduced. Mutations are changes in the sequence of nucleotides in a gene, and they occur with time in all organisms.

If two species arose from a common ancestor only recently, there will not have been much time for each to acquire mutations independently. Their genetic material will therefore still be very similar. But if two species are only distantly related, the number of differences in their DNA will be correspondingly greater. Because more time has passed since they shared a common ancestor, each species will have accumulated mutations on its own for a longer period. Taxonomists use the number of differences seen in a gene (mutations in the nucleotide sequence) as an indicator of relatedness.

CASE: STREAMS, SNAILS, AND SCHISTOSOMES

Schistosomiasis is a debilitating disease caused by parasitic flatworms in the genus *Schistosoma* (**Figure 4.4a**). Over 200 million people are infected worldwide, primarily in the tropics. In Latin America, the causative agent is *Schistosoma mansoni*. These parasites develop as larvae in freshwater snails in the genus *Biomphalaria* (**Figure 4.4b**). After larval development and reproduction, the parasites leave the snail and enter the water. Humans become infected when they enter contaminated water bodies (**Figure 4.4c**). The larval flatworm burrows through the skin and migrates to the veins surrounding the intestine. Following mating, eggs are produced that pass into the intestine and are released with the feces. Upon hatching, the larval worm seeks out an appropriate snail, and the cycle is complete. Adult worms can survive for many years in the human host, producing millions of eggs. Many of the eggs fail to pass into the intestine and get swept by the circulatory system to the liver, where they become lodged. Inflammation caused by the eggs can eventually lead to severe liver disease. Interestingly, not all *Biomphalaria* snails are susceptible to infection. The most susceptible species in South America is *Biomphalaria glabrata*. Others, such as *B. obstructa*, are resistant. It is unknown whether or not *B. temascalensis*, found in the area of Temascal, Oaxaca, in southern Mexico, is subject to infection.

Dr Randall DeJong, a biologist at the National Institutes of Health, has examined the taxonomic relationship of *Biomphalaria* snails in Latin America. He proceeded by extracting DNA from the three species above and determining the nucleotide sequence for an rRNA gene from each type of snail. He then found that the sequences for *B. obstructa* and *B. temascalensis* were almost identical. The sequence for *B. glabrata* was significantly different from that of the other two species.

1. What is an rRNA gene and why is it useful in this analysis? How does a comparison of genes between species help deduce their taxonomic relationship?
2. Which two of these three snail species share a more recent common ancestor?

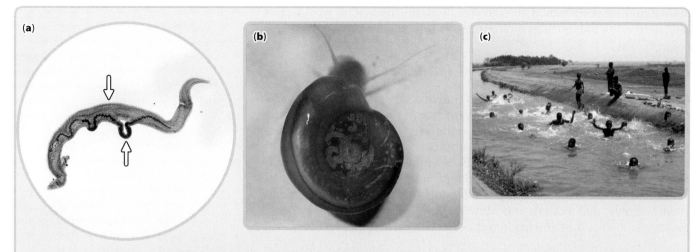

Figure 4.4 Schistosomiasis. (a) Adult male and female schistosomes. The slender female worm (indicated by the lower arrow) lies in a groove along the surface of the male's body (indicated by the upper arrow). In infected humans, these paired schistosomes are found in the mesenteric veins surrounding the intestine. (b) *Biomphalaria glabrata*. These aquatic snails are the most important schistosome larval host in South America. (c) Humans become infected when they enter water contaminated with schistosome larvae.

3. In the area of Temascal, *B. temascalensis* is the only *Biomphalaria* species found. Do you think there is a likelihood of schistosomiasis in these areas?

4. How would the rRNA gene Dr DeJong studied in this snail genus compare with the same gene in other taxonomic groups of animals?

Cells typically possess structures called ribosomes (see Chapter 3, p. 57), which are used to assemble amino acids into proteins. Part of each ribosome is composed of RNA, and the genes that code for this RNA are called rRNA genes. These genes are very useful in taxonomic analysis for several reasons. First, they are found in all living things, so they can be used to compare very different organisms. Second, rRNA genes mutate slowly and at a more or less constant and predictable rate. Separate species accumulate unique gene mutations and the longer ago they diverged, the more different their genes will be. Because the number of mutations is time-dependent, scientists like Dr De Jong can actually deduce how long ago separate species were formed. DNA can thus serve as a molecular clock, which can be used to construct phylogenetic trees.

So, is schistosomiasis likely in the Temascal, Oaxaca, area? Dr DeJong's data indicate that *B. obstructa* and *B. temascalensis* diverged only recently, while the ancestor of *B. glabrata* branched off from the others relatively early in the evolution of this snail genus. Because the closely related *B. obstructa* is resistant to infection, it is likely that schistosomes cannot infect *B. temascalensis* either. Streams in this part of southern Mexico are therefore probably schistosome-free. Other *Biomphalaria* species extend into the southern United States, but they are also resistant, and an analysis of rRNA genes places them closer to *B. obstructa* and *B. temascalensis* than to *B. glabrata*.

If the rRNA gene of *Biomphalaria* snails were compared with that of other snails in different genera but the same family, there would no doubt be an increased number of sequence differences. Snails are members of the phylum Mollusca, which is divided into eight classes. Snails are in the class Gastropoda. Other familiar molluscan classes include Bivalvia (clams and oysters, for example) and Cephalopoda (squid and octopuses). Any two snails would be more similar to each other in terms of rRNA gene sequence then they would be to a member of a different molluscan class. A snail and an octopus, however, would have greater similarity than either would to an animal in a different phylum such as a beetle (phylum Arthropoda).

Kingdoms are organized among three domains

Originally, all life was divided into two principal kingdoms: Animalia and Plantae (animals and plants). However, many types of living things did not fit easily into either of these categories, and the recognition of major structural differences between eukaryotic and prokaryotic cells (see Chapter 3) led to important changes in classification. Eventually, three more kingdoms were added—Protozoa (single-celled organisms but with eukaryotic cells similar to those of animals and plants), Fungi (eukaryotes such as mushrooms, molds, and yeasts), and Monera (the prokaryotes). This scheme became known as the five-kingdom classification system.

In the 1980s, researchers used the powerful new techniques of molecular biology to dramatically change the way we look at the relatedness of living things. By comparing rRNA genes from a very large number of organisms, these scientists found that living things could actually be divided into three groups. These very large divisions, encompassing all five kingdoms, are termed **domains**.

Molecular data showed that the prokaryotes (formerly the phylum Monera) actually had two different evolutionary origins. The single kingdom was therefore reorganized into two very distantly related domains. Most of the former members of the kingdom Monera with which we are most familiar belong to the domain **Bacteria**. Other prokaryotes once considered to be bacteria are different enough to merit their own domain, **Archaea**. Additional study of archaea has revealed other important structural and biochemical differences from the bacteria. Animals and plants, along with protozoa and fungi, are grouped together in the domain **Eukarya** (**Figure 4.5**). All of these organisms have eukaryotic cells, and rRNA analysis suggests that they originally evolved from a common ancestor.

Having reviewed the fundamentals of classification, it is time to turn our attention to each of the principal groups of microorganisms. A complete review of all species is well beyond our scope. Rather, we will consider the important characteristics and the major divisions within each important lineage, providing representative examples where appropriate. Think of this section, then, as your roster of microbial players.

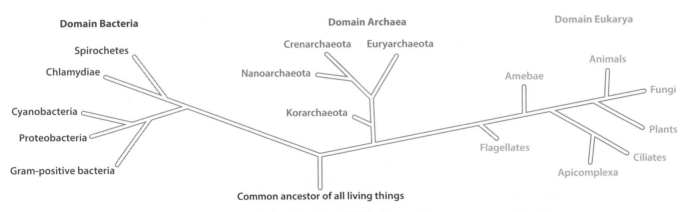

Figure 4.5 The three domains of life. A phylogenetic tree based on ribosomal RNA similarity, showing some of the principal groups in each of the three domains. The exact details of trees such as this one are still under investigation, and alternative trees exist. Note that domain Archaea is somewhat closer to domain Eukarya than either is to domain Bacteria.

Microorganisms are found in all three domains of life

As previously noted, we can only guess how many species of microorganisms actually exist. No one knows, because so few have been described so far. There are currently just over 9000 described species of prokaryotes (bacteria and archaea). The real number of species no doubt numbers in the millions or even up to a billion. A single ounce of sea water may contain thousands of species, most of them unknown to science. As for the archaea, scientists are only just beginning to understand their diversity. Until recently, it was believed that archaea were largely limited to extreme environments, such as very hot, salty, or acidic habitats (see Figure 1.1c). We now know that in many common environments, a large proportion of the microorganisms found turn out to be archaea. The sea, for example, teems with them.

The eukaryotic microorganisms are also a diverse lot. While on a hike, for example, if you pause to drink from an apparently clear freshwater stream, you may become infected with *Giardia intestinalis* and develop a case of what is commonly called hiker's disease (**Figure 4.6**). The water becomes contaminated when certain mammals pass *Giardia* cysts into the environment while defecating. If you enjoy drinking beer, on the other hand, you may be interested to know that different styles of beer are made by using different species of yeast in fermentation. To produce an English-style ale, the brewmaster adds *Saccharomyces cerevisiae* to a filtrate of the mixed malt and hops. A lager is the result if *Saccharomyces carlsbergensis* is used. Both *Giardia* and the two yeast species are examples of microorganisms in the domain Eukarya.

We will begin by reviewing the two prokaryotic domains, first the bacteria, followed by the archaea. We will then move on to examine those microorganisms in the domain Eukarya.

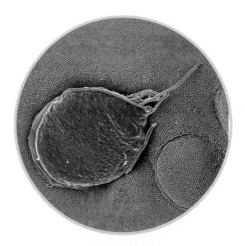

Figure 4.6 *Giardia intestinalis.* These flagellated protozoans use a ventral sucker to attach to the lining of the digestive tract, blocking proper nutrient absorption and causing intestinal distress. Note the circular mark on the intestinal lining left by a ventral sucker.

The domain Bacteria consists of numerous lineages

Genetic analysis indicates that the domain Bacteria can be divided into more than a dozen major phylogenetic lineages. A phylogenetic tree showing the relationship of several of the major groups is provided in **Figure 4.7**.

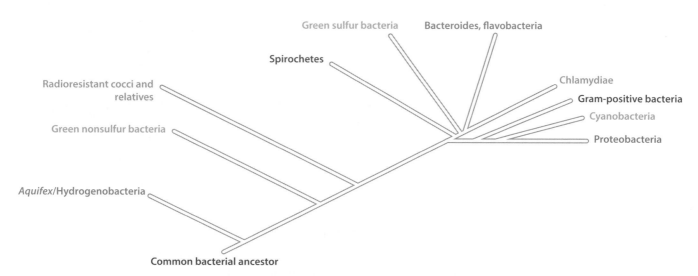

Figure 4.7 Phylogenetic tree of the domain Bacteria based on comparisons of rRNA. There are over a dozen distinct lineages in the domain. This phylogenetic tree shows a proposed relationship between several of the principal groups. While not comprehensive, this tree highlights those lineages that best illustrate the evolution of this domain or those that are of particular medical or ecological importance. The precise relationship between the various lineages is still under investigation, and other possible trees have been proposed.

As discussed earlier, phylogenetic trees are often constructed by comparing rRNA genes. The closer two groups are on the tree, the more similar are their rRNA sequences. Branches in the tree indicate where an ancestral lineage diverged into separate groups. Trees may sometimes differ based on the specific data a scientist uses to construct a tree or on the exact manner in which the tree is produced. For bacteria, in particular, there is still no precise and clear-cut agreement as to how the various lineages are related. Furthermore, there is no firm consensus on which bacterial lineages should be grouped together as kingdoms. Depending on the source, each group has been referred to as a separate kingdom, a phylum, or even a class. Consequently, until this issue is resolved, it is probably preferable to simply refer to each of the large groups as distinct lineages, each one united by a common evolutionary history.

The earliest bacteria were adapted to life on the primitive Earth

CASE: THE HEAT WAS ON!

Earth is estimated to have formed about 4.5 billion years ago, yet it would be another 400 million years before there was even a solid piece of ground to stand on. Prior to that time, Earth's surface was an inferno of molten rock. About 4.1 billion years ago, Earth had cooled enough for the crust to solidify. Another 700 million years later, the first cells appeared (**Figure 4.8**). However, Earth was still an inhospitable place. Temperatures were much higher than they are today, and without an ozone layer, the surface was bombarded by ultraviolet (UV) radiation.

1. **What characteristics would have been typical of the earliest cells?**
2. **Are any of these characteristics present in the modern ancestors of these first prokaryotes?**

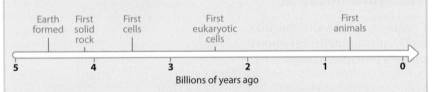

Figure 4.8 Origin of Earth and appearance of the first cells. Earth is thought to have formed approximately 4.5 billion years ago. The first cells appeared about 1 billion years later, about 3.5 billion years ago. Eukaryotes were first present approximately 2.5 billion years ago.

Let's consider the bacteria living today that most resemble those ancient pioneers of 3.5 billion years ago. As seen in Figure 4.7, the *Aquifex/Hydrogenobacter* group were the first to branch off the main lineage within the domain Bacteria. As such they are the most primitive members of this domain, closest to the original ancestor of all bacteria. The term "primitive," when used in this context, does not imply that these organisms are somehow less complex or that they are poorly adapted to their environment. Rather, it simply means that this group has changed the least since the evolutionary origin of the domain.

These bacteria grow at very high temperatures (up to 95°C or 203°F) and are therefore known as **thermophiles** (from the Greek for "heat-loving"). Their ability to tolerate high temperatures, along with their phylogenetic position near the base of the bacterial tree, makes sense in light of prevailing opinion that the early Earth was very hot. Continuing up the

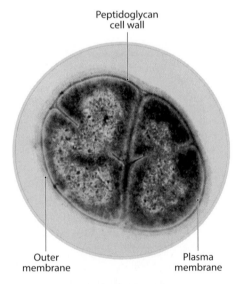

Figure 4.9 *Deinococcus radiodurans*. This species is highly resistant to radiation, in part because of its ability to repair radiation-induced damage to its DNA. Each cell is surrounded by a peptidoglycan cell wall typical of Gram-positive bacteria, as well as an outer membrane seen in Gram-negative species.

evolutionary tree, we next encounter the **green nonsulfur** bacteria, another group of thermophiles. Interestingly, although they grow well at elevated temperatures, they do not survive the extreme heat tolerated by members of the *Aquifex/Hydrogenobacter* lineage. This suggests that although Earth was still quite warm when this group evolved, it had already cooled to some degree. The two lineages discussed so far are composed entirely of thermophilic species. Thermophiles are found in other groups of bacteria, but only in these most primitive groups is the characteristic universal. Green nonsulfur bacteria are also photosynthetic, using sunlight as an energy source to produce sugars.

The **radioresistant cocci** and relatives are noteworthy because of their ability to tolerate radiation. One species, *Deinococcus radiodurans*, is even more radiation-resistant than the bacterial endospores discussed in Chapter 3. These microbes can survive up to 3 million rads of radiation. For a human, exposure to fewer than 500 rads can be lethal. Not surprisingly, a number of unusual adaptations make this possible. Although unrelated to other Gram-positive bacteria, *D. radiodurans* has a multilayered Gram-positive cell wall. It also has an outer membrane, typically found only in Gram-negative bacteria (**Figure 4.9**). High doses of radiation are usually lethal primarily because they cause mutations in a cell's DNA, but *D. radiodurans* has a remarkable ability to repair its DNA, far exceeding our own. Such a characteristic would have been useful when life first evolved because of the high levels of ambient radiation. The resistance seen in species such as *D. radiodurans* may be a legacy of those times.

The tree spreads out

At this point, our phylogenetic tree diverges into a number of distinct lineages. The **spirochetes** are a group of bacteria having an unusual helical and flexible shape (**Figure 4.10a**). They are found in many environments, and several species infect humans and animals, causing a handful of important diseases. Perhaps the most significant of these is syphilis, caused by *Treponema pallidum* (**Figure 4.10b**).

Green sulfur bacteria, like the green nonsulfur bacteria described previously, are photosynthetic. As their name suggests, however, sulfur-containing compounds are used in the photosynthetic process. Notice in Figure 4.7 that the green sulfur and green nonsulfur bacteria are only distantly related. This suggests that photosynthesis evolved more than once in taxonomically distinct groups. Indeed, it seems to have evolved several times in different lineages scattered across the bacterial domain.

If your dentist has ever told you that you are at risk for gum disease or gingivitis, you can probably blame *Bacteroides gingivalis*, a member of the next lineage on our tree, **flavobacteria**. Because *B. gingivalis* cannot grow in the presence of oxygen, it is called a **strict anaerobe**. Consequently, it flourishes in the oxygen-free environment under the gum line, where it causes the inflammation that results in disease. When you floss your teeth, you actually allow oxygen into this space, inhibiting the growth of the bacteria and reducing the likelihood of gingivitis.

Chlamydiae are interesting in several respects. First, they cannot reproduce unless they enter the cytoplasm of an appropriate host cell. Therefore, they are examples of **obligate intracellular parasites**. Although they have a Gram-negative outer membrane, they are unrelated to other Gram-negative species. Unlike more familiar Gram-negative bacteria, they have

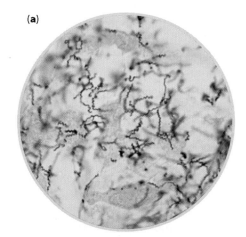

(a)

(b)

Figure 4.10 *Treponema pallidum.*
(a) *Treponema pallidum*, a representative spirochete, is the causative agent of syphilis.
(b) Syphilitic lesions caused by *T. pallidum*.

no peptidoglycan layer between the plasma and outer membranes. *Chlamydia trachomatis* inhabits the human urogenital tract, and infections caused by this species are perhaps the most common of all sexually transmitted infections. This species can also infect the eye. The scarring that occurs on the cornea can result in trachoma, the leading cause of blindness in humans.

The most recently evolved lineages are found at the farthest branches of the tree

There are three remaining evolutionary lineages in our bacterial tree (see Figure 4.7). The **Gram-positive bacteria** contain many familiar genera, such as *Staphylococcus* and *Streptococcus*. Because members of this lineage are so common and because they affect human health and welfare in so many ways, they will reappear often throughout this text.

We owe a large debt of gratitude to the **cyanobacteria** (**Figure 4.11**). Before their evolution, approximately 2.4 billion years ago, the atmosphere contained about 0.1% free oxygen. Today the air we breathe is about 20% oxygen, much of it due to the photosynthetic activity of cyanobacteria. These bacteria, therefore, helped change Earth's environment in a way that made the later evolution of aerobic (oxygen-requiring) life possible. Today they continue to play a major role in ecological processes. Cyanobacteria release large amounts of oxygen, especially in aquatic and marine environments. Additionally, many species remove nitrogen from the atmosphere and convert it to a form that can be used by plants—a process called nitrogen fixation. The bluish-green coloration of cyanobacteria comes from their photosynthetic pigments.

The final branch on our phylogenetic tree represents the **proteobacteria** (see Figure 4.7). This large and varied group contains most of the Gram-negative species with which we are familiar (*Escherichia coli*, for example). Although they are all Gram-negative, proteobacteria span the gamut in terms of cell shape and metabolism. They may be cocci or rods, motile or nonmotile, aerobic or anaerobic. While some species are photosynthetic, the majority are not. The name proteobacteria refers to the Greek god Proteus, who could change his shape, alluding to the great diversity in this group. DNA analysis, however, confirms that these bacteria share a common origin. Like the Gram-positive bacteria, many proteobacteria impact human or animal health or play important roles in ecological processes, so we will encounter many members of this group in subsequent chapters.

Although they are prokaryotic, members of domain Archaea are distinct from bacteria

If you visited a very salty (or hypersaline) pool in Death Valley, or a hot spring in Yellowstone where the water is near boiling, you could be excused for thinking that nothing could survive in such a place. But you would be wrong. Such extreme environments are the habitat of choice for many species of archaea, often called extremophiles (see Figure 1.1c). Their adaptations to harsh environments suggest that they also, like the thermophilic bacteria discussed earlier in our case, were among Earth's earliest inhabitants. As previously discussed, it was not until recently that these organisms were considered to be anything other than unusual bacteria. Although archaea superficially resemble bacteria, phylogenetic analysis indicates that these

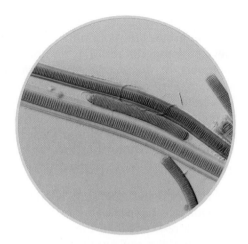

Figure 4.11 Cyanobacteria. Photosynthesis in the cyanobacteria evolved independently from photosynthesis in other bacterial lineages. Oxygen, released by cyanobacteria as a photosynthetic waste, helped to create an oxygen-rich atmosphere.

two prokaryotic groups are only remotely related. Indeed, rRNA comparisons place domain Archaea somewhat closer to domain Eukarya than either group is to domain Bacteria.

The structure of archaea varies considerably. As we discussed in Chapter 3, cell walls, although usually present, never contain peptidoglycan. Many archaea are rods or cocci, but others have highly unusual shapes (**Figure 4.12a**). They may or may not require oxygen, and some are photosynthetic.

The archaea differ from bacteria in other important ways. They have different lipids, for example, in their cell membranes than bacteria. In Chapter 3 we learned that these archaeal lipids are often more stable than those of bacteria, which is thought to be an adaptation to the high temperature and other extreme environments in which they are often found. The manner in which they make proteins is actually more like that of eukaryotes than it is to that of bacteria. This observation supports the hypothesis mentioned above that archaea share a common ancestor with eukaryotes more recently than either group does with bacteria.

Phylogenetically, domain Archaea is currently divided into four main lineages (**Figure 4.13**), although their taxonomy remains controversial. Part of the problem is that most species have never been studied in the laboratory and are known only from their DNA found in the environment.

We still know little about Korarchaeota, presumably the most primitive lineage, and therefore assumed to be those most similar to the original archaeal ancestor. Crenarchaeota are a lineage of extreme thermophiles and are thought to be the most heat-loving of all prokaryotes. Some members of this lineage, as mentioned in Chapter 3, have a phospholipid monolayer rather than a typical bilayer in their cell membranes. The phospholipids on one side of the membrane are covalently linked to those on the opposite side, forming one continuous layer. This is thought to represent an adaptation to extreme high temperatures, because it further enhances membrane stability.

Other Crenarchaeota are extreme acidophiles, living in environments with pH below 1.0. That means that if it were hot enough, and with proper nutrients, they could thrive in battery acid! The Nanoarchaeota, consisting of only a single known species, *Nanoarchaeum equitans*, are unusual in several respects. First of all, they are among the smallest (perhaps the very smallest) of all living things, measuring only 400 nm in diameter (see **Figure 4.12b**).

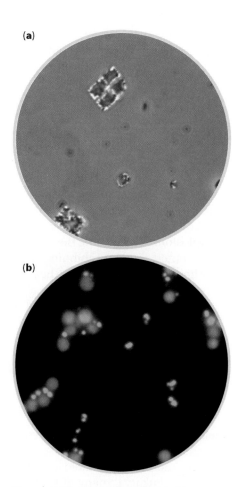

Figure 4.12 Unusual Archaea. (a) *Haloquadratum walsbyi*. This oddly shaped species is almost perfectly square. It lives in very salty (hypersaline) pools, and consequently is considered an extremophile. The genus name reflects its preference for high salt environments ("halo" from the Greek for "salt") and its shape ("quadra" from the Latin for "square"). (b) *Nanoarchaeum equitans* is the only known member of the Nanoarchaeota. The tiny cells (the small bright spheres, only about 0.4 micrometers in diameter) are about five times smaller than the host cells (the larger, fuzzier cells, about 2 micrometers in diameter) to which they are attached. *N. equitans* is the only known parasitic Archaea, relying on the host cells for the lipids it cannot synthesize.

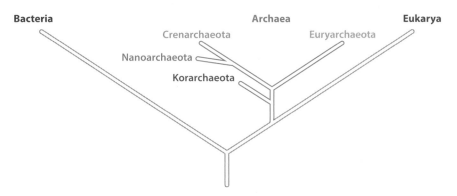

Figure 4.13 Phylogeny of the four principal lineages of Archaea. Korarchaeota are thought to be the most primitive lineage, most similar to the original ancestor of all Archaea.

Furthermore, they are the only known species of parasitic archaea. *N. equitans* cannot make its own lipids; it essentially steals them from the host cell (also an archaea), without which it cannot live. Both host and parasite live in scalding-hot, sulfur-rich water. The fourth lineage of Archaea is Euryarchaeota. The extreme halophilic (salt-loving) archaea are members of this group; some grow in salt deposits or other environments where salt concentrations reach seven times that found in the ocean. The **methanogens**, which thrive in the oxygen-free environments found in sewage, swamps, and the intestines of many animals, are also Euryarchaeota. The name of this group refers to the fact that they release methane as a metabolic waste product. Methane gas is flammable; because it is also odorless, it presents a particular hazard. If you collect compost, you may know that it is advisable to keep your compost heap away from structures and closed spaces to prevent an explosion. Also, turning the compost heap from time to time allows oxygen to enter, thereby inhibiting methanogen growth. Methane is also a greenhouse gas, helping to trap heat in the atmosphere. The methanogens in a single cow's intestines produce about 10 cubic feet of the gas each day. Billions of cows, all releasing tons of methane, are thought to be part of the global warming problem.

Eukaryotic microorganisms include the protozoa and the fungi

Notice in **Figure 4.14** that initially life branched off in two directions. One lineage gave rise to domain Bacteria, whereas the other lineage later branched into domains Archaea and Eukarya. These last two domains are therefore more closely related to each other than either is to domain Bacteria. In domain Eukarya, many of the more ancient lineages are placed in the kingdom Protozoa, which consists of single-celled eukaryotes that are abundant as both parasites and free-living forms. Later in the tree, we see that modern animals, plants, fungi, and some protozoa were the result of a more recent burst of diversity.

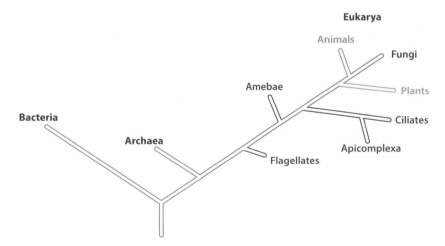

Figure 4.14 Domain Eukarya. A phylogeny of the Eukarya, based on rRNA similarity. Note that animals (in orange) and fungi (in black) are more closely related to each other than either group is to plants (in green). The four lineages in blue are all currently classified as members of the kingdom Protozoa. The exact relationship between these lineages is still under investigation.

Protozoa are single-celled organisms within the domain Eukarya

CASE: THE CAT'S OUT OF THE BAG ON A PROTOZOAN PARASITE

Donna and Ben have been married for 8 years. They have just learned that Donna is pregnant. Anxious to provide proper prenatal care, they read extensively on the subject of how to best care for the developing baby. In a parenting magazine, they are horrified to read that cats pose a risk to a developing fetus. Cats are often infected with the protozoan parasite *Toxoplasma gondii*, which can be transmitted to a pregnant woman and across the placenta to the fetus. Birth defects due to *Toxoplasma* infection in a newborn can include deafness, vision problems, and other neurological abnormalities. Both Donna and Ben had cats growing up and they adopted their current cat, Woody, shortly after they got married. They hate the idea of parting with their feline friend, but they certainly don't want to endanger their baby. They wisely put off making any important decisions about their cat, however, until they consult with their doctor. While at the doctor's office, Donna and Ben ask the following questions:

1. **Is it true that Woody poses a risk to the baby?**
2. **Can Donna be tested for *Toxoplasma* infection? If she is infected, can antibiotics protect both her and the baby?**
3. **Should they find a new home for Woody?**

Protozoa are microscopic, unicellular eukaryotes that lack cell walls. Their lack of photosynthetic chloroplasts distinguishes protozoa from the unicellular algae, with whom they are often categorized in a superkingdom called Protista.

Protozoa are found in many environments, including fresh and salt water and moist terrestrial habitats. They feed by ingesting other organisms or organic material. Protozoa often have complex life cycles. While many are free-living in the environment, others are parasites, and some have great medical and veterinary significance. When conditions permit, they exist in an active feeding and reproducing stage as **trophozoites**. However, if environmental conditions deteriorate, many protozoa form metabolically inactive, protective structures called **cysts**. In parasitic species, cysts are most common among those that spend a portion of their life cycle outside the body of a host. Often these cysts are excreted in feces into the environment, where they might ultimately be ingested by a new host. Once the cyst is ingested, it will develop into the active trophozoite, completing the life cycle. Other protozoa reach new hosts by way of **vectors** such as mosquitoes, which ingest the parasite when they feed on blood. The parasite is then transferred to the next host when the vector feeds a second time. Because protozoa that are carried by vectors are never exposed to the external environment, there is no cyst formation.

Phylogenetic analysis shows that the protozoa are not necessarily closely related to each other (see Figure 4.14). In this sense, the kingdom Protozoa can be compared to a grab bag where organisms that do not seem to fit elsewhere are placed. It is likely that further research may reorganize this kingdom, and what we now refer to collectively as kingdom Protozoa will one day be divided into several distinct lineages. By convention, the four main groups of protozoa are often recognized on the basis of their mode of locomotion.

Flagella are used by the flagellates for locomotion

Flagellates have one or more flagella, which they use for locomotion (**Figure 4.15a**). Some are important pathogens, including *G. intestinalis*, the cause of hiker's diarrhea mentioned earlier (see Figure 4.6). Several species within the genus *Leishmania* (see Figure 1.4b) can cause the disease leishmaniasis. This disease is widespread in Africa, South America, and the Middle East. Humans can become infected after being bitten by sand flies, which serve as vectors. The symptoms of leishmaniasis include severe ulceration and lesions on the skin (see Figure 1.4c). Because the *Leishmania* protozoa are found in the blood, military personnel who have served in the Middle East are not permitted to donate blood upon their return.

Amebae move as they extend parts of their cytoplasm

Amebae are protozoa that move by extending portions of their cytoplasm called pseudopodia (**Figure 4.15b**), changing shape as they move. Most amebas are free-living. Others can cause severe dysentery. *Entamoeba histolytica*, for example, is thought to infect close to half a billion people and to cause up to 100,000 deaths annually. Amoebic dysentery is mainly a problem in parts of the world where sanitation is poor. Cysts are passed in the feces of infected individuals, where they may subsequently come in contact with food or water, to infect new hosts. Access to clean drinking water vastly reduces the risk of infection.

The beating of numerous cilia permits locomotion in the ciliates

The ciliate protozoa are recognized by the presence of cilia, which beat in a coordinated manner, directing locomotion. Perhaps your first view of the microbial world was a drop of pond water under the microscope. If so, you may have observed free-living *Paramecium* swimming furiously across the slide in search of prey (**Figure 4.15c**). Most ciliates are free-living and do not cause disease. One exception known to tropical fish hobbyists is *Ichthyophthirius multifiliis*. These parasitic protozoans penetrate the mucous coat on a fish's skin or gills and feed on epithelial and blood cells. The result is the fatal skin condition called ich or white spot.

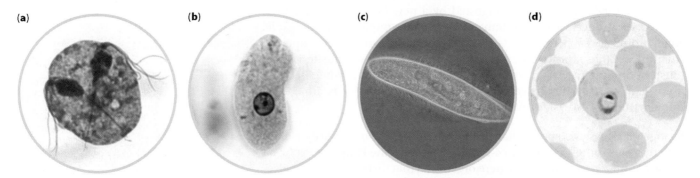

(a)   **(b)**   **(c)**   **(d)**

Figure 4.15 Representative protozoans. (a) A flagellate, *Trichomonas vaginalis*, the most widespread agent of any protozoan disease, causes itching and inflammation of the female reproductive tract. It is transmitted via sexual contact. (b) *Entamoeba histolytica*, an ameba, is a common cause of amebic dysentery. It is transmitted through contaminated food and water and is a problem wherever adequate sanitation is not assured. (c) The common freshwater ciliate *Paramecium* is a free-living (nonparasitic) protozoan. (d) *Plasmodium*, the apicomplexan that causes malaria. The parasite in this photograph is the darkly stained ring like structure, inside a human red blood cell.

Apicomplexa typically do not possess structures related to locomotion

All protozoa in the fourth group are parasitic and most are intracellular parasites, spending at least part of their life cycle inside host cells. They are named for a complex of structures found at the apex of the cell. There are usually no structures related to locomotion, and most are nonmotile.

Some of the most lethal human pathogens are **apicomplexans** in the genus *Plasmodium* (**Figure 4.15d**), which cause malaria. *Plasmodium* infects over 200 million humans worldwide each year, killing over 0.5 million annually. The parasite is transmitted by the bite of specific vector mosquitoes. Many other animals besides humans are also affected. If you visit Hawaii, for instance, almost all of the beautiful birds you see near the coast are non-native species. Many native birds can now persist only at elevations above 1500 m, where malaria-bearing mosquitoes are uncommon. Hawaii did not originally have the mosquitoes necessary to transmit malaria. They first arrived in about 1850, probably on a ship, and they brought *Plasmodium* with them. Indigenous birds, with no previous exposure, had no resistance to the disease and died in large numbers.

Toxoplasma gondii, introduced in our case, is also an apicomplexan. Like many parasites, *T. gondii* has a complex life cycle, involving more than one host (**Figure 4.16**). As trophozoites, *T. gondii* live in the intestine of cats,

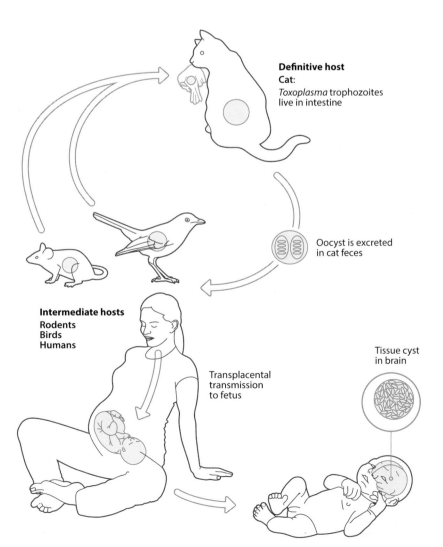

Definitive host
Cat:
Toxoplasma trophozoites live in intestine

Oocyst is excreted in cat feces

Intermediate hosts
Rodents
Birds
Humans

Transplacental transmission to fetus

Tissue cyst in brain

Figure 4.16 Life cycle of *T. gondii*.
Sexual reproduction takes place in the cat, which therefore is called the definitive host. Cyst-like oocysts are passed in the cat's feces, where they may be later consumed by a rodent or other small animal, which serves as the intermediate host. Here the oocysts form tissue cysts, which usually remain dormant until they reenter the definitive host, when a cat eats the intermediate host. Humans can also become infected if they consume oocysts. If a woman first becomes infected during pregnancy, there is a danger that the parasite may cross the placenta and infect a fetus before it is contained by the woman's immune system. Such infection can cause serious birth defects.

where they feed and reproduce sexually. Because sexual reproduction takes place in the cat, cats serve as the **definitive host** for this parasite. Reproduction results in the production of **oocysts**, resistant cyst-like structures that pass out of the intestine with the cat's feces. Rodents or birds may accidentally consume these oocysts. Following ingestion, the oocyst ruptures, releasing the parasite into the intestine. The organisms penetrate the gut wall and spread throughout the body, eventually forming dormant tissue cysts in various organs. Infected rodents or birds, capable of re-infecting cats, are known as **intermediate hosts**. The cycle is completed when a cat kills and eats an infected animal, reestablishing the infection in the cat's intestine.

Humans can become infected if they inadvertently consume oocysts. Tissue cysts then form in the infected human. In someone with a healthy immune system, these cysts usually remain dormant for life, without causing problems. Most longtime cat owners, including Donna, will probably test positive for *T. gondii*, but will not have symptoms because the immune system usually keeps the parasite from causing illness. In an immunocompromised individual, such as an AIDS patient, *Toxoplasma* may cause severe inflammation, which can result in tissue damage and death.

Donna and Ben asked their doctor about antibiotics. Antibiotics would not cure a *Toxoplasma* infection because antibiotics are useful only to treat bacterial infections, and most function by attacking unique prokaryotic features, such as the cell wall or bacterial ribosomes. They have no effect against eukaryotic pathogens. When a *Toxoplasma* infection must be treated, as it would in someone with AIDS, entirely different drugs capable of inhibiting eukaryotic cells must be used.

So, is it time for Woody to go? Since Donna has had cats for years, it is likely that she is already infected with *T. gondii*. In her case, Donna's immune system will have contained the infection as dormant tissue cysts, preventing any spread to her fetus. She will also be resistant to new infections. There is no need to find a new home for Woody.

The real danger is if a woman becomes infected for the first time during pregnancy. The parasite may then spread across the placenta before it is contained. Because a fetus does not have a fully functioning immune system, the infection can cause severe damage, resulting in birth defects. Consequently, a kitten is a poor choice for a baby shower gift, if the mother-to-be has never owned cats. A good precaution in any event is for Ben (not Donna) to change the cat litter frequently during the pregnancy. When oocysts pass out of an infected cat, they need to be exposed to oxygen for 24–48 hours before they can infect another animal. Fresh litter each day greatly reduces the risk of exposure to infective oocysts.

Many fungi are involved in disease or ecological processes or are useful for industrial purposes

Fungi are perhaps most familiar to us either as food or because of their use in baking and brewing. But the importance of fungi far exceeds their gastronomic value. If you have never seen a Costa Rican golden toad, for instance, you have probably missed your chance. This spectacularly colored amphibian is presumably extinct. If so, this would be the first known vertebrate extinction caused by a parasite. The culprit is thought to be an unusual fungus known as *Batrachochytrium dendrobatidis*. You might still find an American chestnut tree to admire, but the vast stands of chestnuts that once blanketed eastern North America are gone. Again, the reason is a fungal pathogen that was first introduced into New York in about 1900.

These examples illustrate the substantial impact that fungal parasites can have on the welfare of animals and plants. The vast majority of fungi are *not* pathogenic, however, and the role played by many fungi in ecological processes can hardly be overstated. Fungi are ubiquitous in the soil, and many fungi extract nutrients from dead or decaying organic material in the soil. They are thus important as **decomposers**—organisms that break down and recycle nutrients in the environment (**Figure 4.17**). The fungi have no equal when it comes to converting large molecules like proteins and complex carbohydrates into their constituent building blocks. These small precursor molecules are recycled back into the environment, where they become available to growing plants. Other fungi form important **mutualistic** relationships with specific living plants; the fungus and the plant live together in an association that benefits both partners. The fungi absorb food from the living plants, while in return they provide the plant with crucial nutrients. Forests would indeed be barren places were it not for the fungi unobtrusively recycling nutrients and augmenting plant nutrition.

About 100,000 species of fungi have been described, but the actual number of fungal species is no doubt many times greater than this. Hundreds of new species are discovered each year. Originally, fungi were believed to be primitive or degenerate plants that lacked chlorophyll and thus could not perform photosynthesis. We now recognize fungi as a separate lineage that differs from other eukaryotes in important ways. Ironically, a comparison of rRNA places fungi closer to the animals than to the plants (see Figure 4.14). These molecular data are supported by other features shared by fungi and animals but not by plants. For example, all fungi synthesize a resilient structural carbohydrate called chitin that is often found in fungal cell walls. Plants do not produce this molecule, but most animals do. The exoskeleton of insects and crustaceans, for example, is composed of chitin. Additionally, plants store glucose in the form of starch to meet future energy demands (see Chapter 2, pp. 30–31). Both animals and fungi assemble individual glucose molecules into a storage polysaccharide called glycogen.

Photosynthetic organisms, like plants, are termed **autotrophs** (self-feeders) because they can synthesize organic molecules from inorganic starting material. Fungi, like animals, are **heterotrophs**, which must obtain organic compounds from other organisms. Unlike animals, however, which typically ingest their food, fungi absorb nutrients from the environment. Fungi secrete digestive enzymes into the environment, where they degrade organic matter into a form that the fungi can absorb. Those fungi that grow on dead material are termed **saprophytes**. Parasitic fungi absorb nutrients directly from a living host organism. Mutualistic fungi live within and are supported by the living tissue of a host, and they provide nutrients or other digestive services in return. Although there are a few aquatic species, most of the known fungi are found in terrestrial habitats.

Fungi have one of two body plans and can be unicellular or multicellular

The fungi have one of two basic body plans. While some are microorganisms, others can be quite large. The first group are single-celled fungi such as yeast (**Figure 4.18a**). The second group are the **filamentous fungi** (**Figure 4.18b**), which are made up of an extensively branched system of tube-like

Figure 4.17 Fungi are important decomposers. This fungus (the irregularly shaped growth in the middle of the photograph) is growing on a forest floor. Fungi secrete enzymes, which digest organic molecules in the plant material. The fungi then absorb the digested components to meet their own nutritional needs.

(a)

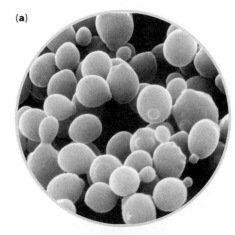

(b)

Figure 4.18 Fungal body plans. (a) Single-celled yeast. (b) A filamentous fungus. The many tube-like hyphae collectively make up the fungal body, called the mycelium.

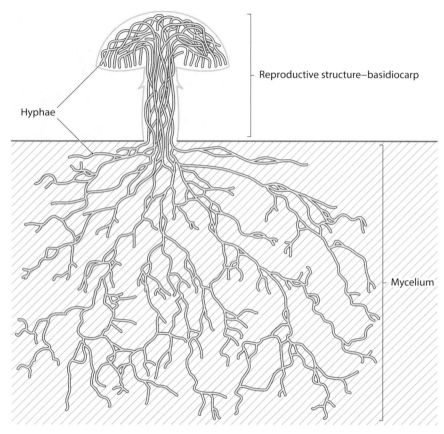

Reproductive structure–basidiocarp

Hyphae

Mycelium

Figure 4.19 A fleshy fungus. The "mushroom" is simply the reproductive portion of this otherwise filamentous fungus. The tightly packed hyphae of the mushroom form the above-ground basidiocarp. Hyphae involved in feeding make up the below-ground mycelium.

filaments called **hyphae**. The entire network of hyphae making up a fungus is collectively called the **mycelium**. In the **molds**, the hyphae of the myce-lium are loosely intertwined throughout the fungus. This gives molds their characteristic fuzzy appearance. **Fleshy fungi**, such as mushrooms, on the other hand, produce reproductive structures made of tightly packed hyphae. The remaining nonreproductive portion of the body may be a loosely orga-nized mycelium, as seen in molds (**Figure 4.19**).

As more and more fungal species have been described, the taxonomy of this group has been repeatedly revised. Currently there are eight recognized phyla. Here we review some of the most important of these groups.

The Chytridiomycota represent a primitive phylum of fungi

Chytridiomycota (chytrids) are a relatively primitive phylum; that is, of the phyla we discuss here, they are believed to be the most similar to the original fungal ancestor (**Figure 4.20**). Chytrids are unusual in that they are primarily aquatic. Most members of this phylum are saprophytic and function eco-logically as decomposers, but a number of them are known parasites of invertebrates.

Until recently it was believed that the chytridiomycota did not infect verte-brates. Unfortunately, we now know this is not the case. *Batrachochytrium dendrobatidis*, mentioned earlier in this section, is a chytrid parasite of amphibians. The fungus infects skin cells of frogs and other amphibians

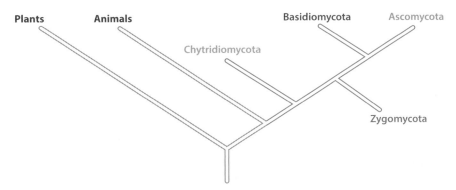

Figure 4.20 Phylogeny of the fungi. Kingdom Fungi is composed of eight phyla. Four especially prominent phyla are depicted here, each indicated in a different color. Chytridiomycota is the most primitive phylum. Basidiomycota and Ascomycota are more related to each other than they are to the other two phyla.

(**Figure 4.21**), although some amphibians are apparently resistant to infection. Enzymes released by the fungus on a susceptible species digest molecules in the skin, which are then absorbed as a nutrient source. Ultimately the amphibian in unable to breathe properly or maintain a proper balance between water and salts in its body. As noted earlier, golden toads are now extinct in Central America, presumably due to a massive *B. dendrobatidis* infestation. It remains unclear why this problem has cropped up only recently, but some research implicates climate change. Freshwater habitats in the Costa Rican highlands have become scarcer in recent years due to reduced rainfall. Consequently, frogs are forced to crowd into the few remaining water bodies. Such crowding increases the likelihood of parasite transmission from infected to uninfected individuals.

Zygomycota often form mutualistic relationships with plants

Most zygomycota are saprophytic, but some species participate in mutualistic associations with the roots of plants. Others are parasitic in plants or animals. Perhaps the most familiar fungus in this phylum is the common black bread mold, *Rhizopus stolonifer* (**Figure 4.22**). When *R. stolonifer* spores land on a piece of bread or other suitable substrate, the spores may germinate and produce specialized hyphae that spread across the bread, absorbing nutrients. Tons of food must be thrown away each year due to damage caused by *R. stolonifer* and other food-decomposing fungi.

What the zygomycota take away in food spoilage, they more than give back as plant mutualists. Many members of this phylum grow on plant roots and actually extend hyphae into root cells. In return for sugars provided by the plant, the fungus supplies the plant with phosphorus. Such associations between fungi and plant roots are called **mycorrhizal associations**, and they are found in up to 80% of all land plants. They are especially common in grasslands and tropical forests, where phosphorus is often scarce.

Like the Zygomycota, many Basidiomycota are important plant mutualists

Basidiomycota include the familiar mushrooms and puffballs. "Mushroom" is actually the common name for the spore-producing structure called the **basidiocarp** (see Figure 4.19). This structure is composed of tightly packed

Figure 4.21 A golden toad. These amphibians are now believed to be extinct, the result of infection with the chytridiomycotan *Batrachochytrium dendrobatidis*. If true, this would be the first known case of a vertebrate extinction due to a parasite.

Figure 4.22 Zygomycota. The common bread mold, *Rhizopus stolonifer*. This saprophytic fungus is a common cause of food spoilage. Hyphae, extended into the bread, release digestive enzymes, which digest carbohydrates in the bread. The digested products are then absorbed by the fungal cells.

hyphae and is produced solely for reproduction. The much larger mycelium of the fungus remains below ground.

Like zygomycota, many species of basidiomycota form mycorrhizal associations with plants. In the basidiomycota, however, the fungal hyphae do not actually invade the root cells. Rather, a dense network of hyphae surrounds the roots. Once again, the fungus benefits by absorbing sugars from the plant. Mycorrhizal basidiomycota, however, typically provide the plant with nitrogen rather than phosphorus as seen in zygomycota. This type of mutualism is most common in colder northern and temperate forests. In such environments the growing season is short and decomposition of plant material is slow. Consequently, nitrogen tends to remain bound up inside plant material. Very little is available in the environment. Without the mutualistic fungi, plant growth would be severely limited by a lack of nitrogen. In warmer climates, where zygomycota provide plants with phosphorus, the longer growing season ensures more rapid organic breakdown and ample nitrogen availability.

Other members of this phylum are important plant parasites, such as the smuts and rusts. Rust infections appear as numerous colored spots on the surface of a plant (**Figure 4.23**). The best known rusts cause stem rust on wheat and other grains. Smuts primarily attack the ovaries of grasses and eventually destroy the grain kernels. Many are important parasites of corn, cereals, and other crops. Wooden structures often fall prey to wood-rotting basidiomycota that can digest the cellulose or other complex carbohydrates found in wood.

While many are decomposers, other members of the Ascomycota cause serious disease

CASE: BAD NEWS FOR BATS

Bats in North America are under siege. White nose syndrome, caused by the fungus *Pseudogymnoascus destructans,* was first identified in New York in 2006. The disease is named for the distinctive fungal growth that occurs on the muzzles of hibernating bats. Since its arrival, the highly lethal fungus has spread to 25 states and five Canadian provinces, where it has killed approximately 7 million bats. Eleven species of bats, including three species that are already endangered, are being affected by this plague, identified by the New York Department of Conservation as "the gravest threat to bats ever seen." In some locations, bat populations have declined by over 90%.

1. **How, precisely, does *Pseudogymnoascus destructans* kill bats?**
2. **Where did this fungal pathogen come from?**
3. **Are there any preventative measures or treatments for affected bats that can be used?**

Most Ascomycota are saprophytic and serve as important decomposers. The majority produce hyphae, but some species are unicellular yeasts. These include the common brewer's yeast *S. cerevisiae*, used to make bread, beer, and wine.

A variety of important plant parasites are members of Ascomycota. An example is the fungus responsible for chestnut blight, mentioned earlier in this chapter (**Figure 4.24**). *Cryphonectria parasitica* reached New York on plants imported from Asia in about 1900. The spores of *C. parasitica* are spread by wind, rain, or birds. If they land in a crack or wound in the bark of

Figure 4.23 Parasitic rust. A representative basidiomycotan, orange rust, growing on a blackberry plant. Like other fungi, this rust secretes enzymes that digest organic molecules in the surrounding environment, which it then absorbs. In this case, those molecules are found in the tissue of its host plant.

Figure 4.24 Chestnut tree stem infected with chestnut blight. This fungal plant pathogen, responsible for the decline of chestnut trees in North America, is a member of the phylum Ascomycota.

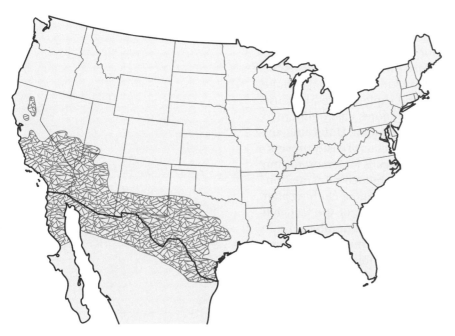

Figure 4.25 Distribution of valley fever in North America.
Valley fever is caused by the soil fungus *Coccidioides immitis*. The fungus
thrives in dry, desert soils of the American Southwest and northern Mexico.

a chestnut tree, the spores germinate, and as the hyphae grow, they form a
lesion which may expand to the point of girdling the tree limb. If this occurs,
everything above the lesion is killed. By 1940, over 3.5 billion American
chestnut trees had died. Today, only a very few remain.

Fungal infections in animals are called **mycoses**, and several important mycoses in humans are caused by members of Ascomycota. Vaginal yeast infections are the result of an overgrowth by *Candida albicans*. In the throat, a
C. albicans infection is called thrush. Interestingly, *C. albicans* is generally
present in small numbers even in healthy individuals. Its numbers are usually
held in check by various bacteria. However, when these bacteria are disrupted,
as they may be during antibiotic therapy, the yeast is able to proliferate.

Coccidioides immitis is another potentially serious pathogen of interest. This
soil fungus is endemic to the American Southwest and northern Mexico
(**Figure 4.25**). Infections are acquired by inhalation of spore-laden dust.
Most cases are asymptomatic, but respiratory symptoms resembling influenza or bronchitis are common. Occasionally the disease can develop into a
far more serious progressive form, with the fungus spreading to other organs,
including bones, skin, viscera, and brain. Those most at risk include farmers,
highway workers, and others regularly exposed to large amounts of dust. A
notable outbreak occurred in the wake of the 1994 Northridge earthquake in
southern California. Over 200 people were diagnosed with valley fever, as
this fungal disease is commonly called. Three people died. During earth-
quakes seemingly everyone is at elevated risk, as fungal spores along with
tons of soil are catapulted into the atmosphere. Unfortunately, short of
avoiding dust, little can be done to reduce the risk of infection. The fungus is
ubiquitous in the soil, and eradicating it is out of the question. With earth-
quakes a part of life on the West Coast, valley fever is no doubt here to stay.

Finally, there is *Pseudogymnogymnoascus destructans*, the Ascomycota that
causes the white nose syndrome described in our case (**Figure 4.26**). The
infection causes the bats to come out of hibernation, which raises their
metabolism. This forces them to use up the fat they had stored for the winter,

Figure 4.26 White nose syndrome. A
little brown bat (*Myotis lucifugus*) infected with
Pseudogymnoascus destructans. The fungal
growth on the face is typical. The fungus is
believed to have been introduced into North
America from Europe, where it has been found
on healthy bats. *P. destructans* grows well at low
temperatures and proliferates during the winter,
causing the bats to awaken from hibernation.
The now-metabolically active bats use up their
fat reserves. Besides the millions of bats that have
succumbed to this fungal infection, innumerable
pounds of insects have gone uneaten, due to
drastically reduced numbers of the insect-eating
mammals.

resulting in their starvation. The fungus is found in Europe, but European bats are apparently resistant to the disease. It is likely that it was inadvertently carried to North America sometime prior to the initial New York outbreak in 2006, perhaps on the clothes or shoes of a traveller who had visited a bat habitat in Europe. Since that time it is believed to have spread in large part by people who visit abandoned mines or who explore caves, because these are favored bat hibernation sites. The fungal spores easily adhere to equipment or shoes, and may then detach in a new location. Unfortunately, there are no obvious treatments for affected bats, or means of easily preventing the further spread of the fungus. There have been calls for a moratorium on visiting caves in affected areas in an attempt to reduce transmission. Certainly it is a good idea to decontaminate any equipment or clothing after visiting a cave or mine. Research is ongoing and in spite of the fact that *P. destructans* continues to spread and to kill bats in large numbers, there have been some signs of progress. For instance, in 2015 it was announced that the soil bacterium *Rhodococcus rhodochrous* can inhibit the growth of *P. destructans.* If the antifungal substance produced by the bacteria can be characterized or if bats could be inoculated with live bacteria, it may eventually be possible to treat infected bats or prevent infection in the first place.

Looking back and looking forward

Now that we know how organisms are classified, how a well-constructed taxonomy reflects evolutionary relationships, and what some of the principal groups of microorganisms are, we are more fully prepared to explore other aspects of their biology, including the impact they have on both their environment and other living things, including humans.

But we are not quite done with our microbial who's who. It is now time to investigate those entities that hover in the shadows, at the border between the living and nonliving worlds—namely, the viruses and the even more bizarre prions. Neither viruses nor prions are composed of cells, nor do they display many of the other characteristics we normally associate with living entities, yet they certainly impact our health and wellbeing. Influenza, herpes, and the infamous Ebola virus are just a few examples of viruses that cause more than their share of misery. Then there are the even simpler prions, which cause a peculiar set of diseases that includes the notorious mad cow disease. In Chapter 5, we'll see how they do it.

Garland Science Learning System

- http://garlandscience.rocketmix.com/students/

- Discover the scientist's answer to Juliet's question "What's in a name?"

- Test your knowledge of this chapter by taking the quiz

- Familiarize yourself with the terminology used in this chapter by using the vocabulary review

- Get help with the answers to the Concept questions

Concept questions

These questions are designed to help you start thinking like a microbiologist. The answers are not always simply found in the text. Instead, you will need to take the concepts about which you have learned and apply them to new situations. Some of the questions may not even have just a single correct answer. Help is provided as part of the GSLS resources, which can be accessed through http://garlandscience.rocketmix.com/students/.

1. If your dog has fleas, a likely culprit is the dog flea, otherwise known as ctenocephalides canis. Rewrite the scientific name of the dog flea in proper form.

2. Lobsters and crabs are in the class Crustacea. Grasshoppers are in the class Insecta, as are fruit flies and mosquitoes. Both fruit flies and mosquitoes are in the order Diptera. All of these animals are in the phylum Arthropoda. Construct a phylogenetic tree that shows the evolutionary relatedness of these five types of arthropods.

3. As seen here in Chapter 4, more than one lineage of bacteria is photosynthetic, but these groups are not necessarily closely related. On the basis of this information, and by looking at the position of the photosynthetic bacteria relative to other bacteria as illustrated in Figure 4.7, does it appear that the process of photosynthesis evolved once or more than once?

4. Suppose that some sort of bacteria contaminated a spacecraft on the launch pad and that the spacecraft was sent into orbit. When the spacecraft returns to Earth, amazingly, some bacteria are still alive. Considering the various lineages of bacteria discussed here in Chapter 4, which would you suspect might survive their journey into space?

5. Why is penicillin ineffective against eukaryotic microorganisms that infect humans?

6. *Pneumocystis jirovecii* is a fungal pathogen that can cause serious respiratory disease in AIDS patients. It was once classified as a protozoan. What sort of evidence may have resulted in the reclassification of this organism.

7. Suppose that all fungi on Earth instantly went extinct. What would be the environmental consequences?

8. *Cryptosporidium parvum* is a protozoan parasite that often infects humans, causing watery diarrhea that lasts for several days. Humans become infected when they drink contaminated water. The parasites live in the intestines, and infective stages are released with the watery feces. Do think that this parasite would have both a trophozoite stage and a cyst stage, or only the trophozoite stage?

9. Some researchers have suggested that the current taxonomic system, which organizes life into three domains, should actually be reduced to a two-domain system. After looking at Figure 4.5, which of the following ideas do you think these scientists are proposing? (a) Unite the bacteria and archaea into one domain. The second domain would remain domain Eukarya. (b) Unite the archaea and the eukarya into one domain. The second domain would remain domain Bacteria. (c) Unite the bacteria and eukarya into one domain. The second domain would remain domain Archaea.

10. What is meant by the term "extremophiles"?

11. It was discussed here in Chapter 4 how the taxonomy of the protozoa is still under investigation. One point that was not mentioned is that the protozoan with rRNA most similar to the animals and the fungi is a type of flagellate called a choanoflagellate. Look at Figure 4.14 and then answer the following questions: Do you think choanoflagellates should be grouped with other flagellates? If not, where would you place a branch labeled choanoflagellate on this tree?

LIFE'S GRAY ZONE: VIRUSES AND PRIONS

It almost seems that not a day goes by without news about some new disease that threatens humanity. Ebola and Zika virus are just two of the more recent entries into the rogues' gallery of dangerous maladies. And it is not just humans that are vulnerable to such new threats. Elephant endotheliotropic herpesvirus (EEHV) is killing baby Asian elephants. First documented in 1995, there have been over 50 cases in zoos across North America and Europe, only nine of which were successfully treated. Wild Asian elephants are also at risk. Although there have also been serious cases reported in adults, most infected older elephants do not develop life-threatening illness. In affected elephants, extensive damage to the cells lining the capillaries causes blood loss and hemorrhaging, which ultimately leads to shock. The disease can be treated with the rapid application of appropriate drugs, but this has only been successful in around a third of cases. Young elephants apparently contract the disease through physical contact with infected adults.

Meanwhile, deer and elk across North America are succumbing to chronic wasting disease (CWD). First recognized in 1967 in Colorado, this invariably fatal disease has spread to 23 US states and two Canadian provinces—most recently to Maryland in 2011. Michigan, with a single case on a captive breeding farm in 2008, reported its first case in a wild deer in 2015. The disease is characterized by weight loss ending in death, and behavioral symptoms such as repetitive walking in a set pattern. It is thought to spread from animal to animal through saliva. There is no treatment.

None of the diseases mentioned above are caused by prokaryotic or eukaryotic microbes. EEHV, for instance, is a virus, related to those herpesviruses that plague humans. CWD is a prion disease. Neither viruses nor prions are composed of cells, nor do they display many of those characteristics of living things discussed in Chapter 1. Rather, they occupy a nebulous position, not really alive in the conventional sense, but certainly not the same as nonliving matter either. Here in Chapter 5, we will explore exactly what they are, what characteristics they possess, and how they affect our wellbeing in so many ways.

All case studies have a few questions at the end, the answers to which will become apparent as you read the sections following the case.

Viruses are acellular parasites, requiring a host cell in which to replicate

CASE: "DEATH" IN THE RUE MORGUE

Edgar Allan Poe died October 7, 1849, at the age of 39. His death has always been attributed to the effects and complications of alcoholism and drug abuse. Michael Benitez at the University of Maryland, on the other hand, believes Poe died of rabies. Benitez came to this conclusion by studying records of Poe's symptoms at the time of his death. He was first delirious with tremors and hallucinations and then slipped into a coma. He emerged from his coma and was calm and lucid before he lapsed back into delirium. After 4 days, he died. These symptoms, say Benitez, are tell-tale for rabies but are not typical for alcoholism. Additionally, records show that Poe had abstained from drinking for 6 months prior to his death. We will never know for certain if rabies killed one of the greatest American authors (**Figure 5.1**), but can we continue to definitively blame Poe's death on the bottle? "Nevermore."

1. **What type of infection is rabies? How is the agent that causes rabies different from the other prokaryotic or eukaryotic organisms we have already discussed?**
2. **Why are the symptoms associated with a rabies infection primarily neurological?**

Figure 5.1 **Edgar Allan Poe.**

Rabies, caused by rabies virus, has been recognized as a serious disease since ancient times. As far back as 4000 years ago, the Mesopotamians had written laws detailing the responsibilities of dog owners, designed to reduce the risk of this dreaded illness. But the doctors who treated Edgar Allan Poe can be forgiven for not understanding what was affecting the author. At the time of Poe's death, it would still be over 40 years before anyone even suggested that something called a virus existed.

What exactly are viruses and how do they differ from other infectious agents? Nobel Prize-winning immunologist Sir Peter Medawar described them as "bad news wrapped in protein," which is a fairly good definition.

First of all, most viruses are small (**Figure 5.2a**). A typical bacterium might have a length of approximately 5 micrometers (μm; 1,000,000 μm = 1 meter). Viruses range in size from about 20 to several thousand nanometers (nm) in length (1000 nm = 1 μm, so 1,000,000,000 nm = 1 meter). The largest human virus, ironically the smallpox virus, is approximately 300 nm long and is barely visible with the strongest light microscope. The rabies virus is 170 nm long and about 70 nm wide. Not all viruses are quite so tiny, however. In the past 15 years or so, several giant viruses, all of which infect protozoa, have been discovered (**Figure 5.2b**).

Second, unlike prokaryotes and eukaryotes, viruses are acellular—they are not composed of cells. At its simplest, a viral particle consists of only nucleic acid, enclosed in a protein coat, much as Medawar described them. As we will see shortly, many viruses are somewhat more complex than that, but viruses are far simpler than the cell-based microorganisms we have discussed so far.

Third, we have previously noted that all living things use DNA to code for the proteins they need to survive. Some viruses also encode genetic information

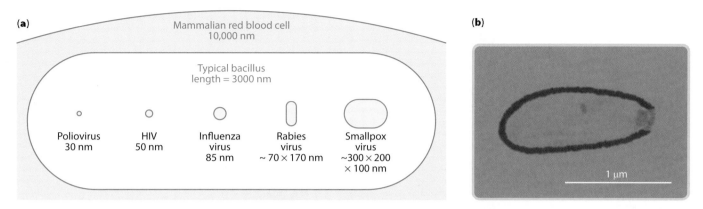

Figure 5.2 Sizes of viruses. (a) Most viruses are significantly smaller than typical bacterial and animal cells shown for comparison. (b) In recent years, several very large viruses have been discovered. Pithovirus is the largest known virus, 1.5 micrometers in length and 0.5 micrometers in width. It is as large as some small bacteria, and it is larger than the smallest eukaryotic cells, found in a type of alga. Discovered in Siberia in 2014, pithovirus infects certain amebas.

in DNA, but many rely on RNA as genetic material. Consequently, viruses are often broadly classified as either DNA or RNA viruses.

Fourth, all viruses are obligate intracellular parasites; they can reproduce only if they invade a host cell. A viral particle has only a small number of genes. This genetic material codes primarily for the structural proteins of the coat and other required proteins, including some enzymes needed for replication. Everything else is supplied by the host cell. Outside of a host cell, viruses cannot replicate and they show few if any of the properties commonly attributed to living things. However, once inside an appropriate cell, viruses essentially commandeer the host cell's machinery and use it to replicate themselves. In this manner, an infected cell becomes a viral factory. The virus uses host enzymes and other resources, including amino acids, as well as host structures like ribosomes, to make copies of itself. Newly assembled viral particles are released from the host cell, which frequently kills the cell or seriously compromises normal functioning.

Because they are acellular, and because they lack many fundamental characteristics of living things, are viruses even alive? The answer to this question, however, really depends on how we define "life." Viruses certainly replicate and evolve as other living things do, and the way we treat viral disease is the same whether or not we consider them to be alive. Perhaps the most accurate way to view viruses is to think of them, as implied in this chapter's title, as straddling the border between the living and nonliving world.

Most viruses have one of a few basic structures

An individual, complete viral particle is called a **virion**. As previously stated, all viral particles are composed of genetic material surrounded by a protein coat. Depending on the type of virus, the genetic material may be DNA or RNA. The protein coat is called the **capsid**, and the individual subunits that make up the capsid are called **capsomeres**. Each capsomere is composed of one or more proteins. Collectively, the nucleic acid and the capsid are called the **nucleocapsid**.

Most capsids have one of two basic shapes. Many capsids are composed of 20 capsomeres, all in the shape of equilateral triangles. Such viruses have the appearance of geodesic spheres and are known as **icosahedral** (20-sided) viruses (**Figure 5.3a**). **Helical** viruses (**Figure 5.3b**) have capsomeres that fit together to form a spiral around the enclosed nucleic acid. Apart from these

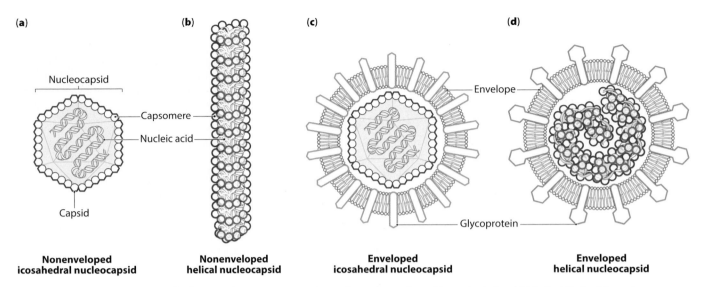

(a) Nucleocapsid, Capsomere, Nucleic acid, Capsid

Nonenveloped icosahedral nucleocapsid

(b)

Nonenveloped helical nucleocapsid

(c)

Enveloped icosahedral nucleocapsid

(d) Envelope, Glycoprotein

Enveloped helical nucleocapsid

Figure 5.3 Viral structure. The two main shapes for viral capsids are (a) icosahedral and (b) helical. Both of these shapes may be enveloped, (c) and (d), surrounded by a phospholipid bilayer that is embedded with glycoproteins.

two most common shapes, some viruses have a **complex** structure. Smallpox virus, for example, has a capsid that is composed of multiple layers of protein, neither icosahedral nor helical in shape.

Many viruses are surrounded by an **envelope** that is composed of a phospholipid bilayer in which viral proteins called **glycoproteins** are embedded (**Figures 5.3c** and **5.3d**). Other viruses are **nonenveloped** and lack this structure. As we will see, the presence or absence of an envelope and its associated glycoproteins tells us a great deal about how a particular virus enters and leaves a host cell. Many viral nucleocapsids contain a small number of enzymes. These are enzymes lacking in the host cell, which are required for successful viral replication.

Specific viruses usually infect only certain hosts and certain cells within those hosts

An important characteristic of a virus is its **host range**—the spectrum of hosts that it is able to infect. Some viruses infect only animals, while others infect only plants, fungi, and so forth. Even within a group such as animal viruses, however, a given virus cannot infect all animal species. For example, cats are susceptible to feline leukemia virus (FLV). While dangerous to cats, this virus is incapable of infecting humans. On the other hand, while humans are susceptible to Epstein–Barr virus, the virus responsible for mononucleosis (mono), cats are not.

Even within an appropriate host, only certain types of cells can be infected. For example, the rabies virus that may have killed Edgar Allan Poe can also kill dozens of other mammalian species. However, rabies virus is able to infect only certain cell types within these hosts. The **specificity** of the virus describes the cell types a particular virus can successfully infect. What determines specificity? Rabies virus is an enveloped virus. The glycoproteins that are embedded in the viral envelope can bind to proteins found in the plasma membrane of nerve cells (neurons). The protein in the plasma membrane of the neuron is similar to a lock, while the glycoprotein is like a key. Only cells that have the proper membrane protein can be infected by this virus (**Figure 5.4**). This explains why the symptoms accompanying rabies are largely neurological. Hepatitis B virus, on the other hand, lacks

(a)

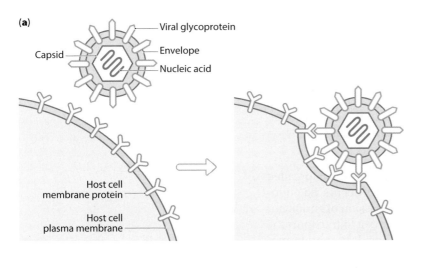

Viral glycoprotein

Capsid

Envelope

Nucleic acid

Host cell
membrane protein

Host cell
plasma membrane

(b)

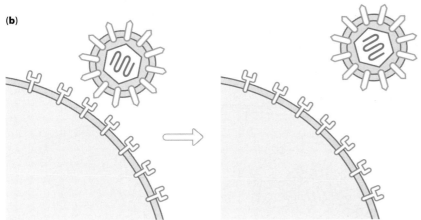

Figure 5.4 Viral host-cell specificity.
The ability of a virus to infect specific host cells is due in large part to the ability of the virus to bind to proteins found in the host-cell membrane. (a) The virus has glycoproteins that match the three-dimensional structure and chemical properties of the host-cell membrane protein. Chemical attraction between the virus and the host proteins binds the virus to the host cell. The virus can then enter the cell via one of several mechanisms. (b) The same virus cannot bind a different cell type, because proteins found on this cell are not complementary to the viral glycoproteins. Consequently, this cell type is resistant to infection by this specific virus.

the necessary glycoproteins to bind to neurons and therefore cannot infect this cell type. Hepatitis B virus has a different envelope glycoprotein, one that allows it to bind to proteins on the surface of liver cells. While some viruses are very specific, others can infect many cell types; the host molecule to which they attach is widespread and found on a variety of different cells. Nonenveloped viruses attach to target host cells via components of their capsid. In poliovirus, for instance, the point where three capsomeres come together forms a region called a canyon. Cells lining the intestine have proteins in their membranes that can fit snugly into these canyons, permitting infection by the virus.

Replication of animal viruses proceeds through a series of defined steps

The infection of a susceptible animal cell by a virus can conveniently be divided into several steps, which we will consider in sequence. Many of the differences that occur in the replicative cycle of different viruses depend on whether or not the virus has an envelope, and whether its genetic material is DNA or RNA.

Viruses must first attach to and enter an appropriate host cell

The first step in the infection process, called **attachment**, is the contact between a virion and its target host cell. A virion is not motile. It is merely transported passively until it happens to interact with a cell bearing the proper receptor. For enveloped viruses, the glycoproteins in the envelope bind the proper receptor in the host-cell plasma membrane. In nonenveloped viruses, proteins forming part of the capsid often function as attachment sites. The receptor molecule to which a particular virus attaches is typically a membrane protein of some sort. Only cells bearing the correct receptor are vulnerable to a specific virus. For instance, human immunodeficiency virus (HIV) can infect only cells that bear proteins called CD4 on their surface.

Following attachment, the virion must enter the host cell in a step called penetration (**Figure 5.5**). For nonenveloped viruses, entry occurs via endocytosis (**Figure 5.5a**) (see Chapter 3, pp. 63–64, for a discussion of endocytosis). The binding of the virion to the receptor causes the cell to transport the virus across the membrane and into the cytoplasm inside a membrane-bound vesicle. Enveloped viruses may also enter a cell via endocytosis, but many employ a different strategy called **fusion** (**Figure 5.5b**). During fusion, viral attachment brings the plasma membrane and the viral envelope into close proximity. The two lipid bilayers actually fuse together, releasing the nucleocapsid into the cell cytoplasm. Many of the details of fusion remain unclear, but it appears that when **viral glycoproteins** bind to their receptors,

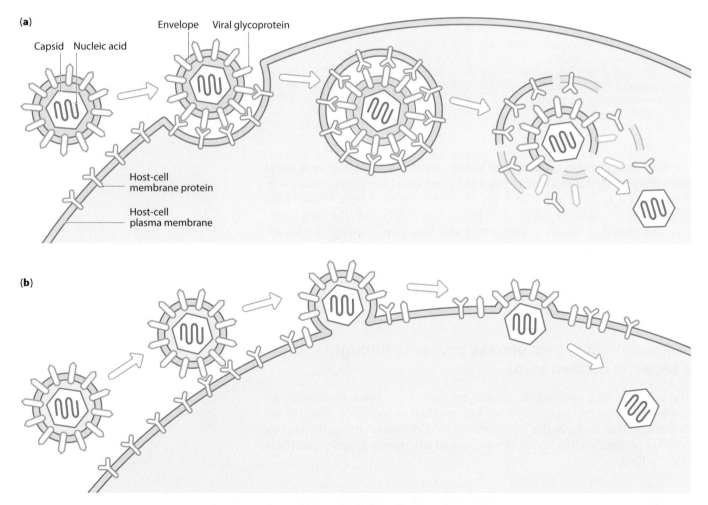

Figure 5.5 Entry of an animal virus into a host cell. (a) Entry via endocytosis. Following viral attachment to host-cell membrane receptors, the host cell transports the virus into its cytoplasm. Once endocytosis is complete, the viral particle is released into the cytoplasm. Both enveloped and nonenveloped viruses may enter a cell via endocytosis. (b) Entry via fusion. Following viral attachment, the host-cell plasma membrane and the viral envelope are brought into close proximity. Following fusion of the two lipid bilayers, the viral nucleocapsid is released into the host-cell cytoplasm. Only enveloped viruses enter host cells by fusion.

the glycoproteins change their shape due to the interaction. This disrupts the envelope and plasma membrane, allowing them to fuse.

Once they have entered the cell, viruses release their protein coat prior to replication

Before it replicates, the virus must shed its capsid (**Figure 5.6**). This process, called **uncoating**, may occur in endocytotic vesicles, where the pH is low. In other cases, host enzymes called proteases digest the viral protein coat. Once a virus has uncoated, we say that it has entered the **eclipse phase**. An intact viral particle no longer exists; the eclipse phase ends only when newly replicated virions are assembled at the end of the replicative cycle.

Following uncoating, two important tasks must be completed before new viral particles are assembled. First, the nucleic acid must be replicated, and second, new structural proteins must be produced (see Figure 5.6). These two important processes are collectively called the **synthesis** stage. Different viruses achieve these two goals in various ways that primarily reflect the type of nucleic acid they carry. We will return to this topic shortly, following our completion of the viral replicative cycle.

The final two stages are the assembly of new virions and their release from the host cell

Once new viral nucleic acid and structural proteins are made, new virions are formed in a process called **assembly** (see Figure 5.6). In some viruses, the capsomeres first assemble to form the capsids, and the genetic material is then inserted. In others, the capsomeres latch onto the genetic material, ultimately producing the new viral particle.

The final step in the viral replicative cycle is the **release** of newly formed virions from the infected cell. Nonenveloped viruses are generally released by **cell lysis**. The host cell literally explodes, releasing the newly assembled virions. The released virions may now contact other host cells, to begin a new round of viral replication. Lysis invariably kills the host cell.

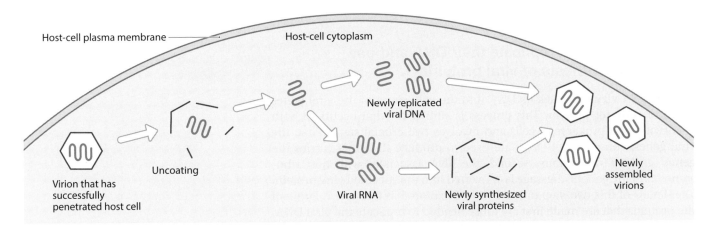

Figure 5.6 Uncoating, synthesis, and assembly of a DNA virus. Once a virion successfully penetrates a host cell, it must release its nucleic acid. This may occur either due to the acidity of an endocytic vesicle or the activity of host enzymes. The subsequent synthesis stage consists of two events: The viral DNA is replicated into many new DNA copies and the DNA is transcribed into viral RNA, which in turn is used to produce new viral proteins. The new viral proteins then assemble around the newly synthesized DNA, producing new progeny virions.

(a)

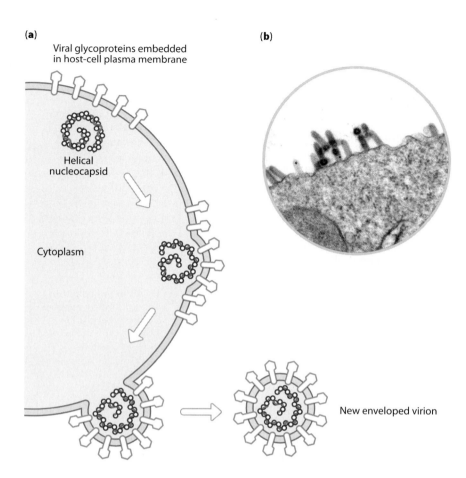

Viral glycoproteins embedded
in host-cell plasma membrane

Helical
nucleocapsid

Cytoplasm

(b)

New enveloped virion

Figure 5.7 Viral release by budding.
(a) Stages in the release of an enveloped helical virion are shown. Enveloped viruses acquire their envelope as they bud through the plasma membrane of the infected host cell. The envelope is actually a portion of the host-cell plasma membrane. Viral glycoproteins, already produced during the synthesis stage (see Figure 5.6), become embedded in the plasma membrane and are acquired along with the lipid bilayer as newly assembled viral particles leave the cell. (b) Electron micrograph of several vesicular stomatitis virus (VSV) particles budding from the surface of an infected cell. Closely related to rabies virus, VSV affects cattle, horses, and pigs as well as humans. It is transported by biting insects.

Enveloped viruses, on the other hand, are typically released from infected cells by **budding (Figure 5.7)**, wherein the assembled nucleocapsid pushes through the plasma membrane. In doing so, it becomes coated with plasma membrane material that now forms the viral envelope. The virus has already coded for envelope glycoproteins, which are already present in the plasma membrane, and the new virus will acquire them along with its envelope. Unlike lysis, budding does not necessarily result in cell death.

The synthesis step described above is substantially different for DNA and RNA viruses. In the next sections we consider some of the details in the synthesis stage for these two viral groups.

DNA viruses must replicate their DNA and use it to direct synthesis of viral proteins

Once a **DNA virus** releases its DNA into the cell, some of the viral genes begin to code for protein. This process is similar to what occurs in both eukaryotic and prokaryotic cells and involves two crucial steps. First, the viral genes, composed of DNA, are used to produce RNA that carries the genes' genetic instructions. Second, the RNA associates with host ribosomes, and the genetic message is converted into the specified viral protein. The details of this two-step process will be explored in Chapter 7. Some of the proteins that are made first are those needed to replicate the viral DNA. The viral DNA is then replicated, and hundreds or even thousands of new DNA molecules are produced. Finally, another round of protein synthesis takes place, primarily using host enzymes and resources. This time the proteins being produced are the structural proteins that will make up the completed virions. Examples of DNA viruses that replicate in this manner include

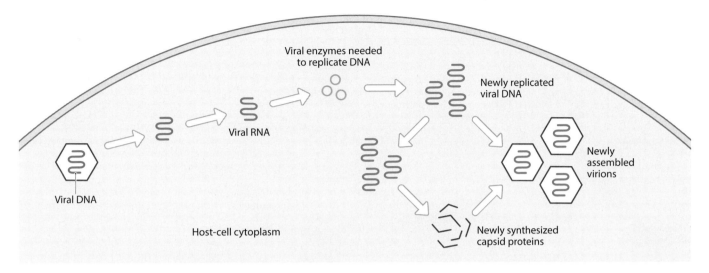

Figure 5.8 Synthesis in a DNA virus. To produce new progeny viral particles, the infecting virion must both replicate its DNA and synthesize necessary proteins. Initially, enzymes that the virus needs to replicate its DNA are produced. DNA replication then follows. Additional structural proteins including capsid proteins are then produced, permitting the assembly of new virions.

the herpesviruses and the papillomaviruses that cause warts. The synthesis stage of a DNA virus is illustrated in **Figure 5.8**.

RNA viruses rely on RNA rather than DNA as their genetic material

Unlike any prokaryote or eukaryote, **RNA viruses** encode their genetic information in RNA. For some RNA viruses, DNA plays no role whatsoever in their synthesis stage. Although many of the details vary for different RNA viruses, a general overview of their synthesis stage is provided in **Figure 5.9a**. For these viruses, the RNA is replicated many times to provide the genetic material for newly synthesized virions. The RNA also guides the production of viral proteins, including capsid proteins, which will assemble around the newly replicated RNA genetic material.

RNA viruses are unique in their ability to make copies of their RNA in this way. To do so, they require an enzyme called **RNA-dependent RNA polymerase**, which is unique to RNA viruses. A few familiar RNA viruses include poliovirus, rhinoviruses (which cause colds), measles virus, and influenza virus. The rabies virus introduced in our case is likewise an RNA virus.

Retroviruses use their RNA to produce DNA

Although they are RNA viruses, **retroviruses** use a very different replication strategy in their synthesis stage. Retroviruses also carry RNA in their nucleo-capsid, but instead of replicating it into new RNA molecules, they convert their RNA to DNA in a process called **reverse transcription** (**Figure 5.9b**). Reverse transcription requires an enzyme called **reverse transcriptase**. The newly synthesized DNA is then transcribed back to many copies of the viral RNA. This RNA serves as the genetic material for new virions and is translated into structural proteins. By far, the most important human retrovirus is HIV, the causative agent of AIDS.

Because reverse transcriptase is not used by our cells, it makes an attractive target for antiviral drugs. Ideally, interfering with this enzyme should have no adverse consequences for the host cell. Many of the important currently available anti-HIV medications act by targeting this enzyme. The topic of antiviral drugs will be more fully explored in Chapter 13.

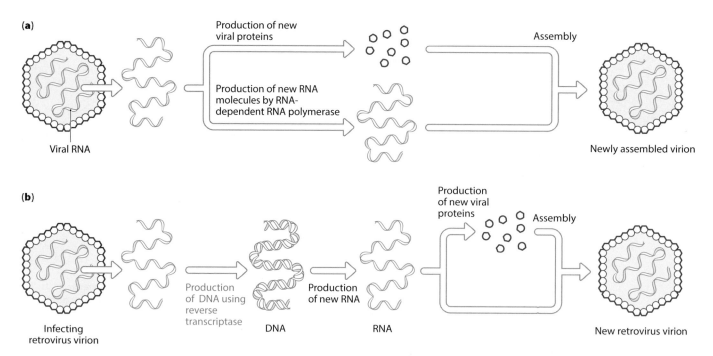

Figure 5.9 Replication of RNA viruses. (a) RNA viruses use RNA, both to copy into new RNA molecules and to guide the production of new viral proteins. These newly produced proteins assemble around the newly synthesized RNA molecules to form new virions. (b) In retroviruses, RNA is converted into DNA, which is then used to make many copies of RNA. The newly synthesized RNA is used both for protein production and for incorporation into new progeny virions.

Not all viral infections cause symptoms or kill host cells

Not all viral infections affect the host in the same way. If the virus undergoes repeated rounds of replication, the infection is termed **acute**; infected cells essentially devote themselves to viral production and ultimately die as a consequence. Some viruses, such as influenza and rabies viruses, cause acute infections only. The viruses keep reproducing and infecting new cells until either the host immune system eliminates the intruder or the host dies (**Figure 5.10**).

Chronic infections are characterized by a slow release of viral particles that may or may not kill the host cell; if such infections do result in individual cell death, host-cell replication is able to keep up with cell death, and there may be few or no signs and symptoms of disease in an infected individual. A good

Time highlighted in pink corresponds to when signs and symptoms of disease are apparent.

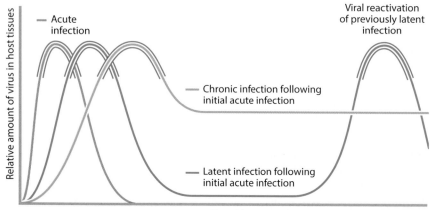

Figure 5.10 Types of viral infections. Symptoms of acute infections are associated with that period of time when levels of virus are high. Chronic infections typically start with an acute infection; viral levels then diminish, and viral replication continues at a lower rate. The amount of virus produced during the course of a chronic infection may or may not be sufficient to cause symptoms. Latent infections also typically begin as acute infections. Following the acute infection phase, the virus enters a period of dormancy (the latent period), during which it does not replicate. This latent period lasts indefinitely; reactivation does not necessarily occur. If reactivation does occur, there will be a new acute episode.

example is a hepatitis B virus (HBV) infection. A person infected with HBV initially suffers through an acute phase with severe illness. In many individuals, the virus is destroyed by the immune system and the patient returns to the uninfected state. In others, the immune system can eventually suppress but not entirely eliminate the virus. Consequently, low levels of virus continue to be produced over a prolonged period of time, sometimes lasting many years.

Some DNA viruses and retroviruses cause **latent** infections. Latent infections begin as acute infections, but then the virus enters a quiescent period during which there is no additional production of viral progeny (see Figure 5.10). During this latent phase, the viral DNA codes for few if any viral proteins. Furthermore, it does not replicate independently of the host DNA. If the host cell divides, however, the viral DNA is duplicated along with the host DNA prior to cell division. Consequently, if a cell infected with a latent virus divides, the two newly produced daughter cells will both be infected with the virus.

Viruses may persist in this latent state for many years or in some cases for the entire life of the host. However, a sudden change in the health of the host or certain environmental factors can cause viral **reactivation**. A reactivated virus is a previously latent virus that has resumed replication. During this time, symptoms may become apparent in an infected individual.

Herpesviruses undergo replication cycles of this type. For instance, human herpesvirus type 1 (HHV-1) is the causative agent of cold sores. When a person is first infected, an acute infection causes cell death, resulting in the oral lesion. An effective immune response eventually destroys replicating viruses. However, the virus will remain latent and may be reactivated at any time. The reappearance of the cold sore reflects such reactivation. If you are infected with HHV-1, you will remain infected for life, even if you never experience any symptoms after the initial infection. If you get a cold sore a second time in the same spot, it is a consequence of the same initial infection event. This is quite unlike an infection with influenza virus; once you recover from a bout of the flu, the virus is gone. If you suffer through another case of the flu in the future, you can blame it on a second infection. EEHV, the virus discussed in the introduction, is a herpesvirus specific to elephants. The baby elephants that die due to EEHV succumb during the early, acute phase. If they survive that period, the elephants remain infected for life, with the virus entering a latent phase. Young elephants apparently become infected when they interact with adults who carry the latent virus. In these adults, the virus may reactivate in a few cells—not enough to cause symptoms in the adult, but enough to infect the young elephants.

We are only just beginning to understand the factors that might cause a latent virus to reactivate. Anything that suppresses the host immune system, such as cancer chemotherapy or HIV infection, makes reactivation more likely. In the case of HHV-1, factors such as exposure to UV light seem to play a role.

Viruses damage host cells in several ways

Disease-causing viruses create problems for cells by interfering with their normal functions. Any such interference is known as the virus' **cytopathic effects**. One important cytopathic effect is the ability of many viruses to interfere with the host-cell plasma membrane. Recall from Chapter 3 that the plasma membrane maintains cell integrity and is important in regulating transport into or out of the cell. In virally infected cells, the budding of enveloped viruses can significantly compromise the ability of the plasma membrane to carry out these functions. Viruses can also damage cells by interfering with the ability of the cell to carry out its own protein synthesis or by simply using resources that the cell requires for its own needs. Some viruses prevent the host cell from copying its DNA. Others can initiate a process that ultimately leads to cancer.

Let us return to Edgar Allan Poe and the rabies virus that may have killed him. Oddly, in spite of the extreme symptoms exhibited by rabies victims, examination of infected neurons reveals little damage. There are, however, large, prominent clumps of viral proteins visible in the cytoplasm. These Negri bodies, as they are called, are often the only visible sign of infection. Presumably, Negri bodies interfere with normal neuron function, resulting in the neurological symptoms typical of rabies.

Like animals, plants are susceptible to many viral infections

Unlike animal cells, plant cells are surrounded by a rigid cell wall and plant viruses are unable to reach the plasma membrane when the cell wall is intact. Consequently, plant viruses infect cells through breaks or wounds in the cell wall. In many cases this damage is caused by insects, bacteria, or fungi.

In other respects, plant viruses are similar to animal viruses. Like animal viruses, they may carry DNA or RNA. Their mode of replication is essentially the same, as is their overall morphology. In one important respect, however, viral infections in plants are quite different. Plant cells are connected to one another via small junctions, which essentially form connections between the cytoplasm of adjacent cells. In effect, all cells linked by these junctions have a common cytoplasm. Progeny of plant viruses can spread to new host cells through these junctions and through the phloem, which functions as the plant's circulatory system. Because of this, plant infections are often systemic. In other words, whereas animal viruses usually affect only specific organs or tissues, plant viruses can affect cells throughout the entire organism.

Many important plant diseases are caused by viruses. Plant viruses take a large economic toll in crops such as corn and wheat, but they cause even greater problems in perennial plants like potatoes. Plants suffering from viral infections are often stunted (**Figure 5.11a**). Irregular coloration often appears on leaves and stems, and green pigments may be lost. Unlike animals, plants rarely recover from a viral infection because, as mentioned previously, the entire plant is often infected. They also lack the ability to mount a highly specific immune response. In a few happy cases, viral infection in plants is actually desirable. The attractive variegated color of some tulips is due to a viral infection that is transmitted through the bulbs (**Figure 5.11b**).

Bacteriophages are viruses that infect bacteria

Even bacteria do not get a free ride when it comes to pathogens. They are subject to infection with bacteria-attacking viruses called **bacteriophages** or phages for short.

The most intensively studied phages are those that infect *E. coli*. A specific phage known as T4 provides an illustrative example. A T4 virion has a complex structure, with an icosahedral head containing DNA and a hollow, helical tail (**Figure 5.12**). The tail region is associated with other characteristic structures called the tail fibers, tail pins, and base plate.

Figure 5.11 Plant viruses. (a) Tobacco mosaic virus. The leaf on the right is from a tobacco plant infected with tobacco mosaic virus. Note its stunted and discolored appearance compared with the leaf from an uninfected plant on the left. (b) The attractive variegated color in some tulips is the result of a viral infection.

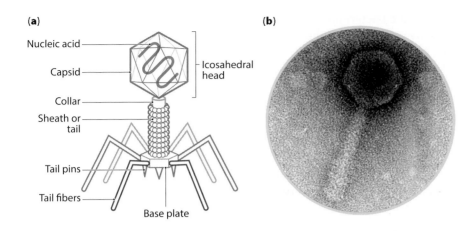

(a)

Nucleic acid

Capsid

Collar

Sheath or tail

Tail pins

Tail fibers

Base plate

Icosahedral head

(b)

Figure 5.12 Structure of T4, a representative bacteriophage. (a) Note the complex structure, including the icosohedral head encompassing the nucleic acid, and the helical tail, which makes initial contact with the host cell. (b) An electron micrograph of a bacteriophage.

The initial step in the infection cycle of bacteria involves the contact and adherence to the bacterial surface by the phage (**Figure 5.13a**). This process, called **adsorption**, cannot occur on just any bacterial cell. To successfully adhere, molecules on the phage tail and tail fibers must match specific molecules on the bacterial surface that serve as receptors. A bacterium lacking these molecules is resistant to infection.

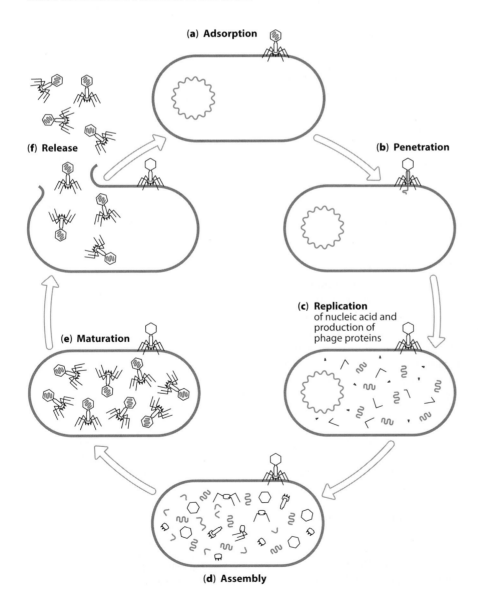

(a) Adsorption

(f) Release

(b) Penetration

(c) Replication of nucleic acid and production of phage proteins

(e) Maturation

(d) Assembly

Figure 5.13 Replicative cycle of bacteriophage T4. This bacteriophage is able to infect *E. coli* cells. (a) Adsorption: The phage makes contact and adheres to the bacterial surface. (b) Penetration: The helical sheath contracts, forcing the viral DNA into the cell cytoplasm. (c) Replication involves both replication of the viral DNA and inhibition of host-cell activity. (d) Assembly of new phage particles. (e) Maturation of phages into newly produced infective viral particles. (f) Lysis of the host cell and release of newly produced, mature phages.

Once the phage has attached, it injects its DNA into the bacterium in a step called **penetration** (**Figure 5.13b**). This involves the contraction of the helical sheath, which forces the hollow tube into the cell cytoplasm, much like a microscopic syringe. In the process, the viral DNA is released into the cell's interior. The viral capsid does not enter the cell. It remains as an empty shell, attached to the cell exterior.

Once DNA has entered the host cell, the metabolic activity of the cell is blocked and the cell begins producing proteins coded for by the viral DNA (**Figure 5.13c**). These include proteins that block normal host-cell activity, enzymes required for viral replication, and structural proteins needed to construct new capsids. The viral DNA is also repeatedly copied. All energy required for these processes is provided by the host cell.

As new viral DNA and structural proteins are produced, they spontaneously assemble into new virions (**Figure 5.13d**). Up to several hundred new phages may ultimately be produced (**Figure 5.13e**). Eventually, the host cell bursts, releasing the progeny phages (**Figure 5.13f**). These newly released phages may then contact and infect other susceptible cells.

The T4 replicative cycle we have just described is termed a **lytic cycle**, because it results in bacterial cell lysis and death. Not all phages, however, undergo a lytic cycle. Infection of *E. coli* with phage lambda, for instance, does not necessarily cause lysis. After releasing its DNA into a susceptible bacterium, lambda does not immediately reproduce. Rather, it enters an inactive period called the **prophage** state (**Figure 5.14**). In the prophage state, the viral DNA is actually incorporated into the bacterial DNA. If the bacterium itself divides, the viral DNA will also replicate along with it, but otherwise, the infected cell remains largely unaffected. This state of viral inactivity inside an infected cell is called **lysogeny**. Under certain conditions, however, the virus may begin to replicate, returning to the lytic cycle. Ultraviolet light or exposure to certain chemicals is known to promote a return to the lytic cycle in phage lambda.

Phages can influence the number of bacteria and even the diseases that they cause

Although they do not infect animals or plants, bacteriophages affect humans in several important direct and indirect ways. First, the number of phages in the environment is almost inconceivable. A single milliliter of sea water may contain close to 50 million of them, and virologists have found similar numbers in soil. According to some estimates, in the oceans alone, up to 40% of all bacteria may be destroyed each day by phages. Consequently, they may influence the entire world's food supply by limiting the numbers of bacteria available for other organisms to eat. The vast numbers of killed bacteria no doubt impact nutrient cycling in the ocean and other environments.

Of perhaps more immediate concern to humans is the effect lysogenic phages have on the virulence of some bacteria. For example, cholera is caused by *Vibrio cholerae*, a Gram-negative species (**Figure 5.15**). This often deadly disease is caused by a toxin produced by the bacteria that results in massive diarrhea. However, the bacterium cannot make the toxin by itself. Rather, lysogenic phage genes inside the cell instruct the host to create the toxin. Bacteria that are uninfected with these phages do not cause disease.

Yet other phages may actually provide us with an invaluable service—namely, protection from potentially pathogenic bacteria. It has recently

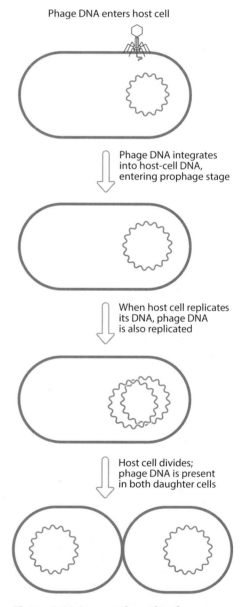

Phage DNA enters host cell

Phage DNA integrates into host-cell DNA, entering prophage stage

When host cell replicates its DNA, phage DNA is also replicated

Host cell divides; phage DNA is present in both daughter cells

Figure 5.14 Lysogenic cycle of bacteriophage lambda. Once phage DNA is released into the cytoplasm of the host cell, it incorporates into the host DNA. During the lysogenic cycle, no new progeny phages are produced. If the bacterial cell replicates, the phage DNA, called the prophage, is replicated along with the host DNA. Under appropriate conditions, prophages in the lysogenic cycle may return to the lytic cycle.

been proposed that phages in our intestine help keep intestinal bacteria in check, and thus unable to easily cause disease. These phages typically adhere in almost astronomical numbers to the mucus lining the intestine. Researchers have speculated that these phages stand ready to attack potentially disease-causing bacteria that attempt to penetrate the mucus to reach the underlying intestinal lining. The phages benefit from this relationship because of their access to a steady stream of bacteria in which to reproduce. In other words, this may be an odd and ancient partnership between animals and phages that is beneficial to both.

Finally, because they destroy specific bacteria, some have suggested that bacteriophages could be utilized therapeutically to treat bacterial infections. This idea was first proposed in the early twentieth century. Spotty results, however, combined with the development of powerful antibiotics dampened interest in phage therapy. However, in recent years, with the development of antibiotic resistance by many bacteria, some researchers are taking a fresh look at this approach. We will return to this topic in Chapter 13.

Figure 5.15 Cholera. This 1912 painting by the French painter Jean-Loup Charmet depicts the horror associated with cholera epidemics.

Prions are infectious proteins

CASE: LAST LAUGH FOR THE "LAUGHING DEATH"

Up through the 1950s, 2–3% of the Fore people of New Guinea mysteriously died of the "laughing death" each year. The disease, also called kuru, started with increased clumsiness, headaches, and a tendency to giggle inappropriately. Within a few months, victims could no longer walk. Total incapacitation, followed by death, occurred within a year. Curiously, the disease affected women and young children almost exclusively (**Figure 5.16**). Older boys and men were spared. The reason for this was unknown until 1957, when Carleton Gajdusek, of the US National Institutes of Health, determined how the disease was transmitted. The Fore people practiced ritual cannibalism as part of their burial ceremony. When a tribal member died, it was customary for female family members to prepare the body and eat parts of the deceased's brain. Small children also consumed brain tissue, but older boys and adult men did not participate. Gajdusek postulated that an infectious agent was transmitted through the brain tissue, explaining why only those who had consumed such tissue were affected. Kuru has now been eradicated, because cannibalism has not been practiced since the 1960s.

1. What exactly causes kuru?
2. How is this infectious agent unlike any other that we have thus far discussed?
3. What other diseases are caused by similar agents?

Figure 5.16 A young Fore child infected with kuru. This photo was taken in the early 1950s, before measures to eradicate kuru were initiated.

Kuru may be gone, but other diseases, such as Creutzfeldt–Jakob disease (CJD) and bovine spongiform encephalopathy (BSE), also known as mad cow disease, are still with us. Likewise, chronic wasting disease (CWD), discussed in the introduction, continues to plague deer and elk in North America. These diseases all cause neurological symptoms and they are all uniformly fatal. They are all caused by highly unusual infectious agents called **prions**.

Prions are infectious proteins. Their very existence was not proposed until 1982, and it is only since the 1990s that most of the scientific community has accepted the notion that a protein could cause infection. The concept of prions was hotly contested because their existence violates a basic tenet of biology—that only nucleic acids can replicate themselves and serve as genetic material.

But prions do not replicate in the traditional sense. Prions seem to be abnormally folded forms of a protein that already exists in the brain (**Figure 5.17**). The best-studied prion disease is scrapie, which affects sheep (see Figure 1.6b). It is named for the behavior of infected animals, which includes scraping themselves bare against fence posts or other solid objects. On the surface of the sheep's brain is a protein called PrP^c. Occasionally this protein folds up incorrectly. It is then called PrP^{Sc}. This aberrant protein is able to interact with normal PrP^c proteins, forcing them to refold in the abnormal three-dimensional shape (**Figure 5.18**). In other words, PrP^{Sc} acts like a mold or template, which forces PrP^c to take on the disease-causing shape. Eventually, the PrP^{Sc} proteins build up to such an extent that they begin to form abnormal plaque on the surface of brain cells. Whether or not it is these plaque deposits that are the direct cause of the disease remains to be seen. Other prion diseases, including kuru, CJD, and CWD are also believed to be caused by improper folding of PrP^c and the ability of the aberrant protein to cause normal proteins to adopt the abnormal shape.

Prion disease can be caused by mutation of the gene coding for PrP^c, but it can also be transmitted between individuals. Creutzfeldt–Jakob disease, for instance, can be transmitted via transplants or surgical instruments. Ninety percent of all prion disease in humans has been identified as CJD. Typical age of onset is approximately 65, and the disease is characterized by progressive dementia. Like all known prion diseases, it is always fatal.

Bovine spongiform encephalopathy (BSE) is a prion disease of cattle. The disease is named for the brain's spongy appearance in affected cows (**Figure 5.19**). It is also popularly called mad cow disease because of the erratic behavior of affected cattle.

It is believed that the disease was spread by the previously common practice of adding bone meal to cattle feed. Bone meal often has nerve tissue attached to it, and such tissue may contain infectious prion particles. The disease came to the attention of the general public when it suddenly appeared in British cattle in 1980. There were over 180,000 cases and millions of cattle were slaughtered in an effort to contain the epidemic. The use of animal products in cattle feed was banned. Many were concerned that humans who consumed meat from infected cows might develop a similar disease. There was also speculation that CJD was a human form of mad cow disease. This is probably not the case, but disturbingly, in 1996, a new form of CJD called variant CJD (vCJD) was first reported in Great Britain. Unlike CJD, vCJD has an average age of onset of 27. Furthermore, the disease kills its victims in just over a year, as opposed to about 3 years for CJD. About 230 people have died of vCJD, mostly in Great Britain.

Could vCJD be the human equivalent of mad cow disease, and did British victims become infected by eating contaminated meat? Certainly the sudden appearance vCJD in 1996 is consistent with such a link.

Looking back and looking forward

We have now completed our examination of our microbial roster. Viruses and prions are far simpler than the cellular life forms reviewed in Chapter 4. Viruses, the ultimate parasites, cannot replicate unless they infect appropriate host cells. Once inside, viruses commandeer the cell, diverting host resources to their own replication, often with dire consequences for the host

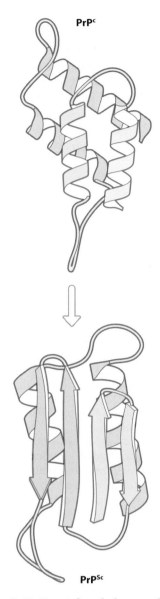

Figure 5.17 Normal and abnormal forms of the PrP protein. The PrP^c protein is normally found on brain cells. The normal PrP^c sometimes takes on the form of the aberrant PrP^{Sc}. The normal protein in its correct three-dimensional structure has four spiral-like regions called alpha helices. In the abnormal form, two of these alpha helices, shown in red, lose their helical form and take on a different, elongated structure, indicated by the colored arrows.

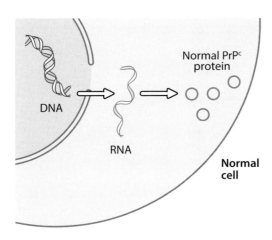

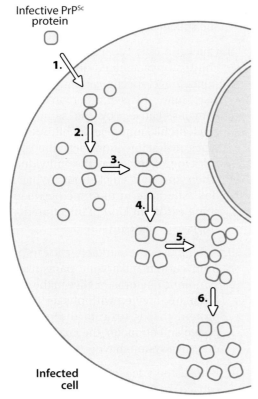

Infected
cell

Figure 5.18 Conversion of the normal PrP protein to the abnormal form. The aberrant PrP^{Sc} can force normal PrP^c proteins to take on the aberrant shape. (1) PrP^{Sc} (the red square) may be acquired by consuming contaminated animal products or may be due to a mutation in the gene coding for the protein. (2) The aberrant protein is able to interact with the normal protein (blue circles) and force it to take on the aberrant shape. (3)–(6) As this process continues, more and more of the normal PrP^c is converted to PrP^{Sc}. High levels of the aberrant protein form abnormal plaque deposits on the surface of brain cells, which may be involved in the nervous disorders seen in individuals suffering from prion disease.

cell. Prions, even less complex than viruses, are essentially rogue proteins that force other host proteins into an aberrant shape. As these abnormally folded proteins build up, symptoms of disease appear.

So now, the stage is set. We know who the microorganisms are, and we are starting to understand what they do. But before we continue to investigate them, we will consider what they did. In Chapter 6, we examine microbes with a historian's eye and investigate some of the astounding ways in which they have influenced human affairs throughout history. We will see that empires have fallen and societies altered, all because of microorganisms. We will also review the colorful history of microbiology itself, looking at some of the important discoveries that have advanced the science of microbiology to where it is today.

Figure 5.19 Stained cross section of the brain from a cow that died of BSE. The white areas scattered across the brain tissue are places where neurons have died due to the buildup of the prion protein. Brain tissue preparations from individuals that have died of other prion diseases, including CJD and CWD appear similar.

Garland Science Learning System

- http://garlandscience.rocketmix.com/students/

- Investigate viral factories and their assembly lines

- Test your knowledge of this chapter by taking the quiz

- Familiarize yourself with the terminology used in this chapter by using the vocabulary review

- Get help with the answers to the Concept questions

Concept questions

These questions are designed to help you start thinking like a microbiologist. The answers are not always simply found in the text. Instead, you will need to take the concepts about which you have learned and apply them to new situations. Some of the questions may not even have just a single correct answer. Help is provided as part of the GSLS resources, which can be accessed through http://garlandscience.rocketmix.com/students/.

1. What is fundamentally different about the way viruses reproduce compared with cellular forms of life?

2. Referring back to Chapter 1, in which the properties of living things were discussed, make an argument that (a) viruses are not living things or (b) viruses are living things.

3. An anti-HIV drug called Enfuvirtide binds to and interferes with the glycoproteins found in the viral envelope. Which stage of the viral replicative cycle is Enfuvirtide interrupting?

4. RNA-dependent RNA polymerase is a unique viral enzyme produced by many RNA viruses. It is not produced by any prokaryotic or eukaryotic cells. Would this enzyme make a reasonable target for drugs active against the RNA viruses that produce it? Why or why not?

5. If you know that a particular virus is enveloped, what does this information alone suggest about its replicative cycle?

6. What is meant by the "eclipse phase" in the viral replicative cycle? Referring back to the steps described in the replicative cycle here in Chapter 5, after which step does it begin? After which step does it end?

7. Suppose that a particular virus has a wide host range but narrow tissue specificity. What precisely does this mean? What about a narrow host range and broad tissue specificity?

8. Chickenpox is caused by the chickenpox virus. If a child is unvaccinated and exposed to the virus, she might develop clinical signs of this disease. Many years later, this same person may develop a case of shingles, caused by the same viral infection that caused chickenpox as a child. Referring back to the types of viral infections described in this chapter, state the type of infection the individual is experiencing when she had chickenpox. What type of infection does she have in the time between chickenpox and shingles? What kind of infection does she have while she is experiencing shingles?

9. Aphids use a remarkable trick to protect themselves from certain eukaryotic parasites. The aphids have symbiotic bacteria that live in their gut. These bacteria often are infected with phage. The phages produce a protein that is toxic to the parasites. What type of phage infection do the bacteria within the aphid's digestive system have?

10. Before the idea of prions was proposed, diseases caused by what we now know to be prions were believed to be caused by as-yet unidentified viruses. Some of the evidence used to show that such diseases were not caused by viruses included the use of DNAses (enzymes that digest DNA) and RNAses (enzymes that digest RNA). Experiments were performed in which brain tissue from infected mice was treated with either DNAse or RNAse. Other, previously healthy mice were then exposed to the DNAse- or RNAse-treated tissue. In either case, these newly exposed mice developed the same disease as the original mice from which the brain tissue had been taken. How do these results help demonstrate that a virus was not involved?

THE MICROBIOLOGY OF HISTORY AND THE HISTORY OF MICROBIOLOGY

6

The first Tsar of All Russia, Ivan IV (aka Ivan the Terrible), was by all accounts a monster whose reign of terror was a sickening tale of mass torture and murder. But he didn't start out that way. Ivan's reign, starting in 1547, began as an enlightened one in which a legal code was established, schools were founded, and the most oppressive nobles were banished. Things changed in 1564, when he abandoned the royal palace for parts unknown. His first message home stated that he was "unable to brook the treachery by which I was surrounded." Russian nobles found him in a distant village and begged him to return. Ivan agreed under the condition that he would be free to execute all traitors. Once he returned, the carnage began.

All evidence points to syphilis as the reason for Ivan's dramatic personality change. When Ivan died, the throne passed to his son Fedor, whose own syphilis probably explained why he was described as a "slow witted idiot." The resulting chaos from his ineffective rule ended when the first of the Romanovs assumed the throne in 1613.

Would Russian leadership have developed differently if Ivan's enlightened period had continued? Was Russian history changed by the mind-altering disease affecting Ivan and his son? We can never definitively answer these questions, yet the questions themselves highlight the often unseen and underappreciated part that microbes have had in shaping human history. Infectious disease, to cite just two instances, hastened the fall of Rome and the abandonment of feudalism in medieval England.

Here in Chapter 6, we consider these and other examples, looking at history with a microbiologist's eye. This is not to suggest that microbes have played the only or even the principal role in determining past events. To do so would be absurd. In many cases, however, our understanding of why certain important historical events occurred as they did can be improved by also considering them from the microbiologist's perspective (**Figure 6.1**). Furthermore, by thinking about how history has been altered by microbes, we gain added insight into the biology of the organisms themselves.

We will then turn our attention to the history of microbiology itself. The study of microorganisms has a rich and colorful past, and we will focus on some of the important people and key discoveries that have led to the development of modern microbiology.

Figure 6.1 The Triumph of Death. This painting by the Flemish artist Pieter Brueghel (1525–1569) depicts the impact of plague on society. During Brueghel's life, Europe was still recovering from the devastation of the Black Death, caused by bubonic plague, almost 200 years earlier.

All case studies have a few questions at the end, the answers to which will become apparent as you read the sections following the case.

If microbiologists wrote history...

CASE: THE SPANISH CONQUEST OF MEXICO

In the early 1500s, the Native American population in what is now Mexico is estimated to have been between 15 and 30 million. The Aztec Empire controlled much of the land. The Aztecs developed a sophisticated society, comparable to the great civilizations of ancient Europe and Asia. In 1519, the Spanish leader Hernán Cortés landed on the Yucatan Peninsula with about 500 soldiers. Captivated by stories of the Aztecs' wealth and the splendor of their capital of Tenochtitlan (now Mexico City), he marched inland, intent on conquest (**Figure 6.2**). Cortés exploited several advantages as he moved toward Tenochtitlan. He obtained the support of many other native tribes by promising to relieve them of Aztec domination. According to Aztec legend, the god Quetzalcoatl was prophesized to return from across the eastern ocean, and the unusual appearance of the white, bearded soldiers in their metal armor convinced many that the Spanish were the fulfillment of that prophecy. If the newcomers truly were Quetzalcoatl and his minions, as divine figures they would be invincible. The Spanish also had horses and weapons, including muskets and cannons, that were unknown in the Americas. Yet after initially appeasing the Spanish, the Aztecs fought fiercely once the Spanish had occupied Tenochtitlan and their intent became obvious. The Aztecs expelled the invaders from the capital but failed to destroy them. Eventually the Spanish returned, and began an extended siege of Tenochtitlan. Meanwhile, a smallpox epidemic raged among the Aztecs, severely crippling their ability to resist. The Spanish eventually conquered the Aztecs, forcing them into servitude. The death toll among the native peoples was staggering. Within 50 years, the native population in Mexico had shrunk to 3 million. By 1600 it was under 2 million, a mere 5–10% of the pre-conquest population. Many of these deaths were due to smallpox.

1. Why did smallpox exact such a heavy toll on Native Americans but not the Spanish?
2. Why were the Spanish aided by smallpox but not decimated by so-called American diseases to which the native peoples were less vulnerable?
3. How was the conquest of the Americas different from the European experience in Africa?
4. Are there other examples of diseases influencing the outcome of major historical events? What characteristics do the involved microorganisms have in common?

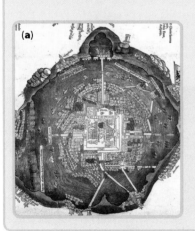

Figure 6.2 The Spanish arrive in Mexico. (a) A map of Tenochtitlan, published in 1524, based on Cortés' memory of the city. (b) Initial meeting between Cortés and the Aztec leader, Montezuma, as depicted in *Lienzo de Tlaxcala*.

As you may have already guessed, the answer to the first part of question 4, above, is emphatically "yes!" Unseen microbial allies and enemies have turned the tide in many epic occurrences. Before we look closer at smallpox in the New World, though, let's go back to even earlier times to examine the part played by infectious organisms in some of history's seminal events.

Disease turned the tide in the Peloponnesian War between Athens and Sparta

Following the defeat of the invading Persians in the epic naval battle of Salamis in 480 BC, the power of Athens was at its height. Thus began the Golden Age, an unparalleled period of creativity and advances in philosophy, architecture, and the arts.

This period of extraordinary enlightenment was all too brief. In 431 BC, the Peloponnesian War broke out between Athens and the rival Greek city state of Sparta. The conflict was initially a standoff. Sparta had a strong army, but no navy. Athens, with a powerful navy but a weak army, fought a defensive war on land and an offensive war at sea. Although the Athenians were crowded behind their city fortifications, they might have lasted indefinitely, because the city was connected to a port and they could bring in supplies by ship. However, crowded conditions increase the likelihood of epidemic disease. Dense aggregations of people often coincide with poor sanitation, which makes the transmission of many pathogens, especially those transmitted in contaminated food or water, more likely. Other forms of transmission, including airborne transmission, are enhanced when people are crammed together in close quarters. Rodents, which carry many diseases transmissible to people, also thrive in congested urban environments. Unless proper preventative measures are taken, a crowded city with poor sanitation is an epidemic waiting to happen. In 430 BC, that is exactly what occurred (**Figure 6.3**). A plague, perhaps arriving in Athens via a trading vessel from Egypt, erupted that killed between one-third and two-thirds of the city's population.

The nature of this disease remains unknown. The historian Thucydides describes it as striking rich and poor alike. Symptoms included rapid onset, high fever, and extreme thirst. Eventually, pustules formed on the skin of the victims. Thucydides observed that those who recovered from the illness were best suited to care for the sick, because they somehow became resistant to a second infection following their recovery. This is one of the first descriptions of acquired immunity, an important aspect of our immune defense against invading microbes and a topic we will explore more fully in Chapter 12.

Many experts feel that the disease described by Thucydides was an unusually virulent form of scarlet fever, caused by the Gram-positive bacterium *Streptococcus pyogenes*. Others have implicated smallpox, measles, or bubonic plague. Alternatively, the plague of Athens was possibly caused by a disease that no longer exists or by one that has changed so much that it is now unrecognizable. Regardless of the cause, the plague undoubtedly contributed to the eventual defeat of Athens and the end of the Golden Age. With the loss of so much of its population, Athens was never able to prevail. The war dragged on for 27 years before Sparta was ultimately victorious in 404 BC.

Figure 6.3 The plague of Athens. Before the plague, Athens' population is estimated to have been about 315,000. The plague is estimated to have killed up to two-thirds of this population. In his chronicle of the plague, the historian Thucydides wrote, "The bodies of dying men lay one upon another, and half-dead creatures reeled about in the streets. The catastrophe was so overwhelming that men cared nothing for any rule of religion or law."

The fall of the Roman Empire was accelerated by several disease outbreaks

In the late second century, the Roman Empire stretched from the British Isles to North Africa and from Portugal to Arabia (**Figure 6.4**). Over the next few centuries, however, Rome and its empire began a long and slow decline. The fall of Rome is certainly one of the most important events in the history of Western civilization. Historians have debated for years why Rome collapsed, and the reasons are undoubtedly complex. Pressure from tribes, such as the Goths, Vandals, and Franks along the northern frontier, was important. The Roman economic system and its inability to sustain itself over the vast distances in the empire are often cited as a key reason for Rome's decline. The general decadence of the wealthy and luxury-seeking Romans is also considered part of the story behind the breakdown of the Roman army. Indeed, by the late fourth century AD, Germanic tribesmen rather than Roman citizens formed the bulk of the Roman army in the West.

Why did the Romans allow the empire's security to fall increasingly into the hands of non-Romans? While moral decay may have been a factor, it is also probable that a decline in Rome's population played at least a part in the changing makeup of the Roman legions. And some of the reasons for the population decline were disease-related. Starting in the first century BC, epidemic malaria became a regular part of Roman life. The problem was especially bad in the agricultural districts surrounding the capital, where mosquitoes thrived. Mosquitoes and other arthropods that transmit pathogens between humans are known as **vectors**, and the diseases they transmit are called **vector-borne diseases**. Malaria is an important example. The problem became so bad that many cultivated areas around Rome were abandoned, resulting in a reduction in food supply.

Other epidemics may have contributed to Rome's slow decline. These included the Great Plague of Cyprian, which began in 251 AD. The cause of

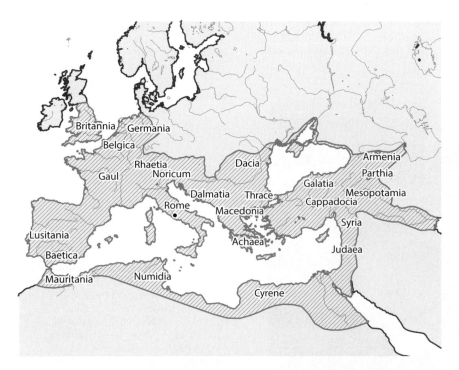

Figure 6.4 The Roman Empire at its height, circa 150 AD. Over the next 250 years, Rome began a slow decline.

this plague is unknown. Cyprian, the Christian bishop of Carthage, described symptoms that included explosive diarrhea, high fever, and gangrene of the hands and feet. The plague lasted at least 19 years, with higher mortality than any previous epidemic known. The disease was thought to spread not only by contact with infected individuals but also via any articles used by the sick, including clothing. The plague ebbed each summer, with fresh outbreaks each fall. Over half of those who contracted the mystery illness died.

By the late fourth century AD, the Roman Empire was divided into Western and Eastern Empires. After the Visigoths sacked Rome in 410, the Western Empire was weakened and disorganized. Those in the Eastern Empire dreamed of recapturing the West and this dream almost became reality under the leadership of the Emperor Justinian. With the goal of resurrecting the grandeur of a united empire, Justinian attacked the former Western Empire in 533. He recaptured Sicily as well as portions of the Italian peninsula, including Rome. After retaking parts of Spain, he made bold plans to lead his forces back into Gaul and even Britain.

In 540, however, disaster struck. Bubonic plague broke out in North Africa and quickly spread to the rest of the known world, reaching the Eastern capital of Byzantium between 541 and 542. The outbreak, now known as the Plague of Justinian, may have been the worst epidemic in human history. Up to 5000 people a day died in the Eastern Empire. So high was the death rate that graves could not be dug in time. Whole buildings were filled with corpses and set ablaze. Ships were loaded with the dead, rowed out to sea, and cast adrift.

Descriptions of the disease allow its clear diagnosis: bubonic plague. Victims experienced high fever, and swollen glands (buboes) in the armpits and groin. Often the buboes ruptured, resulting in gangrenous lesions and causing excruciating pain. Death typically followed within a week. All ages were vulnerable. The plague would strike an area and then recede, only to return again. No town or village was spared. Many cities were wiped out or abandoned, as were agricultural areas. The entire empire descended into panic and confusion.

We can never know to what extent the plague quashed Justinian's ambitions and led to the long, gradual decline of the Eastern Roman Empire. But it seems clear that at the very least, these events were hastened by this disease and that bubonic plague assisted in irrevocably changing the history of the Western world.

The Black Death altered European society forever

In 1346 the Tatars from Central Asia attacked a Genoese trading colony on the shores of the Black Sea. During the siege, bubonic plague broke out among the attackers. In an early example of biological warfare, the Tatars catapulted bodies of dead plague victims over the walls into the city. Plague is another example of a vector-borne disease. Fleas, rather than mosquitoes, serve as the vectors. Fleas abandon a dead host and search for new, living hosts to colonize. This is apparently what happened in the trading colony. Plague soon appeared among the Genoese, some of whom managed to escape the beleaguered colony via ship. A few days after their ship arrived in the home port of Genoa, the first cases of plague in the city were reported.

Thus began the notorious bubonic plague epidemic known as the Black Death (**Figure 6.5**). While Italian merchants brought the disease to Europe, the Tatars and their Mongol allies, in whom the epidemic began, spread

Figure 6.6 Plague victims. A mass grave in Martigues, France containing skeletons of individuals who died during the Black Death.

plague north into Russia, and east into India and China. The devastation caused by this worldwide epidemic, also known as a **pandemic**, is hard to comprehend. Although the exact numbers are unknown, in Europe alone it is estimated that 24 million people may have died from the plague, representing over 25% of the total population at the time.

A mortality rate of such stupendous proportions had major social and political ramifications. Deaths from plague destabilized the economies of major European powers, and even temporarily halted the Hundred Years War between France and England. In Siena, local Catholic authorities were hoping to attract the papacy to the city by tripling the size of Siena's cathedral. With the outbreak of bubonic plague, these plans were permanently shelved. The fate of the Greenland Vikings is also intriguing. Ships from Norway probably carried the infection to the Greenland colonies, which had been founded by Erik the Red. Plague may have weakened the colonists to such an extent that they were either conquered or absorbed by the indigenous Inuit people. Greenland was then forgotten for over 200 years until it was rediscovered in 1585. Because Viking colonists in Greenland were known to have visited the Canadian coast, the extinction of the Greenland settlements may have permanently altered North American history.

The impact on the social fabric of society was particularly acute in England. The epidemic first spread to the British Isles in the summer of 1348. It reached London in November of that year and arrived in Ireland in 1349. By 1350 it had spread to Scotland. At the time, England was still largely an agricultural society, governed by a feudal system of indentured servitude. Prior to the outbreak of plague, England's growing population resulted in a regular supply of cheap labor. The feudal lords were able to keep wages extremely low and restrict the movement of peasants to the cities or other places they might seek better financial opportunity. With bubonic plague, however, came a severe manpower shortage. Peasants could now demand higher wages and were allowed greater freedom of movement to meet labor needs. As they gained unprecedented economic independence, the feudal system began to crumble. Within 150 years it had entirely disappeared.

European explorers imported many diseases into the New World

Rarely in the history of humanity was the role of infectious microorganisms more clear than in the conquest of the Americas (**Figure 6.6**). Cortés, introduced earlier in our case study, was not the only conquistador who was assisted by lethal pathogens. Once established in Mexico, smallpox spread rapidly to South America. Millions of Incas in Peru, including the reigning emperor, died. When Francisco Pizarro arrived with a small band of Spanish soldiers, the Inca Empire was in political and social upheaval. Pizarro was thus able to usurp control without significant military opposition.

Other diseases that Europeans unwittingly introduced into the New World included measles, influenza, and malaria. Native Americans, far more than Europeans, were highly susceptible to these diseases. There are several interrelated reasons for this difference in susceptibility. These imported diseases originated in Europe and Asia, so Europeans already had a long history, stretching back perhaps thousands of years, interacting with the microorganisms causing these diseases. As we will explore more fully in later chapters, all mammals, including humans, eventually evolve some measure of resistance to regularly encountered diseases. Because these diseases were absent in the Americas before the arrival of the Spanish, no such

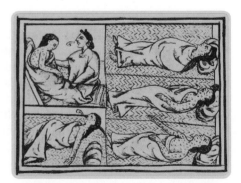

Figure 6.6 Smallpox in the Americas.
A drawing of Aztec smallpox victims, from the *Florentine Codex*.

opportunity was afforded to the Native Americans. Therefore, although the Spanish could still contract and transmit diseases such as smallpox, they were much less likely than the native populace to develop serious illness. The Aztecs watched as their own people died in large numbers while the Spanish remained relatively disease-free.

Why, then, as we asked in our case study, were the Spanish not decimated by so-called American diseases to which the native population was resistant? The conquest of the Americas occurred in a breathtakingly short period of time, in large part because there were few such diseases. This was certainly not the case when Europeans attempted to colonize Africa. A plethora of indigenous African maladies lay in ambush to waylay potential invaders. Consequently, while the Spanish conquest of Mexico and Peru was completed in a few short years, it was not until the twentieth century that the European colonization of Africa was complete.

To answer this question, we must first consider some of the qualities shared by diseases such as smallpox, measles, and influenza. They are all **acute** in nature. Once a host is infected with an acute disease, the pathogen replicates rapidly, and symptoms of disease appear quickly. Furthermore, there are only two possible outcomes—that is, within a short time the host either recovers or dies. If the host recovers, he or she will typically be immune to the same pathogen, at least for a while. We will explore how such immunological memory is developed in Chapter 12.

Because agents of acute disease either rapidly kill their host or are themselves destroyed, they must be transmitted to new hosts quickly and efficiently if they are to survive. Smallpox virus, for instance, is transmitted through the air. A single cough or sneeze by an infected individual can infect many other potentially susceptible hosts. The situation is different for **chronic diseases** (Table 6.1). These are conditions that last indefinitely, sometimes for years or decades. The host immune system is able to control but not eliminate the disease-causing microorganism. For a pathogen causing chronic disease, there is far less need for rapid transmission. It can survive in its current host indefinitely and has a relatively long period of time in which to reach new hosts.

Table 6.1 Characteristics of acute and chronic diseases. Acute diseases generally begin with a rapid onset of symptoms, and the disease resolves rapidly, with either elimination of the pathogen or death of the host. Chronic diseases last indefinitely and may cause symptoms for long periods of time.

CHARACTERISTIC	ACUTE DISEASE	CHRONIC DISEASE
Replication of infectious agent in the host	Very rapid	Generally less rapid
Disease course	Short incubation; rapid onset of symptoms; quick resolution (immunity or death of host)	Longer incubation; slow onset of symptoms; no quick resolution (may persist months, years, or decades)
Immune response	Generally rapid; result is complete elimination of the pathogen, with resistance to subsequent infection by the same pathogen	Incomplete; although the pathogen may be contained, it is not eliminated
Transmission	Rapid and efficient transmission to new susceptible hosts is required for pathogen survival	Rapid and efficient transmission is not required for pathogen survival
Host population size	Large population of susceptible hosts is required for pathogen survival	Large, susceptible host population is not required for pathogen survival
Example	Smallpox, measles	Tuberculosis

Acute diseases need large populations of potential hosts. Consequently, they are often called crowd diseases. Remember that each new host either recovers quickly, and is subsequently immune, or dies. The pathogen, therefore, needs a large supply of susceptible hosts to persist. When an acute disease is introduced into a small isolated population, it quickly uses up all potential hosts and then dies out itself. Studies have shown, for instance, that measles virus will die out in populations under 1 million. When measles was accidentally introduced into the Faroe Islands in 1781, a severe epidemic broke out. The Faroes are an isolated group of islands in the North Atlantic, and after the virus spread through the entire population, it went extinct. The islands then remained measles-free until an infected individual from Denmark brought the virus back to the islands by ship in 1846. In large populations, new hosts in the form of newborns, immigrants, or people in whom immunity has worn off, ensure a constant supply of susceptible individuals, thus allowing the virus to persist indefinitely.

Molecular analysis of pathogens causing acute disease has confirmed that many of them are most closely related to pathogens causing similar crowd disease in animals. Smallpox virus and measles virus, for instance, are most closely related to viruses causing acute diseases in cattle. Most authorities believe that in the 8000 years since cattle were domesticated in western Asia, these animal viruses made the jump from cattle to humans (**Figure 6.7**). Other acute disease agents are similarly related to viruses found in domestic animals, including pigs and chickens that had been raised for thousands of years in Europe and Asia.

In the Americas, however, there was no similar long-standing tradition of animal husbandry. Only dogs, turkeys, guinea pigs, llamas, and ducks had been domesticated prior to European contact. Thus, the opportunities for animal-to-human transmission were greatly reduced. This is not to say that the Americas were disease-free, but the relatively infrequent or even non-existent contact between Native Americans and large herds of animals resulted in far fewer human acute illnesses.

In this manner, Cortés and other conquerors were aided by a one-sided exchange of deadly microbes. Perhaps a dozen or more major infectious diseases were introduced into the New World by Europeans. With the possible exception of syphilis, the origins of which are still debated, no similar lethal disease was transferred from the Americas back to Europe.

Unlike the Americas, Africa was not fully colonized until the 1900s

The rapidity with which the Americas were conquered stands in stark contrast to the European experience in Africa. In spite of numerous efforts, attempts to colonize or even to explore Africa up through the nineteenth century often met with dismal failure. In 1816, for instance, the exploration of the Congo River by Captain James Tuckey was cut short when almost all members of his party were struck by "an intense, remittent fever and black vomit in some cases." In 1841, Captain H. D. Trotter set out to explore the Niger River with 145 Europeans and 133 Africans. After they had travelled 100 miles upstream, severe fever broke out. Among the Europeans, 130 became seriously ill and 50 died. All of the Africans remained healthy.

Native Africans did not have large herds of domestic animals, but unlike the situation in Mexico and South America, the Africans lived in close proximity to enormous herds of wild grazing animals. These animals served as reservoirs for many agents of disease to which the animals themselves, as well as

Figure 6.7 Animal husbandry has a long tradition in the Middle East. Cattle are thought to have been domesticated in the valley of the Tigris and Euphrates Rivers, in what is present-day Iraq, approximately 8000 years ago. This sculpture of a Mesopotamian king with the body of a cow illustrates the importance that cattle had in early Mesopotamian culture.

native Africans, were relatively insusceptible. Sleeping sickness, yellow fever, and other diseases took a heavy toll on both Europeans and their animals. Until the cause and prevention of such ailments was understood, little progress in the colonization of Africa was possible. Indeed, the tsetse fly, which carries the protozoans that cause sleeping sickness, has been called Africa's greatest game warden. Until it became possible to control these flies, large segments of Africa's most fertile lands were off limits to the colonists for raising domestic livestock because these non-African animals were so susceptible to sleeping sickness.

Several important events of the nineteenth century owe their outcomes to microorganisms

The nineteenth century is especially illustrative of the relationship between infectious disease and human history. During this time period, national armies of unprecedented size swept back and forth across continents, providing the large crowds necessary for acute disease, as well as a means for introducing them into new populations. Cities, too, were expanding rapidly and the crowded, unsanitary conditions in these urban centers permitted many diseases to flourish. Furthermore, the nineteenth century saw a dramatic improvement in communication and record-keeping, allowing us to better chronicle events of the day. It is probably no accident that the late nineteenth century gave birth to the modern science of microbiology, as scientists struggled to make sense of disease. Our next few examples are all well-documented instances of the impact infectious disease has had on human affairs during the 1800s, setting the stage for the Golden Age of microbiology.

Napoleon's dreams of empire were compromised by disease

Napoleon is one of the major historical figures of the nineteenth century. Decisions made by the French emperor dramatically altered the political landscape in North America and Europe. In at least two of his most monumental endeavors, disease loomed large.

At the end of the eighteenth century, France controlled an enormous tract of land in North America, stretching from the Mississippi River to the Rocky Mountains and from the Gulf of Mexico to Canada. The French also possessed several Caribbean islands, including the colony of Saint-Domingue (now Haiti), where large sugar plantations were established.

In 1801, after more than a decade of revolt and turmoil in Haiti, Napoleon sent over 20,000 troops to the colony. Before long, most of these soldiers were laid low by yellow fever. Well over half eventually died. These heavy losses influenced Napoleon's decision not to risk even more French soldiers in defense of French territory in North America. Napoleon reasoned that these troops could be better used extending his empire in Europe. In 1803, he sold the vast Louisiana Territory to the United States (**Figure 6.8**).

In the early summer of 1812, Napoleon turned his attention east and invaded Russia. Initially his plans went well, but as his army of over 500,000 men crossed through Poland, typhus erupted among his troops. Typhus is caused by the bacterium *Rickettsia prowazekii* and is transmitted by body lice (**Figure 6.9**). Typhus most often strikes when people are crowded together under unsanitary conditions, because failure to bathe or to wash clothing allows the lice to flourish. History is filled with examples of armies being

Figure 6.8 The United States, 1803. Territory purchased by the United States from France as the Louisiana Purchase is highlighted in red.

plagued with typhus because soldiers often find themselves in situations that are ideal for typhus transmission.

By the time Napoleon's army reached Moscow on September 14, only 100,000 of his troops remained. Typhus, combined with starvation of the ill-equipped soldiers, accounted for most of the deaths. Ultimately, Napoleon was forced to withdraw from Russia, defeated by typhus, hunger, and the Russian winter. By the end of the year, approximately 20,000 French troops staggered into Vilnius on the Baltic Sea. Fewer than 1000 of the once mighty army were ever again fit for duty. Napoleon's dream of conquering Russia had ended for good.

Irish history was irrevocably changed by a disease affecting potatoes

The development of agriculture, starting in about 7000 BC, constituted one of the greatest revolutions in the history of mankind. Farming allowed people to live in large, settled communities for the first time and gave them some measure of control over their food supply. Such progress, however, comes at a cost. Agriculture carries with it certain inherent risks. Oftentimes, when one crop is especially well suited to a particular region or when there is an overwhelming preference for a specific crop, there is a tendency to cultivate only this one plant species. The result is a **monoculture**—an agricultural area dominated by a single crop. Reliance on monocultures is a dangerous gamble. During favorable conditions, farmers may count on bumper harvests. If crop pests such as insects, plant viruses, or fungi invade, however, the results can be disastrous. The pest can increase in numbers rapidly as it feeds on its favored food source and quickly spreads through and destroys the entire crop. If the population depends too heavily on this monoculture for sustenance, the result is famine. Such devastation is rare in natural ecosystems consisting of many plant species. It is unlikely that any plant pest can attack more than a few plant types, so a variety of plants can severely curtail the damage to the overall environment caused by any one pest.

The Irish potato famine provides a startling example of how a crop disease can affect a society. It also warns against relying too heavily on a single food source. It is rumored that the potato was first introduced into Ireland by Sir

Figure 6.9 The human body louse, *Pediculus humanus humanus.* Typhus is a bacterial, vector-borne disease transmitted by body lice. These lice thrive in crowded conditions where people are unable to bathe or wash clothes regularly. Consequently, soldiers through the mid-twentieth century were especially plagued by typhus. Note the prominent front legs, with which the louse holds tightly onto clothing.

Walter Raleigh in about 1588 when he planted potatoes in his garden. The crop thrived in Ireland, and by the late 1600s, potatoes had become a staple in the Irish diet. In 1846 disaster struck in the form of a plant disease called potato blight (**Figure 6.10**). The disease is caused by *Phytophthora infestans*. These organisms appear fungus-like but are more closely related to algae. The disease spread rapidly, causing almost total crop failure. Because the Irish depended so heavily on potatoes, the consequences were severe. Within only a few months the entire economy collapsed. Eventually, 750,000 to a million of Ireland's approximately 8 million inhabitants starved to death or died of diseases related to malnutrition. Many more emigrated to other countries.

Today the population of Ireland is about 6.4 million, still well below the pre-famine total. The large Irish communities in cities like Chicago and Boston owe their existence in some measure to the fungus-like organism that precipitated the disaster.

French efforts to build a canal in Panama were thwarted by mosquito-borne disease

Before 1914, the only sea route between the Atlantic and Pacific Oceans was via a long and hazardous trip around Cape Horn or through the Straits of Magellan at the tip of South America. As early as the sixteenth century, the Spanish proposed building a canal across the Isthmus of Panama to solve this problem, but no serious attempt to undertake this task was made until the late nineteenth century. In 1879, the Frenchman Ferdinand de Lesseps secured government backing for what he called "la grande entreprise." The project would be the biggest financial undertaking of all time and would enhance French prestige and influence tremendously.

Yet it was not to be. Mosquitoes thrived in the lakes and swamps through which the canal was to pass, and the French suffered enormously from two mosquito-borne diseases—malaria and yellow fever. In 1882, a year after digging began, approximately 20% of the French workforce had died. By 1883, the death toll averaged 200 each month. When 17 newly graduated French engineers arrived in 1885, only one remained alive 1 month later. As word of the horrendous death rate filtered back to France, it became increasingly difficult to find employees willing to risk death by disease in the Panamanian jungle. By 1889, the project was abandoned.

It was not until 1897 that the English scientist Ronald Ross obtained evidence that mosquitoes transmitted the malaria protozoan *Plasmodium*. Although the Cuban Carlos Finlay had suggested that mosquitoes transmitted yellow fever in 1881, this idea was not confirmed until the 1890s.

This newly discovered information about malaria and yellow fever proved invaluable when, in 1904, the American president Theodore Roosevelt decided to tackle the challenge of linking the Pacific and Atlantic (**Figure 6.11**). William Gorgas, who had helped to control mosquito-transmitted diseases in Havana, was put in charge of health and sanitation in Panama. Gorgas employed the same techniques that had worked so well in Cuba. Buildings for workers were made mosquito-proof with fine-gauge copper mesh. Standing water was drained wherever possible. Water bodies that could not be drained were sprayed regularly with kerosene to kill mosquito larvae. When digging began in 1907, yellow fever was already under control, and deaths from malaria were significantly lower. When the canal was completed in 1914, the annual death rate among American workers in Panama from all causes was 6 in 1000. In comparison, the death rate in New York and London was about 14 in 1000.

Figure 6.10 Potato infected with *Phytophthora infestans*. The brown streaks growing through the potato are caused by the infection. Though fungus-like in appearance, this plant pathogen, the cause of the Irish potato famine, is more closely related to algae.

Figure 6.11 Theodore Roosevelt confronts "yellow jack." A cartoon from the early 1900s, first published in *Harper's Weekly*, depicting President Roosevelt and Uncle Sam squaring off with yellow jack, as yellow fever was often called. Because of recent discoveries about the transmission of yellow fever, Roosevelt knew that successful construction of the Panama Canal required that disease control measures first be implemented.

Gorgas became a hero. He was appointed Surgeon General in the United States and was knighted by the King of England. Meanwhile, the United States gained unprecedented influence in Central America that continues until the present day.

A bacterial metabolic process aided the British navy in World War I

Microbes are not important just because of the diseases that they cause. Many common food items are made with the help of microorganisms, and industry relies on microbial assistance for many routine industrial processes. Here we describe one such industrial use of microorganisms, which in a small way influenced the outcome of the First World War.

As World War I wore on, the British Navy began to suffer from a cordite shortage. Cordite was used on naval vessels as smokeless gunpowder. With cordite, a ship could fire its weaponry without billows of smoke giving away its position. Acetone is a carbon compound required for cordite production and it can ordinarily be cheaply produced by heating wood. Acetone was obtained at that time as a by-product of charcoal production. So great was the demand for cordite, however, that supplies of acetone became severely limited.

Looking for an alternative means of acetone production, David Lloyd George, Director of Research for Explosives, travelled to Manchester, where he met with Professor William Perkin, a leading organic chemist. While there, he also met Perkin's student, Chaim Weizmann, who was interested in fermentation (**Figure 6.12**). Weizmann had already discovered that *Clostridium acetobutylicum*, a Gram-positive bacterium, produced a mixture of butyl alcohol and acetone from sugar. When Lloyd George heard of this, he immediately dispatched Weizmann to London to consult with Winston Churchill, who was then First Lord of the Admiralty.

Weizmann explained to Churchill that the bacterium could produce the required acetone in abundance, if a large enough fermentation and distillation plant could be found. Churchill used his authority to commandeer a gin

Figure 6.12 Chaim Weizmann and David Lloyd George. (a) Weizmann was born in what is now Belarus in 1884. After studies in Switzerland, he moved to Britain, where he developed a technique for acetone synthesis, using bacteria. (b) David Lloyd George, after a successful military career, went on to become Britain's first, and to date only, prime minister of Welsh ancestry.

factory and have it converted to acetone synthesis. Weizmann was appointed chief scientist of the Nicholson Gin Company.

The cordite shortage was abated. Acetone production was eventually shifted to Canada, where an abundance of corn ensured a constant supply of sugar for the fermentation process. Lloyd George went on to become Prime Minister of Great Britain, while Weizmann eventually became president of the World Zionist Organization, which was seeking the creation of a Jewish homeland in the Middle East. Weizmann sought the Prime Minister's support, and in 1917 Britain issued the Balfour Declaration, instrumental in the establishment of such a homeland and the eventual founding of the State of Israel. Weizmann was elected as Israel's first president in 1948.

Polish doctors spared their villages from World War II's devastation with microbial assistance

A little-known event in World War II illustrates how ingenious people can be when it comes to survival under difficult circumstances. The incident also offers yet another example of how a microorganism can impact human affairs in unexpected ways.

Normally when the immune system is confronted with a particular pathogen, proteins called **antibodies** are produced that react with and neutralize the microorganism. Antibodies are usually very specific—each type of antibody binds to only one type of microbe. In some cases, however, antibodies against one microorganism will bind to a second organism. When an antibody against one microorganism can also bind to a second microbe, the antibody is said to be **cross-reactive**.

Earlier in the twentieth century, Polish doctors discovered that antibodies against *Rickettsia prowazekii*, the causative agent of epidemic typhus, were cross-reactive with *Proteus vulgaris* OX19, a harmless strain of Gram-negative bacteria. They used this phenomenon to develop a reliable test for typhus (**Figure 6.13**). Blood was drawn from a suspected typhus patient and mixed on a glass slide with *Proteus* OX19 cells. If the patient was infected with typhus, anti-*Rickettsia* antibodies present in his or her blood would bind to the *Proteus* and cause easily observed clumping on the slide.

World War II began in 1939 when the Germans invaded Poland. In the Polish villages of Rozvadow and Zbydniowie, about 200 km southwest of Warsaw, two Polish physicians decided to try a clever deception to spare their villages from the Nazi onslaught. Drs Eugeniusz Lazowski and Stanislav Matulewicz injected people in their villages with *Proteus* OX19. These villagers remained healthy but produced anti-*Proteus* antibodies. The Polish doctors took blood samples from these individuals and sent them to the German occupation authorities for testing. When the tests came back positive, the Germans became convinced that a typhus epidemic was raging in the region. The Germans were so afraid of typhus that they declared an epidemic zone in the area, and German oppression in at least one part of occupied Poland was dramatically decreased.

Some infectious disease is caused by non-microorganisms

Our emphasis in this text is microorganisms—living things ordinarily too small to see with the unaided eye. In addition to microorganisms, however, humans are vulnerable to infection with a variety of much larger parasitic worms known as **helminths**. A **parasite** is any organism, microscopic or not,

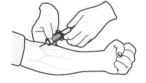

1. Take blood sample from suspected typhus patient.

2. Mix blood sample on glass slide with *Proteus* OX19.

3. Examine slide for evidence of cross-reactivity between anti-*Rickettsia* antibodies and *Proteus* OX19.

No anti-*Rickettsia* antibodies in patient's blood	Anti-*Rickettsia* antibodies in patient's blood

No cross-reactivity between anti-*Rickettsia* antibodies and *Proteus* OX19	Antibody cross-reactivity causes clumping of *Proteus* OX19
Patient does not have typhus	**Patient has typhus**

Figure 6.13 The Weil–Felix test for typhus. If a patient is infected with *R. prowazekii*, he or she will have antibodies to this bacterial species circulating in the blood. These antibodies will also bind to *Proteus* OX19, a harmless soil bacterium. Because of this cross-reactivity, a diagnosis of typhus can be easily confirmed by mixing the patient's blood serum with *Proteus* OX19. Clumping of the *Proteus* bacteria indicates that the patient is positive for typhus. A lack of clumping rules out a typhus infection.

that lives in or on another organism, from which it obtains a place to live, reproduce, and obtain nutrients. Our final example of how history can be shaped by disease involves such a helminth parasite. Specifically we will discuss how an outbreak of schistosomiasis may have influenced the geopolitical map in the early days of the Cold War.

Following the defeat of Japan in 1945, the island of Taiwan reverted to Chinese administration, yet the late 1940s were a time of great unrest in China. The Chinese Communists, led by Mao Zedong, were at war with the Kuomintang led by General Chiang Kai-shek. As the civil war progressed, the Kuomintang nationalists were forced to retreat. Finally, in late 1949 Chiang and his government abandoned the Chinese mainland and retreated to the island of Taiwan, establishing a new capital in Taipei.

The Chinese Communists began immediate preparations to invade Taiwan. Thousands of troops moved into southern China to prepare for an attack across the Taiwan Strait. Because invasion of the island would require an amphibious assault, the Communist troops were taught to swim in the rivers and irrigation canals in the region. Many of these water bodies harbored schistosome-infected snails. As we learned in Chapter 4 (see pp. 70–71), schistosome worms develop as larvae in specific freshwater snails. When larval development is complete, the worms leave the snails and penetrate directly through the skin of their human hosts, where they mature to adulthood. Between 30,000 and 50,000 Chinese soldiers were infected and an epidemic of acute schistosomiasis broke out. Although we may never know to what extent schistosomiasis influenced the thinking of the Communist authorities, the invasion was postponed. Meanwhile, a US naval fleet arrived off the Taiwanese coast to protect the island against invasion. With the threat of American involvement now looming large, plans to invade Taiwan were shelved indefinitely. In 1954, the United States and Taiwan signed a mutual defense treaty, assuring the survival of the Nationalist government.

The science of microbiology has a vibrant history

CASE: THE ASSASSINATION OF PRESIDENT GARFIELD

On July 2, 1881, only 6 months into his term, President James A. Garfield was shot twice by an assailant in Washington, DC. The first shot caused a superficial wound in the president's arm. The second entered Garfield's chest. The immediate concern of doctors was to find the second bullet, and even before Garfield was moved, they began digging in the wound. In the late nineteenth century, any bullet wound was extremely serious, and if the bullet could not be extracted at once, the prognosis for recovery was poor. Garfield himself realized this, and when an attending physician tried to comfort him, the President replied, "I thank you Doctor, but I am a dead man." Two unsuccessful operations to find the bullet were attempted. Alexander Graham Bell, the inventor of the telephone, was summoned to look for the bullet with a crude metal detection device, but he too was unsuccessful (Figure 6.14). Finally, 80 days after he was shot, Garfield died. Unfortunately, if the incident had occurred only a few years later, or if doctors attending the stricken president had utilized recently discovered principles of infection control, Garfield might have survived. There is a sad grain of truth to a statement made by the assassin, Charles Guiteau, before

he was hanged. Upon learning that Garfield had died, Guiteau replied, "The doctors killed him. I simply shot him."

1. **What discoveries in the emerging field of microbiology might have saved President Garfield's life?**

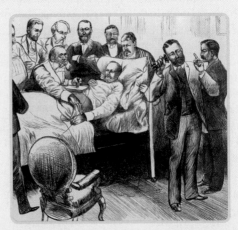

Figure 6.14 Alexander Graham Bell and President James Garfield. An 1881 illustration from *Harper's Weekly*, showing Alexander Graham Bell looking for a bullet in President Garfield with his magnetic induction device.

Garfield died of a bacterial infection. Antibiotics to treat bacterial infections were still a half century in the future and even simple infection control procedures such as hand washing were rarely if ever utilized. The doctors who first probed President Garfield's wounds did so with unsterilized instruments and with fingers still soiled from work on other patients earlier in the day. During the two surgeries, no attempt would have been made to prevent infection of the surgical incisions.

The emerging science of microbiology led to vast improvements in the treatment of wounds and the prevention of infection during surgery. To understand how, we need to examine the origins of microbiology itself (Table 6.2).

Table 6.2 Some key highlights in the history of microbiology. A timeline showing a few of the major milestones in microbiology over the last 350 years. The period between about 1850 and 1910 is often called the Golden Age of microbiology, because of the unprecedented number of breakthroughs that occurred during this time.

YEAR	EVENT
1674	Van Leeuwenhoek views "animalcules" with his microscopes.
1721	Lady Montagu learns of smallpox variolation, introducing the technique into Europe.
1796	Jenner develops an anti-smallpox vaccine.
1847	Semmelweis introduces hand washing into Vienna hospital, reducing incidence of "childbed fever."
1854	Nightingale introduces modern nursing techniques, including routine sanitation.
1857	Pasteur demonstrates fermentation in wine.
1858	Pasteur disproves spontaneous generation.
1864	Pasteur develops pasteurization.
1865	Lister develops aseptic technique.
1876	Koch develops "germ theory of disease" and identifies causative agent for anthrax.
1881	Koch develops pure culture technique.
1882	Hesse suggests agar as bacterial medium; Koch identifies causative agent for tuberculosis.
1884	Koch's postulates formally established; Gram stain introduced.
1885	Pasteur develops concept of attenuation and develops rabies vaccine.
1892	Ivanowski describes "filterable agent" causing tobacco mosaic disease.
	Smith and Kliborne describe the first vector-borne disease: Texas red water fever, transmitted by ticks.
1894	Yersin identifies bacteria responsible for plague.
1896	Ross demonstrates mosquito transmission for malaria.
1898	Loeffler and Forch describe filterable agent for hoof-and-mouth disease.
1900	Reed identifies role for mosquitoes in yellow fever transmission.
1908	Ehrlich develops salversan for treatment of syphilis.
1929	Fleming discovers penicillin.
1933	Electron microscope is developed.
1935	First sulfa drugs; viruses are first crystallized.
1944	DNA identified as the genetic material.
1946	Transfer of DNA between bacterial cells is first described.

(continued)

Table 6.2 Some key highlights in the history of microbiology. A timeline showing a few of the major milestones in microbiology over the last 350 years. The period between about 1850 and 1910 is often called the Golden Age of microbiology, because of the unprecedented number of breakthroughs that occurred during this time (*continued*).

YEAR	EVENT
1953	*Watson and Crick publish their paper on the structure of DNA.*
1960	*Gene regulation in bacteria is described.*
1973	*First recombinant DNA is developed.*
1979	*First pharmaceutical (insulin) is produced in bacteria via recombinant DNA technology.*
1986	*First approval of a vaccine (hepatitis B) is created with recombinant DNA technology.*
1995	*First entire genome, the bacterium* Haemophilus influenzae, *is sequenced.*
	First "designer drug" against a microorganism, protease inhibitors used against HIV, is introduced.
2006	*The vaccine against cervical cancer, caused by the human papillomavirus, is introduced.*
2013	*Technology derived from an understanding of the prokaryotic CRISPR/Cas system is used to alter DNA in plants and animals.*

Microorganisms were first discovered in the seventeenth century

In 1674, Anton van Leeuwenhoek first observed pond water with a simple microscope he had constructed (**Figure 6.15**). It had been known for some time that lenses could be used to magnify objects, but van Leeuwenhoek was the first to combine precision lens grinding with an intense curiosity about things nobody had thought to look at before. And what he saw astounded him: an entire universe of previously unknown living things, invisible to the naked eye. Van Leeuwenhoek not only saw but also carefully described bacteria and protozoa. When he looked at other samples, including small soil particles and even his own feces, he continued to find more of what he called "animalcules." While looking at organisms from teeth scrapings in his mouth, he noted that they were apparently killed if he drank hot tea before taking a sample. His findings were made public through a series of letters he sent to the Royal Society of London until his death in 1723.

Van Leeuwenhoek's discoveries started an intense scientific debate that lasted nearly 200 years. Where did these microscopic organisms come from? Many believed that they could somehow arise spontaneously from non-living substances. Nine years before van Leeuwenhoek's first descriptions of microorganisms, spontaneous generation had been disproved for animals by the Italian Francesco Redi (**Figure 6.16**). Many people, however, refused to accept Redi's conclusions, and for the newly discovered microorganisms in particular, spontaneous generation still had many adherents.

The science of microbiology was born in the second half of the nineteenth century

Microbiology truly emerged as a distinct scientific discipline after 1850. One of the first important breakthroughs occurred in 1859 when Louis Pasteur conducted a simple yet elegant experiment, which finally settled the question of microbial spontaneous generation. In his experiment, Pasteur demonstrated that a liquid broth could not give rise to microorganisms spontaneously (**Figure 6.17**). Pasteur believed that the microorganisms apparently arising spontaneously in a liquid broth were merely carried into the broth through the air. To test this idea, Pasteur sterilized meat broths by

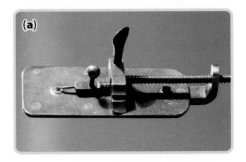

(a)

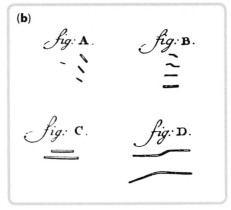

(b)

Figure 6.15 Anton van Leeuwenhoek was the first to describe microorganisms. (a) One of van Leeuwenhoek's early microscopes. (b) Drawings of "animalcules" from one of van Leeuwenhoek's notebooks.

Figure 6.16 Redi's experiment refuting spontaneous generation. The experiment, conducted in 1665, showed that maggots developed on rotting meat only if flies had access to the meat first. When flies could not reach the meat, no maggots developed.

Meat is exposed in open container. Flies enter and lay eggs on meat, which hatch into maggots.

When flies are prevented from reaching meat by fine gauze, no maggots appear in meat.

boiling them in flasks. Half of his flasks had standard straight necks, while the other half had necks bent into an S-shape. Both types of flasks were open to the air. Microorganisms soon appeared in the straight-necked flasks, but the S-necked flasks remained sterile. Pasteur reasoned that although both types of flasks were open to the air, microorganisms were unable to enter the S-necked flasks because they were trapped in the bend of the neck. When Pasteur snapped off the necks of these flasks, microorganisms quickly appeared in these broths as well. These results made it clear that organisms could not be created spontaneously, even if air entered the flasks freely. Life could only arise from previous life. The notion of spontaneous generation was discredited for good.

Pasteur advanced the science of microbiology in numerous ways, making him, by any measure, one of the giants of nineteenth-century science (**Figure 6.18**). At about the same time that he was disproving the idea of spontaneous generation, Pasteur was also making important contributions to the science of winemaking. The wine industry has always been an important part of

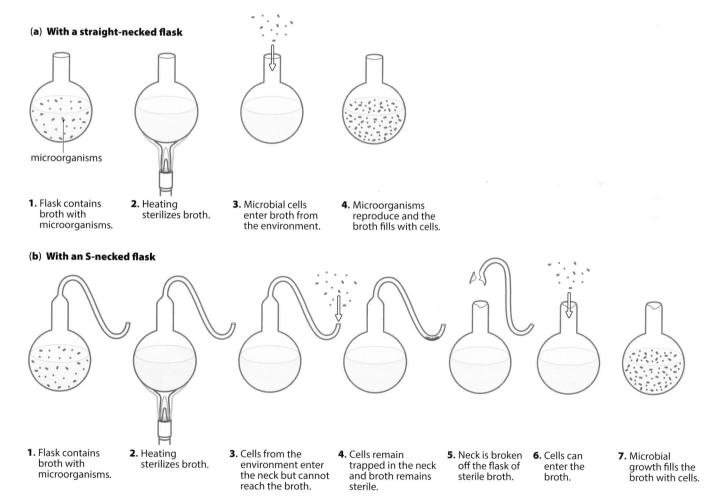

(a) With a straight-necked flask

microorganisms

1. Flask contains broth with microorganisms.

2. Heating sterilizes broth.

3. Microbial cells enter broth from the environment.

4. Microorganisms reproduce and the broth fills with cells.

(b) With an S-necked flask

1. Flask contains broth with microorganisms.

2. Heating sterilizes broth.

3. Cells from the environment enter the neck but cannot reach the broth.

4. Cells remain trapped in the neck and broth remains sterile.

5. Neck is broken off the flask of sterile broth.

6. Cells can enter the broth.

7. Microbial growth fills the broth with cells.

Figure 6.17 Pasteur's experiment disproving spontaneous generation of microorganisms. (a) If microorganisms had access to a sterilized broth through the air, eventually they eventually colonized the broth. (b) If, however, the broth-containing flask was constructed in such a way that microorganisms could not reach the broth, no such colonization occurred.

the French economy. Prior to the mid-1800s, however, wine was too often spoiled when it mysteriously turned sour.

In 1857, a group of French wine merchants asked Pasteur to investigate the problem. At that time, it was unclear how sugars in wine or other alcoholic beverages were converted into alcohol. Pasteur found that yeast, a microscopic fungus, released alcohol as a metabolic waste product in the absence of oxygen in a process called **fermentation**. Furthermore, he discovered that if certain bacteria had contaminated the wine, these bacteria competed with the yeast for sugar, converting it into acetic acid (the key ingredient in vinegar). It was this acid that gave the spoiled product its sour taste.

Pasteur found that he could eliminate these bacteria by temporarily heating the wine. This process, now called **pasteurization**, is still routinely used to remove unwanted microorganisms from many beverages, including beer, milk, and fruit juice. This link between microorganisms and food spoilage suggested that microorganisms might play a role in other important processes as well, including disease.

The germ theory of disease was advanced by the understanding of the need for hospital sanitation

In the second century AD, the Greek physician Galen attributed good health to a proper balance of the body's four elemental fluids or so-called humors. When these four fluids—blood, phlegm, black bile, and yellow bile—were not present in the correct proportions, disease resulted. Galen's ideas influenced medical thought for hundreds of years. Others believed that various illnesses were caused by foul odors, by vapors emanating from swamps, or by evil spirits. Up through the mid-nineteenth century, few had considered the idea that organisms too small to see could cause certain diseases.

Things began to change in the mid-nineteenth century. In 1847, the Austrian physician Ignaz Semmelweis (**Figure 6.19**) made an important discovery that paved the way for the **germ theory of disease**—the concept that microorganisms could cause disease. Semmelweis was concerned by the many women who died shortly following childbirth in Vienna's largest hospital. All too frequently, women in the hospital delivered healthy babies, only to die within a few days of what was called childbed fever. Curiously, only women who were admitted to the hospital's First Clinic developed the illness. When the First Clinic was full, obstetric patients were admitted to the Second Clinic and these women did not contract childbed fever. Semmelweis observed that the two clinics were the same except for one thing; in the First Clinic, medical students delivered the babies, whereas the Second Clinic was staffed by midwives.

Semmelweis got a clue to the mystery when a medical student cut his finger during an autopsy and died of an illness identical to childbed fever. Medical students in the hospital performed autopsies in the morning and later delivered babies in the First Clinic obstetric ward. Semmelweis reasoned that the students picked up something deadly on their hands during the autopsies and carried it to the new mothers in the First Clinic.

In 1847, hand washing was not routine in hospitals, but Semmelweis ordered the students to wash their hands with a chlorine solution before entering the clinic. The number of childbed fever cases in First Clinic immediately plummeted. Unfortunately, Semmelweis' discovery was ridiculed and the practices he instituted were not widely adopted. We now know that childbed fever is caused by the bacterium *Streptococcus pyogenes*, which can easily be carried on the hands between patients.

Figure 6.18 Louis Pasteur (1822–1895). Pasteur is regarded as one of the founders of modern microbiology. His contributions include the first vaccine for rabies and the development of pasteurization, still used to remove potentially harmful microorganisms from many food products.

Figure 6.19 Ignaz Semmelweis (1818–1865). When Semmelweis proposed that childbed fever was caused by a common agent transmitted on the hands of attending doctors, he was ridiculed by others in the medical profession. He was eventually dismissed by the hospital, whereupon he suffered a mental breakdown. Today, however, he is regarded as the "savior of mothers" in Austria.

Fortunately Semmelweis' work did not go entirely unnoticed. During the Crimean War of 1854–1856, thousands of lives were saved due to the ground-breaking work of Florence Nightingale (**Figure 6.20**), widely recognized as the founder of modern nursing. Nightingale introduced the basic concepts of hospital cleanliness to British military hospitals and insisted that filthy clothes and bandages be regularly replaced or sanitized. Her Nightingale School of Nursing, founded after her return to England, was the first school in the world to specifically train nurses and educate them about the importance of sanitation in hospitals.

The English surgeon Joseph Lister (**Figure 6.21**) was also familiar with Semmelweis' work, as well as with Pasteur's studies linking wine spoilage to bacteria. Furthermore, Pasteur had also demonstrated that a protozoan could cause disease in silkworms. Could microorganisms also cause disease in humans? Lister knew that carbolic acid, also called phenol, killed bacteria. He reasoned that if bacteria did in fact cause human disease, treating wounds and bandages with phenol might reduce such disease. He successfully tested his ideas in 1865. As word of Lister's **aseptic technique** spread, it gradually became accepted around the world.

Unfortunately, in 1881 neither routine hand washing, first championed by Semmelweis, nor the use of aseptic technique as developed by Lister was commonplace in the United States. Had these practices been more readily adopted in this country, it is likely that the story of President Garfield would have had a happier ending.

The development of pure culture technique allowed rapid advances in microbiology

The definitive proof that a specific bacterium caused a particular disease was finally provided by the German microbiologist Robert Koch in 1876 (**Figure 6.22**). Crucial to this endeavor was the development of **pure culture technique**. Koch realized that if a specific microorganism was causing disease, any host would also harbor many other microorganisms that may have nothing to do with the disease in question. Unless he could isolate these various microorganisms, he would have no means to determine exactly which species was actually the pathogen. Koch found that he could take samples from a host and spread them out on solid nutrients such as potato slices. Different microbial species grew into different shaped and colored colonies. He then inferred that each of these colonies had arisen from a single bacterial cell and consequently represented a pure culture. Cells could then be extracted from each of these colonies for further testing.

Potato slices, however, were crude and did not support the growth of all microbes. Koch experimented with gelatin as a growth medium, but gelatin, which melts at temperatures required for maximum bacterial growth, turned out to be a poor substitute. Fortunately, Fannie Hesse, the wife of one of Koch's co-workers, suggested agar, which she regularly used as a solidifying agent in her fruit preserves. Agar turned out to be an ideal growth medium and substantially assisted Koch in his isolation of pathogenic bacteria.

Koch was the first to link a specific microorganism to a particular disease

Koch was interested in determining the cause of anthrax, a serious disease of domestic livestock. He inspected the blood of animals that had recently

Figure 6.20 Florence Nightingale (1820–1910). Called "the lady with the lamp," Nightingale is regarded as the founder of modern nursing, training hundreds in sanitary technique at her Nightingale School of Nursing, the world's first such institution. She was also an accomplished writer and helped to found the modern discipline of epidemiology.

Figure 6.21 Joseph Lister (1827–1912). Lister, a surgeon at the Glasgow Royal Infirmary in Scotland, found that infection of surgical wounds could be prevented if wounds and bandages were treated with phenol.

died of anthrax and found characteristic rod-shaped bacteria. When he isolated these bacteria and then injected them into healthy animals, the animals developed anthrax. Koch found bacteria in the blood of these animals as well. He isolated these bacteria and compared them with the bacteria he had isolated initially. They were identical. Koch thus confirmed that this organism, later named *Bacillus anthracis*, was the causative agent of anthrax.

On the basis of this work, Koch developed a protocol for demonstrating that a particular disease was caused by a specific microorganism. These investigative steps, called **Koch's postulates** (**Figure 6.23**), led to remarkable advances in the field of medical microbiology. Koch used his postulates to identify *Mycobacterium tuberculosis* as the causative agent of tuberculosis, and by 1900 the causes of many diseases, including typhus, diphtheria, and gonorrhea, had been identified. As the nineteenth century ended, the causative agent of bubonic plague was finally discovered as well. Alexander Yersin, a Swiss scientist, traveled to China to track down the source of this perennial scourge of humankind. Using pure culture techniques, he isolated bacteria from the inflamed lymph nodes or buboes of plague victims and then injected these bacteria into rats. Not only did the rats develop disease, but Yersin was able to isolate more of the same bacteria from the rats, satisfying Koch's postulates. This bacterial species was subsequently named *Yersinia pestis* in Yersin's honor. Koch's postulates are still valuable today and are still employed to track down the cause of apparently new diseases, a topic we will return to in Chapter 14.

The germ theory led to major advances in disease control

Once the link between microorganisms and specific diseases was firmly established, it became possible to develop ways to control these microorganisms. Such control was realized in various ways. Improved hospital care reduced the likelihood of infection. The pioneering work of individuals such as Semmelweis, Nightingale, and Lister was crucial in this regard. People also began to appreciate the need for clean food and water. Before the acceptance of the germ theory, drinking water was frequently contaminated with raw sewage (**Figure 6.24**). One could argue that safe drinking water and indoor plumbing have done more to rein in infectious disease than any other single innovation.

Nevertheless, clean food and water cannot control all diseases. In the late 1800s, evidence began to accumulate that insects and other arthropods could transmit certain diseases. In 1878 it was demonstrated that elephantiasis, caused by the accumulation of specific parasitic worms in lymph nodes, was transmitted through the bite of an infected mosquito. In 1892, Theobald Smith and Frederick Kilbourne determined that Texas red water fever, a serious disease of cattle, was caused by a protozoan parasite that was transmitted by ticks, and in 1896, Ronald Ross, an officer in the British Army, proved that malaria was mosquito-transmitted.

The demonstration that yellow fever was a mosquito-borne disease is particularly illustrative. Until the role of mosquitoes was clearly elucidated, control of the disease was essentially impossible. You will recall that the Cuban Carlos Finlay was the first to argue that yellow fever was transmitted via the bite of infected mosquitoes. In 1898, with the outbreak of the Spanish–American War, the US Army became especially interested in yellow fever. Consequently, the Yellow Fever Commission, headed by Walter Reed, went to Cuba to determine the source of this disease. To test Finlay's mosquito

Figure 6.22 Robert Koch (1843–1910). Along with Pasteur, Koch is regarded as one of the founding fathers of microbiology. He was the first to show that a bacterial species (*Bacillus anthracis*) could cause a specific disease (anthrax), and he was awarded the Nobel Prize for his work demonstrating that tuberculosis was caused by the bacterium *Mycobacterium tuberculosis*. The protocol that he developed to link a specific microbe with a particular disease, called Koch's postulates, is still used today.

transmission hypothesis, the commission carefully reared mosquitoes in the laboratory to rule out any previous exposure to humans. The mosquitoes were then allowed to feed on patients with yellow fever. Subsequently, these same mosquitoes were permitted to feed on human volunteers, including commission members James Carroll and Jesse Lazear. Both developed yellow fever, and while Carroll survived the ordeal, Lazear vomited blood, became delirious, and died 12 days following infection. Although Lazear paid with his life, the way was now paved for the control of yellow fever, which, as discussed previously, contributed to the construction of the Panama Canal.

Even though yellow fever could now be managed by controlling mosquitoes, it would still be a number of years before the exact infectious agent causing this disease would be discovered. Yellow fever is caused by the yellow fever virus. As discussed in Chapter 5, viruses are very small—too small, in fact, to be seen by even the best microscopes available at the start of the twentieth century. The first indication that there might be something smaller than bacteria that could cause disease came in 1892. The Russian Dmitri Ivanowski was trying to determine the cause of mosaic disease in tobacco plants. He found that whatever caused the disease was so small that it could pass through filters fine enough to prevent the passage of all known bacteria. When he injected healthy plants with the filtered fluid, they too developed mosaic disease. In 1898, a similar filterable agent was found to cause hoof-and-mouth disease in cattle. In the 1930s, it finally became clear that these were not merely unusually small bacteria but were a fundamentally different type of microbe. Wendell Stanley showed that these mysterious viruses ("virus" from the Latin for "poison") were so simple that they could be crystallized like a chemical compound. In the 1940s, the development of the electron microscope gave microbiologists their first look at viruses.

With the development of the first vaccines, the field of immunology was born

Until recently, smallpox was one of the great scourges of humanity. Yet over 1000 years ago, medical practitioners in China and India inoculated healthy people with material taken from smallpox pustules. Inoculated individuals generally developed only very mild cases of smallpox and they would be immune to future infection with the disfiguring and often lethal form of the disease. This practice was not without risk. Occasionally the inoculated person developed a full-blown case of virulent smallpox. The procedure, known as **variolation**, was unknown in the West until 1721. In that year, Lady Montagu, wife of the British Ambassador to the Ottoman Empire, observed the practice and became an enthusiastic supporter. She had her own children inoculated and introduced variolation to England.

Although variolation became popular in England, many people remained unprotected, and in the 1790s a smallpox epidemic erupted in western England. Edward Jenner, a young physician, was dispatched by the Royal Society of Medicine to the scene of the outbreak. In the course of his investigations, Jenner learned that the young girls who milked cows were unafraid of smallpox. Among these milkmaids it was common knowledge that

1. Suspected organisms are isolated from an infected animal and grown in pure culture.

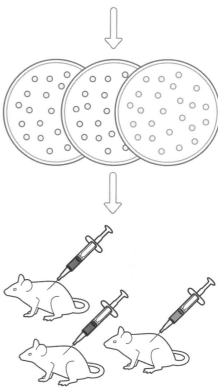

2. Each suspect organism is injected into healthy but susceptible animals.

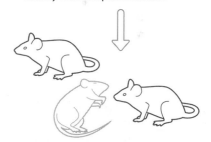

3. All experimental animals remain healthy except one, which sickens and dies.

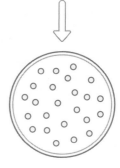

4. Organisms isolated from the dead experimental animal are confirmed as being one of the original suspect organisms.

Figure 6.23 A demonstration of Koch's postulates. Developed by Robert Koch, Koch's postulates are used to demonstrate that a specific microorganism causes a particular disease. All four postulates, numbered 1–4 in the figure, must be satisfied to prove a definitive link between the disease and the microbe.

although they developed a mild condition called cowpox on their hands, they rarely contracted smallpox. Jenner wondered if infection with cowpox somehow protected against the far more serious smallpox. In 1796 he inoculated fluid from a cowpox lesion into the skin of an 8-year-old boy (**Figure 6.25**). The boy developed cowpox as expected. He then inoculated the boy with fluid from a smallpox lesion. The boy remained healthy. Inducing immunity to smallpox in this way became known as vaccination. The word "vaccination" derives from the Latin word "vacca," meaning "cow."

Louis Pasteur is credited with the quotation, "chance favors the prepared mind," and he demonstrated this to good effect in the mid-1880s. Fresh from his assistance to the French wine industry, Pasteur was asked by agricultural officials in 1885 to investigate a serious disease in chickens. Using the postulates developed by Koch, Pasteur identified the bacteria causing fowl cholera. He then began work on a vaccine for the disease. At one point he mistakenly left some of the bacteria unattended while he went on a short holiday. When he returned to work, he inoculated chickens with these same bacterial cultures and found that while the chickens became ill, they did not die. He attributed this to the fact that the cultures were old. When he inoculated these same chickens with fresh bacteria, however, the birds developed no illness whatsoever. Pasteur correctly reasoned that the weakened bacteria stimulated resistance to a later challenge with the more virulent bacteria. Nowadays, microorganisms are often purposely weakened for use as vaccines against unweakened, still virulent strains. The process of weakening microbes is called **attenuation**. Pasteur had discovered the concept of an attenuated vaccine, in part because of his earlier error. He went on to develop similar attenuated vaccines against anthrax and rabies.

Antimicrobial drugs are a twentieth century innovation

Killing microorganisms is easy. The trick is to do so without harming the patient. Early attempts at antimicrobial therapy often did as much harm

Figure 6.24 London's water. A cartoon from an 1827 London newspaper describing the Thames River, the source of the city's water, as a "monster soup."

Figure 6.25 The first smallpox vaccine. (a) In this painting, Edward Jenner is inoculating a young boy with material from a smallpox lesion. The boy, who had earlier been inoculated with cowpox, failed to develop smallpox. (b) Cartoon by the British satirist James Gillray suggests the fear felt by many about the new technique of smallpox vaccination. The cartoon implies the vaccine, made from cowpox, will turn people into part-human and part-cow.

as good because substances were used that were toxic to both host and microbial cells.

In 1906 Paul Ehrlich, a German chemist, found that certain histological dyes would stain bacterial but not animal cells. Ehrlich reasoned that if these chemicals bound only to bacterial cells, there must be fundamental differences between the two cell types. Furthermore, he thought that if these dyes could be made in a toxic form, the dyes might bind to and kill bacterial cells while leaving animal cells unaffected. He thus began the search for what he called his "magic bullet"—a compound that would selectively kill only bacterial cells.

Ehrlich's quest for such a magic bullet bore fruit in 1908 when he discovered salvarsan, a compound that was effective against syphilis. Unfortunately, it would be almost 30 years before a second reliable antimicrobial became readily available. In 1935, researchers at the German company I.G. Farben discovered that the reddish dye prontosil could be used effectively to treat streptococcal infections. The chemist Gerhard Domagk played the leading role in the drug's development. Further work demonstrated that prontosil was converted in the human host to sulfanilamide, the first sulfa drug. There are now hundreds of sulfa derivatives, and they are still frequently used today.

Sulfa drugs are considered to be synthetic drugs because they are organic chemicals manufactured in a laboratory. **Antibiotics**, alternatively, are antibacterial compounds that are produced by microorganisms. The first true antibiotic was discovered in 1929. In another example of how mistakes can sometimes result in good fortune, Alexander Fleming saw that his *Staphylococcus* cultures had been contaminated with mold. As discussed in Chapter 1 (p. 10), instead of merely disposing of his apparently ruined cultures, Fleming noticed that the bacteria failed to grow in an area close to the contaminant. Subsequent analysis showed that the mold, *Penicillium notatum* (now called *Penicillium chrysogenum*), was releasing a compound that was toxic to the bacterial cells. Fleming called the substance penicillin. Although he understood that it might be useful for combating bacterial infections, penicillin proved difficult to purify in large quantities. It was not until the 1940s that production problems were overcome and penicillin became widely available for clinical use.

Into the modern era

In the last 75 years microbiology has continued to advance, sometimes at blinding speed. In the late 1940s and early 1950s, biologists began to understand such fundamental concepts as the structure of DNA, the mechanism of enzyme activity, and the basis of metabolism. As it became increasingly understood that these processes were essentially the same in all living things, scientists began to use microorganisms as **model organisms**. In other words, important concepts first discovered in microorganisms were found to have broad application in other living things as well. Microorganisms are especially well suited to serve as model organisms because they are relatively easy to rear in the lab and they reproduce so quickly. Consequently, large numbers necessary for study can be grown in a short time. Furthermore, as we saw in Chapter 3, their cell structure is simple compared with that of eukaryotes, making it easier to examine fundamental biological processes.

In 1944, DNA was established as the genetic material, and in 1946 it was shown that DNA could be transferred between bacterial cells, a process we will explore more fully in Chapter 7. In 1953, the American James Watson and the Englishman Francis Crick published their model for the structure of DNA. In the 1960s, we began to understand how proteins were produced and how the activity of DNA was regulated, again through the use of microbial model systems. As the 1970s began, it was demonstrated that pieces of animal DNA could be inserted into bacterial DNA, resulting in hybrid, **recombinant DNA**. The stage was now set for the ongoing molecular revolution—our ability not only to decode the genetic material but also to manipulate it. One of the most recent, potentially earth-shaking developments has been the discovery that prokaryotes have a unique type of immune system, protecting them against phage. This system, called CRISPR/Cas, not only protects bacteria, but as we will see in Chapter 15, has provided us with potentially game-changing technology to alter the genes of other living things, including humans, breathing new life into the field of gene therapy (see Table 6.2). Other ramifications of the molecular revolution, including the production of pharmaceuticals by bacteria, and the creation of pest-resistant agricultural crops, will likewise be investigated. We must remember, however, that went it comes to microorganisms, the future is far from uniformly rosy. In 2016 the National Academy of Medicine issued a report stating that future epidemics could kill millions and result in economic losses measured in trillions of dollars. Among the many reasons for this sober assessment is the problem of drug resistance. Ongoing research into new means to avoid such calamity will no doubt keep investigators busy for the foreseeable future.

And let's not forget the momentous changes that have occurred in our understanding of the how our normal bacteria affect our wellbeing. It has been understood for a while that our resident microbes, our microbiota, help to digest our food and to protect us from pathogens. Now it seems that much of who we are is dictated by the microbiota. They help to determine our body weight, our likelihood of developing allergies, and even our mood. We will have occasion to investigate these phenomena further in subsequent chapters, as we continue on our exploration of the microbial world.

Looking back and looking forward

By examining the often-unappreciated part that microorganisms have had on human affairs throughout our history, we achieve two objectives. First, we gain clearer perspective on some of the important biological factors that have impacted the arc of our history and how they have done so. Second, we gain added understanding of the organisms involved and how they influence us in many subtle and not-so-subtle ways. Furthermore, by examining the lively history of microbiology, we better grasp how this science developed and how we have arrived at our current understanding of microorganisms.

After our detour into the past, we return to the present. We have already commented on the remarkable diversity of microorganisms. In Chapter 7 we will explain the biological basis of that diversity. We also provide a scientific explanation for the common observation that offspring tend to resemble their parents. In other words, we turn our attention to genetics—the study of heredity. Although we will focus on microorganisms, we will find, as we recently learned in our discussion of model organisms, that basic genetic processes are similar in all living things.

Garland Science Learning System

- http://garlandscience.rocketmix.com/students/

- Discover how self-experimentation and a peptic ulcer won the Nobel Prize in Medicine

- Test your knowledge of this chapter by taking the quiz

- Familiarize yourself with the terminology used in this chapter by using the vocabulary review

- Get help with the answers to the Concept questions

Concept questions

These questions are designed to help you start thinking like a microbiologist. The answers are not always simply found in the text. Instead, you will need to take the concepts about which you have learned and apply them to new situations. Some of the questions may not even have just a single correct answer. Help is provided as part of the GSLS resources, which can be accessed through http://garlandscience.rocketmix.com/students/.

1. Alternative history has been quite a popular genre for fiction writers (for example, a fictional history in which the Roman Empire had persisted until modern times). Suppose you wished to write an alternative fiction novel in which the Aztec Empire had developed ocean-going ships and used them to invade Europe in the early 1500s, instead of Europeans having invaded the Americas. Provide a short, one-paragraph summary of your novel's plot.

2. What is the difference between an acute and a chronic disease? Why are small isolated populations at less risk of acute disease, unless such diseases are introduced from the outside?

3. In writing to his wife on November 29, 1812, Marshal Ney, top commander of Napoleon's army, said, "General Famine and General Winter, not Russian bullets, have conquered the Grand Army." Why should he have included "General Louse" and "General Typhus"?

4. During the French attempt to construct the Panama Canal, bedposts were often placed in small bowls of water to prevent ants from crawling up the bedposts and into the beds. How might this practice have contributed to the French failure to complete the canal?

5. Suppose you were working in Vienna's Central Hospital, along with Ignaz Semmelweis, in the mid-1800s. Semmelweis shares with you his ideas regarding the cause of childbed fever. He asks you to conduct an experiment in the hospital to see if he is correct. How would you proceed?

6. Here in Chapter 6, we learned about the experiments of Redi and Pasteur that disproved the notion of spontaneous generation. Design another experiment that would likewise support the conclusions of Redi and Pasteur—that life can come only from preexisting life.

7. One of the necessary conditions, when using Koch's postulates to demonstrate that a particular microorganism causes a specific disease, is that it must be possible to grow the microorganism in pure culture. Why is this?

8. What makes microorganisms especially useful as model organisms?

9. Why is the late nineteenth century often referred to as the Golden Age of microbiology?

10. By far, the most commonly sold type of banana in the world is the Cavendish banana. In fact, in 2016 the BBC stated that "practically every banana consumed in the western world is directly descended from a plant grown in a Derbyshire hothouse 180 years ago." Currently, Cavendish bananas are being devastated by a fungal disease called black leaf streak disease. The problem is so bad that Cavendish bananas may be commercially unviable within a decade or two. In light of the discussion of the Irish potato famine here in Chapter 6, why do you think this is such a problem for Cavendish bananas, while in most cases, plant diseases do not lead to such eradication of the affected plants?

MICROBIAL GENETICS

In the spring of 2015 came the disturbing news that a woman infected with a highly drug-resistant strain of tuberculosis boarded a plane in India in April and flew to Chicago. Once there, she traveled by car to visit relatives in three states before seeking medical attention in May. It was unknown how many people were exposed to this especially dangerous strain of *Mycobacterium tuberculosis*, and health officials began the laborious task of trying to track down all of her contacts for TB testing. Amid all the anxiety and media coverage, a fundamental question remained. Some strains of *M. tuberculosis* are easily treatable. Others are not. Why, if they are all members of the same species, do they differ in this way?

The answer to this question comes down to genetics, the study of biological inheritance. Most of the variation that we observe between organisms and most of the similarities between parents and offspring can be explained in terms of differences and similarities in their genetic material. With the exception of the RNA viruses discussed in Chapter 5, the genetic material is deoxyribonucleic acid (DNA).

Such variation is taken for granted by just about everyone in familiar species like dogs or cats. Nobody, for instance, is likely to confuse a poodle and a Pekingese, even though they are both easily recognizable as dogs. Microbes are no less variable. Why do some strains of *Escherichia coli* cause serious disease, while most others live peacefully in your intestines without causing problems? Some bacteria have extracellular structures, such as capsules, while others of the same species do not. What explains this difference? Why do all living things, microbial or otherwise, tend to resemble their parents or siblings more than they do unrelated individuals? Here in Chapter 7, we aim to find out.

We will begin by examining the structure and function of DNA. We will then learn how this most remarkable of molecules influences the characteristics we observe in any organism and how it can change over time.

DNA is an information molecule

DNA is found in the cells of all living things, and all DNA does the same thing—it determines which proteins a particular cell can make. In other words, DNA, as an information molecule, encodes the information necessary to produce various proteins.

In Chapter 2 (pp. 38–39), we began to understand how DNA accomplishes this task. DNA is composed of individual building blocks called **nucleotides**. Part of each nucleotide is composed of a nitrogen-containing base, of which

All case studies have a few questions at the end, the answers to which will become apparent as you read the sections following the case.

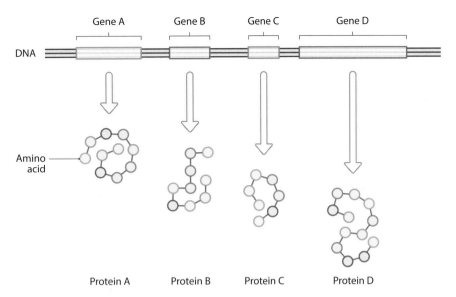

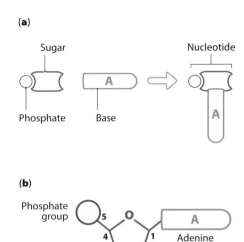

Figure 7.1 **Genes and proteins.** Genes code for proteins. Each gene contains the genetic information, encoded in the sequence of nucleotides, for the production of a protein. The nucleotides "spell out" which of the 20 amino acids (indicated by the colored circles in the protein) is required at a particular point in the protein. A small protein may consist of a few dozen amino acids. A large protein may be composed of 1000 or more amino acids.

there are four types: adenine, thymine, guanine, and cytosine. Consequently, there are four types of nucleotides. A strand of DNA may be composed of thousands or millions of nucleotides and nucleotides with different bases may be aligned in any possible order. Much of an organism's DNA is organized into **genes** (**Figure 7.1**). A gene is a sequence of DNA nucleotides that codes for a single protein. In any gene, the sequence of nucleotide bases specifies the exact amino acid sequence needed to make a specific protein (see Figure 2.29). The nucleotides therefore represent a code, utilized for protein production.

One way to think about this is to picture the four types of DNA nucleotides as letters in a genetic alphabet. These letters spell out words, each of which represents one of the 20 amino acids found in proteins. If, for example, we know that a particular protein consists of 300 amino acids (a fairly typical size), we know that the gene that codes for this protein must contain 300 genetic words, each composed of DNA nucleotides. A single bacterial cell may have several thousand genes, indicating that it can produce several thousand proteins.

Although DNA is composed of genes and contains the exact instructions for protein production, DNA does not actually link the amino acids together to form the protein. Remember that DNA is an information storage molecule; it is like a cookbook with thousands of recipes for different proteins. It is not the cook. As we will see, other molecules are responsible for reading these protein recipes and assembling the specified amino acids, as per DNA's instructions.

The structure of DNA is the key to how it functions

To fully appreciate how DNA serves as the genetic material, we first need to become familiar with its unique structure. To begin, let's take a closer look at an individual nucleotide (**Figure 7.2a**). Each nucleotide consists of three

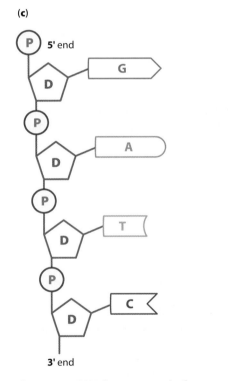

Figure 7.2 **DNA is composed of nucleotide building blocks.** (a) Simplified diagram of an individual nucleotide, in this case, A. All nucleotides contain identical phosphates and sugars, but the base may be adenine, thymine, guanine, or cytosine. (b) A more detailed diagram, depicting the same nucleotide. Note that the carbons on the deoxyribose sugar are numbered 1–5. The base is always attached to carbon number 1, and the phosphate is always attached to carbon number 5. (c) A schematic diagram showing a short length of one DNA strand. Note that one end of a DNA strand, the one with a free number 5 carbon, is called the 5′ end. The end with a free number 3 carbon is called the 3′ end.

parts: a five-carbon sugar called deoxyribose, a phosphate, and a nitrogen-containing base. The sugar and the phosphate are identical in all nucleotides. Differences in the associated base make a nucleotide either A, T, G, or C.

Notice in **Figure 7.2b** that each of the five carbons in the deoxyribose sugar has a specific number. The phosphate is always attached to carbon number 5, and the base (adenine, thymine, guanine, or cytosine) is always attached to carbon number 1. In a strand of DNA, nucleotides are linked together into long chains (**Figure 7.2c**). The phosphate on each nucleotide (attached to carbon 5) is linked to carbon 3 on the adjacent nucleotide. The end of the DNA molecule on which a phosphate is found is termed the 5′ (5-prime) end. The other end, with an unlinked carbon number 3, is called the 3′ (3-prime) end. A one-letter code is used to describe the nucleotides in the DNA strand. In Figure 7.2c, the code in this short DNA strand is written G-A-T-C (the code is always read from the 5′ end). The larger **purines**, adenine and guanine, are double-ringed structures, while thymine and cytosine are smaller, single-ringed structures called **pyrimidines** (see Figure 2.27b).

We have now examined the structure of a DNA strand, but a complete molecule of DNA is usually composed of two strands. In other words, DNA is a double-stranded molecule (**Figure 7.3**). The sugars and phosphates on each strand form a backbone with the bases pointing inward. The two strands are held together by hydrogen bonds that form between bases on opposing strands. The shape and chemical properties of the bases are such that the base pairing between opposite strands is highly specific: T pairs with A, and C pairs with G. You might think of the bases as pieces of a jigsaw puzzle to help you visualize this point. The structure of A is such that it typically can form hydrogen bonds only with T. Likewise, the specific shape and chemical properties of G and C ensure that they hydrogen-bond only with each other. Notice that, as Figure 7.3 shows, a T-A pair is held together by two hydrogen bonds, whereas a C-G pair forms three hydrogen bonds. Furthermore, the two strands are positioned opposite each other in an inverted orientation, known as **antiparallel**—that is, one strand runs in the 5′ to 3′ direction, while the opposing strand is in the 3′ to 5′ direction. The two DNA strands wind around each other in a spiral configuration. Because there are two such strands and because a spiral configuration is known as a helix, DNA is said to adopt a structure known as a **double helix**.

If the sequence of nucleotides on one of the two DNA strands is known, the sequence on the opposite strand can instantly be inferred because of the specificity of base pairing. As we will see, this observation about DNA's structure provided a powerful clue as to how the genetic material is replicated.

DNA is found on chromosomes

Storing DNA within a cell is no trivial matter. The DNA of a single bacterium, for example, if stretched out, is approximately 1000 times longer than the cell itself (**Figure 7.4a**). Such a large molecule cannot be dispersed at random throughout the cell. Rather, it must be neatly packaged in a manner that allows it not only to fit within the cell but also to carry out its functions. In both prokaryotic and eukaryotic cells, DNA is organized into discrete units called **chromosomes**. A chromosome consists of both DNA and associated proteins. These proteins provide a scaffolding around which the DNA coils (**Figure 7.4b**). Such packaging allows the DNA to be condensed many times. In a typical bacterial cell such as *E. coli*, for instance, the condensed DNA takes up only about 10% of the cell's total volume.

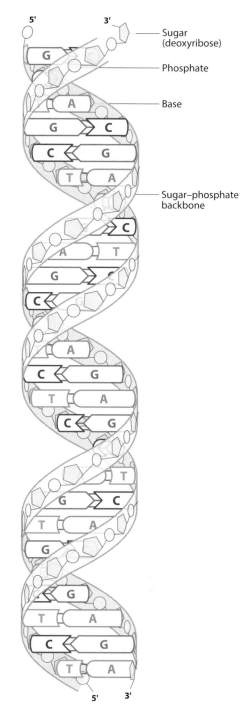

Sugar (deoxyribose)

Phosphate

Base

Sugar–phosphate backbone

Figure 7.3 The DNA double helix. A DNA molecule is composed of two DNA strands held together by hydrogen bonds (in red) between the paired bases. The 5′ and 3′ ends of each strand are indicated. The two complementary strands are antiparallel to each other. The pairing of the nucleotide bases is specific, such that adenine bonds with thymine and guanine bonds with cytosine. Note that there are two hydrogen bonds between an A-T pair and three between a G-C pair.

(a) **(b)**

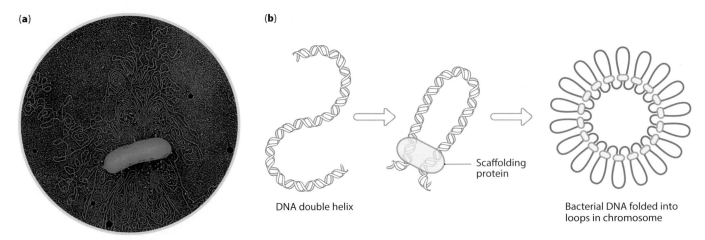

Scaffolding protein

DNA double helix

Bacterial DNA folded into loops in chromosome

Figure 7.4 **The prokaryotic chromosome.** (a) In this photograph, DNA, seen here as a mass of squiggly lines, is emerging from a ruptured bacterial cell. (b) DNA wraps around proteins, forming loops, which help to compact DNA within the cell.

In prokaryotic cells, which lack a membrane-bound nucleus, the chromosome is found in the cell's cytoplasm. In eukaryotic cells, chromosomes are located within the nucleus. Prokaryotic and eukaryotic chromosomes differ in several other notable ways. First, a prokaryote's DNA is usually circular (**Figure 7.5a**). This circular DNA, associated with its scaffolding proteins, forms the circular prokaryotic chromosome. In eukaryotes, linear DNA and its associated scaffolding proteins form linear chromosomes (**Figure 7.5b**).

Second, although there are exceptions, a prokaryotic cell generally has one chromosome. This means that all of a prokaryote's genes are found on this single chromosome. In contrast, the DNA of eukaryotes is organized into a number of chromosomes. The exact number is species-specific. Human cells contain 46 chromosomes. The cells of fruit flies, horses, and onions contain 8, 64, and 16 chromosomes, respectively. Eukaryotic chromosomes often come in pairs, so the 46 chromosomes of humans consist of 23 pairs. Each member of the pair carries the same genes, meaning that a eukaryotic cell usually has two copies of each gene: one on each chromosome in the pair. Prokaryotes, with only one chromosome, usually have only a single copy of each gene. There are, however, examples of bacteria with multiple copies of a particular gene on the same chromosome

Each DNA molecule can be accurately copied into two new DNA molecules

CASE: DNA: FORM SUGGESTS FUNCTION

The short, 900-word paper published in the British journal *Nature* by James Watson and Francis Crick on the structure of DNA in April 1953 is one of the most important publications in the history of biology. With the molecule's structure now elucidated, it became clear how DNA was able to code for characteristics and how genetic information was transmitted from generation to generation. The structure of DNA also provided important clues as to how it could be duplicated prior to cell division. Indeed, in a masterpiece of understatement, Watson and Crick concluded their paper with the following sentence: "It has not escaped our notice that the specific pairing we have postulated immediately suggests a possible copying mechanism for the genetic material."

1. **What did Watson and Crick mean by this statement?**

(a)

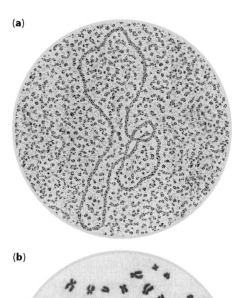

(b)

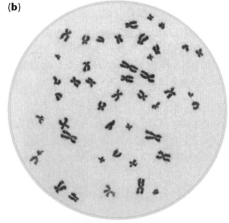

Figure 7.5 **Organization of eukaryotic and prokaryotic DNA.** (a) Prokaryotic DNA is organized into a circular chromosome. This photograph, taken through an electron microscope, shows a bacterial plasmid (see Chapter 3, pp. 56–57), represented as a long, twisted loop. A bacterial plasmid such as this may contain fewer than a dozen genes. Larger bacterial chromosomes, also circular, typically carry a few thousand genes. (b) Eukaryotic DNA is organized into linear chromosomes. The human chromosomes seen here are an example.

DNA replication is the process in which a DNA molecule is copied into two identical DNA molecules. Such replication is critical. When cells divide, each newly formed cell must receive a complete set of genetic information if it is to produce proteins and carry out its various functions. As the quote in our case suggests, the secret to DNA replication lies in the specific base pairing of the double helix.

Each of the two original, **parental DNA** strands is used as a template for the synthesis of a new, complementary **daughter DNA** strand (**Figure 7.6**). In other words, the base sequence on each parental strand dictates the complementary sequence on the newly synthesized daughter strands. If the specific base on the parental strand is guanine, then, as the growing daughter strand is synthesized, a nucleotide containing cytosine will be added opposite this guanine. If the next base on the parental strand is a thymine, the next nucleotide added to the daughter strand will contain an adenine. If a short section of the parent strand consists of the nucleotides TGATCAGTGGTA, the specified nucleotides on the daughter strand will be the complementary nucleotides ACTAGTCACCAT. This continues until every nucleotide on the parental strands has been successfully paired and replication is complete. There are now two complete molecules of DNA, each consisting of one original parent strand and one newly synthesized daughter strand. This is exactly what Watson and Crick alluded to in our case's quotation.

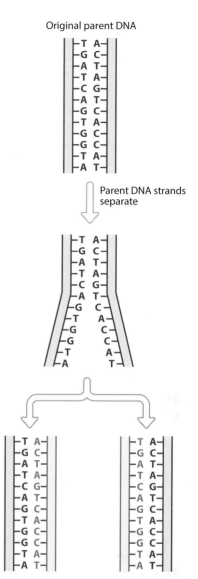

Two new molecules of daughter DNA result, each consisting of one of the parent strands and one newly synthesized daughter strand

Figure 7.6 DNA replication. The two strands of a DNA molecule wind around each other in a double helix, but are shown as two straight lines in this diagram for simplicity. The sequence found on the parental strand acts as a template for the synthesis of a new daughter strand. Each of these two, identical molecules consists of one original parent strand (in gray) and one newly synthesized daughter strand (in red).

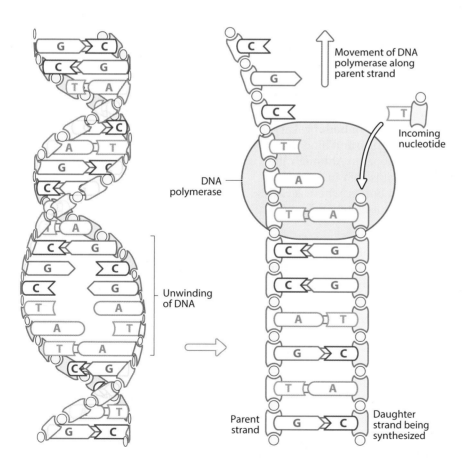

Figure 7.7 DNA replication. DNA replication begins when the DNA double helix unwinds and the bases separate. As the strands separate, DNA polymerase moves along the resulting single parental strands (in gray), using its nucleotide sequence as a template for the synthesis of the daughter strands (in red). The diagram shows the process as it takes place along one parental strand; a similar process occurs simultaneously opposite the other parental strand.

DNA replication is mediated by more than a dozen proteins and enzymes, one of which is **DNA polymerase**. During replication, the two strands of the parental DNA double helix separate from each other and unwind. As the two parental strands are separating, DNA polymerase moves along them toward the progressing separation point, reading the nucleotide sequence on the parental strand and adding new nucleotides to the growing daughter strands (**Figure 7.7**).

Exact points where nucleotides are added to the growing daughter strands are called **replication forks** (**Figure 7.8**). As the parental DNA continues to unwind, nucleotides are added individually at a progressing replication fork. On one of the parent strands, nucleotides are added one after another, until the daughter strand is complete. Along the other parent strand, the nucleotides are added until they form a short fragment of nucleotides. A series of these fragments are joined together at a later time. Each new DNA molecule will consist of one strand of original parental DNA and one strand of new daughter DNA (see Figure 7.6).

DNA polymerase is very accurate but occasionally it makes a mistake. For example, if a base on the parental strand is cytosine, DNA polymerase might erroneously add an opposing adenine instead of guanine. Yet the truly remarkable thing about DNA polymerase is how rarely such mistakes occur; the DNA might be copied perfectly thousands or millions of times before a single error takes place. This level of accuracy is an important feature of DNA replication. As we will discuss later in this chapter, even when it makes a mistake, DNA polymerase can often detect the error and correct it. Between its high precision and its capacity to correct many of the mistakes that sneak though, DNA replication is accurate indeed.

DNA replication begins at a site called the origin

The **origin** is the point on the DNA where the two parental strands first separate and DNA replication begins (**Figure 7.9**). Replication in both prokaryotes and eukaryotes begins when specific proteins recognize and bind to the origin. The origin typically consists of about 250 base pairs. Although the exact sequence can vary, it is generally a location that is rich in A-T pairs. Recall that A and T are held together by two hydrogen bonds, whereas G-C pairs are maintained by three such bonds. The fewer hydrogen bonds in A-T pairs enable strand separation to occur more easily.

In prokaryotes, the DNA is generally a circular molecule. The parental strands separate at a single origin (**Figure 7.10a**), and DNA polymerase binds at each of two replication forks moving in opposite directions around the circular, parental DNA. Because the DNA is a closed loop, the two forks ultimately meet when DNA replication is completed. When the original cell divides into two new cells, each of these newly formed cells obtains one of the newly synthesized, circular DNA molecules (**Figure 7.10b**). In eukaryotes, alternatively, DNA is linear. Moreover, eukaryotic DNA has multiple origins (**Figure 7.11**). This means that replication is occurring simultaneously at several points. Two replication forks form at each of these origins and move in opposite directions.

A cell's characteristics are largely determined by the proteins it produces

What precisely is the connection between a series of nucleotides and a characteristic such as your cat's black fur or whether or not a strain of *Staphylococcus aureus* is resistant to penicillin? How exactly does a specific

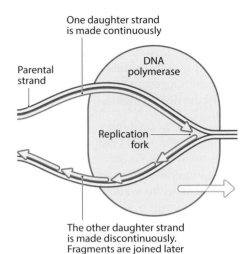

Figure 7.8 A replication fork. A precise point where nucleotides are added to the growing daughter strands (in red) by DNA polymerase is called a replication fork. One daughter strand is synthesized continuously into a single long DNA molecule, while the other daughter strand is synthesized in a series of short fragments, which are later joined together. The replication fork continues to move forward as the parental DNA (in gray) unwinds ahead of the DNA polymerase. The open arrow shows the direction in which DNA polymerase moves.

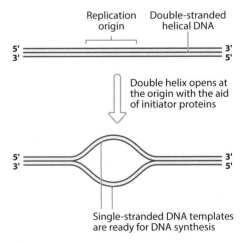

Figure 7.9 Replication starts at an origin. Specific proteins can recognize the base sequence at the origin and separate the parental DNA strands at that point. The now single-stranded parental DNA can serve as a template for synthesis of the daughter strands.

Figure 7.10 DNA replication and cell division in prokaryotes. (a) In prokaryotes, the circular DNA double helix is opened at a single origin (highlighted in green). Two replication forks are formed and move in opposite directions. DNA polymerase binds at each replication fork, producing the daughter strands (in red). When the two replication forks meet, DNA replication terminates, and the newly formed DNA molecules separate. (b) Following DNA replication, the parental cell divides and each new daughter cell receives one complete DNA molecule.

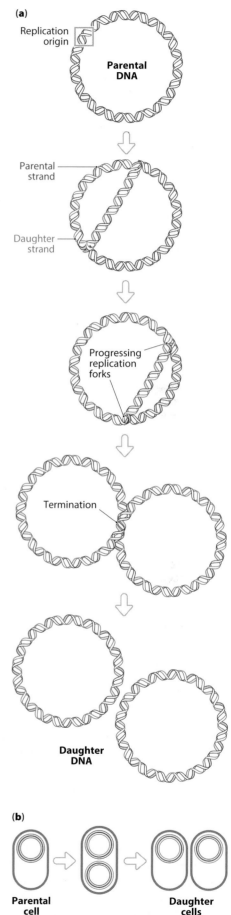

gene influence a particular characteristic? We will start to answer these questions by describing the actual manner in which DNA's code is interpreted by cells.

It may be surprising to learn that DNA, as an information molecule, primarily codes for proteins and that a fundamental difference between different organisms is simply the proteins that they make and when they make them. After all, cells also contain many nonprotein components and these also vary in different organisms. Recall from Chapter 2, however, that many proteins are enzymes, and enzymes catalyse biological reactions, including the synthesis of molecules such as carbohydrates and lipids. Although DNA does not code for specific nonproteins directly, it does code for the enzymes that synthesize these molecules. In this way, a cell's DNA determines all of the processes that the cell can carry out.

As we have recently learned, each protein produced by a cell is encoded in a gene within that cell's DNA. Furthermore, we have seen that a gene is a linear sequence of nucleotides, and the sequence of bases in these nucleotides encodes the information necessary to produce the specified protein (see Figure 7.1). Consequently, genes determine exactly what proteins a cell can synthesize, and proteins are ultimately responsible for the characteristics of a cell. The sum total of an organism's genes is called its **genome**. The genome of a bacterial species may consist of 4000–5000 genes, whereas the human genome is composed of 20,000–25,000 genes. Because each gene

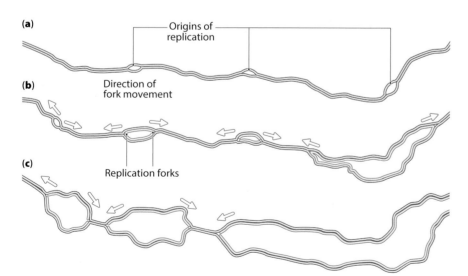

Figure 7.11 Multiple origins in eukaryotic DNA. In eukaryotes there are multiple origins along the length of the linear DNA molecule. As the DNA strands separate at each origin, two replication forks are formed. The replication forks at each origin move in opposite directions, until the entire parental DNA molecule is replicated.

codes for a single protein, geneticists regard the gene as the basic unit of heredity.

What exactly does it mean when we say that a gene codes for a protein? Recall that although DNA contains information specifying protein structure, DNA itself does not link the required amino acids together. Rather, a gene's encrypted instructions must be read by other molecules and ultimately translated into the series of amino acids that form the specified protein. The conversion of a gene's nucleotide sequence into a protein is called **gene expression**.

Gene expression consists of two basic steps, the first of which is called **transcription**. In this step, the DNA in a specific gene is used to guide the synthesis of RNA (see Chapter 2, p. 39). The genetic information in the gene has now been encoded in the molecule of RNA. In the second step, called **translation**, the DNA code that has been encoded or transcribed into the RNA is used to link the specified amino acids in the proper order to form a protein (**Figure 7.12**). Like DNA replication, protein synthesis is similar in all cells, but we will highlight a few of the important differences in the ways that bacteria, archaea, and eukaryotes transcribe and translate DNA's genetic blueprint.

In transcription, a gene's DNA is used to produce complementary RNA

In the first step of protein synthesis, transcription, the genetic information encoded in the nucleotides of a gene is transferred to RNA through complementary base pairing. Briefly, a specific enzyme, RNA polymerase, recognizes the beginning of a gene, and once it binds at this site, the enzyme unwinds the DNA sequence to be transcribed. It next assembles complementary RNA nucleotides into a molecule of messenger RNA (mRNA; **Figure 7.13**). If a short series of nucleotides in the DNA is GCGCT, for example, the corresponding nucleotides strung together by RNA polymerase will be CGCGA. When RNA polymerase reaches the end of the gene, indicated by a special termination site, the RNA polymerase detaches and the newly synthesized messenger RNA is released.

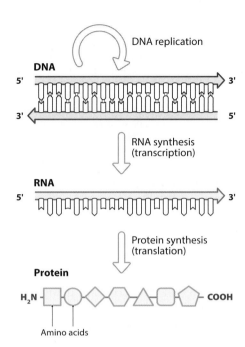

Figure 7.12 Gene expression. In addition to its ability to replicate, DNA encodes genetic information. The flow of genetic information from DNA to RNA (transcription) and from RNA to protein (translation) occurs in all living cells. This flow of information is so critical to our understanding of life that it is often called the central dogma of genetics.

CASE: CHRISTMAS TREES AND TRANSCRIPTION

If you take electron-microscopic images of DNA in cells that are actively producing protein, you will likely see structures that look very much like Christmas trees. There are central fibers corresponding to the "trunk" and attached "branches," which are short at the top and longer at the bottom of the "tree." If such cells are treated with an enzyme that digests DNA but not RNA (that is, a DNase), the trunks of the trees disappear, leaving only the branches. If similar cells are treated with an enzyme that digests RNA only (that is, an RNase), the trunks remain but the branches disappear.

1. **What is the nature of these "Christmas trees"?**
2. **What is the explanation for the results observed following treatment with DNase or RNase?**

Like DNA, RNA consists of a series of nucleotides (**Figure 7.14**). Unlike DNA, however, which is a double-stranded molecule, RNA is typically single-stranded, and in place of the deoxyribose sugar that is an essential part of DNA, a slightly different sugar, ribose, is found in RNA. Furthermore, RNA does not contain the base thymine. In its place, RNA uses a different

Figure 7.13 An overview of transcription. RNA polymerase recognizes the beginning of a gene, and once it binds at this site, it separates the DNA strands. RNA polymerase then moves along the gene, separating the DNA strands as it goes. Using the nucleotide sequence of one of the DNA strands as a template, RNA polymerase synthesizes a complementary molecule of RNA, using available RNA nucleotides. As RNA polymerase moves down the length of the gene, the DNA double helix closes behind it. When RNA polymerase reaches the termination site, it releases the RNA and detaches from the DNA, allowing the DNA to regain its original double helix structure.

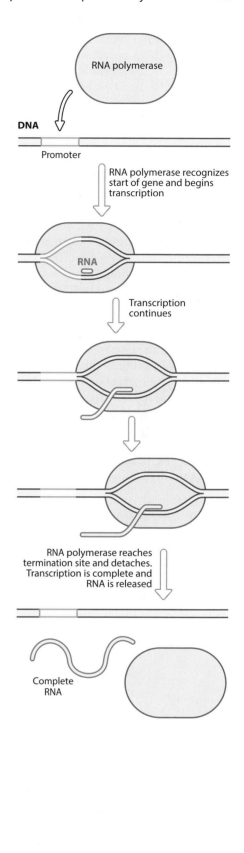

single-ringed pyrimidine called uracil. The other three bases found in DNA (guanine, cytosine, and adenine) are also found in RNA.

During the first step in protein synthesis, transcription, the code in DNA guides the synthesis of a complementary sequence of RNA nucleotides. For example, if a short segment of DNA consists of the bases GGCATC, the complementary bases on the newly synthesized RNA will be CCGUAG. Note that adenine in the DNA is transcribed into uracil in the RNA instead of thymine. However, not all the DNA in an entire DNA molecule is transcribed at once. Genes are transcribed either individually or in small groups as needed, meaning that each RNA molecule is transcribed from only a small portion of the entire DNA molecule.

In bacteria, transcription of a particular gene begins when a specific short base sequence found near the beginning of a gene is recognized and bound by an enzyme called **RNA polymerase**. This base sequence, called a **promoter**, essentially acts as a "start" signal (**Figure 7.15**).

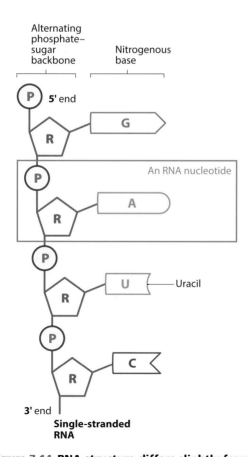

Figure 7.14 RNA structure differs slightly from that of DNA. RNA contains the sugar ribose and the base uracil in place of thymine. It is typically single-stranded.

A given promoter typically controls the transcription of only one or a few genes. Once RNA polymerase binds the promoter, it unwinds the DNA to expose the DNA nucleotide bases. Incoming RNA nucleotides that are complementary to the DNA are added, forming a growing RNA chain. In this way, the sequence of bases in the RNA is dictated by the base sequence in the DNA. Transcription continues until RNA polymerase reaches a series of DNA nucleotides called the **termination sequence**. At this point the RNA is released and RNA polymerase detaches from the DNA, allowing the DNA double helix to re-form. Because only a few genes at most are transcribed into one RNA molecule, RNA is far shorter than DNA.

For a particular gene, only one DNA strand (the **coding strand**) is transcribed (**Figure 7.16**). How does RNA polymerase distinguish between the coding and the opposite, noncoding strand? Recall that transcription begins at a specific promoter region. These promoters are found only on the coding strand. The noncoding strand, with a sequence complementary to the promoter, is not recognized by RNA polymerase.

The "Christmas trees" described in our case are actually genes caught in the act of transcription (**Figure 7.17**). Each gene is being transcribed by many RNA polymerase enzymes at once. Some of the RNA polymerases are almost finished transcribing the gene and are approaching the termination sequence. Their attached RNA molecules represent the longer branches at the bottom of the tree. At the top of the tree, other RNA polymerases are just starting out at the beginning of the gene. Since they have not yet transcribed many nucleotides, their attached RNA molecules are shorter and represent the shorter branches. DNase digests the DNA trunk of our trees, leaving only the branches. RNase has the opposite effect; the trunk remains, stripped of its RNA branches.

While very similar, transcription in eukaryotes varies from that of bacteria

Transcription is much the same for eukaryotes. One important difference, however, is the manner in which a gene's promoter is recognized. In bacteria, recall that RNA polymerase recognizes and binds to promoter regions to initiate transcription. In eukaryotes, promoters are first bound by a complex of proteins called **transcription factors**. RNA polymerase recognizes and binds to these proteins rather than directly to the promoter. Interestingly,

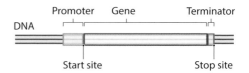

Figure 7.15 Gene organization. A length of DNA showing the respective locations of a promoter, a gene, and a termination site.

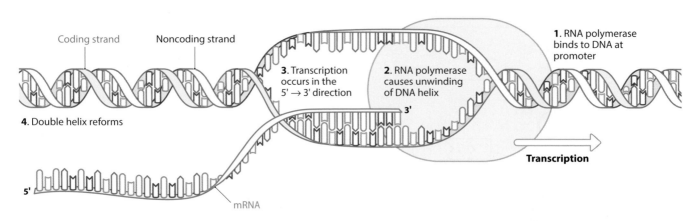

Figure 7.16. Transcription. Transcription begins when RNA polymerase binds the promoter of a gene. RNA polymerase unwinds the DNA and, using the DNA as a template, moves along the DNA, adding the appropriate, complementary RNA nucleotides. As transcription continues, the DNA double helix re-forms behind the RNA polymerase. Only one of the two DNA strands (the coding strand) is transcribed by RNA polymerase.

archaea also use transcription factors and are consequently more like eukaryotes than they are like bacteria in this regard.

Many eukaryotic genes are interrupted by intervening noncoding sequences called **introns** (**Figure 7.18**). The coding regions that are actually translated into a protein are called **exons**. The number of introns found in eukaryotic genes ranges from zero to close to 100. Introns are less common in unicellular eukaryotes and are even rarer in bacteria. Some genes in archaea also contain introns.

To help visualize this concept, consider the sentence "THE RAT SAW THE CAT." If this sentence were equivalent to a genetic message in bacterial DNA, the same message in a eukaryote might read "THE XQXY RAT SAW VRVNGT THE CAT." The nonsense letters represent two introns, interspersed between three exons. Both introns and exons are initially transcribed, but following eukaryotic transcription, introns are removed from mRNA in a process called **RNA splicing**. The function of introns is still under investigation, but they may participate in regulating the rate of protein synthesis.

Amino acids are assembled into protein during translation

In the second step of protein synthesis, translation, the mRNA formed during transcription is used to specify the amino acids that are linked together to form a protein. Briefly, a structure called a ribosome binds at the end of the mRNA molecule (**Figure 7.19a**). The ribosome moves along the mRNA, reading the nucleotides and translating the information contained in the nucleotide sequence into a series of amino acids. Each set of three nucleotides specifies a particular amino acid (**Figure 7.19b**). For example, if the first three nucleotides read by the ribosome contain the bases AUG, a particular amino acid called methionine will be the first amino acid in the protein. If the next three nucleotide bases are CGU, a different amino acid, arginine, will next be linked to the growing protein chain (**Figure 7.19c**). This process continues as the ribosome progresses along the mRNA, until it reaches a three-base sequence at the end of the mRNA that does not specify any amino acid. At this point, the ribosome detaches from the mRNA, the newly completed protein is released, and protein synthesis is complete (**Figure 7.19d**).

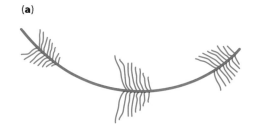

(a)

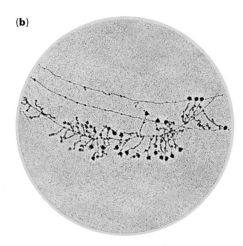

(b)

Figure 7.17 Treelike appearance of genes being transcribed. (a) The "branches" (in blue) are RNA molecules being transcribed from the DNA (in red). Shorter branches indicate an earlier point in the transcription of each of the three genes illustrated here. Longer branches, representing longer RNA molecules, are nearing completion of transcription. (b) An actual electron-microscopic image of transcription "Christmas trees." Experiments similar to the one described in the case, conducted in 1970, offered powerful clues regarding RNA's role in gene expression.

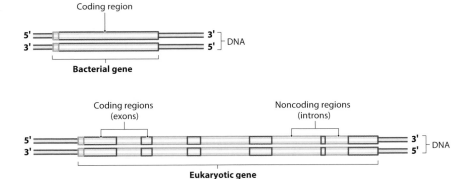

Figure 7.18 Bacterial and eukaryotic genes are organized differently. Bacterial genes usually consist of a single stretch of uninterrupted nucleotides that encodes the amino acid sequence of a protein. In contrast, many eukaryotic genes are interrupted by noncoding sequences called introns. The nucleotide sequences in DNA that are ultimately translated into a protein are called exons.

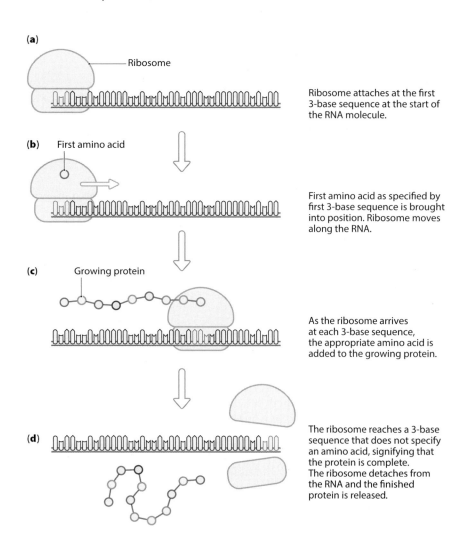

(a)

Ribosome

Ribosome attaches at the first 3-base sequence at the start of the RNA molecule.

(b) First amino acid

First amino acid as specified by first 3-base sequence is brought into position. Ribosome moves along the RNA.

(c) Growing protein

As the ribosome arrives at each 3-base sequence, the appropriate amino acid is added to the growing protein.

(d)

The ribosome reaches a 3-base sequence that does not specify an amino acid, signifying that the protein is complete. The ribosome detaches from the RNA and the finished protein is released.

Figure 7.19 An overview of translation. (a) Translation begins when a ribosome binds to the beginning of the RNA nucleotide sequence. (b) Each set of three nucleotide bases on the RNA specifies a particular amino acid; the first amino acid is brought into position according to the first three-base sequence. (c) The ribosome moves along the RNA molecule reading the nucleotides, three at a time, and adding the amino acids in the order specified by RNA. (d) Finally, the ribosome reaches a three-base sequence that does not specify an amino acid, indicating that the protein is complete. At this point, the ribosome disassociates from the RNA, and the now-completed protein is released.

CASE: PROTEIN SYNTHESIS: DEAD IN ITS TRACKS

When she begins to experience vague symptoms that include coughing, tightness in the chest, and fatigue, Leslie starts taking some amoxicillin that is left over from an old prescription. After a week she has not improved, so she visits the doctor who, following an examination, prescribes a different antibiotic—erythromycin. The doctor tells Leslie that he suspects that she has a case of atypical or "walking pneumonia," caused by the bacterium *Mycoplasma pneumoniae*. He also tells her that the amoxicillin that she took has no effect on this bacterial species and he uses the incident to discourage her from such self-treatment in the future. Within a few days of starting erythromycin treatment, Leslie feels much better. By the end of the 10-day prescription, her symptoms have completely resolved.

1. **How did the erythromycin cure Leslie?**
2. **Why didn't the amoxicillin work?**

Cells produce not just one, but a number of different RNA types, several of which are needed to translate RNA into protein. The RNA that encodes DNA's specific instructions for the production of a protein is called **messenger RNA (mRNA)**. As the name implies, mRNA carries DNA's message to the site in the cell where amino acids are actually assembled into protein.

The mRNA sequence of bases specifies the precise amino acids that make up a protein. These bases are recognized in groups of three, called **codons**, and

Ala	Arg	Asp	Asn	Cys	Glu	Gln	Gly	His	Ile	Leu	Lys	Met	Phe	Pro	Ser	Thr	Trp	Tyr	Val	Stop
	AGA									UUA					AGC					
	AGG									UUG					AGU					
GCA	CGA						GGA			CUA				CCA	UCA	ACA			GUA	
GCC	CGC						GGC		AUA	CUC				CCC	UCC	ACC			GUC	UAA
GCG	CGG	GAC	AAC	UGC	GAA	CAA	GGG	CAC	AUC	CUG	AAA		UUC	CCG	UCG	ACG		UAC	GUG	UAG
GCU	CGU	GAU	AAU	UGU	GAG	CAG	GGU	CAU	AUU	CUU	AAG	AUG	UUU	CCU	UCU	ACU	UGG	UAU	GUU	UGA

Ala	Arg	Asp	Asn	Cys	Glu	Gln	Gly	His	Ile	Leu	Lys	Met	Phe	Pro	Ser	Thr	Trp	Tyr	Val	Stop
A	R	D	N	C	E	Q	G	H	I	L	K	M	F	P	S	T	W	Y	V	

Figure 7.20 The genetic code. The three-base sequences represent the 64 different codons, specifying the 20 different amino acids, each of which is given in both its three-letter and one-letter abbreviation. Most amino acids are represented by multiple codons. Three codons do not specify any amino acids. Rather, they function as "stop" codons, signaling the end of the protein-coding sequence.

each three-base codon specifies a particular amino acid. For example, UCU (transcribed from AGA in the DNA) codes for the amino acid serine (Ser), while GGA (transcribed from CCT) codes for glycine (Gly). **Figure 7.20** provides the complete genetic code. Note that of the 20 different amino acids, only tryptophan (Trp) and methionine (Met) are specified by only a single codon. Other amino acids correspond to as many as six different codons (for example, leucine, Leu). Three of the possible codons do not specify any amino acid. Rather, they function as stop codons, to indicate when a protein has been completed.

Amino acids are linked together into a growing protein on **ribosomes** (see Chapter 3, p. 57). Each ribosome is a complex assembly of many proteins along with a second type of ribonucleic acid, called **ribosomal RNA (rRNA)**. A ribosome attaches to the end of an mRNA molecule to initiate translation. The ribosome then moves toward the other end of the mRNA, one codon at a time, reading the codons from one end to the other. At each codon, the appropriate amino acid is brought into position by a third type of ribonucleic acid, called **transfer RNA (tRNA)** (**Figure 7.21**). Each tRNA has a three-base sequence called the **anticodon** that is complementary to a specific codon on the mRNA. When a ribosome arrives at a particular codon, a tRNA bearing the appropriate amino acid temporarily binds the codon via its anticodon. A peptide bond is then formed between the newly arriving amino acid and those amino acids already incorporated into the growing protein chain. The previously arriving tRNA is then released. In this way, codons are read sequentially along the mRNA until the ribosome arrives at a stop codon indicating that the protein is complete. At this point, the ribosome releases its protein and detaches from the mRNA (**Figure 7.22**).

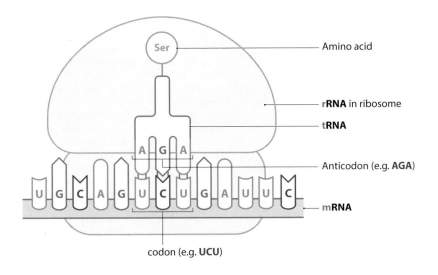

codon (e.g. **UCU**)

Figure 7.21 Ribosomal RNA (rRNA), transfer RNA (tRNA), and messenger RNA (mRNA). Ribosomes, the site of amino acid assembly into proteins, contain both protein and rRNA. Transfer RNA (tRNA) carries the correct amino acid (serine in this example) into a growing protein chain by binding temporarily to specific codons on the mRNA. Note the position of the anticodon, complementary to a specific codon on the mRNA.

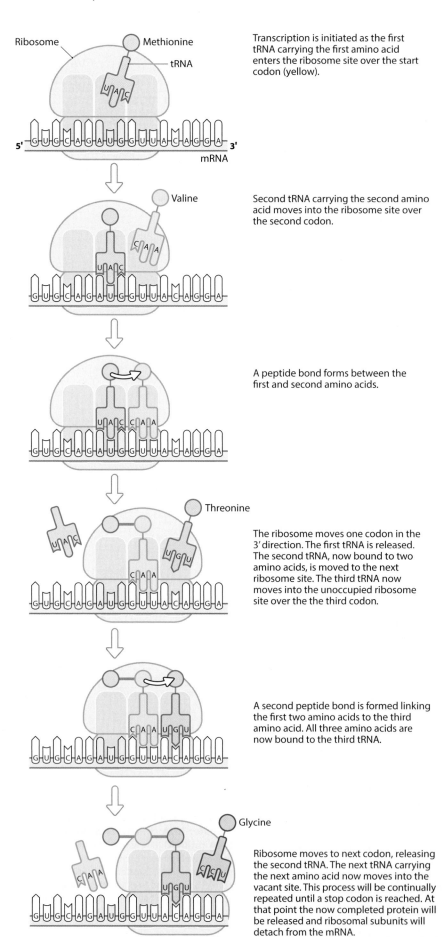

Transcription is initiated as the first tRNA carrying the first amino acid enters the ribosome site over the start codon (yellow).

Second tRNA carrying the second amino acid moves into the ribosome site over the second codon.

A peptide bond forms between the first and second amino acids.

The ribosome moves one codon in the 3' direction. The first tRNA is released. The second tRNA, now bound to two amino acids, is moved to the next ribosome site. The third tRNA now moves into the unoccupied ribosome site over the the third codon.

A second peptide bond is formed linking the first two amino acids to the third amino acid. All three amino acids are now bound to the third tRNA.

Ribosome moves to next codon, releasing the second tRNA. The next tRNA carrying the next amino acid now moves into the vacant site. This process will be continually repeated until a stop codon is reached. At that point the now completed protein will be released and ribosomal subunits will detach from the mRNA.

Figure 7.22 The process of translation. The first codon (the start codon, yellow) on the mRNA is almost always AUG, coding for methionine (Met). The first tRNA, with an anticodon complementary to the first codon, brings the first amino acid (Met) into position. The second tRNA, bearing the amino acid specified by the second codon, likewise binds, and a peptide bond forms between the first two amino acids. The ribosome then moves, one codon at a time, and the correct amino acid is brought into the growing peptide chain by the appropriate tRNA. Ultimately, the ribosome reaches a stop codon, causing both the ribosome and the now-completed peptide to be released.

The erythromycin that Leslie took to treat her infection acts by blocking translation of bacterial mRNA. Specifically, erythromycin binds to the bacterial ribosome and prevents it from moving along the mRNA molecule. Unable to translate mRNA, the bacteria fail to produce protein, and reproduction grinds to a halt. The bacteria are then generally eliminated by the immune system within a short time. Many other antibiotics interfere with bacterial translation in different ways, and such drugs might be used as alternatives to erythromycin to treat a case of atypical pneumonia such as Leslie's. But amoxicillin will not work because, as a penicillin-type drug, it acts by inhibiting the synthesis of new cell wall material (see Chapter 3, p. 51). As you may recall, *M. pneumoniae* is unusual because it lacks a cell wall. With no cell wall synthesis to inhibit, amoxicillin is useless against *M. pneumoniae*. We will examine antibiotics and when to use (or not use) them in Chapter 13.

Like transcription, translation is similar in bacteria, archaea, and eukaryotes, but we will mention two important differences. First, in eukaryotic cells, transcription occurs in the nucleus. Before translation can occur, a molecule of mRNA must pass through pores in the nuclear membrane to reach the ribosomes in the cytoplasm (see Chapter 3, pp. 61–62). In bacteria and archaea, there is no nuclear membrane, so in prokaryotic cells, ribosomes can associate with the end of an mRNA molecule before transcription is even completed. In other words, the two processes of transcription and translation, completely separate and discrete in eukaryotes, overlap in prokaryotes (**Figure 7.23**). This is part of the reason why proteins are produced more quickly in most prokaryotic cells, and it is another example of how the elegant simplicity of prokaryotes often gives them an edge in terms of efficiency.

Another difference with great practical significance involves the ribosome. Although they fulfil the same function, ribosomes in eukaryotes contain more protein and rRNA than bacterial ribosomes. Their mass is therefore greater. This difference may sound trivial, but the relative safety of drugs such as erythromycin is based on this fact. Although erythromycin can bind to and interfere with bacterial ribosomes, the antibiotic cannot bind the larger, more massive ribosomes found in eukaryotes. Because human or other animal ribosomes are unaffected, these drugs can be used to control bacterial growth without unacceptable side effects. The ribosomes of archaea are more like those of eukaryotes than those of bacteria—another reminder of how different the two domains of prokaryotes are from each other.

Gene activity is often carefully regulated

Arriving home on a cold winter night, your first action might be to flip on some lights and turn up the heater. Later you might cook dinner or watch TV. These uses of energy might be considered essential or at least enjoyable. Suppose, however, that the next morning you left home quickly, forgetting to turn off the lights and leaving the coffee maker and maybe the radio on. Now you are clearly wasting energy and you will pay the price when the utility bills arrive.

Cells must make similar decisions regarding their energy budgets. At certain times, specific proteins are required and must be synthesized, in spite of the energy needed to make them. As conditions change and less of a particular protein is needed, cells need a way to repress protein production in order to conserve energy. Then, should the protein be needed later, cells must be able to switch protein synthesis back on. Those that cannot regulate protein

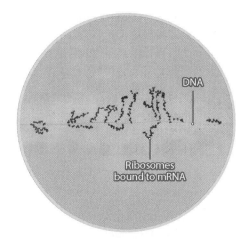

Figure 7.23 In prokaryotes, transcription and translation are overlapping. Because prokaryotes lack a nuclear membrane, translation of an mRNA can begin even before transcription of that mRNA is completed. This electron micrograph shows a gene being transcribed from left to right. The darkly stained bodies are ribosomes attached to mRNA. The mRNA becomes progressively longer as transcription occurs, allowing greater numbers of ribosomes to bind and initiate translation.

synthesis in this way pay their own price in terms of wasted energy, inefficiency, and reduced growth. Genes that are turned on and off as conditions warrant are called **regulated genes**. Other, nonregulated genes are always being expressed. Nonregulated genes code for proteins that the cell needs all the time—there are no situations in which it makes sense to shut them off. Nonregulated genes are called **constitutive genes**. Examples include genes coding for crucial structural proteins that are always needed in constant, large amounts.

Many bacterial genes are regulated simultaneously in groups of genes called operons

Let's now examine how the expression of regulated genes is controlled. In prokaryotes, many such genes are organized into gene clusters, called **operons**. Operons are groups of genes the products of which are all involved in the same overall process, even though each one codes for its own protein (**Figure 7.24**). An operon also contains control regions that regulate gene activity. Genes are either activated or repressed as a unit—they are either all on or all off. The genes coding for lactose-digesting enzymes are organized in an operon called the *lac* operon. Several different proteins, each coded for by a different gene in the operon, are required if a bacterium is going to digest lactose. There is never a time when a cell would want only some of these genes active. It is therefore efficient to regulate them as a group. Operons are common in bacteria but are not found in eukaryotes, where most genes are regulated individually.

Consider the conditions under which the genes of the *lac* operon should be expressed or repressed. When lactose is present in the environment, it is to a cell's advantage to digest it. Bacteria respond to the presence of lactose by activating the genes of the *lac* operon. In the absence of lactose, expression of these genes is energetically wasteful, so these same genes are repressed.

Figure 7.25 illustrates a portion of bacterial DNA showing the three protein-coding genes involved in lactose digestion. Adjacent to these genes are

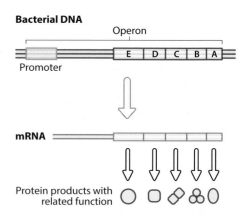

Figure 7.24 Operons are gene clusters whose expression is controlled as a unit. The letters A–E identify different genes in an operon, all of which code for proteins involved in a related process. They therefore all use a single promoter and are activated or inhibited as a unit.

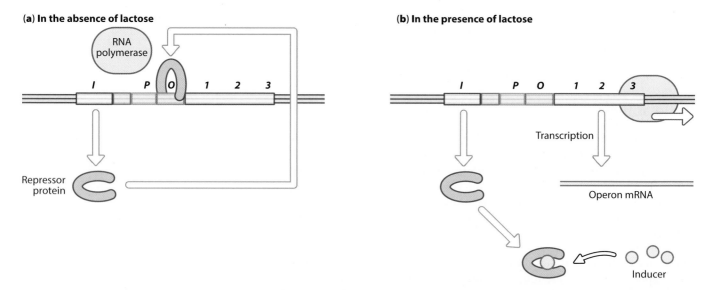

Figure 7.25 The *lac* operon. To digest lactose, three different proteins are required. The genes coding for these proteins (the structural genes) are numbered 1–3. (a) When lactose is absent from the environment, the repressor is bound to the operator (O), preventing RNA polymerase from binding to the promoter (P) and thereby blocking transcription of the structural genes. (b) In the presence of lactose, a lactose molecule binds to the repressor, which causes the repressor to come off the operator. Lactose, thus, functions as the inducer (I). Once the repressor is removed, RNA polymerase can bind the promoter and transcribe the structural genes.

crucial control regions. The **promoter**, as described previously, is a DNA base sequence recognized by RNA polymerase. The **operator** is found between the promoter and the three protein-coding genes. Protein-coding genes are often called structural genes, to distinguish them from control regions of the operon. An additional regulator gene is found at some distance from the control regions. It codes for a large protein called the **repressor**, which is able to bind to the operator when lactose is absent.

When lactose is absent from the environment, the repressor binds tightly to the operator (**Figure 7.25a**). With the repressor in place, RNA polymerase cannot bind the promoter. Consequently, no transcription of the genes in the operon can occur. Alternatively, if the environment changes and lactose becomes available, some lactose enters the cell. The repressor protein has a binding site on its surface for lactose. The binding of lactose causes the shape of the repressor to change slightly. In its new shape, the repressor cannot bind the operator (**Figure 7.25b**). This allows RNA polymerase to bind the promoter. Transcription of the genes ensues, and following translation, lactose digestion can begin. Because lactose itself activates the *lac* operon, lactose is termed the **inducer**.

This mechanism allows the cells to produce lactose-digesting enzymes only under the proper environmental conditions, minimizing energy waste. The *lac* operon will be repeatedly induced or inactivated in response to changing amounts of lactose in the environment. Many other operons function in a similar manner. Operons such as these are termed **inducible**, because the substance to be digested actually activates or induces transcription.

Some operons are regulated through repression

Other operons utilize a slightly different type of regulation called **repression**. In this case, the operon is turned off when a crucial substance is available in the environment, as opposed to the inducible *lac* operon, where we saw that the operon was turned on in the presence of lactose. A good example is provided by those bacteria that can make their own tryptophan (Trp), a necessary amino acid. Such bacteria utilize the *trp* operon to regulate synthesis of the amino acid. When a bacterial cell has sufficient tryptophan, it is in the cell's interest to suspend tryptophan synthesis until levels of the amino acid again fall to low levels. Consequently, abundant tryptophan acts not to induce but to inhibit the operon.

Repressible operons like the *trp* operon have similar structure and work in a similar way to the *lac* operon, with one big difference: RNA polymerase can bind the promoter only when tryptophan is *not* present and the cell needs to make more of the amino acid. Unlike lactose, which removes the repressor from the operator, tryptophan, when present, binds the repressor and allows the repressor to bind. Once the available tryptophan is used up and there is none left to bind the repressor, the repressor leaves the operator. RNA polymerase can once again transcribe the genes in the operon, resulting in increased production of the needed amino acid.

Bacteria often coordinate their gene expression through chemical communication

Bacteria are not always as independent agents as you might guess. Rather, many bacteria use chemical signals to keep tabs on their local population density and regulate gene expression accordingly. This phenomenon, called **quorum sensing**, allows bacteria to synthesize certain proteins only when the bacteria sense that their population has reached a certain threshold level.

To illustrate how quorum sensing works, consider the marine bacterial species *Aliivibrio fischeri*. This bioluminescent species can either exist as free living cells or symbiotically in the light-producing organ of the Hawaiian bobtail squid (Figure 7.26a). The bacteria are able to produce an enzyme called luciferase, which is involved in the production of light. The production of luciferase requires an operon consisting of several structural genes. When *A. fischeri* cells are living independently in the ocean, RNA polymerase cannot bind the promoter and no luciferase is produced. An individual cell would gain no advantage from producing this enzyme; a single cell cannot produce enough light to even be visible and producing luciferase in such situations would be a waste of energy. When the bacteria colonize the squid's light producing organ, however, they all activate their operons at the same time, producing the observed bioluminescence in the squid. This association is beneficial for both the squid (they use the light to hide their silhouette and are less likely to be preyed upon) and for the bacteria (the squid provide bacteria in the light organ with sugar and amino acids). But how do the bacteria know their density is high enough to make luciferase production worthwhile?

The bacteria all produce a chemical messenger called the **autoinducer**. Furthermore, they all have receptor molecules in their cytoplasm that can bind the autoinducer. When the bacteria are at low density, the only autoinducer in the immediate vicinity of a given cell is that which the cell itself makes. This molecule quickly dissipates in the ocean and little if any of it reenters the cell and binds the receptor (Figure 7.26b). If the cells are at high concentration, however, as they would be in the light organ, the amount of autoinducer surrounding each cell goes up. Some of this molecule diffuses into each cell and binds receptor molecules. The autoinducer/receptor complex can now bind the promoter of the luciferase operon, facilitating binding of RNA polymerase, and allowing transcription to commence (Figure 7.26c). Transcription begins at about the same time in all cells, permitting the production of visible light.

The list of quorum sensing examples continues to grow. Bacteria may even become more or less virulent due to this phenomenon, or they may assemble into complex aggregations, consisting of many cells.

Eukaryotes use various mechanisms to regulate gene expression

Eukaryotes utilize a large number of mechanisms to activate or inhibit their genes, many of which are still being investigated. One important mechanism depends upon how tightly the DNA on a chromosome is wound around its scaffolding proteins. Inactive genes tend to coil more tightly around these proteins, making the DNA much denser at these points. Genes that are actively being expressed are wound in a looser configuration, and as inactive genes become active, the DNA making up these genes becomes less dense. An exciting, recent discovery is that RNA, long thought to function only in transcription and translation, has been found to play an important role in gene regulation. Small RNA molecules called **microRNAs** now seem to be crucial in determining when genes are expressed. This is certainly one of the most important biological discoveries in recent years and is certain to be a major research focus for the foreseeable future.

The environment influences the nature of genetically determined characteristics

The specific genetic makeup of an individual for a particular trait is termed that individual's **genotype** for the trait. The expression of that genotype,

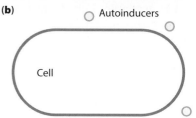

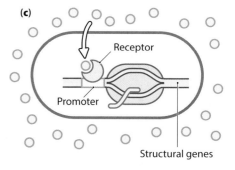

Figure 7.26 Quorum sensing. (a) The Hawaiian bobtail squid, *Euprymna scolupes*. The light organ or photophore of the squid contains *Aliivibrio fischeri*, symbiotic bacteria responsible for the squid's bioluminescence. (b) At low cell density, although each cell produces some of the autoinducer (green circles), this molecule quickly dissipates in the surrounding sea water, and little or none reenters the cell. (c) At high concentration within the squid's light organ, levels of the autoinducer are high and some diffuses back into each cell. Inside the bacterial cell, the autoinducer can bind its receptor (purple). The receptor then binds the promoter (green box) for the luciferase operon, inducing transcription of the structural genes.

which results in the outward appearance of the individual, is called the individual's **phenotype**.

To illustrate the difference between genotype and phenotype, consider one genetic trait in humans—freckles. Two people with the same genotype will have the exact same genetic information for this trait. Even these two individuals, however, may end up appearing somewhat different—one person may have more numerous and/or more prominent freckles. If so, their phenotypes are different. Various factors might explain this difference. Exposure to sunlight, for one thing, can cause more pronounced freckling, demonstrating the impact that environmental factors can have on genetically based characteristics. This example highlights an important concept in genetics: Genes for a characteristic (the genotype) strongly influence, but do not wholly shape, the final observed outcome (the phenotype). An individual's potential for a certain characteristic is determined by genotype. The actual outcome of that potential, the phenotype, is influenced by various other, often environmental, factors. This important principle applies to all living things, from amebas to zebras.

Genetic variation can arise through diverse mechanisms

We opened this chapter with a discussion of the airline passenger who may have spread a drug-resistant strain of tuberculosis-causing *Mycobacterium tuberculosis* to others as a consequence of her travels. We then posed the question: Some strains of *M. tuberculosis* are easily treatable. Others are not. Why, if they are all members of the same species, do they differ in this way?

We have already learned part of the answer to this question. All organisms, even those of the same species, vary in terms of their genetic material. When it comes to genetics, variety is indeed the spice of life. Here we will find out where that variety comes from.

Sexual reproduction permits the generation of new genetic combinations

Genetic information is passed from parents to offspring when an organism reproduces. Reproduction in bacteria is a relatively simple affair; a single-celled bacterial cell divides into two new cells. The original, reproducing cell is called the **parent cell**. The newly formed progeny cells are called **daughter cells**.

Each daughter cell contains an exact copy of the parent cell's DNA. The DNA of a parent cell is copied prior to actual cell division and each daughter cell receives a single copy as the cell divides (**Figure 7.27**). This type of reproduction, in which a single parent cell gives rise to two identical daughter cells, each receiving the same genetic information, is called **asexual reproduction**.

Reproduction in many multicellular organisms, including many plants and most animals, is more intricate. This **sexual reproduction** involves the mixing of DNA from two parent organisms to create offspring with a novel combination of genetic information. Unlike the simple division of a parent cell in asexual reproduction, sexual reproduction involves the union of specialized sex cells called **gametes**—one female and one male. In many familiar organisms, the female and male gametes are called the

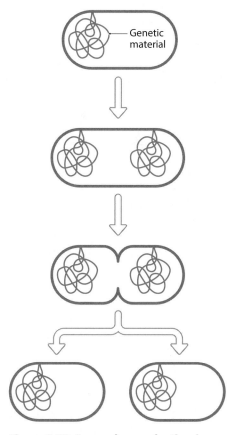

Figure 7.27 Asexual reproduction in bacteria. Following replication of the genetic material into two copies, the bacterial cell divides, forming two new individual bacteria.

egg and the **sperm**, respectively. In most cases, each gamete contains exactly half of the parent organism's DNA. When an egg and a sperm combine, the resulting offspring obtains a full complement of DNA, half from each parent. Because the offspring has genetic information from both the male and female parents, it is genetically distinct from either parent. This ability to generate new combinations of genetic material is called **genetic recombination**.

We have learned that the gene is the basic unit of heredity. And even though all members of a given species have the same genes, they do not necessarily have the same forms of those genes. For example, all humans have genes that code for eye color, but the specific form of those genes determines whether the eyes will be brown or blue. Such alternative forms of the same genes are called **alleles** of that gene. In most cases, a eukaryotic cell carries two copies of each gene—one obtained from the female parent and one from the male. The two gene copies may or may not be the same allele. Prokaryotes, on the other hand, usually have only a single copy of each gene.

Sexual reproduction provides a mechanism to greatly increase the amount of genetic variability in a population. During gamete formation, each egg or sperm receives only one copy of each gene, which explains why each gamete ends up with only 50% of the genetic material. Exactly which of the two copies a specific gamete receives, however, is largely random. That means that different gametes end up with different combinations of the parents' alleles (**Figure 7.28**). The alleles are therefore randomly shuffled with the formation of each gamete, increasing genetic variation enormously. During sexual reproduction, a male gamete unites with a female gamete, in both of which the parent's alleles were scrambled. The result is that no two offspring are ever likely to be identical.

Horizontal gene transfer allows for genetic recombination in asexually reproducing organisms

While sexual reproduction has an obvious, built-in mechanism for generating genetic variation, this is not the case for asexual reproduction. In Figure 7.27 we saw that asexual reproduction should normally result in offspring with genetic material identical to that of the parent cell. Yet prokaryotes also display considerable genetic variability, relying on a variety of mechanisms collectively called **horizontal gene transfer**. Horizontal gene transfer refers to the transfer of genes between organisms in some way other than traditional reproduction. Reproduction, sometimes called vertical gene transfer, refers to genes passed from parent to offspring.

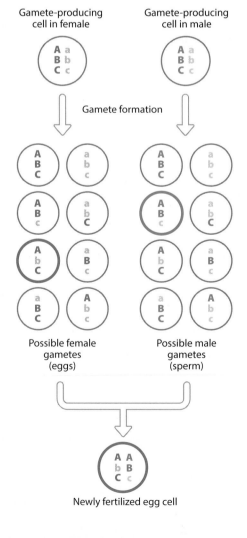

Figure 7.28 Potential for genetic variability in sexually reproducing organisms. In this example, only three genes are considered. Each parent has two copies of each gene. **A** and a, **B** and b, and **C** and c represent different alleles for each of these three genes. Each parent donates only one of each gene pair to each gamete. Consequently, each parent can form eight possible different gametes. Any female gamete can potentially be fertilized by any male gamete, resulting in 64 combinations. Suppose the egg (highlighted in red) containing **A**, b, and **C** is fertilized by the sperm (highlighted in blue) containing **A**, **B**, and c. The result is a fertilized egg with the genotype **AA**, b**B**, and **C**c. In typical sexually reproducing organisms with thousands of genes, there are almost limitless new gene combinations.

CASE: GRIFFITH'S TRANSFORMING FACTOR

In the 1920s, the English microbiologist Frederick Griffith was studying *Streptococcus pneumoniae*, a bacterial species that sometimes causes pneumonia. Griffith knew that only those *S. pneumoniae* strains with a polysaccharide capsule exterior to their cell wall could cause disease. Strains lacking capsules were harmless. Griffith wanted to understand how the capsule contributed to disease. To do so, he performed the experiment shown schematically in **Figure 7.29**. He first inoculated mice with bacteria lacking capsules. As expected, these mice remained disease-free. When he inoculated mice with encapsulated bacteria, the mice sickened and died. Griffith next killed encapsulated bacteria by exposing them to heat. When mice were exposed to these dead cells, the animals remained healthy. None of these results were particularly surprising. But when Griffith exposed mice to a mixture of killed encapsulated bacteria and living *S. pneumoniae* lacking capsules, he was astounded when

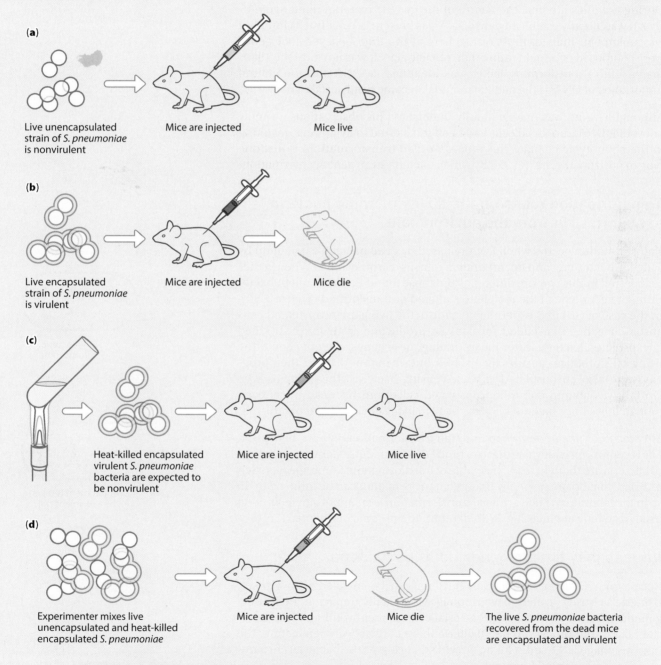

(a)

Live unencapsulated strain of *S. pneumoniae* is nonvirulent

Mice are injected

Mice live

(b)

Live encapsulated strain of *S. pneumoniae* is virulent

Mice are injected

Mice die

(c)

Heat-killed encapsulated virulent *S. pneumoniae* bacteria are expected to be nonvirulent

Mice are injected

Mice live

(d)

Experimenter mixes live unencapsulated and heat-killed encapsulated *S. pneumoniae*

Mice are injected

Mice die

The live *S. pneumoniae* bacteria recovered from the dead mice are encapsulated and virulent

Figure 7.29 Griffith's transformation experiment. (a) When mice were infected with live, unencapsulated bacteria, the mice remained healthy. (b) Mice infected with live, encapsulated bacteria developed disease and died. (c) When encapsulated bacteria were heat-killed and then injected into the mice, the mice remained disease-free. (d) If live unencapsulated bacteria were mixed with heat-killed encapsulated bacteria, mice later injected with these bacteria were killed. Bacteria isolated from these dead mice were also encapsulated.

the mice developed pneumonia and died. When he cultured bacteria from these dead mice, he was further amazed to find live, encapsulated bacteria. Griffith hypothesized that a "transforming factor" of some sort had been released by the dead encapsulated bacteria and that this factor had been absorbed by the living, previously harmless bacteria. The transfer of this factor permitted the living cells to now produce the capsule and cause disease.

1. What was Griffith's transforming factor?

2. How did this factor pass from dead to living bacteria, and once it had passed, how did it permit the unencapsulated bacteria to start forming capsules?

3. How common is this phenomenon? How is it similar to or different from other mechanisms of genetic exchange in prokaryotes?

During Griffith's lifetime, the nature of the genetic material remained elusive. It was already established that DNA was present in cells, but its function was unknown. Indeed, most researchers at the time believed that genes were composed of protein rather than nucleic acid. It was not until the 1940s that Griffith's transforming factor was identified as DNA and the central importance of DNA as the genetic material became firmly established.

Although Griffith was unable to fully understand his observations, we now know that DNA from dead cells had been transferred to living cells, resulting in the phenotypic change. This process, called **transformation**, is just one type of horizontal gene transfer by which bacteria swap genetic information.

Transformation relies on the ability of some bacteria to absorb DNA from the environment

Transformation occurs when DNA released by dead bacteria is taken up by a living bacterium and incorporated into its chromosome. When genes transferred in this way are different from those found in the recipient bacterium, the genotype of the recipient is altered or transformed (**Figure 7.30**). In the case of Griffith's experiment, some of the living unencapsulated cells absorbed genes that coded for capsule production from the heat-killed encapsulated bacteria. After incorporating these genes, not only could the living cells produce capsules themselves, but when they divided, their progeny were also encapsulated, disease-causing bacteria. This explains why Griffith was able to culture encapsulated cells from his dead mice, even though these mice had not been exposed to living encapsulated organisms.

Streptococcus pneumoniae can undergo transformation, in part, because it has receptors on its surface that can bind DNA fragments. Many other groups of bacteria are thought to be incapable of transformation. Some species such as *E. coli* that do not normally undergo transformation can be induced to do so in the laboratory. Such artificial transformation is an important technique used in genetic engineering. We will return to this topic in Chapter 15.

Viruses may shuttle genes between bacteria

Recall from Chapter 5 that even bacteria can be infected with viruses. Bacterial viruses, called bacteriophages, normally replicate their own genetic information within the bacterial cell, frequently killing their bacterial host in the process. When bacteriophages are actively replicating, the bacteriophage DNA codes for an enzyme that cleaves the bacterial chromosome into many small fragments.

Occasionally, during the assembly of new viral particles, a small amount of bacterial DNA is incorporated into the phage head with the viral nucleic acid. The resulting virion will be unable to replicate. If it is released, however, it may still be able to inject its DNA, including the mistakenly acquired

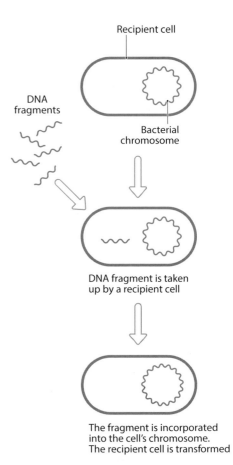

Figure 7.30 Transformation in bacteria. Donor DNA fragments may be taken up by certain bacterial species. If the donor DNA is incorporated into the chromosome of the recipient cell, the recipient becomes transformed.

bacterial DNA, into a new bacterial cell. Occasionally, genes transported in this manner become incorporated into the chromosome of the recipient bacterial cell.

In this process, known as **transduction**, the bacteriophage has essentially acted as a device to shuttle bacterial genes between donor and recipient cells (**Figure 7.31**). Like transformation, transduction may alter the genotype and phenotype of the recipient bacterium. Transduction is a common form of genetic exchange between bacteria and it occurs in many species.

In conjugation, small portions of DNA are transferred directly from one cell to another

In addition to the main bacterial chromosome, some cells have an additional circular loop of DNA called a **plasmid**. Like the chromosome, the plasmid carries genes that code for specific proteins, although the number of genes is small. A typical plasmid may even code for fewer than a dozen proteins, while a bacterial chromosome typically codes for a few thousand. Plasmids can usually replicate independently of the main chromosome, and cells may have more than one of these structures.

Many bacteria with plasmids can transfer these plasmids to other cells. This process of genetic exchange is called **conjugation** (**Figure 7.32**). The best-studied example of such conjugation involves the plasmid found in *E. coli*. Cells that carry the plasmid are called F⁺ cells (fertility-positive), whereas those lacking it are called **F⁻ cells**. A gene on the plasmid allows **F⁺ cells** to produce a structure called the **sex pilus**. The sex pilus, constructed of protein, physically connects the F⁺ and F⁻ cells and pulls them together. This contact induces the F⁺ cell to replicate its plasmid. The two strands of plasmid DNA separate, and one of the resulting single strands is transferred to the F⁻ cell. The single strand of plasmid DNA in both cells now gives rise to their respective complementary strands. Essentially, the F⁻ cell is now converted into an F⁺ cell. Such a cell has been genetically altered and can now produce any new products coded for by genes on the plasmid.

In conjugation, unlike transformation and transduction, the donor and recipient cells must come into actual physical contact. Furthermore, conjugating cells must ordinarily be of opposite types, with only F⁺ cells serving as donors and only F⁻ cells acting as recipients.

Transposons are short DNA sequences that can move from one site in the genome to another

Short sequences of DNA can be found in the genomes of both eukaryotes and prokaryotes that are able to move from one site to another. In prokaryotes this movement might be from the chromosome to a plasmid or from one position on the chromosome to another (**Figure 7.33**). These so-called "jumping genes" are more formally known as **transposons** and they represent yet another way that genetic diversity can be generated.

All transposons encode information for the production of transposase, an enzyme that both cuts and reseals the DNA, allowing the movement of the transposon from one site to another. Some larger transposons also carry genes for other characteristics such as drug resistance or toxin production. If such a transposon moves from the main bacterial chromosome to a plasmid, the genes associated with that transposon might then pass to another cell via conjugation.

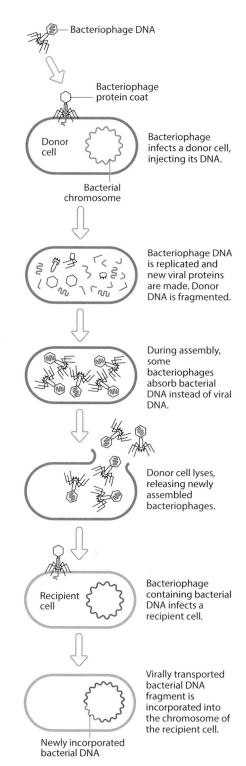

— Bacteriophage DNA

Bacteriophage protein coat

Donor cell

Bacterial chromosome

Bacteriophage infects a donor cell, injecting its DNA.

Bacteriophage DNA is replicated and new viral proteins are made. Donor DNA is fragmented.

During assembly, some bacteriophages absorb bacterial DNA instead of viral DNA.

Donor cell lyses, releasing newly assembled bacteriophages.

Recipient cell

Bacteriophage containing bacterial DNA infects a recipient cell.

Virally transported bacterial DNA fragment is incorporated into the chromosome of the recipient cell.

Newly incorporated bacterial DNA

Figure 7.31 Transduction in bacteria. A bacterial virus or bacteriophage mistakenly incorporates a fragment of bacterial DNA into its capsid during assembly. If the bacteriophage is released and infects a second bacterial cell, the bacterial DNA it contains is transferred to the recipient cell, where it may be incorporated into the recipient cell's chromosome.

Figure 7.32 Bacterial conjugation. (a) The process of conjugation as it might occur between an F⁺ donor *E. coli* cell and an F⁻ recipient cell. Following conjugation, the F⁻ cell is converted to an F⁺ cell and may now serve as a donor. (b) Notice that the upper cell (the F⁺ cell) has numerous fimbriae. If the ability to produce these structures is encoded by a plasmid gene, the F⁻ cell below may obtain it and consequently undergo a phenotypic change. Because fimbriae help a cell to adhere to surfaces inside a host, cells with fimbriae are sometimes more virulent than those without fimbriae. Thus, in this photograph, the bottom cell may be in the process of acquiring a new virulence factor.

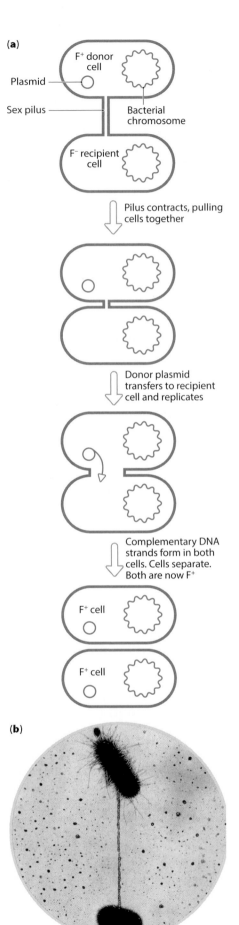

Genetic recombination in prokaryotes has great significance for humans

Microbial gene exchange might seem to have little practical importance for humans. However, the consequences of such seemingly trivial events can be monumental. To illustrate, consider a modern hospital where large amounts of antibiotics are being used. Within the hospital environment, there are often strains of bacteria that develop resistance to one or more antibiotics. Without gene transfer, if a particular bacterial strain became drug-resistant, such resistance would remain confined to that strain. Because of mechanisms such as those discussed in this section, however, resistance genes can be exchanged between different strains and even different species of bacteria. As a result of such wholesale genetic transfer, strains of bacteria resistant to all antibiotics can arise. It is an unfortunate fact that people often develop serious illness and sometimes die from infections that were previously easy to treat. Drug resistance (to be explored in more detail in Chapter 13) is not the only dangerous characteristic that bacteria can swap by gene transfer. For instance, we now occasionally hear about outbreaks of disease caused by food items contaminated with the "killer *E. coli*" (*E. coli* O157:H7). Symptoms of infection include bloody diarrhea and kidney failure. Among the virulent characteristics of this strain is the ability to produce a protein toxin. It is believed that the toxin gene was acquired by transduction from an entirely different species of bacteria (*Shigella dysenteriae*).

Not all microbial gene exchange is cause for alarm. Soil bacteria able to degrade environmental pollutants have transferred this capability to other bacteria, which can then also break down dangerous chemical contaminants. These newly created bacteria may eventually be used to degrade pollutants in a process called bioremediation. Bioremediation will be fully examined in Chapter 17.

We have now seen that genes can exist in different forms, known as alleles, and that both prokaryotes and eukaryotes have various mechanisms to mix and match these alleles, generating novel genetic combinations. Yet one big question regarding genetic variation remains—namely, how do various

Figure 7.33 Transposons in bacteria. Transposons are short DNA segments with the ability to move from site to site within the genome. The cell in the center of the figure has a transposon (in green) integrated into its chromosome (in red). This cell also has a plasmid (in blue). Bacterial transposons might move from one site on the bacterial chromosome to another (left arrow) or they might move from the chromosome to a plasmid (right arrow). If this plasmid is subsequently transferred to another cell via conjugation, the new cell will also acquire whatever genetic information is encoded by the transposon DNA.

forms of a gene arise in the first place? In the next section we will provide this important missing piece to the puzzle of genetic variability.

Mutations are the original source of genetic variation

CASE: END OF THE LINE FOR A LAST-LINE DEFENSE?

Vancomycin has been used against Gram-positive bacterial infections since the 1950s. It has been especially valuable against bacteria that are resistant to other commonly used antibiotics. For some bacterial strains, such as certain strains of *Staphylococcus aureus* that are resistant to a variety of drugs, vancomycin is the only drug that is still effective. The antibiotic works by disrupting assembly of the bacterial cell wall.

In recent years, however, many strains of bacteria have acquired resistance to vancomycin as well. In these bacteria there is a mutation that changes the composition of the cell wall. Vancomycin cannot attach to the modified cell wall in the mutant strain. Many patients infected with vancomycin-resistant bacteria are essentially untreatable, because these bacteria are often resistant to other antibiotics as well.

1. **What exactly is a mutation?**
2. **What type of mutation would cause the vancomycin resistance seen in the case?**
3. **How and why do mutations occur?**

Nothing in this world always goes right. In our discussion of DNA replication and gene expression, we described things as they are supposed to happen. Although the mechanisms we have discussed are remarkably faithful, they do not work perfectly every time. Damage to or mistakes in DNA base sequences can occur and such changes in the genetic material are called **mutations**. Most mistakes are immediately repaired by various mechanisms and therefore do not persist. We will discuss some of those repair mechanisms later in this section.

Many mutations have no effect on an organism's phenotype, and cells with such mutations look and behave normally. Other mutations may alter a cell's appearance or ability to function. If a mutation is severe enough, it may be lethal and in rare situations, as in our case, a mutation can even provide a new advantage. In other words, the consequences of a change in a gene's base sequence can span the gamut from deadly to neutral to beneficial. If a mutation kills the cell, there will be no transmission to progeny cells. Other nonlethal mutations may become a permanent part of the cell's genetic material and will thereafter be transmitted to subsequent generations. In this section we will consider some of the ways that DNA can be altered and why the effects of mutations on phenotype can vary so greatly.

As we have stated, many mutations have no effect whatsoever on phenotype. Although the base sequence of a particular gene is changed, the protein coded for by this gene remains the same. To understand how this is possible, recall that a gene's DNA code is transcribed into mRNA, which is translated one codon at a time into protein. Imagine that a particular codon, transcribed from normal, nonmutated DNA was UU**U**. Further imagine that, due to a mutation in the DNA, the codon that results is now UU**C** (**Figure 7.34a**). This would be caused if an adenine, at the appropriate spot in the DNA, were erroneously replaced by a guanine.

In the genetic code (see Figure 7.20), you will see that both the UUC and UUU codons code for the same amino acid, phenylalanine. Consequently

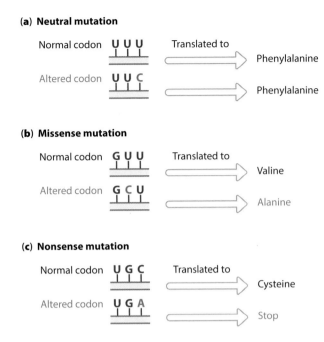

(a) Neutral mutation

Normal codon U U U   Translated to → Phenylalanine

Altered codon U U C   → Phenylalanine

(b) Missense mutation

Normal codon G U U   Translated to → Valine

Altered codon G C U   → Alanine

(c) Nonsense mutation

Normal codon U G C   Translated to → Cysteine

Altered codon U G A   → Stop

Figure 7.34 Single-base mutations. (a) In a neutral mutation, the amino acid sequence in a protein is unchanged. (b) A missense mutation alters the amino acid sequence of a protein. Such mutations, depending on how they affect protein function, may range from lethal to harmful to beneficial. (c) Nonsense mutations, in which a readable codon is converted into a stop codon, are almost always highly detrimental. Neutral, missense, and nonsense mutations are caused when DNA polymerase makes an error during DNA replication, inserting an incorrect nucleotide on the daughter strand.

the resulting protein will be unchanged. Because the specified protein is unaltered, such a mutation has no effect, either detrimental or beneficial, on phenotype. Mutations of this sort are often called **neutral mutations**.

Other mutations can result in a change in a protein's amino acid sequence, and such a change can be harmful or beneficial. For example, consider a mutation that changes a normal codon reading **GU**U into **GC**U (**Figure 7.34b**). While GUU codes for valine (see Figure 7.20), GCU codes for alanine. Consequently, as a result of this mutation, the translated protein will have a different amino acid sequence. If the mutated form of the protein works less efficiently than the nonmutated form, the mutation is likely to be harmful. If the mutated protein is completely unable to carry out its specific function, the mutation may be lethal, especially if the protein carries out an essential task. Rarely, the mutated protein actually works more efficiently or can do something new that the normal protein cannot. If this happens, we may observe a rare, beneficial mutation. Mutations such as these, which change the amino acid sequence of a protein, whether harmful or beneficial, are often called **missense mutations**, because they change the so-called sense of the DNA.

Frameshift mutation (insertion)

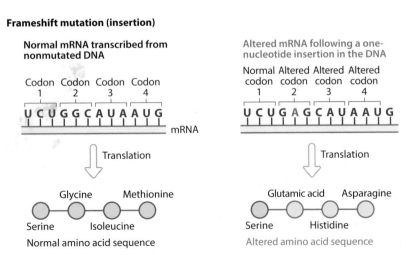

Normal mRNA transcribed from nonmutated DNA

Codon 1 Codon 2 Codon 3 Codon 4

U C U G G C A U A A U G   mRNA

↓ Translation

Serine — Glycine — Isoleucine — Methionine

Normal amino acid sequence

Altered mRNA following a one-nucleotide insertion in the DNA

Normal codon 1 Altered codon 2 Altered codon 3 Altered codon 4

U C U G A G C A U A A U G

↓ Translation

Serine — Glutamic acid — Histidine — Asparagine

Altered amino acid sequence

Figure 7.35 A frameshift mutation. In this case, an extra base has been inadvertently inserted. Other frameshift mutations represent deletions, in that a base has inadvertently been left out.

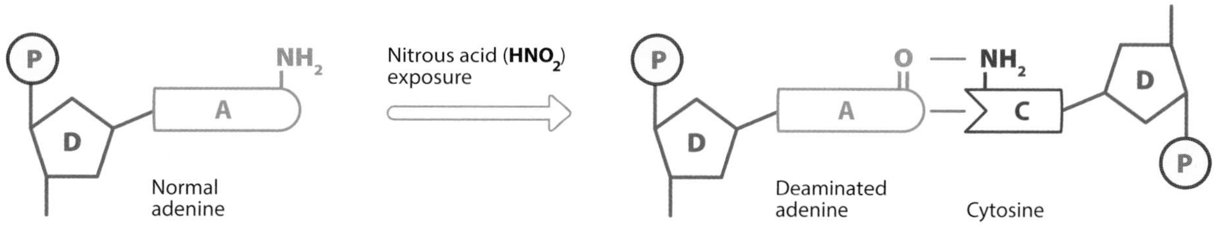

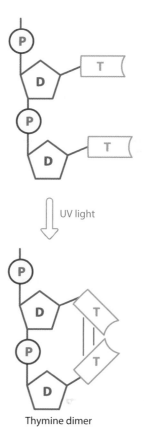

Figure 7.36 Nitrous acid, an example of a chemical mutagen. Exposure to nitrous acid (HNO_2) may cause deamination of nitrogenous bases. In this example adenine has been deaminated. It subsequently base pairs with cytosine rather than thymine. When this occurs during DNA replication, it can lead to a neutral, missense, or nonsense mutation.

A particularly harmful mutation can occur when a readable codon is converted into a stop codon. In the example provided in **Figure 7.34c**, the codon UG**C** normally codes for cysteine. If, due to a mutation in the DNA, the codon is now UG**A**, the normal codon has been converted into a stop codon. During translation, protein synthesis will be prematurely terminated at this point, resulting in an incomplete protein. These kinds of mutations almost always have a severely negative impact on phenotype. A mutation such as this, in which a readable codon is converted into a stop codon, is called a **nonsense mutation**.

Harmful mutations also occur when an extra base is either mistakenly inserted into or deleted from newly synthesized daughter DNA by DNA polymerase during DNA replication. Such a mutation is called a **frameshift mutation** (**Figure 7.35**). Frameshift mutations generally have a very severe impact on the resulting protein, because many amino acids are changed in the protein chain. If a base is inserted or deleted, all codons after the mutation are altered. In other words, during translation, the codons are shifted by one nucleotide, altering all the remaining amino acids in the protein.

The mutation in our case that resulted in resistance to vancomycin is almost certainly a missense mutation, because only missense mutations can alter the amino acid sequence in this manner. This particular mutation is an example of a rare beneficial mutation. In a patient treated with vancomycin, such mutant bacterial cells have an enormous advantage over sensitive cells. Because they continue to replicate while sensitive cells are being destroyed, resistant cells quickly come to predominate.

Mutations are caused by many factors

What causes mutations in the first place? Remember that enzymes such as DNA polymerase do not work perfectly every time. When DNA is replicated, there is a chance that a mistake will occur and a mutation may result. These random errors are called **spontaneous mutations** and they tend to be rare. A given gene might mutate only once for every several million times it is replicated, although certain genes tend to be more or less prone to mutation.

Exposure to certain chemicals and some forms of radiation can increase the frequency of mutations significantly, often up to 1000 times. Such agents are called **mutagens**.

Chemical mutagens can alter DNA in various ways, increasing the likelihood of replication errors. Nitrous acid (HNO_2), for instance, removes amino groups ($-NH_2$) from nitrogenous bases. This may cause the base to subsequently base-pair incorrectly during DNA replication. For example, if adenine loses its amino group, it is likely to base pair with cytosine rather than thymine (**Figure 7.36**).

Figure 7.37 Formation of a thymine dimer. UV light can break the hydrogen bonds between thymine and adenine. Two adjacent thymine bases, disrupted in this way, may then covalently bond to each other, forming a dimer.

Another group of chemical mutagens are the **intercalating agents**. These three-ringed molecules are about the same size as a pair of DNA nucleotides. They cause problems by inserting themselves at a replication fork during DNA synthesis. As the nucleotides are pushed apart, an extra base may be added to or left out of the new DNA strand, resulting in a frameshift mutation. The mold *Aspergillus flavus* produces aflatoxin, which can induce frameshifts in this manner. Because *A. flavus* often contaminates peanuts, this food is sometimes associated with aflatoxin-induced mutations.

Radiation can also damage DNA and thus act as a mutagen. Gamma radiation and X-rays are powerful forms of radiation that can break bonds in molecules. The remaining atoms or molecules often form **free radicals**. These are highly reactive compounds that latch onto and damage other molecules, including DNA. Such damage can cause errors in DNA replication or even cause breaks between a nucleotide sugar and its base.

Ultraviolet (UV) radiation can break the hydrogen bonds between thymine and adenine nucleotides. If two such thymines are adjacent to each other on the same DNA strand, they may then covalently bond with each other. Two thymines covalently linked together in this way are called a **thymine dimer** (**Figure 7.37**). The dimer distorts the shape of the DNA molecule, preventing further DNA replication. UV radiation can damage DNA in human skin cells and that is why we are often warned about the danger of overexposure to sunlight. Unlike gamma radiation and X-rays, UV radiation cannot penetrate below the skin's surface and consequently poses no risk to deeper tissues. Most microorganisms, however, are easily penetrated by UV radiation, making UV light an effective means of control. Ultraviolet lighting is sometimes installed in places like hospital operating rooms, where contaminating microorganisms can be a serious problem. In other situations, such as prisons, where unnatural crowding provides increased opportunities of disease transmission, UV lights can help to minimize such risks. To prevent human DNA damage, however, such lights are turned on only when people are not actually in the area to be decontaminated.

The first bacterium to develop resistance to vancomycin may have been briefly exposed to a mutagen similar to those described. Alternatively, this now-major medical problem may have started out as a random spontaneous mutation in a bacterium living in the intestine of a hospitalized patient. We will never know. The consequences of this seemingly minuscule event, however, will be with us for years to come.

Many mutations are repaired before they can affect phenotype

Part of the reason we see so few mutations is that most of them are quickly repaired. Both prokaryotic and eukaryotic cells have a sizable repertoire of repair enzymes that guard against mutations. Only those few mutations that manage to slip by these repair mechanisms persist.

In our discussion of DNA replication, we mentioned that not only does DNA polymerase incorporate new nucleotides during DNA synthesis, but also it can check its work and repair any mistakes. This self-checking capacity is called **proofreading**. If DNA polymerase inserts the wrong nucleotide, the enzyme often cannot proceed because the aberrant base cannot properly hydrogen-bond to the nucleotide on the parental strand (**Figure 7.38**). DNA polymerase then actually moves in reverse, cutting out the last nucleotide that it added. Following removal of the incorrect nucleotide, the enzyme gets

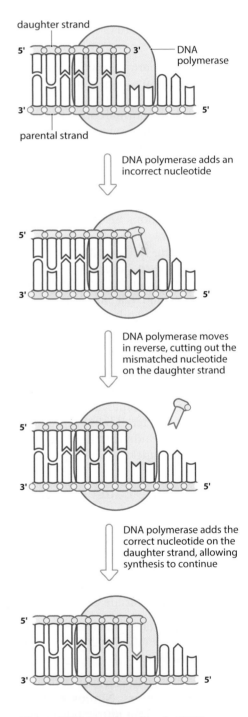

Figure 7.38 Proofreading by DNA polymerase. If an incorrect nucleotide is added to a growing strand, the DNA polymerase can cleave it from the strand before continuing.

a second chance to insert the correct nucleotide into the growing daughter strand.

One way that dimers can be repaired is by **DNA mismatch repair enzymes**, which recognize and cut the damaged strand and remove a small section of DNA on both sides of the dimer in the process. A molecule of DNA polymerase replaces the removed bases, restoring the DNA to its mutation-free state. This same mechanism can be used to remove and replace mismatched bases that were able to avoid repair by DNA proofreading. As a result, this **mismatch repair** functions as a backup to the proofreading ability of DNA polymerase (**Figure 7.39**).

Mutations are the raw material of evolution

Although mutations are rare, random events, they are the only way that new genetic information can initially come into being. If mutations never occurred, all members of a species would not only have the same genes, they would also have precisely the same alleles of those genes; they would be genetically identical. Genetic recombination, whether sexual or asexual, would accomplish nothing, since new gene combinations would not be possible. If the environment changed in any way, as it does for bacteria such as *S. aureus* when vancomycin is being used, there would be no lucky few who just happened to have the right form of a gene to survive and thrive in the face of this changing environment.

In other words, there would be no **evolution**—the genetically based change in the phenotype of a population over successive generations. Such change can happen over time precisely because of genetic variability in a population. Because of such variation, under any specific environmental conditions, some individuals are better adapted for survival than others. Those whose genetic makeup makes them more suited to a particular environment are more likely to survive and tend to produce more offspring. These offspring inherit those genetic traits that result in better adaptation, and over many generations, the population as a whole comes to take on the phenotype of the better-adapted individuals. The diversity of living things we see today, both microbial and nonmicrobial, owes its existence to this never-ending interplay between genes and the environment. We will revisit the topic of evolution, especially as it applies to microorganisms, in Chapter 9.

Looking back and looking forward

As microbiologists we are confronted by issues every day that only make sense in light of genetics. With an understanding of basic genetic principles, we are well positioned to understand phenomena such as drug resistance and the appearance of pathogenic strains of certain microbes that are typically not thought of as especially pathogenic. We also begin to comprehend the enormous diversity found within the microbial world and the underlying genetic mechanisms that explain it.

In no way is that diversity more on display than in an investigation of metabolism—the means by which cells use energy. The sky's the limit when it comes to the various ways that microbes meet their energy needs, and understanding how they do it will help us to appreciate how, among other things, bacteria are used to make yogurt, why hydrogen peroxide is a good antiseptic to keep in your medicine cabinet, and why rotting fish stink.

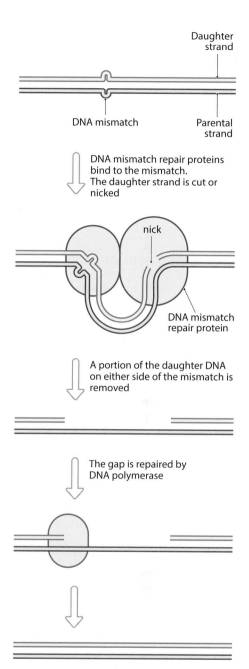

Figure 7.39 Mismatch repair. DNA repair enzymes can correct errors that occur during DNA replication.

Garland Science Learning System

- http://garlandscience.rocketmix.com/students/

- Discover how bacteria turn genes on and off

- See sexual contact between bacteria

- Test your knowledge of this chapter by taking the quiz

- Familiarize yourself with the terminology used in this chapter by using the vocabulary review

- Get help with the answers to the Concept questions

Concept questions

These questions are designed to help you start thinking like a microbiologist. The answers are not always simply found in the text. Instead, you will need to take the concepts about which you have learned and apply them to new situations. Some of the questions may not even have just a single correct answer. Help is provided as part of the GSLS resources, which can be accessed through http://garlandscience.rocketmix.com/students/.

1. If a particular DNA molecule consisted of 28% guanine-containing nucleotides, what percentages of nucleotides would contain adenine, cytosine, and thymine?

2. DNA is described as having an antiparallel structure. What exactly does this mean?

3. In Chapter 7, we learned about the "proofreading" ability of DNA polymerase. But as we learned in Chapter 5, some viruses rely on RNA rather than DNA as their genetic material. Furthermore, when these viruses replicate their RNA genome, they rely on other enzymes, such as RNA-dependent RNA polymerase, which do not have proofreading ability. On the basis of this information, how do you think the mutation rate in RNA viruses compares with that in DNA viruses?

4. A stretch of DNA has the following sequence:

   5′-GTAATCGTAGCTTAC-3′
   What would the complementary DNA strand be? (Be sure to label the 5′ and 3′ ends, remembering that nucleic acids are always drawn in the 5′ to 3′ direction.)

5. A stretch of mRNA has the following sequence:

   5′-AGUUCUAGGCCC-3′
   Using Figure 7.20, translate this sequence into the proper amino acids.

6. Suppose that, in a particular strain of bacteria, the *lac* operon is mutated in such a way that it can never be repressed. In other words, it is permanently in the induced state. What problems would you expect this to cause for the bacteria?

7. A particular species of bacteria has an operon called the *ara* operon. When it is active, enzymes are produced that allow the bacteria to digest the sugar arabinose. Does this sound like an inducible system (like the *lac* operon) or a repressible system (like the *trp* operon)? Under what environmental conditions would you expect the operon to be actively transcribed?

8. *Aliivibrio fischeri* cells are added to a series of test tubes containing appropriate growth solution in increasing concentration. No bioluminescence is seen at lower concentrations. At a particular intermediate concentration, cells are observed to glow. Do you think tubes containing cells at even higher concentration would also glow? Why or why not?

9. A new form of prokaryote is discovered that has its own, unique genetic code. Unlike any other form of life, this new organism uses only a single type of codon for each amino acid. Compared with other organisms, do you think that harmful mutations would be more or less common in this newly discovered prokaryote?

10. In a large colony of bacteria, all individual cells have identical genotypes. You remove members of this colony and place them in different environmental conditions. For example, some are placed at relatively high temperature, while others are placed at relatively low temperature. Some are placed in a nutrient-rich environment, while others are placed in a nutrient-poor environment. After several generations in these new environments, you inspect the cells in each. Based on what you have learned about genetics here in Chapter 7, do you expect them to all have the same phenotype? Why or why not?

METABOLISM AND GROWTH

8

On the evening of March 20, 1966, Washtenaw County, Michigan, was the scene of one of the best documented UFO sightings of all time. Local residents observed flying saucers hovering above and even landing in a local swamp. Several photographs were taken before the mysterious objects vanished into the night sky (**Figure 8.1**).

In the 1960s, the US Air Force still administered Project Blue Book, which investigated purported UFO sightings. As part of this project, J. Allen Hynek arrived on the scene to look into the mystery. Although UFO buffs contend to this day that Hynek, who later coined the phrase "close encounters of the third kind," was pressured by his superiors, the investigator's conclusion was that the sightings were the result of swamp gas—a term that has subsequently often been used to dismiss UFO sightings.

Washtenaw County may or may not have been visited by extraterrestrials, but one thing is certain; countless microorganisms thrive in the oxygen-depleted, organically rich water of swamps. Many of these microbes produce swamp gas—waste products such as methane gas that can sometimes produce eerie glows. And such swamp gas may be responsible for at least some of the odd things, including UFOs, that people sometimes report seeing in swamps.

There is nothing very unusual about what these microorganisms are doing. All cells, microbial or otherwise, break down molecules to provide the energy needed for survival and growth. As cells digest these molecules, waste products are released. In the case of swamp-dwelling prokaryotes, that waste is swamp gas. As we will see, other organisms release different wastes, depending on exactly how they go about the business of breaking down large molecules.

Once cells digest these molecules, the energy they obtain can then be used in other reactions that require energy rather than release energy. For instance, all cells must construct larger biological molecules, including proteins or nucleic acids, from smaller building blocks such as amino acids or nucleotides. Such synthesis requires an input of energy. When we discuss all the ways that cells use energy, both how they get it and how they use it, we are discussing **metabolism**.

Often, in daily conversation, we use "metabolism" only as it applies to weight gain or the rate at which we burn calories (for example, "Brenda never gains weight because she's got a rapid metabolism"). But that is only part of what metabolism is all about. Metabolism is the sum total of our energy budget. It is somewhat analogous to how we use money. We don't just "earn it" (by

Figure 8.1 Alien visitors or microbial metabolism? March 20, 1966, Washtenaw County, Michigan.

All case studies have a few questions at the end, the answers to which will become apparent as you read the sections following the case.

eating); sometimes we save it in the form of polysaccharides and fats (when we "earn" more then we need), and we also need to "spend" it (to synthesize needed cell components).

Metabolism is similar in all living things

One important point should be emphasized because it simplifies our task in this chapter considerably. The basics of metabolism are similar, whether we are talking about archaea growing in a hot spring, a lilac bush growing in your garden, or a rabbit sneaking in at night to eat the lilac. You may recall from Chapter 1, in listing the properties that distinguish life from nonlife, that all organisms need an energy source, which they convert to other usable forms of energy. This means that in a broad, general sense, what applies to one type of living thing applies to others as well. This is not to say all living things are identical in every detail. To highlight just one difference, while animals, fungi, and many microorganisms have to eat or absorb organic compounds, plants and some microorganisms make their own biological molecules. The next time you feed your house plants, look closely at the ingredients in the plant food; you will see no amino acids, lipids, or sugars listed. Plants make all these things themselves as long as they are supplied with a few crucial raw materials, such as carbon dioxide and certain minerals.

Here in Chapter 8, we will start by discussing the basics of metabolism, especially as they apply to microorganisms. We next examine some of the metabolic differences that distinguish microorganisms from each other. These differences explain not only why some microorganisms are the basis of alien sightings but also why cheddar cheese is different from Gouda or how beer is made (**Figure 8.2**). Another reason for studying metabolism is that it is linked closely to growth. By understanding how microorganisms utilize energy, we will be ready to reflect on why various cells reproduce at very different rates, or how a change in the environment might cause a change in a particular organism's growth rate.

Energy released from food molecules is used for other processes that require energy

When a cell digests any biological molecule—a long chain of sugars (a polysaccharide), for instance—energy is released. In this instance, the polysaccharide contains more energy than the simpler sugars that remain. This is an example of an **exergonic** (energy-out) reaction. What happens to the released energy that is no longer part of the sugars? Much of it is lost as heat or, in some interesting cases, as light (**Figure 8.3**). Some, however, may be captured by cells for energy-requiring reactions. Reactions that need an input of energy to occur are called **endergonic** (energy-in) reactions.

Exergonic and endergonic reactions are intimately linked in living things. To visualize this, think of a waterwheel in a river. For the wheel to move, an input of energy is required. In other words, the turning of the wheel is endergonic. The energy is provided by the energy released as the water flows downstream. Likewise, the synthesis of simple sugars into a polysaccharide is endergonic. As the sugars are being linked together to form the polysaccharide, another molecule is being digested simultaneously to provide the necessary energy. Put simply, the endergonic reaction needs the exergonic reaction as an energy source. Reactions that are combined in this way are called **coupled reactions** (**Figure 8.4**).

Figure 8.2 Courtesy of microbial metabolism. The cowboy in this wall painting in Tijuana, Mexico, may not know it, but the beer he is enjoying relies on single-celled fungi (yeast), which break down the sugars in grains to obtain energy. When the yeast cells are denied oxygen, they release ethyl alcohol as a metabolic waste.

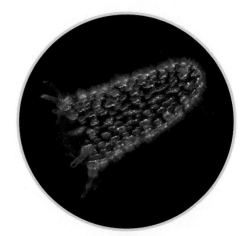

Figure 8.3 Light emitted by bacterial metabolism. This mysterious sea creature, called a pyrosome, is related to the common sea squirt. Individuals can be up to several feet long. Light-emitting bacteria in their tissues cause them to pulsate eerily at night. The light is due to the release of energy from exergonic reactions carried out by the bacteria. It is believed that the intermittent glow of the bacteria attracts food for the pyrosome. It might also start a war. In an odd historical note, some have suggested that the 1964 Gulf of Tonkin torpedo attack on an American ship, which led to American military involvement in Vietnam, was actually pyrosome flashes.

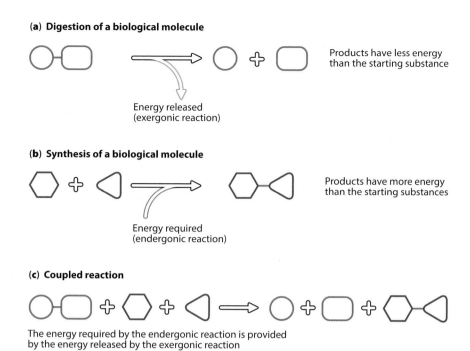

(a) Digestion of a biological molecule

Products have less energy than the starting substance

Energy released (exergonic reaction)

(b) Synthesis of a biological molecule

Products have more energy than the starting substances

Energy required (endergonic reaction)

(c) Coupled reaction

The energy required by the endergonic reaction is provided by the energy released by the exergonic reaction

Figure 8.4 Exergonic, endergonic, and coupled reactions. (a) When a biological molecule is digested, energy is released. The bonds of the larger molecule have more total energy than the bonds of all the smaller substances that remain. Such reactions, in which energy is released, are termed exergonic. (b) Synthesis reactions require an input of energy, because the bonds of the initial starting substances have less total energy than those of the newly synthesized, larger molecule. Reactions that require an energy input are called endergonic reactions. (c) In a coupled reaction, an endergonic and an exergonic reaction occur simultaneously. The exergonic reaction provides the energy necessary for the endergonic reaction to proceed.

Cells convert the energy in biological molecules into ATP

For most endergonic reactions, the required energy is supplied by **adenosine triphosphate** (**ATP**). You can think of ATP as a cell's energy currency. In a nutshell, cells try to capture energy released from exergonic reactions, so they can use it to make ATP. Cells attempt to make as much ATP as they can, to keep their energy "bank account" topped up as much as possible. The stored ATP can then be spent as needed to power endergonic reactions, such as the synthesis of a protein or a large polysaccharide.

ATP consists of three component parts: a molecule of the sugar ribose covalently bound to a molecule of adenine and three phosphate groups (**Figure 8.5**). The phosphates represent the "business end" of the molecule, because they provide the large amount of energy found in ATP. When one of these phosphates is removed, a large amount of energy is released, which is available for use in endergonic processes. After loss of a phosphate, the remaining molecule, with only two phosphates left, is called ADP (adenosine diphosphate). To have enough energy on hand, cells must replenish their ATP stock by adding phosphates back onto ADP.

Why is ATP so energy-rich? Each phosphate group carries a negative charge and the three negative charges strongly repel each other. This makes the bonds between the phosphates unstable. The chemical energy in these bonds is released when a phosphate is cleaved, much as the energy in a compressed spring is liberated when the spring is released.

Cells need a lot of ATP, and cells that are more active metabolically have even greater ATP demands. Making ATP requires rather than releases energy, and cells go to a lot of trouble to make sure they do not run out. The cells of prokaryotes and eukaryotes alike usually convert the energy in organic molecules into ATP, although there is variation in how they do it. Some cells, for example, need oxygen to synthesize ATP. Others make their ATP in the absence of oxygen, and for some of these, oxygen is a deadly poison.

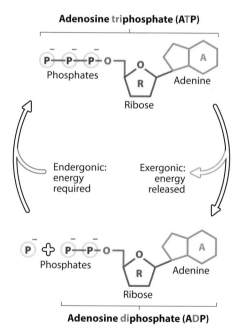

Adenosine triphosphate (ATP)

P—P—P—O
Phosphates   R   Adenine
Ribose

Endergonic: energy required

Exergonic: energy released

P   P—P—O
Phosphates   R   Adenine
Ribose

Adenosine diphosphate (ADP)

Figure 8.5 Interconversion of ATP and ADP. ATP stores large amounts of chemical energy in the bonds linking the phosphate groups. When the bond between the second and the outermost phosphate group is broken, energy is released as the terminal phosphate separates from the remaining molecule of ADP. The synthesis of additional ATP from ADP and phosphate is an endergonic (energy-requiring) process.

Microorganisms fall into all of these categories. Furthermore, not all microbes use organic molecules as an energy source for ATP production. Some use inorganic compounds as an initial energy source, while some use light for this purpose. A full discussion of all the diversity found in microbial metabolism is well beyond the scope of this text, but it is worth remembering that alternative energy is par for the course for many microorganisms.

Energy is transferred from one molecule to another via oxidation and reduction

How is the energy in biological molecules converted into ATP? Chemical energy is actually a property of electrons. Recall, however, that not all electrons in a molecule have the same amount of energy. As discussed in Chapter 2, electrons that are held more tightly by the nucleus are more stable and therefore have less energy. Electrons held less tightly by the nucleus are less stable and consequently have more energy. Energy is released when electrons at higher energy are removed from one atom or molecule and transferred to a different atom or molecule to which they are bound more tightly and thus have less energy. The atom or molecule losing electrons is said to be **oxidized**, while the one gaining electrons is **reduced** (**Figure 8.6**). Oxidation and reduction do not happen independently; if one atom or molecule is losing electrons, then another must be simultaneously gaining them. We refer to the paired oxidation and reduction as a **redox reaction**. Cells break down food molecules through a series of redox reactions.

Something similar happens, for instance, when we burn a piece of wood (**Figure 8.7a**). Wood is composed of long-chain organic molecules containing many carbon–hydrogen bonds, and the electrons forming these C–H bonds are at relatively high energy. As these bonds are broken, the electrons are transferred to oxygen, where they are more stable and at lower energy; the carbon atoms in these compounds are oxidized and the oxygen is reduced. We can further illustrate this process by looking at methane (**Figure 8.7b**), the simplest hydrocarbon. Note in Figure 8.7b that when a methane molecule is fully oxidized, the result is no longer a hydrocarbon. Ultimately all that remains is carbon dioxide; no C–H bonds remain at all. Meanwhile, additional oxygen is reduced and converted into water. The energy that was lost by the electrons during their transfer from the hydrocarbon to oxygen is released as light and heat.

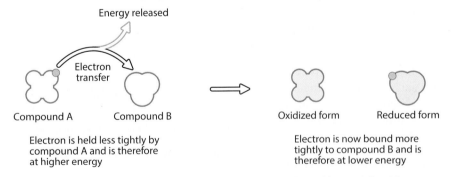

Figure 8.6 Oxidation–reduction reactions. In this example, the compound that loses one or more electrons (compound A) is being oxidized, while the compound gaining those electrons (compound B) is reduced. Because the electron is held less tightly by compound A, it is less stable and at higher energy. Following transfer to compound B, the electron is bound more tightly and is therefore more stable and at lower energy. Thus, this is an exergonic reaction, with energy being released by the electron during its transfer from compound A to compound B.

a biological molecule like a sugar or fat is oxidized or reduced, oxidation results in the loss of carbon–hydrogen bonds (see Figure 8.7b), whereas reduction results in the gain of carbon–hydrogen bonds.

When wood or a hydrocarbon fuel burns, a very large amount of energy is released in a single step—the transferred electrons go from very high to very low energy and the result is fire. But such an explosive release of energy would never work in cells. When cells oxidize biological molecules, they do so in a large number of small steps. Not only would a single enormous release of energy be inefficient, it would no doubt kill the cell. By releasing the energy gradually, the cell can use it efficiently, while remaining at temperatures that are compatible with life.

To visualize this more easily, think about the way that octane (a hydrocarbon) in gasoline is used to power an automobile. In the engine, the octane is oxidized by oxygen slowly and in small increments. In this way, the energy that is released can be efficiently utilized to move the car. You could release exactly the same amount of energy from a tank of gas, simply by throwing a lighted match into the gas tank, oxidizing all the octane at once. This certainly would move the car, but hardly in an efficient way or in a manner that would permit the survival of either the car or the driver.

Oxygen is not the only molecule that can accept electrons during redox reactions. In living things, several other molecules act as electron acceptors when food molecules are oxidized. **Nicotinamide adenine dinucleotide** or **NAD⁺** is an important example (**Figure 8.8**). This molecule is present in one of two states. In its oxidized state, it exists as NAD⁺. When reduced, NAD⁺ picks up two electrons from the molecule being oxidized, as well as a single proton, to become NADH. A second proton is released as an H⁺ ion. NADH can later be oxidized with a different molecule being reduced, releasing still more energy. The now-oxidized NAD⁺ is again ready to accept another pair of electrons. Thus, NADH acts as a sort of electron carrier, first accepting and later donating these electrons. As we will see, the energy released as these electrons are passed along provides the energy needed for a cell to produce ATP.

Cell respiration refers to the oxidation of biological molecules

When they hear the term "respiration," most people immediately think of breathing. Furthermore, we know that we breathe to get enough oxygen, which is absolutely necessary for us to survive. Far fewer people, however, know exactly why we need oxygen. We have already alluded to the reason. Many living things, including animals and many microorganisms, use oxygen to oxidize biological molecules in order to obtain the energy these biological molecules contain. Not all organisms, however, use oxygen for this purpose. As we will see, many microorganisms use other molecules precisely the way we use oxygen—to oxidize biological molecules. No matter which molecules are used, the slow incremental breakdown of biological molecules necessary for ATP production is called **cell respiration**. Essentially, breathing is simply the way animals get the oxygen needed to produce ATP. And without ATP, life is impossible.

Carbohydrates are the primary source of energy used in cell respiration

Although many different biological molecules may be digested to meet energy needs, carbohydrates are generally the molecule of first choice. Most

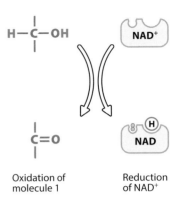

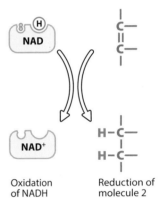

Figure 8.8 Nicotinamide adenine dinucleotide. This electron carrier can exist in either an oxidized (NAD⁺) or reduced (NADH) state. When NAD⁺ is reduced, a biological molecule (molecule 1) is oxidized, losing a carbon–hydrogen bond. The now-reduced NADH may itself become oxidized, reducing a different molecule (molecule 2), which consequently gains carbon–hydrogen bonds.

(a)

Wood   + O_2 →   Smoke + ashes + heat + CO_2 + H_2O

(b)

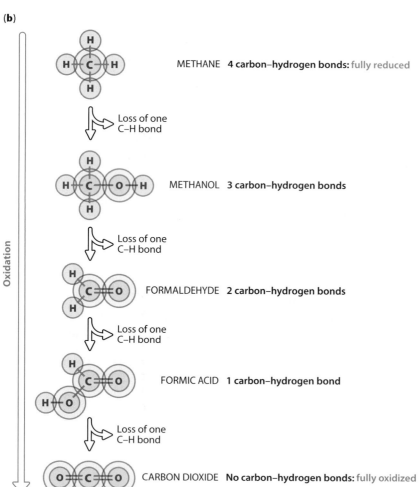

METHANE   **4 carbon–hydrogen bonds:** fully reduced

Loss of one C–H bond

METHANOL   **3 carbon–hydrogen bonds**

Loss of one C–H bond

FORMALDEHYDE   **2 carbon–hydrogen bonds**

Loss of one C–H bond

FORMIC ACID   **1 carbon–hydrogen bond**

Loss of one C–H bond

CARBON DIOXIDE   **No carbon–hydrogen bonds:** fully oxidized

Oxidation

Figure 8.7 Oxidation of carbon fuels. (a) When electrons are transferred to oxygen from the carbon atoms in a fuel such as wood, energy is released in the form of light and heat. Carbon dioxide is the completely oxidized form of carbon, while water is the reduced form of oxygen. (b) Often when biological molecules are oxidized, they lose carbon–hydrogen bonds. When they are reduced, they gain carbon–hydrogen bonds. In this example, methane (CH_4) is fully reduced, because a carbon atom can form a maximum of four bonds. Here, methane is being progressively oxidized as C–H bonds are replaced by bonds between carbon and oxygen. In its fully oxidized form (CO_2), no C–H bonds remain.

Biological molecules are typically oxidized in a series of small steps

As seen in the wood-burning example, when organic molecules are oxidized, they often lose hydrogen atoms, while molecules being reduced often gain them. When an electron is transferred, it often picks up a proton (H^+) at the same time. In biological reactions these protons are freely available in water. Consequently, when a redox reaction occurs, the net effect is that one molecule loses a hydrogen atom while a different molecule gains one. When

commonly, glucose is the starting material of cell respiration. Because glucose is so important for cell respiration, cells tend to store up as much of it as possible, generally by stringing many glucose molecules together into long polysaccharides.

Depending on the organism in question and the environmental conditions in which that organism finds itself, glucose may be completely or only partially broken down. If glucose is completely digested, all carbon–hydrogen bonds are oxidized and the released energy is used for ATP production. If glucose is only partially broken down, some carbon–hydrogen bonds remain. Because less energy is released, the yield of ATP is lower when glucose is only partly digested. We will first consider the complete digestion of glucose. We will then turn our attention to those situations in which glucose digestion can only proceed to a certain intermediate point and no further.

Cell respiration occurs in three stages

Respiration, the oxidation of biological molecules and the conversion of the released energy into ATP, takes place in a series of small steps. These steps can be grouped into three principal stages (**Figure 8.9**). In the first stage, **glycolysis**, a molecule of glucose is partially digested into two smaller molecules and some of the released energy is captured as NADH. In the second stage, the **Krebs cycle**, the oxidation of these smaller molecules is completed, and more energy-carrying molecules such as NADH are formed. In both glycolysis and the Krebs cycle, a small amount of ATP is also made, but most ATP is made in the third and final stage, called **electron transport**. Here, all the energy-carrying molecules formed in the first two stages (mostly NADH) are oxidized, molecules such as oxygen are reduced, and the released energy is used to produce the bulk of a cell's ATP.

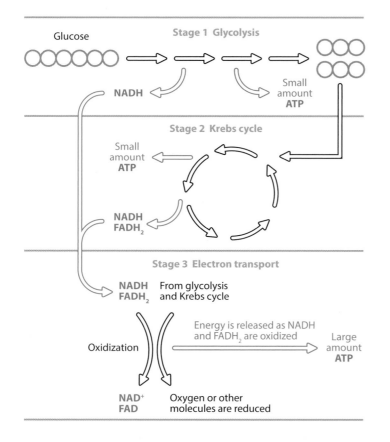

Figure 8.9 Cell respiration occurs in three stages. In stage 1, glycolysis, glucose (a six-carbon compound, represented by six gray circles) is partially oxidized to two smaller glucose derivatives (two three-carbon compounds, represented by three gray circles). As glucose is oxidized, NAD^+ is reduced to NADH. The oxidation of glucose is completed in stage 2, the Krebs cycle, as glucose derivatives from glycolysis are oxidized in a series of steps. Each oxidation results in the reduction of NAD^+ or a similar molecule called FAD, forming NADH and $FADH_2$. Small amounts of ATP are made in these first two stages, but most ATP is synthesized when NADH and $FADH_2$ formed in glycolysis and the Krebs cycle are oxidized in stage 3, electron transport. Energy released from these oxidations provides the energy needed for ATP synthesis in electron transport.

In glycolysis, glucose is partially oxidized, forming two smaller molecules called pyruvate

Glycolysis, stage 1 in cell respiration, is where glucose oxidation begins (see Figure 8.9). By the end of glycolysis, glucose has been split into two molecules of pyruvate. A small amount of ATP is generated in glycolysis, and some of the electrons originally with the glucose are transferred to NAD$^+$, forming NADH. We will now consider these events in more detail.

Glycolysis is common to prokaryotic and eukaryotic cells alike. Glycolysis literally means "the splitting of glucose," which accurately describes what happens, as shown in **Figure 8.10**. One molecule of glucose (a six-carbon molecule) is broken down in a series of steps, into two molecules of **pyruvate** (each with three carbons). Along the way, several notable events occur. First, early in glycolysis there are energy-requiring (endergonic) steps. ATP provides the necessary energy. Although the goal of cell respiration is to make ATP, a small amount of ATP must be consumed in these early steps.

Second, the six-carbon compound is cleaved into two three-carbon compounds. Each of these three-carbon compounds is ultimately converted into pyruvate, the end product of glycolysis. Before that happens, however, each of these three-carbon molecules is oxidized. Recall that when one molecule is oxidized, a different molecule is reduced. In this case, the molecule being reduced is NAD$^+$, which is consequently converted to NADH. These NADH molecules, along with others that will be produced later, will ultimately be used in the final stage of cell respiration, where most ATP production takes place.

Some ATP, however, is directly produced in glycolysis. Two steps in particular are highly exergonic. The energy that is released at these steps is coupled to the synthesis of ATP from ADP and phosphate.

To summarize, each molecule of glucose that enters glycolysis is broken down into two molecules of pyruvate. NADH molecules are formed when intermediate compounds in the pathway are oxidized, and exergonic steps are coupled to the synthesis of a small amount of ATP. The pyruvates remaining at the end of glycolysis will next be oxidized in stage 2 of cell respiration: the Krebs cycle.

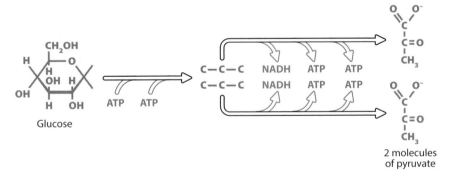

Figure 8.10 Glycolysis. Glucose (containing six carbon atoms) is cleaved into two molecules, each with three carbons, which requires the input of energy in the form of ATP. As each of these three-carbon compounds is oxidized, a molecule of NAD$^+$ is reduced to NADH. The now-oxidized three-carbon compounds continue through a series of steps, ultimately giving rise to two molecules of pyruvate at the end of glycolysis. Highly exergonic steps in the pathway are coupled to the synthesis of ATP.

In the Krebs cycle, pyruvate from glycolysis is completely oxidized and the energy released is transferred as electrons to NAD⁺ and FAD

In stage 2 of cell respiration, the Krebs cycle (see Figure 8.9), pyruvate is completely oxidized in a series of steps. At some of these steps, the energy-carrying compound NAD^+ and a similar compound, FAD, are reduced to NADH and $FADH_2$, respectively. These energy carriers, along with those produced in glycolysis, will provide the energy for ATP synthesis in the third and final stage of cell respiration. We will now take a more in-depth look at what takes place in the Krebs cycle.

At the end of glycolysis, two pyruvate molecules remain from each molecule of glucose. An examination of pyruvate structure (see Figure 8.10) should convince you that there are still several carbon–hydrogen bonds available for oxidation. These bonds are oxidized in the Krebs cycle.

Figure 8.11 summarizes the Krebs cycle. Notice that pyruvate is converted from a three-carbon to a two-carbon molecule before the cycle actually begins. The third carbon is released as CO_2 and a molecule of NAD^+ is reduced to NADH.

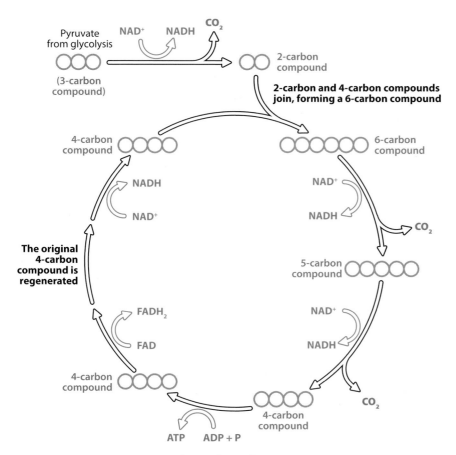

Figure 8.11 The Krebs cycle. Gray circles represent carbon atoms. Pyruvate is first oxidized to a two-carbon compound, which enters the Krebs cycle. This two-carbon molecule joins a preexisting four-carbon compound to form a six-carbon compound. This compound is subsequently oxidized in a series of steps back to the four-carbon compound. Carbons are released in the form of CO_2, and NAD^+ and FAD are reduced to NADH and $FADH_2$, respectively. ATP is generated by coupling its synthesis to an exergonic step of the cycle.

Next, the newly formed two-carbon compound joins an already existing four-carbon compound to form a six-carbon compound. As illustrated in Figure 8.11, this six-carbon compound is progressively oxidized in a series of steps back to the four-carbon compound, losing two additional carbons as CO_2 along the way. The fact that the starting four-carbon compound is ultimately regenerated is why the process is called a cycle. This newly regenerated four-carbon compound is now ready to join another two-carbon compound to begin another round of the cycle.

Figure 8.11 highlights several key events that occur during the cycle. First, as mentioned, two molecules of CO_2 are released. By the end of the Krebs cycle, all of the carbon atoms that were originally found in pyruvate have been released as CO_2. The hydrogens that were originally bound to carbon, however, were transferred to NAD^+ in a series of redox reactions. In one step, a molecule similar to NAD^+ called FAD is reduced instead, forming a molecule of $FADH_2$. FAD has the same function as NAD^+: it serves as an electron carrier, becoming reduced to $FADH_2$ as another molecule is being oxidized. Additionally, note that in one highly exergonic step of the Krebs cycle a molecule of ATP is made, similarly to what we observed in glycolysis. Finally, remember that at the end of glycolysis, two pyruvates remained from the initial glucose. Each of these pyruvates is oxidized in the Krebs cycle, meaning that for each molecule of glucose, the Krebs cycle takes place twice.

To summarize, by the end of the Krebs cycle, glucose has been completely oxidized. The carbon atoms originally present in the glucose have been released as CO_2. A small amount of ATP has been formed in coupled reactions, but much of the energy that was in the carbon–hydrogen bonds of the glucose molecule is now present in the reduced forms of NAD^+ and FAD—namely, NADH and $FADH_2$. These electron carriers will now be oxidized, in the third and final stage of cell respiration. As NADH and $FADH_2$ are oxidized, the energy released will be used to produce most of the cell's ATP.

NADH and $FADH_2$ are oxidized in electron transport, providing the energy for ATP synthesis

Most ATP is synthesized in electron transport, the third and final stage of cell respiration (see Figure 8.9). Briefly, the NADH and $FADH_2$ produced in glycolysis and the Krebs cycle are oxidized. The released electrons are passed between a series of electron carriers and are ultimately passed to oxygen or other compounds. These now-reduced compounds represent the final reduced waste of cell respiration. The energy released as electrons pass from carrier to carrier is used to pump protons across a membrane. As these protons build up on one side of the membrane, they provide a source of potential energy. The protons then flow back to their original location through an enzyme complex. Like water flowing over a dam that is used to generate electricity, the flow of protons through this enzyme complex is utilized to synthesize ATP. Let's now take a closer look at this remarkable process.

A large number of NADH and a smaller number of $FADH_2$ molecules were produced in glycolysis and the Krebs cycle, due to the step-by-step oxidation of glucose. In electron transport, these reduced molecules are oxidized, and the electron pair released from each NADH or $FADH_2$ is passed through a series of electron carriers known collectively as the electron transport chain.

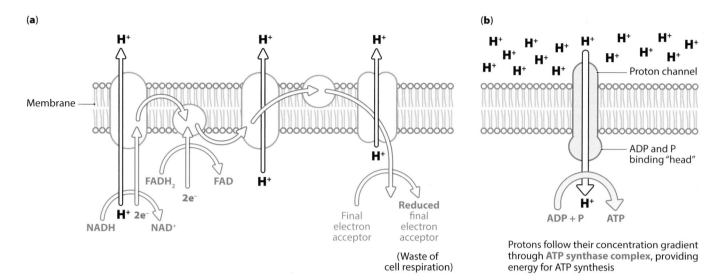

Figure 8.12 The electron transport chain. In prokaryotes, the components of electron transport are embedded in the plasma membrane. In eukaryotes, these components are found in the inner mitochondrial membrane (a) NADH or FADH$_2$, formed in glycolysis and the Krebs cycle, is oxidized by components of the chain in a series of redox reactions, until they are passed to a final electron receptor. As electrons are passed from component to component, the energy lost by the electrons is used to pump protons across the membrane. (b) The buildup of protons supplies energy for ATP synthesis. The protons flow back to either the cytoplasm (prokaryotes) or the mitochondrial matrix (eukaryotes) through the ATP synthase proton channel. As they do so, these protons release energy, which causes the head portion of the ATP synthase to spin rapidly. This spinning provides the force to join ADP and P, forming ATP.

The components of the electron transport chain are embedded in membranes: the plasma membrane of prokaryotes and the inner mitochondrial membrane of eukaryotes (**Figure 8.12**).

As we see in Figure 8.12, the first components of the electron transport chain oxidize NADH or FADH$_2$, converting either of these electron carriers back to NAD$^+$ or FAD, which can now be reused in glycolysis or the Krebs cycle. Electrons are subsequently passed through the entire chain in a series of redox reactions that ends when the electrons are ultimately passed to a **final electron acceptor**. The final electron acceptor, now reduced by the electrons that have passed down the chain, is released as a waste product. As electrons are passed from one component of the chain to the next, these electrons lose energy. Thus, the redox reactions of the electron transport chain are exothermic. As we will shortly see, this energy release is used to generate the lion's share of a cell's ATP.

The components of the electron transport chain fall into only a few categories of molecules. Some are membrane-bound proteins, which contain associated molecules derived from B vitamins. Other protein components contain metallic ions such as iron. Still others are nonprotein, hydrophobic molecules.

Different organisms use different molecules in their electron transport chains. Bacteria, archaea, and eukaryotes all typically differ in their electron transport components. Within the domain Bacteria alone, there is significant variety. In fact, the identification of pathogenic bacteria in a diagnostic laboratory is sometimes based on the identification of electron transport components. *Pseudomonas* bacteria, for example, have an electron chain component called cytochrome oxidase, which is lacking in many other species. A simple test for the presence of this compound can help to distinguish *Pseudomonas* from other organisms.

Different microorganisms may utilize different final electron acceptors

In many prokaryotes, and even more commonly in eukaryotes, oxygen is the final electron acceptor at the end of electron transport. As oxygen accepts the electrons that were released by NADH or $FADH_2$ at the beginning of the electron transport chain, the oxygen is reduced to H_2O. By the time the electrons unite with oxygen, they have lost a great deal of their energy. The bonds between the hydrogens and the oxygen are very stable and at low energy. Water, in this case, represents a waste product. This is the reason that so many living things need oxygen to survive; we need it as a final electron acceptor in electron transport. Without oxygen to remove electrons at the end of the electron transport chain, respiration grinds to a halt. Organisms that utilize oxygen in this way are relying on **aerobic respiration**.

CASE: NO FISH FOR YOU!

Anya loves seafood, and with company coming for dinner she goes to the fish market and purchases several fresh halibut filets. After buying the fish, she runs several errands, and by the time she arrives home, over an hour has passed. When she unpacks the fish in the kitchen, Anya notices that it smells funny. Pressing it to her nose, she now cannot avoid the fact that her planned entrée absolutely stinks, emitting a strong fishy odor. Confused and angry, she discards the fish in the trash. The halibut was odor-free and fresh when she selected it in the fish market. In any event it looks like fish is off the menu this evening.

1. **What happened to the fish? Why does unfrozen fish so quickly turn fishy?**
2. **Why don't beef and chicken start to smell bad unless they've been left unfrozen much longer?**

We are so used to thinking of oxygen as essential for life that it can be surprising to learn that many microorganisms do just fine without it. These **anaerobic** microbes rely on final electron acceptors other than oxygen. Some anaerobes never utilize oxygen. Others use oxygen when it is available but readily switch to a different final electron acceptor when oxygen is absent.

Examples of final electron acceptors used by anaerobes include sulfate (SO_4^{2-}) and nitrate (NO_3^-). When sulfate accepts electrons at the end of electron transport, the now-reduced molecule, hydrogen sulfide (H_2S), is released as a waste. Some organisms use carbon dioxide as a final electron acceptor, reducing it to methane (CH_4). Methane is a greenhouse gas, so-called because of its ability to trap heat in the atmosphere. Consequently, **methanogens** (generators of methane) contribute to the problem of global warming. Methane is also the primary component of swamp gas. Some would argue that fantastic sightings of aliens, swamp creatures, and the like have more to do with methanogens busily making ATP than with anything else.

And Anya's ruined fish dinner? The flesh of saltwater fish contains a substance called trimethylamine oxide. This odorless substance is used by certain species of anaerobic bacteria as a final electron acceptor in electron transport. Once it accepts electrons, however, the now-reduced trimethylamine is anything but odorless. It has the strong fishy smell that we associate with fish that is not fresh. Because mammals and birds do not have trimethylamine oxide in their flesh, beef and chicken will not develop this unpleasant odor. Any bacterial growth via anaerobic respiration is enough to make fish smell bad. If fish is very fresh or frozen there will be no foul smell, because it contains very few bacteria. That is why, when purchasing seafood, it is important to get it straight home and into the refrigerator.

Protons that are pumped across a membrane in electron transport provide an energy source for ATP synthesis

The yield of ATP obtained directly in glycolysis and the Krebs cycle is modest. Small amounts of ATP were produced in these earlier stages when ATP synthesis (an endergonic reaction) was coupled to exergonic reactions. This relatively small ATP harvest represents about 4% of the energy found in a molecule of glucose. Consequently, by the end of the Krebs cycle, only about 4% of glucose's energy has been converted into a form that a cell can directly use to meet its own energy needs.

Although we have discussed how the energy in electrons is slowly released in electron transport, we still have to explain how this exergonic release of energy translates into ATP production. In Figure 8.12a, notice that as electrons pass through the various components of the chain, the energy released by these electrons is used to pump protons (H^+) across the membrane. This results in a buildup of protons on one side of the membrane (outside the plasma membrane in prokaryotes, outside the inner mitochondrial membrane in eukaryotes). This unequal distribution of protons (high concentration of H^+ on one side of the membrane and low on the other) is called a **proton gradient** (Figure 8.12b). The energy from the electrons has thus passed to the proton gradient, where it is temporarily stored.

The protons that have been pumped across the membrane cannot simply diffuse back. They can, however, flow down their concentration gradient through an enzyme complex called **ATP synthase** (see Figure 8.12b). This remarkable enzyme complex consists of two important and related parts. First, there is a channel-forming portion, which permits protons to move across the membrane. Second, there is a head portion that binds onto ADP and phosphate. As the protons that have built up on one side of the membrane stream back across the membrane through the ATP synthase channel, they release energy. This energy causes the ADP and phosphate-binding head to spin rapidly. This rotation provides the force to link ADP and phosphate, forming ATP, much as the spinning of a windmill provides energy to pump water or perform other tasks.

In aerobic respiration, each glucose molecule digested in glycolysis and the Krebs cycle generates enough NADH and $FADH_2$ to make about 35 ATP molecules in electron transport. Earlier we stated that the ATP synthesized in glycolysis and the Krebs cycle represented about 4% of the energy originally contained in a molecule of glucose. With the addition of ATP produced in electron transport, about 40% of glucose's energy has now been converted into ATP. The amount of ATP produced in anaerobic respiration is variable, but since there are usually fewer carriers in the electron transport chain, there are fewer protons pumped, and ATP yield is often correspondingly less than it is for aerobic respiration. Because they produce less ATP, many anaerobes tend to grow slowly relative to aerobes.

Having discussed the three stages of cell respiration, we should now have a good idea about how cells convert energy in the carbon–hydrogen bonds of glucose into ATP. Remember, however, that sometimes cells cannot break glucose down completely. Instead, glucose is only partially digested. We will now turn our attention to those situations in which cells call it quits before cell respiration is complete.

Incomplete glucose oxidation results in fermentation

> ### CASE: DRIVING UNDER THE INFLUENCE (OF YEAST!)
>
> In 2016 a woman in the upstate New York town of Hamburg employed an unusual defense, which resulted in her drunk driving charge being dismissed; she claimed that she was suffering from "gut fermentation syndrome," in which an abundance of yeast living in her intestine were actually producing the alcohol causing her erratic driving. Tests confirmed that under certain conditions, her blood alcohol levels would skyrocket, even when she had not consumed any alcoholic beverages. After confirming these test results, the judge dismissed all charges.
>
> 1. **Why were yeast in the woman's intestine producing alcohol? Under what conditions are they likely to do so?**
> 2. **What can be done to alleviate this problem?**

The next time you are gasping for air after a particularly strenuous workout, try thinking about what is happening in your cells to take your mind off your discomfort. In a nutshell, your muscle cells are running out of their final electron acceptor—oxygen. That means that respiration can proceed only through glycolysis before it stops. To get back to aerobic respiration and its greatly increased ATP yield, your body responds with increased breathing and heart rate while you sit down to catch your breath. Many prokaryotes, however, can survive indefinitely on glycolysis alone. It should not be surprising that when doing so, such microorganisms grow slowly, owing to their relatively paltry ATP production.

If cells are relying solely on glycolysis to meet their ATP needs, even briefly, what happens to the end product, pyruvate? You might guess that it is released as a waste, but this is not the case. Remember that glycolysis produces a small amount of NADH as well as pyruvate. In the presence of the final electron acceptor, this NADH is oxidized in electron transport, resulting in additional ATP synthesis. Without the final electron acceptor, this cannot occur. Not only is the NADH of no further value, but the cell is now in danger of running out of NAD^+. If the supply of NAD^+ gets too low, even glycolysis stops. Consequently, in the absence of its final electron acceptor, the cell regenerates NAD^+ by oxidizing NADH and reducing pyruvate (**Figure 8.13**). The now-reduced form of pyruvate is released as a waste. This process is known as **fermentation**, and the waste that is released is the **fermentation waste product**.

For many cells, fermentation is like a safety net. Although ATP yield is limited to the low quantities made in glycolysis, you can think of fermentation as a way that such cells get by until the final electron acceptor again

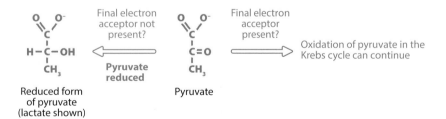

Figure 8.13 The fate of pyruvate. If the necessary final electron acceptor is available, pyruvate will be further oxidized, releasing additional energy to be used in ATP synthesis. If it is unavailable, pyruvate must be reduced in order to regenerate NAD^+, which is required in glycolysis. The reduced form of pyruvate (lactate, in this example) is released as a waste.

becomes available. The length of time that cells can persist on such skimpy ATP production depends on how much ATP they need to survive. Most large animals cannot survive like this for long, because their demands for ATP are too high. Deprived of their final electron acceptor (oxygen) for more than a few minutes, they die. Many microorganisms, however, can persist on fermentation alone indefinitely.

The fermentation waste product varies according to cell type. In an animal's muscle cells, pyruvate is reduced to lactate (see Figure 8.13). Microorganisms release a wide variety of fermentation waste products, many of which are useful in food preparations. Pickles and sauerkraut, for instance, are the result when bacteria are used to ferment the sugars found in cucumbers and cabbage, respectively. Different cheeses have different flavors because the various microorganisms used to ferment milk release different fermentation wastes. Other microbial fermentation wastes such as acetone (see Chapter 6, pp. 120–121) are utilized in industrial processes. In Chapters 16 and 17 we will take a closer look at the role of microbial fermentation in the production of many familiar foods and industrial products.

And the woman who apparently produced her own alcohol? Individuals with gut fermentation syndrome (also known as auto-brewery syndrome) suffer from a rare situation in which an abundance of yeast is growing in the intestine. Although several yeast species can cause this problem, the most likely culprit is *Saccharomyces cerevisiae*, the same species that is commonly used to make beer and wine. Yeast cells will readily use oxygen as a final electron acceptor, but the intestine is very low in oxygen concentration. And yeast cells growing in such anaerobic environments reduce pyruvate to ethyl alcohol and CO_2, the exact same fermentation wastes upon which baking and brewing depend (**Figure 8.14**). So, when people with gut fermentation syndrome consume large amounts of sugar, the yeast increase their rates of fermentation, resulting in alcohol production. As this alcohol is absorbed across the intestine, blood alcohol levels go up, sometimes to dangerously high levels. When she was arrested, the woman in our case had a blood alcohol level of 0.3—more than three times the legal limit.

Treatment focuses on the use of antifungal drugs. There are dietary restrictions, too. The diet should be yeast-free, and sugar should be avoided, in order to deny intestinal yeast the raw material needed for fermentation. Other carbohydrates should be consumed in very low amounts for the same reason.

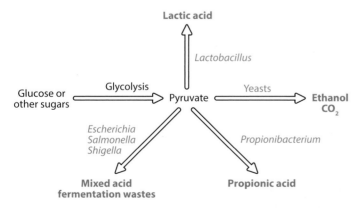

Figure 8.14 Microbial fermentation waste products.
Depending on the microorganism, pyruvate may be reduced to various final fermentation wastes. Many organisms release a combination of waste products, and many fermentation wastes are utilized in the production of various food products.

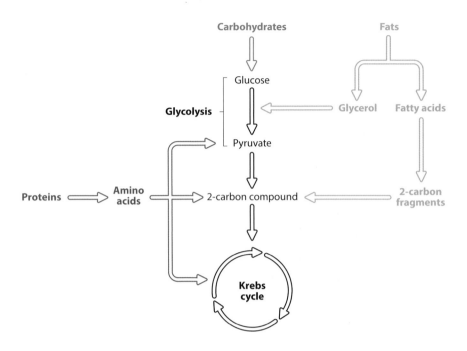

Figure 8.15 Respiration of biological molecules. Carbohydrates, fats, and proteins can all be utilized in cell respiration. Each of these biological molecules can enter the respiratory pathway at different points.

Molecules other than glucose can be used to generate ATP

Although our discussion of cell respiration has focused on glucose, cells can meet their energy needs with various other organic molecules, including other sugars, fats, and proteins. As biological molecules are digested into their component parts, these smaller molecules may enter the cell respiration pathway at various points. Fats, for example, are energy-rich molecules with many carbon–hydrogen bonds available for oxidation. During digestion, fats are first hydrolyzed into glycerol and fatty acids by a group of enzymes collectively called **lipases**. Glycerol and fatty acids are then converted into other intermediate compounds in the cell respiration pathway (**Figure 8.15**). Proteins are first digested into amino acids by enzymes known as **proteases**. Depending on the amino acid in question, it may enter the respiratory pathway at various points.

Not all cells can use all substances. Only those with the required enzymes are able to digest a specific type of biological molecule. To provide just one example, some bacteria can digest the amino acid phenylalanine, because they produce an enzyme called phenylalanine deaminase. Others, lacking this enzyme, cannot use this amino acid as an energy source. A simple laboratory test can determine whether or not a particular species digests phenylalanine (**Figure 8.16**).

Autotrophs produce their own biological molecules

Metabolism includes both the harvesting of energy from biological molecules, such as carbohydrates and fats, and the use of this energy by the cell to carry out various life processes. To obtain these energy-yielding molecules, all living things fall into two broad categories, called heterotrophs and autotrophs. **Heterotrophs** (different feeders) either eat or absorb such molecules. All animals and fungi, along with many microorganisms, fall into this group. Green plants, however, as well as numerous prokaryotes, make their own biological molecules, as long as they are supplied with a few crucial resources such as carbon dioxide and nitrogen. These self-feeders are called **autotrophs**.

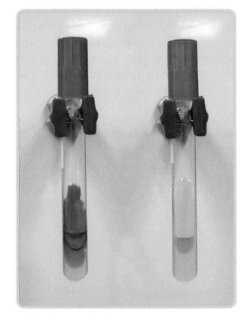

Figure 8.16 The phenylalanine deaminase test. The slant contains the amino acid phenylalanine. Bacteria are streaked onto the slant. Bacteria that digest the amino acid can do so because they produce a specific enzyme called phenylalanine deaminase which permits phenylalanine digestion. An acidic waste is produced as the amino acid is digested. Ferric chloride added to the slant reacts with the acid, producing a green color. If bacteria are negative for the enzyme, no acid waste is produced in the slant. Added ferric chloride remains yellow.

Aerobic cell respiration

$$C_6H_{12}O_6 + 6O_2 \longrightarrow 6CO_2 + 6H_2O + \text{Energy (ATP)}$$

Photosynthesis

$$C_6H_{12}O_6 + 6O_2 \Longleftarrow 6CO_2 + 6H_2O + \text{Energy (sunlight)}$$

Figure 8.17 Summary equations for aerobic cell respiration and photosynthesis. Aerobic cell respiration and photosynthesis in green plants and some bacteria are largely opposites of each other. In respiration, glucose is digested into carbon dioxide and water, and the released energy is used for ATP synthesis. In photosynthesis, energy (sunlight) is used to make glucose from carbon dioxide and water.

The synthesis of biological molecules by autotrophs is endergonic. It therefore requires energy in the form of ATP. Autotrophs get this ATP in one of two ways. **Photoautotrophs** convert the energy in light to ATP, whereas **chemoautotrophs** use the chemical energy in certain molecules to accomplish this same task. We will next briefly consider these two categories of self-feeding organisms.

In photosynthesis, energy in sunlight is used to produce biological molecules

Photoautotrophs—green plants and some microorganisms—have a unique skill; they can create biological molecules out of carbon dioxide and hydrogen-containing compounds, using sunlight as an energy source. This ability is called **photosynthesis**.

Not all photosynthesis happens in precisely the same way, but for green plants and cyanobacteria, photosynthesis is in many respects the opposite of aerobic cell respiration (**Figure 8.17**). We have already seen that, in aerobic respiration, glucose ($C_6H_{12}O_6$) is digested and the carbon atoms that were originally part of the glucose are released as carbon dioxide. Oxygen serves as the final electron acceptor, and once it has accepted electrons, it is released as water. Because this process is exergonic, released energy is used to make ATP.

In photosynthesis, on the other hand, glucose is being made rather than digested (see Figure 8.17). The raw materials needed to make glucose are carbon, oxygen, and hydrogen atoms. The carbon and oxygen come from carbon dioxide, while the hydrogen atoms are supplied by water. When hydrogen atoms are stripped away from water for use in glucose construction, the remaining oxygen is released as a waste. In photosynthesis, energy is required rather than released, because sugar production is endergonic. Sunlight provides the initial energy source for glucose synthesis.

Photosynthesis is basically a two-step process (**Figure 8.18**). In the first step, a series of reactions called the **light reactions**, the energy in sunlight is used to produce ATP. Additionally, a molecule of **NADP⁺** (nicotinamide adenine dinucleotide phosphate) is reduced. The resulting reduced molecule, **NADPH**, serves as a source of hydrogen with which to construct the sugar molecule. NADP⁺ and NADPH are very similar to NAD⁺ and NADH and they serve a similar function: the transfer of electrons from one molecule to another (see p. 170). In step 2 of photosynthesis, ATP and NADPH made in the light reactions are utilized to synthesize sugar. This usually occurs in a set of reactions called the **Calvin cycle**. In the Calvin cycle, CO_2 absorbed from the environment is reduced in a series of steps by NADPH to produce sugar. ATP provides the energy for this endergonic process.

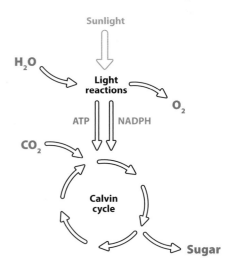

Figure 8.18 Overview of photosynthesis. In the light reactions, energy in sunlight is used to generate ATP, and NADPH is formed. These products are then used in the Calvin cycle, along with atmospheric CO_2, to synthesize sugar.

(a)   **(b)**

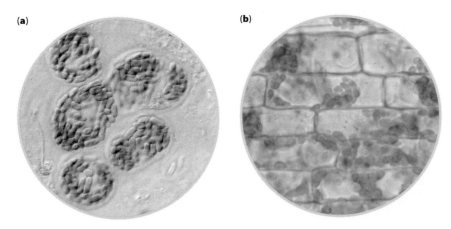

Figure 8.19 Role of membranes in light reactions. Light reactions require the light-absorbing, green pigment chlorophyll. Chlorophyll is embedded in cell membranes. (a) In cyanobacteria, these membranes are called thylakoids, and the chlorophyll in these membranes gives the cells their green color. In many other photosynthetic prokaryotes, chlorophyll is embedded in the plasma membrane. (b) In eukaryotic plants and algae, chlorophyll is found in the membranes of organelles called chloroplasts, seen here in these plant cells as the small green structures within the cell.

The light reactions depend on the presence of the green pigment chlorophyll. The light-absorbing chlorophyll molecules form complexes, which are embedded into membranes. In cyanobacteria, these membranes are called **thylakoids** (**Figure 8.19a**). In other prokaryotes, chlorophyll may be found in the plasma membrane. In eukaryotes, chlorophyll complexes are embedded in membranes found inside organelles called **chloroplasts** (**Figure 8.19b**; also see Chapter 3, pp. 64–65).

Light striking a chlorophyll complex excites electrons in the chlorophyll, elevating them to higher energy. These energized electrons are then captured by components of an electron transport chain (**Figure 8.20**). As

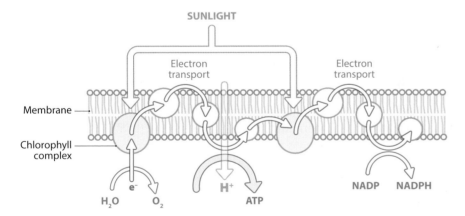

Figure 8.20 Light reactions in green plants and cyanobacteria. Light striking chlorophyll complexes energizes electrons. These high-energy electrons are passed through a series of electron carriers in an electron transport chain. As these electrons are shuttled along the chain, they lose energy, which is used to pump protons across the membrane. As these protons pass back across the membrane through ATP synthase, ATP is generated. The final electron acceptor is $NADP^+$, which is reduced to NADPH. The electrons that reduce NADP to NADPH, originally lost by the chlorophyll complexes, are replaced by electrons removed from water. Oxygen is released as waste.

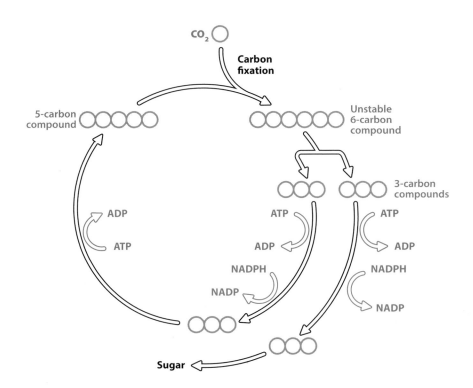

Figure 8.21 The Calvin cycle. In the Calvin cycle, carbon dioxide is fixed into sugar molecules, using NADPH and ATP made in the light reactions as sources of hydrogen atoms and energy, respectively. Carbon dioxide from the environment is first linked to a five-carbon compound (carbon fixation), forming an unstable six-carbon compound that immediately splits into two three-carbon compounds. The resulting three-carbon compounds either remain in the cycle or are used to synthesize sugar.

electrons travel down the chain between the chlorophyll complexes, protons are pumped across the membrane. The return of these protons through ATP synthase results in the production of ATP in a manner similar to that described for cell respiration. At the end of the electron transport chain, the electrons reduce $NADP^+$ to NADPH. In green plants and cyanobacteria, light is also used to split a molecule of water. Electrons extracted from the water replace those lost by chlorophyll at the beginning of the light reactions. Oxygen (O_2) is released at this water-splitting step. In other photosynthetic prokaryotes, other molecules, such as hydrogen gas (H_2) or hydrogen sulfide (H_2S), are split to provide a source of electrons. These photosynthetic microbes do not release oxygen.

The ATP and NADPH produced in the light reactions are then used to produce biological molecules, most commonly in the Calvin cycle. The principal event in the Calvin cycle is **carbon fixation**—the incorporation of atmospheric CO_2 into organic molecules. The Calvin cycle occurs in a series of steps, as illustrated in **Figure 8.21**. The resulting product is glucose or other biological molecules that are required by the cell. The Calvin cycle occurs in the cytoplasm of prokaryotes or inside the chloroplasts of eukaryotes.

Chemoautotrophs use chemical energy in the way that photoautotrophs use solar energy

Chemoautotrophs, like photoautotrophs, make their own biological molecules. They also use CO_2 as a carbon source, but instead of light energy, chemoautotrophs use reduced compounds such as ammonia (NH_3), methane (CH_4), or hydrogen sulfide (H_2S) as an energy source. The energy released when these compounds are oxidized is ultimately converted to ATP, which is used to synthesize biological molecules. We have recently described ammonia, methane, and hydrogen sulfide as wastes released by anaerobic prokaryotes. Chemoautotrophs use these waste products and thus play crucial roles in nutrient cycling and other fundamental ecological processes. We will revisit the topic of how microbes help to recycle nutrients in the environment in Chapter 10.

Microbial metabolism influences growth rate

Having completed our discussion of metabolism, we now move on to microbial growth. The reason why is simple. The rate at which microbial populations grow depends largely on metabolism. Microorganisms that are most efficient at producing and using energy usually grow the fastest. We have already mentioned, for instance, that aerobic organisms typically produce more ATP per molecule of glucose than anaerobic organisms. As a general rule, aerobic microorganisms increase in numbers much faster than anaerobes do.

But things are not quite that elementary. Environmental variables such as nutrient availability or temperature impact metabolic activity, and as metabolism changes, we expect the growth rate to change. Because different species have different environmental requirements, different species respond in various ways to changes in the environment.

Why should we care about any of this? As we have seen throughout this text, some microorganisms pose grave risks to our health and well being. Others cause less critical problems, ranging from food spoilage to underarm odor to unsightly buildups on kitchen drains. Many other microbes are more benign and provide various useful services. Such organisms help us fend off attack by pathogens, produce food and industrial products, and maintain a healthy environment. Knowing which conditions permit optimal metabolic efficiency and thus maximum growth for a given microorganism allows us to either inhibit growth of the "bad guys" or spur on the growth of beneficial microbial species.

Microbial growth refers to population growth

When we talk about microbial growth, we usually mean population growth rather than an increase in size—that is, more cells, not bigger cells. Although individual microbial cells do in fact get larger, this growth is of little significance compared with the astronomical increases in population size that microorganisms can achieve under favorable conditions.

Reproduction, and consequently increase in numbers, is generally a straightforward affair in prokaryotes. Most bacteria and archaea divide by **binary fission**—the simple division of an original parental cell into two new cells (**Figure 8.22**). This process begins with the cell's elongation and the replication of its single chromosome. The replicated chromosomes become attached to the plasma membrane, and new membrane is formed between the two chromosome attachment sites, moving the two chromosomes apart. Subsequently, new plasma membrane and cell wall material forms along the cell's midsection, dividing the cytoplasm into two sections, each with its own chromosome. The separation of these two sections gives rise to two new daughter cells.

Binary fission can quickly give rise to an enormous population, because it can result in exponential growth. As long as conditions remain favorable, the population can double every time the cells in that population divide. A single cell divides in two, the division of two cells results in four, four give rise to eight, and so forth. The time between divisions is called the **generation time**. Some species have a generation time as short as 20 minutes. Others require 24 hours or more. The length of the generation time has important implications for how fast the population grows.

Consider the consequences of exponential growth by binary fission. Suppose we are measuring population size in *Escherichia coli*, which under ideal laboratory conditions can divide every 20 minutes or so. If we start

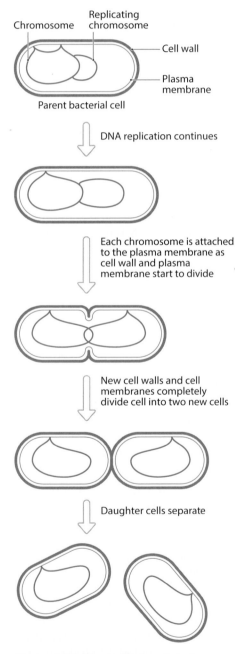

Figure 8.22 Binary fission. A single bacterium gives rise to two daughter cells. Each daughter cell, in turn, can give rise to two further daughter cells, and so forth.

Table 8.1 Exponential growth. These data assume a generation time of 20 min, typical for *E. coli* when grown under ideal conditions.

TIME (MIN)	GENERATION NUMBER	CELLS PER MILLILITER
0	0	100
20	1	200
40	2	400
60	3	800
80	4	1600
100	5	3200
120	6	6400
140	7	12,800
160	8	25,600
180	9	51,200
200	10	102,400
220	11	204,800
240	12	409,600
260	13	819,200
280	14	1,638,400
300	15	3,276,800

with a liquid broth containing 100 cells per milliliter, in 1 hour (three generation times) we would have 800 cells per milliliter (**Table 8.1**). After 2 hours, the population would be 6400 cells per milliliter, and in just 5 hours we would have over 3 million cells per milliliter. Cells are rarely fortunate enough to experience ideal growth conditions for long, however, so population increase of this sort cannot occur forever. As the population continues to increase, the growth conditions deteriorate as resources like nutrients, oxygen, and space are used up. Nevertheless, it is worth emphasizing that, given the opportunity, prokaryotic organisms are capable of astonishing feats of population increase.

A graph showing the number of generations and the number of cells in each generation quickly becomes unwieldy because of the very rapid rate of increase (**Figure 8.23**). Consequently, growth curves are often plotted in terms of logarithms (powers of 10) over a given time period or number of generations. A logarithmic value of 1 is equal to 10 (10^1 or 10×1). A logarithmic value of 2 is equal to 100 (10^2 or 10×10), while a logarithmic value of 3 is equal to 1000 (10^3 or $10 \times 10 \times 10$), and so on. By plotting time versus the logarithm of cell numbers, we obtain a more manageable straight line as the cells grow exponentially (**Figure 8.24**).

Why a straight line instead of a stair-step pattern? Put another way, if all cells waited exactly 20 minutes to divide, wouldn't there be 20-minute periods of no growth, followed by a sudden doubling in the number of cells (see Figure 8.24)? Such a growth pattern would reflect **synchronous growth**—a situation in which all the cells in a population divide at the same time. Bacterial cultures

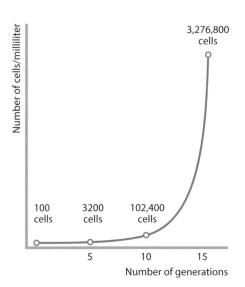

Figure 8.23 Astronomical growth potential of microbes. The graph depicts the potential for exponential growth over a number of generations.

may approximate synchronous growth initially, but different cells very quickly get out of synchrony. Some are dividing while others have just divided. Some take a little longer to divide while others take somewhat less time. The result is a smooth ascent in population size, and the generation time for the population is actually the average generation time for all the cells in the population.

Environmental factors influence microbial growth rate

In 1994, Diana Duyser made herself a grilled cheese sandwich. As she went to take a bite, she noticed a strange image on the sandwich. To the Florida woman, the image looked like the Virgin Mary (**Figure 8.25**). She kept the sandwich in a plastic box for 10 years, and mysteriously no mold ever grew, proof to some this was indeed a miracle of divine intervention. The story drew pilgrims and widespread media attention. In 2004, Ms Duyser sold the sandwich on eBay for $28,000.

We may never know if the lack of fungal growth was due to spiritual or physical causes. Fungi, however, like all living things, require not only adequate nutrition but also a hospitable environment where conditions such as temperature, moisture, and pH are compatible with survival, growth, and reproduction. Throughout this text we have seen numerous examples of how microorganisms may differ in terms of what constitutes a favorable environment. Earlier in this chapter, for example, we learned how some microorganisms depend on oxygen as a final electron acceptor in electron transport, while others use different compounds.

Regarding Ms Duyser's sandwich, microbiologists have speculated that any one of several factors might have prevented fungal growth. These include preservatives in the bread or the sandwich's dried-out condition. The acidity or calcium in the cheese may have helped to inhibit fungal growth. Whatever the cause, divine or otherwise, this example nicely illustrates that in the absence of necessary conditions living things cannot thrive.

In this section we will take a closer look at how oxygen and other crucial physical and nutritional factors can affect the metabolism and consequently the growth of different microorganisms. As we will see, knowing a particular organism's environmental requirements allows us to predict where it lives, and in some cases how it may interact with humans or other living things.

Oxygen affects microbial growth in various ways

CASE: CANINE FIRST AID

Jill has quite a menagerie of pets and she regularly has to deal with various minor animal injuries. So when Johnnie, her rooster, spikes Jessie, her collie, on the muzzle with one of his long leg spurs, Jill knows what to do; she reaches for the hydrogen peroxide and thoroughly floods Jessie's puncture wound with the liquid. Jill observes that as always, the hydrogen peroxide causes heavy fizzing at the site of the wound. She repeats this procedure several times a day for the next week, carefully inspecting the wound for signs of infection. The injury heals without complication, and Jessie is none the worse for wear, although he has a new respect for roosters.

1. What is hydrogen peroxide and how does it reduce the risk of infection?
2. Why is hydrogen peroxide especially useful for treating puncture wounds?
3. What causes the fizzing associated with hydrogen peroxide use?

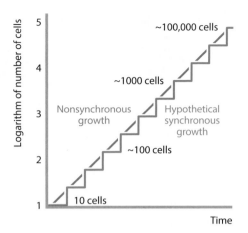

Figure 8.24 Logarithmic microbial growth. If all cells in a growing culture divided at exactly the same time, the result would be synchronous growth and the graph of growth over time would resemble a stair-step pattern. Cells in a growing culture, however, very quickly divide out of synchrony, resulting in a smooth ascent in population size.

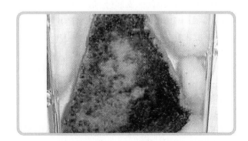

Figure 8.25 Holy toast? The grilled cheese sandwich purported to bear the image of the Virgin Mary. What may have prevented mold from growing on the toast, even after 10 years?

In our discussion of aerobic respiration we learned that atmospheric oxygen is important for many forms of life. Indeed, most multicellular organisms and a significant number of microbial species cannot survive without it. Organisms that require oxygen are called **obligate aerobes**. They need oxygen because it is the only final electron acceptor suitable for electron transport. They are, therefore, obligated to live where oxygen is present.

For other microorganisms, oxygen is important because it's so deadly. It consequently becomes something to be avoided. Why exactly must some species steer clear of oxygen at all costs?

When oxygen takes part in biological reactions, it can be converted into various compounds that have a high likelihood of removing electrons from other biological molecules. Compounds such as this are called **oxidizing agents** and they can be highly damaging to cells. Examples of such oxygen compounds include the superoxide ion (O_2^-) and hydrogen peroxide (H_2O_2). Cells that utilize oxygen in cellular respiration must be able to deal effectively and extremely quickly with these highly reactive and damaging compounds. To do so, aerobic cells have enzymes that rapidly detoxify them. For example, hydrogen peroxide spontaneously breaks down into harmless oxygen and water. However, aerobic cells require an enzyme called catalase to speed up the decomposition before H_2O_2 causes damage. Catalase converts two molecules of hydrogen peroxide into two molecules of water and one molecule of oxygen:

$$2H_2O_2 \rightarrow 2H_2O + O_2$$

Aerobic microorganisms come equipped with enzymes such as catalase to deal with reactive oxygen compounds like hydrogen peroxide, but many anaerobes lack them. These are the **obligate anaerobes**—they are obligated to avoid oxygen, because if oxygen is present, reactive oxygen compounds might be formed, and without detoxifying enzymes, reactive oxygen is lethal. Consequently, obligate anaerobes are found in low-oxygen environments, including lake bottoms or soil, and they rely on compounds such as sulfate (SO_4^{2-}) and nitrate (NO_3^-) as final electron acceptors. The Michigan swamp we described at the beginning of this chapter is a perfect environment for obligate anaerobes, because there is little oxygen in swamp water and because decaying plant material releases other, alternative final electron acceptors.

Because many anaerobes produce little or no catalase, hydrogen peroxide can be used as an antiseptic to help keep wounds infection-free. This is especially true for a puncture wound like Jessie's, because puncture wounds create ideal oxygen-free environments for obligate anaerobes. Because obligate anaerobes introduced into the wound lack catalase, the hydrogen peroxide used to irrigate the injury is likely to kill them, preventing infection. The fizzing that Jill observed is mostly due to the catalase that is present in Jessie's own cells. The enzyme converts some of the hydrogen peroxide to oxygen and water, and the oxygen bubbles that are liberated cause the fizz (**Figure 8.26**).

The oxygen requirements of other microbial species are not so cut and dried and metabolic switch-hitters abound. As previously mentioned, some microorganisms use different final electron acceptors, depending on availability. Many familiar microbes, including *Staphylococcus* and *E. coli*, are **facultative anaerobes** (**Figure 8.27**). They readily utilize oxygen for aerobic respiration when oxygen is available and they usually possess enzymes like

Figure 8.26 Catalase in action. The obvious fizzing observed when hydrogen peroxide is used as an antiseptic is caused by the release of oxygen. In this example, the hydrogen peroxide is being broken down by catalase-positive bacteria. If catalase were placed in contact with catalase-negative, obligate anaerobes, no fizzing would be observed.

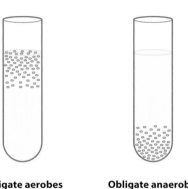
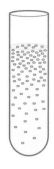

Obligate aerobes
require oxygen and
are catalase-positive.

Obligate anaerobes
cannot tolerate oxygen
and are catalase-negative.

Facultative anaerobes
grow best with oxygen
but can also grow
anaerobically. They are
catalase-positive.

Aerotolerant anaerobes
tolerate but do not use
oxygen. They are
catalase-positive.

Microaerophiles
use small amounts of oxygen.
They produce small amounts
of catalase.

Figure 8.27 Oxygen use by prokaryotes. All organisms able to survive in the presence of oxygen have the catalase enzyme necessary to help detoxify hydrogen peroxide. A lack of catalase and other detoxifying enzymes is the reason why obligate anaerobes cannot tolerate oxygen. Oxygen concentration is high near the surface of the agar tube and low near the bottom. The location of microbial growth reflects how each type of organism responds to oxygen. Obligate aerobes grow near the surface, where oxygen concentration is the highest, while obligate anaerobes grow only at the bottom of the tube. Facultative anaerobes grow best near the surface but can also grow at lower concentrations of oxygen. The growth of aerotolerant anaerobes is independent of oxygen concentration, while microaerophiles grow best slightly below the surface, where oxygen concentration is somewhat less.

catalase. In the absence of oxygen, facultative anaerobes rely on fermentation alone or anaerobic respiration to meet their energetic needs.

Microbial diversity regarding oxygen usage does not stop there. **Aerotolerant anaerobes** can partially break down toxic oxygen compounds, but they do not use oxygen for cell respiration. Consequently, oxygen has no effect, either positive or negative on their growth. They simply ignore it and they grow equally well in environments with or without oxygen (see Figure 8.27). Finally, **microaerophiles** require a small amount of oxygen but they cannot grow at normal atmospheric oxygen concentrations.

Like oxygen, temperature has a major impact on metabolism and growth

A comfortable temperature for one microorganism can be freezing or sweltering for another. As discussed in Chapter 4, some microorganisms are capable of growth at extremely high temperatures and are termed **thermophiles** (heat-loving). Others, called **psychrophiles** (cold-loving), grow only at low temperatures. Many microorganisms grow best at temperatures close to animal body temperature. Because they prefer intermediate or mid-range temperatures, they are called **mesophiles**. Most microorganisms grow over a range of temperatures, allowing the microbiologist to determine a **minimum growth temperature** (the lowest temperature allowing growth) and a **maximum growth temperature** (the highest temperature allowing growth). Population growth will be highest at some intermediate value (the **optimum growth temperature**) between these two extremes (**Figure 8.28**).

It is often relatively easy to predict where best to find thermophiles, psychrophiles, or mesophiles. If you were to sample a hot spring for microorganisms, you would expect to find thermophiles, while cold ocean water is a habitat of choice for psychrophiles. Most of the organisms found in an animal's body are mesophiles.

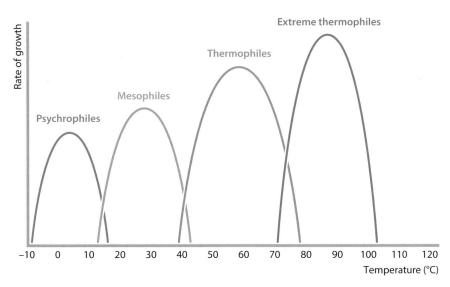

Figure 8.28 Microbial growth rates in response to temperature. For each of the four representative microbial types, there is a minimum growth temperature, below which growth cannot occur, and a maximum growth temperature, above which growth cannot occur. At the minimum and maximum growth temperatures, growth rate is slow and the generation time is relatively long. Each type of organism also has an optimum growth temperature, at which growth is most rapid with the shortest generation time.

These temperature preference categories are not rigidly defined, and many microorganisms are somewhat harder to pigeonhole. For example, food is refrigerated to discourage the growth of mesophiles that can lead to food spoilage. Food kept in refrigerators still spoils, however, meaning that some microbes can grow, even at these chilly temperatures. Perhaps surprisingly, however, most of the common food spoilers are not psychrophilic. Although they can grow at temperatures close to freezing, they grow better between 20 and 40°C and are often called **psychrotolerant**—they tolerate the cold, even though they prefer warmer temperatures. Among mesophiles that infect humans there is often variability in the optimum growth temperature, sometimes with interesting consequences. *Mycobacterium leprae*, the causative agent of leprosy, for example, typically infects fingers, ears, feet, and other cooler regions of the human body. It does not grow well at higher body temperatures found elsewhere on the host. *Treponema pallidum*, which causes syphilis, is especially sensitive to any deviation from its preferred temperature. Indeed, for many years syphilis was actually treated by inducing a high fever to inhibit growth of the bacterium. This was accomplished by deliberately infecting patients with malaria, which causes body temperature to skyrocket.

Various factors interact to determine a microbe's temperature preference. To a large degree, the maximum growth temperature is determined by the temperature at which a cell's enzymes begin to denature (unfold) and lose their ability to function. Lipid composition in the plasma membrane is also different in psychrophiles and mesophiles. Recall from Chapter 2 that saturated fatty acids form straight hydrocarbon chains, while unsaturated fatty acids have twists and kinks formed by carbon–carbon double bonds. This allows saturated fatty acids to be packed together much more tightly. The looser packing associated with unsaturated fatty acids makes them correspondingly more liquid. Consequently, psychrophiles incorporate a large proportion of unsaturated fatty acids in their membrane to help keep their membranes from freezing. Thermophiles, on the other hand, rely on densely packed saturated fatty acids to help prevent their membranes from melting.

Microorganisms have an optimal environmental pH

In Chapter 2, we learned that pH is a measure of the concentration of H^+ ions or acidity, and the lower the pH, the greater the acidity. Like temperature and oxygen, the environmental pH can have a substantial impact on microbial growth. Most species grow best when the environment is neither very acidic nor very basic: in other words, at a pH close to neutrality (pH = 7.0). There are, however, some interesting exceptions. In Chapter 2, for example, we encountered the case involving *Acidithiobacillus ferrooxidans*, which thrive in the highly acidic leachings of a coal mine. **Acidophiles** like *A. ferrooxidans* grow best at low pH. Many fungi are acidophiles, which explains why microbial contaminants of acidic foods like tomatoes and oranges are usually fungal (**Figure 8.29**). Some species thrive in basic environments, such as soda lakes, with pH values above 8.5. As with temperature, some microbes defy easy categorization. *Helicobacter pylori*, for example, thrives in the stomach of many people. Because the stomach is so acidic, you might guess that *H. pylori* is an acidophile. But is it? Recall the case from Chapter 2 in which we learned that *H. pylori*, the cause of most stomach ulcers, produces an enzyme that converts urea in the stomach into ammonia. And because ammonia is basic, it neutralizes the acid in the immediate vicinity of the bacteria.

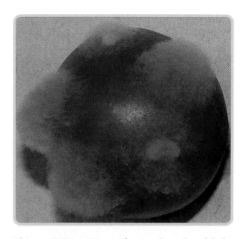

Figure 8.29 pH as a factor in microbial growth. An acidophilic fungus growing on a tomato.

The growth of most, but not all, microorganisms is inhibited by high salt concentration

Before the advent of refrigeration, salt was routinely used as a food preservative (**Figure 8.30**), because it is so good at inhibiting microbial growth. Most microorganisms cannot grow in the presence of high salt concentration, because the salty environment draws water out of the cells, causing them to dehydrate. Yet some microorganisms, the **halophiles** (salt-loving), not only survive but thrive in extremely saline habitats. In Chapter 4, for example, we

Figure 8.30 Salted cod drying at Smith's Wharf, Halifax, Nova Scotia, circa 1950. Salt has been used to prevent microbial growth in food products for thousands of years.

encountered halobacteria, a group of archaea that prefer a salt concentration many times higher than that found in the sea. In some habitats such as the Great Salt Lake in Utah, halophiles account for essentially all the aquatic microbial life.

Like all living things, microbes require certain nutrients to grow efficiently

All living things require carbon-based, biological molecules as building blocks for various cellular components and to produce ATP as previously described. Microorganisms also need phosphorus to synthesize nucleic acids and membrane phospholipids. Protein synthesis depends on a reliable supply of nitrogen. Certain vitamins are needed because they form important parts of enzymes. Small amounts of certain minerals (trace elements) such as zinc or magnesium also interact with proteins to form a complete enzyme. Complexes in the electron transport chain depend upon small amounts of iron or other metallic ions. If these necessary ions are unavailable, ATP production, and consequently growth, is curtailed.

Certain pathogenic microorganisms take drastic measures if the trace elements they require are not present in necessary amounts. *Corynebacterium diphtheriae*, for instance, lives in the throats of many people, often without causing problems. In others it produces a powerful protein toxin that results in the disease diphtheria. Why do these bacteria, which sometimes live peacefully in your throat, suddenly turn vicious? In the presence of abundant iron, the diphtheria-causing toxin is not produced. The toxin is coded for by an operon (see Chapter 7). Operon expression is usually repressed, because the operator is bound by a repressor protein. And to bind to the operator, the repressor must first bind to iron. When iron is present, some iron binds the repressor, and the iron/repressor complex binds the operator, silencing the operon. When iron is lacking, the repressor cannot bind the operator, and the toxin genes are transcribed and translated. The result is diphtheria. As cells in the upper respiratory tract are killed by the toxin, the iron in these host cells is liberated, giving the bacterial cells the iron they need to survive.

Because various microorganisms have particular nutritional requirements, a microbiologist trying to rear them in the laboratory will need to provide different diets to different species. Microorganisms that cannot make many of the biological compounds required for growth are called **fastidious**. Growing fastidious organisms can be tricky, because their nutritional requirements are complex. *Lactobacillus* is a common example. To grow these bacteria in the laboratory, a special **chemically defined medium**, containing specific amounts of purines, pyrimidines, and various vitamins and amino acids, is required. **Nonfastidious** organisms, on the other hand, are easier to grow. Because they can make almost all the biological molecules that they require, they can be grown on **minimal medium**, containing little more than a carbon source and a few other inorganic minerals and salts. Many species, including most medically important bacteria, are grown on **complex medium**, containing a variety of nutrients such as yeast extracts or predigested proteins. Even if a bacterium is capable of synthesizing some of these compounds on its own, growth is accelerated when these biological molecules are provided.

Environmental and nutritional factors, coupled with diversity in terms of metabolism, show just how complicated it can be to coax microorganisms to grow as we would like. Nevertheless, whether or not microorganisms are behaving the way we want, there are certain predictable features to their growth. We will conclude this chapter by looking at the sequence of events as a microbial population grows, levels out, and ultimately decreases.

Microbial populations pass through a sequence of phases called the growth curve

When we are sick with food poisoning or a bacterial respiratory infection, a reduction in bacterial growth, or at least a slowing of population increase, is clearly in our best interest. In other situations, when bacteria are churning out a useful product at our behest, we would like them to grow as quickly as possible. To encourage microorganisms to grow (or not grow) in a manner that suits us, it is useful to understand the phases of growth that all microbial populations go through.

The growth of a bacterial population typically goes through four characteristic phases: the lag, log, stationary, and decline phases (**Figure 8.31**). To illustrate, consider a small number of cells that has just been inoculated onto fresh medium in the laboratory. For the purposes of illustration, assume that initially all conditions are conducive to optimal growth.

Initially, there will be little if any change in the numbers of cells on our growth medium. During this **lag phase**, although the population is not increasing, the cells are preparing to divide. The length of the lag phase depends on various factors, including the normal generation time for the species in question, as well as the environmental conditions the cells were experiencing before they were placed on fresh medium. If the fresh medium contains the same nutrients that the original medium contained, the lag period will be relatively short. If the newer medium contains different nutrients, the lag period will be longer, because the bacteria cannot begin growth until they have synthesized the enzymes needed to digest these new nutrients.

Eventually the cells will begin to divide and enter a period of exponential or logarithmic growth called the **log phase** (see Figure 8.31). Cells in log phase are metabolically at their most active, and they divide at their maximal rate. This is not to say that no cells are dying during this time, but during log phase, the production of new cells by binary fission greatly outpaces any cell death.

Rapid growth cannot continue indefinitely. As cell number increases during the log phase, resources such as nutrients become depleted. Cells often produce waste products that begin to build up, inhibiting growth, and in some cases cells simply fill up all the available space. Consequently, growth begins to slow and the death rate of cells begins to increase. At some point the rate of new cell increase approximates the rate at which cells are dying and we say that the population has entered the **stationary phase**, a period during which the net population remains more or less constant (see Figure 8.31).

Although fewer new cells are produced during the stationary phase, those cells that are produced continue to use resources and produce wastes. Consequently, the conditions for growth deteriorate to such a point that the number of dying cells outnumbers any production of new cells, and the population begins to decrease (see Figure 8.31). Bacterial populations at this stage of growth are said to have entered the **decline phase**. The population may be drastically reduced to only a small number of cells, or it may die out entirely.

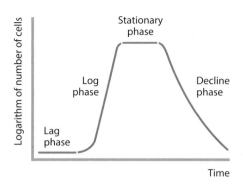

Figure 8.31 Four phases of microbial growth. Bacterial populations typically pass through four characteristic phases in their growth. Initially, there is a period of no growth called the lag phase, during which bacterial cells are preparing to divide. During the log phase, cell division is at its maximum, with the production of new cells exceeding cell death. As conditions deteriorate, growth slows, and eventually reproduction and death are approximately equal. This corresponds to the stationary phase—a time when the population is neither increasing nor decreasing. Finally, populations enter the decline phase, during which cell death exceeds the production of new cells.

As the environment changes, microbial metabolism and therefore the growth curve changes in response

CASE: MAKING YOGURT

Sheri and Jim have always liked cooking and experimenting in the kitchen, so when Sheri sees a recipe for yogurt, they decide to give it a try. They first heat whole milk to about 170°F to reduce the numbers of unwanted bacteria. They then let the milk cool and pour it into a dozen cups. To each cup they add 1 tablespoon of powdered milk and 1 teaspoon of plain yogurt containing active *Lactobacillus acidophilus* cultures (**Figure 8.32**), completely mixing all ingredients. The directions state to carefully seal each cup completely with plastic wrap, but after sealing the first eight cups, they run out of wrap. They place all 12 cups, however, including the uncovered ones, in the oven on the lowest possible setting (120°F). After about 10 hours, they check the progress of their yogurt. The covered cups have all solidified, and Sheri and Jim, after adding some sugar and fresh fruit, enjoy real homemade yogurt. The cups that remained uncovered are still fairly liquid.

1. **Why did only the sealed cups produce yogurt?**
2. **In which of the cups, the sealed or the unsealed ones, was bacterial metabolism more efficient? Which cups contained more bacterial cells? If growth curves could be made for the bacteria in sealed and unsealed cups, how would they differ?**
3. **After each having a cup of yogurt, Jim and Sheri place the remaining cups into the refrigerator. How would this alter the growth curves for the bacteria in the yogurt?**

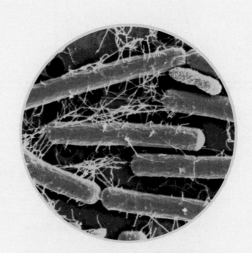

Figure 8.32 *Lactobacillus acidophilus.*
This facultative anaerobe can grow in either the presence or absence of oxygen. When oxygen is limited, the reduction of pyruvate during fermentation results in the production of lactic acid, which is released as a waste product. When these bacteria are used in the production of yogurt, the released lactic acid causes milk proteins to denature, giving yogurt its firm texture. It also gives unsweetened yogurt its characteristic sour taste.

Although microbial populations all go through the same basic stages of growth, the growth curves for different species will appear somewhat different. Consider, for instance, the log phases of an aerobe growing in the presence of oxygen and an anaerobe growing in the absence of oxygen. Generally speaking, we would expect that the rate of increase would be significantly faster for the aerobe for the simple reason that aerobic respiration usually results in a greater ATP yield than anaerobic respiration does (**Figures 8.33a and 8.33b**). Consequently, reproduction for the aerobe is typically faster.

The shape of the growth curve will be impacted by the other factors as well. The lag phase will become longer and the log phase will rise more slowly as physical and nutritional conditions become less ideal. Cells growing under less optimal conditions may enter the stationary phase sooner, and the reduction in population during the decline phase may be steeper. Furthermore, the shape of the curve may change as conditions change. Imagine, for example a thermophile with a minimum growth temperature of 40°C, a maximum growth temperature of 75°C, and an optimum growth temperature of 65°C. If we prepare a culture of these cells, incubate them at 45°C, and then plot their growth curve, we will see that after a prolonged lag phase the cells start to increase in number slowly (see **Figure 8.33c**). If we then change the incubation temperature to 65°C, we should see that the log phase becomes considerably steeper, rising more quickly.

The *L. acidophilus* used by Jim and Sheri to make yogurt provides an excellent way to illustrate some of the key points we have made regarding

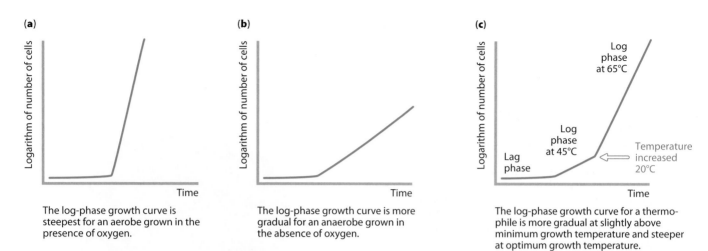

Figure 8.33 Effect of metabolism and growth conditions on rate of growth. Logarithmic-phase growth for (a) an aerobe growing in the presence of oxygen and (b) an anaerobe growing in the absence of oxygen. The faster growth rate for the aerobe reflects the higher ATP yield achieved through aerobic respiration. (c) Effect of temperature on logarithmic growth. For this thermophile, 45°C is just above the minimum growth temperature, so growth is slow. At the optimum growth temperature (65°C), the rate of growth is significantly higher.

metabolism and its effect on growth. This bacterial species is a facultative anaerobe; it can grow in either the presence or absence of oxygen, but metabolism will be more efficient and growth will be faster when oxygen is available (**Figure 8.34a**). To make proper yogurt, anaerobic conditions are required. Without oxygen, the bacteria are forced to rely on glycolysis alone to meet their ATP needs. At the end of glycolysis, pyruvate is reduced to lactic acid via fermentation, which lowers the pH of the milk, causing the milk proteins to denature. In their denatured state, these proteins take on a solid form, giving yogurt its firm texture. The reason the bacteria initially added in the recipe were thoroughly mixed and the cups were tightly sealed was to provide this anaerobic environment. Because the bacteria in the unsealed cups were exposed to increased oxygen, many of the cells in these cups were able to respire aerobically. Less fermentation resulted in less lactic acid, preventing the milk in the uncovered cups from solidifying properly. If we prepared growth curves for the sealed and unsealed cups, we would detect a longer lag period and a less steep log period in the sealed cups that ultimately produced yogurt. The unsealed cups would undoubtedly contain more bacterial cells.

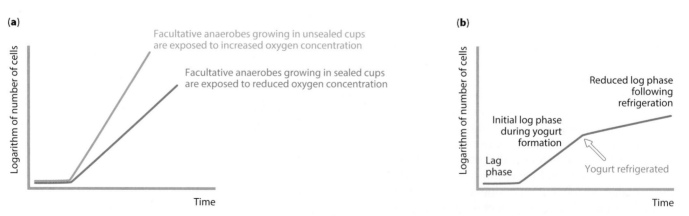

Figure 8.34 *L. acidophilus* growing under different conditions. (a) As a facultative anaerobe, *L. acidophilus* can grow in either the presence or absence of oxygen, but growth is faster when oxygen is present. The bacteria in the sealed cups are exposed to less oxygen, and therefore are depending to a greater extent on fermentation than those bacteria in the unsealed cups. (b) When placed in a colder environment, the growth rate of this psychrotolerant species slows.

Although *L. acidophilus* is mesophilic in terms of its temperature preferences, it is somewhat psychrotolerant. Consequently, when the yogurt was placed in the refrigerator, although log phase may have continued, the slope of the rise would be considerably reduced (**Figure 8.34b**). Growth might continue slowly until other factors, including the digestion of all remaining sugars, moved the bacteria into the stationary phase.

Yogurt making is just one situation in which we might want to influence microbial metabolism and growth. As we continue through this text we will encounter other situations in which we may wish a population of microorganisms to be in one part of the growth curve or another. There is usually an urgency to move disease-causing pathogens into the decline phase, but as we will see in Chapter 13, we do not always want even pathogens to decline in number too rapidly. For other beneficial organisms it may be in our interest to keep them in the log phase as long as possible. As we have seen, much of our ability to do this depends on knowing how a particular species makes and uses energy.

Looking back and looking forward

While all cells utilize cell respiration to convert the energy in biological molecules into ATP, the precise details of cell respiration vary between different organisms. Some, for example, require oxygen, while others do not. Furthermore, not all organisms obtain their biological molecules in precisely the same way. Diversity in terms of metabolism helps explain why different microbes grow at different rates and how different environmental factors affect their growth. Through an understanding of metabolism and growth we are better positioned to control microbial growth to suit our specific goals.

Now that we know a good deal about what microorganisms are and what they do, in Chapter 9 we will learn how they got that way; we will turn our attention to microbial evolution. Through an understanding of the evolutionary process we can begin to fully appreciate how the never-ending interplay between living things and their environment has given rise to the phenomenal diversity of life that surrounds us. Furthermore, an understanding of biological evolution permits us to understand and even to predict phenomena such as bacterial drug resistance and changing levels of pathogen virulence. And remember that we too are evolving. To underscore this point, we will also investigate some of the curious ways that microorganisms have influenced the evolution of humans and other host organisms.

Garland Science Learning System

- http://garlandscience.rocketmix.com/students/

- Discover how you can tell if a bacterium is a vampire

- See the molecular turbine spinning

- Test your knowledge of this chapter by taking the quiz

- Familiarize yourself with the terminology used in this chapter by using the vocabulary review

- Get help with the answers to the Concept questions

Concept questions

These questions are designed to help you start thinking like a microbiologist. The answers are not always simply found in the text. Instead, you will need to take the concepts about which you have learned and apply them to new situations. Some of the questions may not even have just a single correct answer. Help is provided as part of the GSLS resources, which can be accessed through http://garlandscience.rocketmix.com/students/.

1. Electrons are transferred from molecule A to molecule B.

   a. Which molecule has been reduced?

   b. Were the electrons at higher energy with A or B?

   c. Were the electrons held more tightly when they were with A or with B?

   d. With which molecule, A or B, are the electrons more stable?

   e. Is this an endergonic or exergonic reaction?

2. What exactly is meant when we say that, in electron transport, oxygen is the final electron acceptor?

3. Twenty psychrophilic bacterial cells find their way into your milk carton while it's out of the refrigerator. When you return the milk to the refrigerator, the cells begin to grow with a generation time of 2 hours. How many cells, in theory, will be in your milk carton the next time you go to the refrigerator, 10 hours later?

4. A facultative anaerobe suffers a mutation, resulting in its inability to produce catalase. How might this change the environmental conditions under which this organism could survive?

5. Why does oxygen kill obligate anaerobes?

6. Flask A contains yeast in a glucose broth. The flask is exposed to the atmosphere. Flask B also contains yeast in a glucose broth. The flask is tightly sealed, creating an anaerobic environment. Answer the following, assuming that these yeasts are facultative anaerobes.

   a. In which flask would there be more cell division over a period of 48 hours?

   b. In which flask would ethanol be found after 48 hours?

   c. In which flask would the generation time be shorter?

   d. In which flask would the yeast be able to produce greater amounts of ATP?

   e. In which flask would pyruvate get reduced? In which would it be oxidized?

   f. Draw the growth curves for the cells in each flask. How would they differ and why?

7. Draw growth curves for the following:

   a. An obligate anaerobe is placed in optimal environmental conditions. At time x, oxygen is introduced.

   b. A facultative anaerobe is placed in anaerobic conditions. At time x, oxygen is introduced.

   c. An obligate aerobe is placed in optimal environmental conditions. At time x, all oxygen is used up. Conditions are now anaerobic.

   d. You're at a picnic and your potato salad is contaminated with a mesophilic species. After the picnic you take the remaining potato salad home and place it in the refrigerator at time x.

8. Waters in the North Atlantic Ocean are typically low in nitrogen content. Imagine an experiment in which nitrogen was added to a small area in the Atlantic. How might this change the lag phase for microorganisms in the area? The generation time?

9. *Citrobacter freundii* cannot digest the amino acid phenylalanine, whereas *Proteus vulgaris* can. What accounts for this difference? How could you use the test described in Figure 8.16 to distinguish between these two species of bacteria?

MICROBIAL EVOLUTION: THE ORIGIN AND DIVERSITY OF LIFE

In the 1990s, South America was hit by a severe cholera epidemic. The outbreak apparently started when a Chinese ship released contaminated ballast water in a Peruvian port. The bacterium responsible for cholera, *Vibrio cholerae*, is spread through contaminated food and water. The bacteria inhabit the intestine of affected individuals and are released in large numbers in the watery, explosive diarrhea that typifies the disease. Ultimately, about a million cases and almost 10,000 deaths were reported, affecting every country on the continent except Uruguay.

Not every country suffered equally. Ecuador, for instance, had the highest proportion of cases, while Chile had the fewest. Furthermore, the *V. cholerae* infecting Ecuadorian patients caused a more serious form of illness. In Chile, those who contracted cholera suffered milder symptoms.

What explains this difference? Sanitation was relatively poor in much of Ecuador, whereas Chile had relatively sophisticated sanitation. Consequently, for a water-borne pathogen like *V. cholerae*, transmission was much more difficult in Chile. It was therefore in the bacteria's interest to cause milder symptoms and not kill its hosts too quickly, before it had a relatively rare chance to spread. In Ecuador, transmission was more assured, and there was less necessity for the bacteria to become relatively benign.

But exactly how did the bacteria adapt to local conditions in this way? Such a question goes straight to the heart of one of biology's most fundamental questions. What is the mechanism that explains how living things adapt to their environments?

The short answer is that ever since life on Earth began, it has been subject to the rules of biological evolution. Today, after more than 3.5 billion years, we see the result: an almost limitless assortment of organisms, microbial and nonmicrobial, all of which have evolved to survive and reproduce best under specific environmental conditions. And whether we are considering *Escherichia coli* or elephants, the rules governing evolutionary change are the same. Here in Chapter 9 we will make sense of those rules and explain how biological evolution has resulted in the myriad of living things around us. Each of these living things is the product of eons of evolution and each has its own evolutionary story to tell. To fully comprehend those stories, especially as they apply to microorganisms, the best place to start is at the beginning—in this case, the origin of life on Earth.

All case studies have a few questions at the end, the answers to which will become apparent as you read the sections following the case.

Conditions on the early Earth were very different than they are today

Scientists believe that Earth was formed 4.5–4.6 billion years ago (BYA) (**Figure 9.1**). According to accepted theory, our solar system was formed when dust-sized particles accumulated around the sun. As these particles collided, they often stuck together due to electrical attraction. Over time these growing masses attracted still more particles due to their increasing gravity. Ultimately these protoplanets developed into the planets of our solar system.

The young Earth was a vastly different place than it is today. First of all, it was far hotter. Since its formation, the Earth has steadily cooled, losing the heat that was first generated when it was formed. Second, the early Earth's atmosphere contained very little oxygen (O_2). Today, oxygen makes up 21% of our atmosphere. Some O_2 entered the atmosphere by geologic processes, but oxygen did not begin to approach current levels for another 2 billion years or so. At that time, certain photosynthetic organisms evolved the ability to use water as an electron source (see Chapter 8, pp. 181–182), releasing oxygen as a waste product. Consequently, the early atmosphere of our planet was a reducing atmosphere, as opposed to our current oxidizing atmosphere. This fact may have been essential for the initial development of life. Recall from Chapter 8 that when molecules are more likely to be reduced, the synthesis of biological molecules is more likely. In an oxidizing environment, digestion is more likely because molecules tend to be oxidized. Low levels of free oxygen therefore may have set the stage for the sorts of reactions that would ultimately give rise to the first biological molecules.

Hydrogen gas (H_2) was especially abundant in Earth's early, reducing atmosphere. Other reduced molecules containing hydrogen, including methane (CH_4), ammonia (NH_3), and water vapor (H_2O), were also present. These molecules provided the building blocks for the beginnings of life itself.

The first biological molecules were formed from nonbiological precursors

In Chapter 6 we discussed some of the discoveries that disproved the notion of spontaneous generation. Experiments by Redi, Pasteur, and others demonstrated that life could not arise from nonliving material. In the 1920s, however, it was suggested that in the primitive Earth's reducing atmosphere,

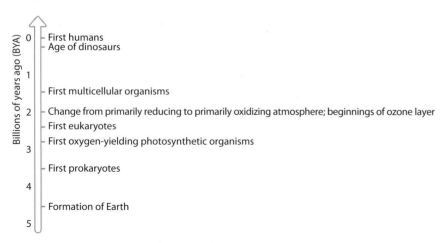

Figure 9.1 Some important events in the history of life. All dates are approximate. Major episodes in the origin of living things will be revisited throughout this chapter.

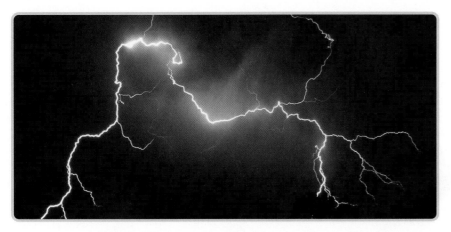

Figure 9.2 Energy for abiotic synthesis. If biological molecules formed in Earth's primitive reducing atmosphere, lightning strikes may have provided the energy necessary for synthesis.

inorganic molecules could assemble into the biological molecules that are the precursors of life. Reduced gases in the atmosphere could donate electrons, meaning that the redox reactions necessary for synthesis could take place. This no longer occurs in our current oxidizing atmosphere, because atmospheric oxygen is such a powerful oxidizing agent.

Even in a reducing environment, an energy source was needed to synthesize large biological molecules from the smaller molecules present in the primitive Earth's atmosphere. Scientists think that UV radiation or lightning could serve in this capacity (**Figure 9.2**). The lack of significant free oxygen meant that no ozone layer would have been present to absorb UV radiation. Consequently, the intensity of such radiation reaching Earth's surface was far higher than it is today.

Thus, it was hypothesized that although spontaneous generation is unlikely today, the unique conditions found on the early Earth meant that the scene was set for **abiotic synthesis**—the building of biological molecules from inorganic precursors.

The hypothesis of abiotic synthesis was tested in 1953 in a brilliant experiment conducted by Stanley Miller, then a graduate student at the University of Chicago. Miller actually recreated Earth's primitive reducing atmosphere in an experimental vessel (**Figure 9.3**). The apparatus consisted of two chambers connected by glass tubing. In one chamber, Miller created a miniature "ocean" of liquid water. A different chamber consisted of a primitive "atmosphere," complete with reduced methane (CH_4), ammonia (NH_3), and hydrogen (H_2) gases. The ocean was boiled with a heating source, such that the water vapor passed into the atmosphere, where in addition to mixing with other atmospheric gases, the molecules were subjected to artificial lightning in the form of electrical discharge from an electrode. Within a day, the mixture took on a reddish hue. After a week, Miller analyzed the chemical composition of the molecules in his apparatus. Amazingly, he found numerous organic compounds, including several amino acids. The possibility of abiotic synthesis in an environment simulating that of the primitive Earth had been confirmed.

Others have proposed that the composition of the primitive Earth's atmosphere was less important than Miller's work suggests. As an alternative hypothesis, it has been suggested that early biochemical reactions first occurred in deep sea vents, where hot water and minerals, along with

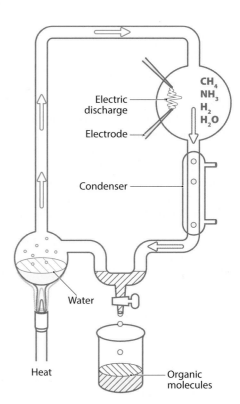

Figure 9.3 Stanley Miller's "origin of life" experiment. Water vapor was passed through a primitive reducing atmosphere consisting of NH_3, CH_4, and H_2, into which energy, in the form of electric sparks, was discharged. Within a few days, amino acids and other organic molecules were formed.

Figure 9.4 Alternative explanations for abiotic synthesis.
(a) A deep ocean, hydrothermal vent called a black smoker. The black color is due to the large amounts of minerals such as iron and carbon that are dissolved as the superheated water leaves the vent. Biological molecules may have first been synthesized at sites such as this. (b) Piece of the Murchison meteorite, which landed in Australia in 1969. Traces of water and amino acids were found on the meteorite, leading some to speculate that Earth was originally seeded with organic molecules formed in space and delivered to Earth on meteorites.

reduced compounds such as methane and hydrogen sulfide, are released into the ocean (**Figure 9.4a**). According to this hypothesis, as reduced compounds rose up through Earth's crust, they provided the raw materials necessary for the synthesis of biological molecules. The superheated water supplied an energy source. Interestingly, these hydrothermal vents provide a suitable environment for many microorganisms today. Many of these bacteria and archaea are extreme thermophiles, able to thrive in the hot water. Others have observed that molecules such as hydrogen cyanide and formaldehyde form in gaseous clouds in space and that these molecules can react to form other more complex biological molecules. It is possible that Earth was originally seeded with these organic molecules by meteorites that bombarded Earth shortly after its formation (**Figure 9.4b**).

Regardless of where the process took place, however, it now appears likely that under certain specific conditions biological molecules, the building blocks of life, can indeed form from abiotic beginnings.

Genetic information may have originally been encoded in RNA instead of DNA

The building blocks of life are not the same as life itself. An essential attribute of living things is their ability to reproduce. Reproduction requires genetic material to transmit information from one generation to the next, and for all modern prokaryotic and eukaryotic organisms, this material is DNA. In our primitive mix of amino acids and other organic molecules it is therefore vital to find a molecule acting as the genetic material, able to both duplicate itself and transmit genetic information. Yet if DNA initially served as the genetic material, we are faced with a "chicken or egg" type conundrum. Although DNA can replicate itself, such replication requires certain enzymes. Enzymes, however, are proteins, coded for by DNA. So, which came first, the DNA that codes for the enzymes or the enzymes needed to copy the DNA?

There was no good answer to this question until the 1980s, when researchers discovered enzymes made of RNA, which they called **ribozymes**. This discovery led to the RNA world hypothesis: the idea that the current DNA world was preceded by an RNA world in which RNA, rather than DNA, was the genetic material.

Why RNA? Prior to the discovery of ribozymes, RNA was thought to be involved only in conveying DNA's message first into mRNA and ultimately into a protein. The discovery of ribozymes, however, demonstrated that RNA could also act as an enzyme, catalyzing certain reactions. In other words, the discovery of ribozymes overturned the long-held assumption that only proteins could serve as enzymes. It has subsequently been found that ribozymes can also catalyze the synthesis of new RNA, and in 2012 it was announced that the active site of ribosomes themselves, so vital to protein translation, contains RNA only and that ribosomes therefore are ribozymes.

RNA can also encode genetic information in its sequence of nucleotide bases, and it can serve as a template for replication. DNA stores information, but it cannot catalyze the replication of that information. Proteins can serve as enzyme catalysts, but they cannot store information. RNA can do both, and proponents of the RNA world hypothesis suggest that RNA served as the initial genetic material and that it catalyzed its own reproduction.

The first cells required a membrane and genetic material

Two important things are needed before we can start to consider any assemblages of biological molecules to be bona fide cells. First, a barrier between the inside and the outside of the cell in the form of a membrane is required. Second, a cell needs genetic information that can replicate and code for essential proteins. We do not know precisely what the first cells looked like, but we do know that when phospholipids are mixed with water, they assemble into small, spherical compartments (**Figure 9.5a**). These compartments, called **liposomes**, are thought to have formed the first cells when they

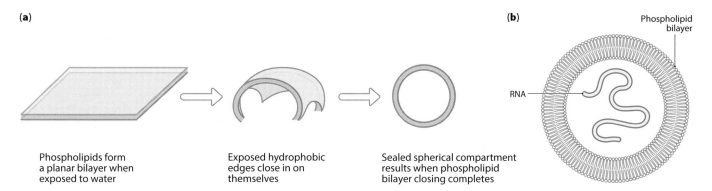

(a)

Phospholipids form a planar bilayer when exposed to water

Exposed hydrophobic edges close in on themselves

Sealed spherical compartment results when phospholipid bilayer closing completes

(b)

Phospholipid bilayer

RNA

Figure 9.5 The first cells. (a) Phospholipid bilayers form spontaneously when phospholipids are placed in water. They then readily close in on themselves to formed sealed compartments. The closed structure is stable because it avoids the exposure of the hydrophobic hydrocarbon tails to water. Such phospholipid spheres, called liposomes, may have constituted the first plasma membranes. (b) The earliest cells may have consisted of liposomes surrounding self-replicating RNA.

incorporated self-replicating RNA (**Figure 9.5b**). Liposomes sometimes grow by engulfing other liposomes and because their membrane is selectively permeable, they can absorb and release certain materials. The enclosed RNA would have allowed these primordial cells to replicate and continue to evolve.

We may never know for sure how the transition to a DNA world took place, but the earliest RNA-based cells would have had an extremely simple metabolism. Compared with proteins, RNA is an inefficient catalyst, and early in life's history, RNA probably catalyzed only a few absolutely critical reactions. As proteins began to appear in cells, it is likely that they took over since they are more efficient catalysts, leaving RNA to mainly encode information (**Figure 9.6**).

DNA may have arrived on the scene as cells gradually became more complex and needed to store increasing amounts of information. RNA may have provided the template for the first DNA molecules, but DNA is inherently far more stable than RNA. DNA is therefore better suited for the transfer of large amounts of information from generation to generation. With more efficient information storage and replication, cells utilizing DNA would eventually replace those relying on RNA. RNA was then available to assume its current task of serving as an intermediate between DNA and protein.

Consequently, somewhere at an early stage in the evolution of life, the transfer of information from DNA to RNA to protein became universal. The precise time when this happened is not known, but it is worth noting that the domains Bacteria, Archaea, and Eukarya all rely on DNA as their genetic material. This fact strongly supports the idea that any transition from an RNA world to a DNA world would have occurred before these three domains of life diverged.

The first prokaryotes are thought to have arisen approximately 3.5 billion years ago

The point at which simple, primordial cells qualified as authentic living things is fuzzy. We can guess that the time-frame is enormous and that the first living organisms were prokaryotes that appeared around 3.5 BYA (**Figure 9.7**). Evidence to support this has come from the recovery of microbial fossils in layered sedimentary rock formations called **stromatolites**

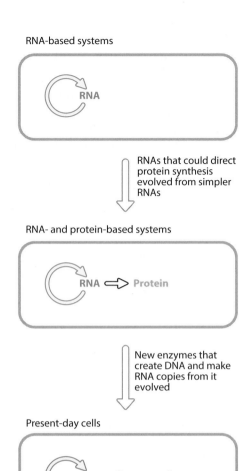

Figure 9.6 Transition from RNA world to DNA. According to this hypothesis, RNA combined genetic and catalytic functions in the earliest cells. Proteins are more efficient catalysts and eventually took on most enzymatic functions. DNA, more stable than RNA, took over the encoding of genetic information as greater amounts of information were required by increasingly complex cells. RNA remained, functioning primarily as a go-between in protein synthesis, while retaining catalytic functions for a few reactions.

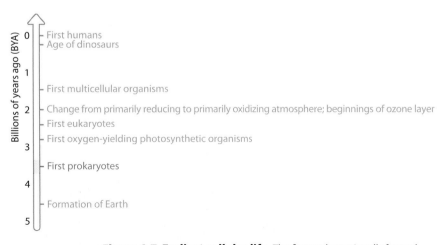

Figure 9.7 Earliest cellular life. The first prokaryotic cells formed approximately 3.5 billion years ago.

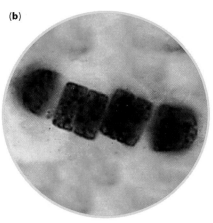

Figure 9.8 The first prokaryotes. (a) The oldest known stromatolites to date have been found in Sharks Bay in Western Australia. The oldest stromatolite fossils are estimated to be approximately 3.5 billion years old. (b) Fossilized ancient bacterium from northern Australia.

(**Figure 9.8**). Many stromatolites contain fossils of filamentous or rod-shaped microorganisms that resemble organisms that are still alive today. These early microbial pioneers were most likely obligate anaerobes and, as discussed in Chapter 4, were probably well-adapted to withstand the intense UV radiation and high temperatures that characterized Earth's environment at that time.

Certain important metabolic pathways evolved in a defined sequence

In Chapter 8 we reviewed the important metabolic processes of glycolysis, cell respiration, and photosynthesis. Most researchers agree that glycolysis was probably the first of these pathways to evolve. Powerful evidence for this is provided by the fact that, unlike photosynthesis or specific types of cell respiration, glycolysis occurs in all cells. This suggests that glycolysis evolved before the diversification of cells into the three domains of life—Bacteria, Archaea, and Eukarya.

Now that cells had a mechanism to harvest the energy in biological molecules such as glucose, it would be advantageous for a cell to develop the capacity to produce its own biological molecules, reducing reliance on environmental sources. Consequently, photosynthesis is thought to have evolved sometime after glycolysis and probably after anaerobic cell respiration. The earliest photosynthetic organisms did not use water as a source of electrons and consequently did not release oxygen. The release of oxygen as a photosynthetic by-product (see Chapter 8, pp. 181–182) is believed to have started about 2.7 BYA (**Figure 9.9a**), and the organisms responsible may have been the distant ancestors of modern-day photosynthetic cyanobacteria (**Figure 9.9b**). Starting at that time, oxygen began its steady buildup in the atmosphere, and by about 2 BYA atmospheric oxygen was common enough that iron-bearing rocks began to undergo oxidation, forming iron oxide (**Figure 9.10**). Iron oxide, commonly called rust, is red in color, and the presence of red beds, reddish-colored iron-rich rocks that were exposed to the atmosphere 2 BYA, attests to the presence of greatly increased levels of oxygen at that time.

This buildup of atmospheric oxygen (O_2) changed the environment in profound ways. When O_2 is bombarded by UV radiation, for example, it can be converted into ozone (O_3). Ozone can absorb UV radiation, and as levels of ozone increased, Earth's atmosphere gained an ozone shield to protect

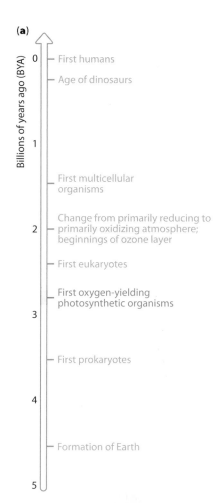

(a)

Billions of years ago (BYA)

0 — First humans
— Age of dinosaurs

1

— First multicellular organisms

— Change from primarily reducing to
2 — primarily oxidizing atmosphere;
beginnings of ozone layer

— First eukaryotes

— First oxygen-yielding
3 — photosynthetic organisms

— First prokaryotes

4

— Formation of Earth

5

(b)

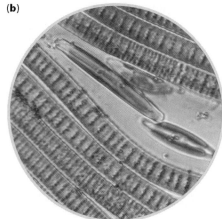

Figure 9.9 The first photosynthetic organisms. (a) Oxygen-yielding photosynthesis is believed to have evolved about 2.7 BYA. (b) The first photosynthetic prokaryotes that released oxygen may be related to modern-day cyanobacteria, seen as the long slender green filaments in the photograph. These bacteria, like green plants, use water as a source of electrons and release oxygen as a waste. Other photosynthetic prokaryotes rely on different electron sources and do not release oxygen.

Figure 9.10 Red beds. Uluru, a well-known rock formation in the Northern Territory of Australia. Iron oxide, formed when iron deposits in the rock were oxidized by atmospheric oxygen, gives Uluru its red color. These rocks were formed about 2 BYA, indicating the approximate time when the atmosphere changed from reducing to oxidizing. Rock strata older than 2 billion years that were not exposed to the atmosphere show no such oxidation.

living things against UV radiation. In Chapter 7 we discussed how radiation acts as a powerful mutagen, and before the development of the ozone layer, most living things could survive only under rocks, in the oceans, or in other habitats protected from direct solar radiation. Following the development of an ozone layer, early microorganisms could spread out over Earth's surface, safe from the damaging effects of UV radiation.

Furthermore, recall from Chapter 8 that oxygen can form powerful and sometimes lethal oxidizing compounds like superoxide ions (O_2^-). Consequently, for many anaerobes, oxygen is highly toxic. Before oxygen levels reached high concentration in the atmosphere, all organisms would have been anaerobic, relying on glycolysis and/or anaerobic cell respiration to meet their energy needs. So, for cells living 2 BYA the choice was simple: They could avoid oxygen by living only in anaerobic environments or they could go extinct. The obligate anaerobes alive today are the descendants of those that dealt with the problem of oxygen by avoiding it. Other primitive prokaryotic lineages perished.

Alternatively, organisms could evolve ways to detoxify oxygen. Some organisms developed the capacity to reduce oxygen to water through a primitive form of electron transport. As we saw in Chapter 8, reducing oxygen in this manner releases energy, and some organisms were able to capitalize on this released energy by using it to produce additional ATP. In other words, the electron transport chain that provides the lion's share of ATP for aerobic organisms first evolved as an oxygen detoxification system. By a happy coincidence, those organisms fortunate enough to develop this strategy received an additional benefit—aerobic respiration.

Eukaryotes evolved from prokaryotic ancestors

CASE: CELLS WITHIN CELLS

As amazing and perhaps as creepy as it sounds, even your own cells are not *completely* your own. As mentioned briefly in Chapter 3, some parts of eukaryotic cells have bacterial ancestors. In the late 1800s, it was first proposed that at least some of the organelles in eukaryotic cells were derived from bacteria that had entered into a symbiotic relationship with the eukaryotic cell. Over time, the two cell types became so interdependent that independent life was no longer an option for either cell. This idea, called endosymbiosis, was largely looked upon as nonsense until the 1980s. At that time, some scientists began to take a fresh look at endosymbiosis, and as they compiled evidence to support it, endosymbiosis gradually gained credibility. Today, most biologists agree that endosymbiosis is the most likely explanation for the origin of at least two important organelles.

1. **Which organelles were most likely acquired by endosymbiosis?**

2. **What evidence exists to support endosymbiosis?**

3. **In what way would endosymbiosis be advantageous to both partners in the symbiotic relationship?**

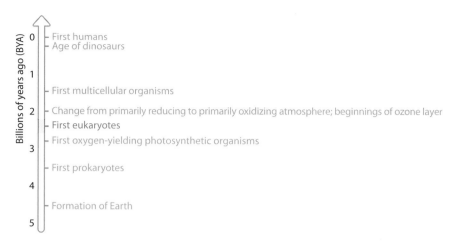

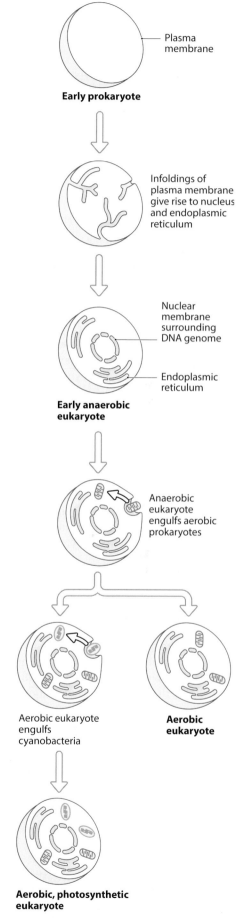

Figure 9.11 The first eukaryotic cells. The earliest fossils that show clear evidence of a membrane-bound nucleus and other distinguishing features of eukaryotic cells first appeared a little more than 2 BYA.

Following the discovery of fossil prokaryotes in the mid-twentieth century, it became apparent that there was an abundant fossil record of early prokaryotic life. Yet nothing that looks like a eukaryotic cell appears in the fossil record until just over 2 BYA (**Figure 9.11**). For the first 1.5–2 billion years of their existence, prokaryotes had the planet to themselves.

The first eukaryotes were probably structurally simple single-celled organisms, resembling prokaryotes, except for the fact that they had certain internal membrane systems such as the nuclear membrane. The nuclear and other internal membranes, including the endoplasmic reticulum and the Golgi apparatus, are thought to have evolved from infoldings of the plasma membrane (**Figure 9.12**). As discussed in Chapter 3, such compartmentalization of the early eukaryotic cell allowed increased intracellular specialization and meant that the cell could carry out more metabolic processes at the same time.

Other organelles, specifically mitochondria and chloroplasts, were still lacking. These organelles were later acquired by **endosymbiosis** (see Figure 9.12). "Symbiosis" refers to a special relationship between two organisms, which is often mutually beneficial. Mitochondria are believed to have evolved when aerobic prokaryotes developed a symbiotic relationship with primitive anaerobic eukaryotic cells. Later, some eukaryotes acquired chloroplasts in a similar manner, allowing them to carry out photosynthesis.

It is generally assumed that mitochondria were acquired before chloroplasts, because of their presence in virtually all eukaryotes (**Figure 9.13**). Later, after eukaryotes had begun to diversify, photosynthetic bacteria were acquired by the branch of eukaryotes that ultimately gave rise to plants.

Figure 9.12 Origin of eukaryotes. The first eukaryotes were probably formed when infolding of the plasma membrane gave rise to membrane-bound organelles, permitting increased subcellular specialization. These early eukaryotes would have been anaerobic and non-photosynthetic. The first aerobic eukaryotes were the result of endosymbiosis with aerobic prokaryotes. These symbiotic aerobes ultimately gave rise to mitochondria. Some of these now-aerobic eukaryotes subsequently developed a second endosymbiotic relationship with photosynthetic bacteria. Increasing interdependence between these bacteria and the eukaryotic cells resulted in the evolution of chloroplasts and eukaryotic cells that were both aerobic and photosynthetic.

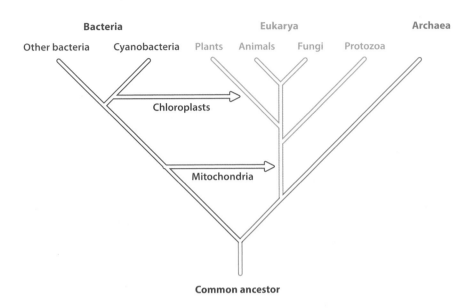

Figure 9.13 Cell evolution. Present-day cells evolved from a common prokaryotic ancestor along three major lines of descent, giving rise to domains Bacteria, Eukarya, and Archaea. Eukarya acquired mitochondria prior to the differentiation of the domain into its various groups. Only the group that ultimately gave rise to plants acquired chloroplasts.

There is no shortage of evidence to support the theory of endosymbiosis:

- Symbiotic relationships between prokaryotic and eukaryotic cells can still be found today. For example, some photosynthetic cyanobacteria reside inside the cells of some unicellular eukaryotes.

- Similarities exist between prokaryotic cells and modern mitochondria and chloroplasts. Like prokaryotes, these organelles can reproduce by binary fission, independently of host-cell division. Both mitochondria and chloroplasts have their own DNA, and like prokaryotic DNA, the chromosomes found in these organelles are circular.

- Using their own DNA, both mitochondria and chloroplasts code for some of their own proteins. Translation of these proteins occurs within the organelle on smaller prokaryotic-like ribosomes, instead of the larger ribosomes found in the cytoplasm of eukaryotic cells.

- Comparison of gene sequences from mitochondrial DNA and certain aerobic bacteria confirms that mitochondria and bacteria are related. Similar evidence confirms the relationship between chloroplasts and photosynthetic cyanobacteria.

The advantages enjoyed by primitive eukaryotes with mitochondria and primitive plant cells with both mitochondria and chloroplasts were straight-forward. As we learned in Chapter 8, aerobic respiration results in higher ATP yields than anaerobic respiration. Eukaryotes harboring aerobic bacterial partners could now partake in this ATP bonus. Early plant cells that acquired photosynthetic bacteria via endosymbiosis no longer needed to depend on an outside source of food molecules. As autotrophs, they could now synthesize their own. And for the once-independent bacteria, now living snugly inside eukaryotic cells, there was less risk of predation as well as the benefits of more stable environmental conditions.

Eukaryotic cells first arose from certain archaeal ancestors

Which prokaryotes actually developed into the first eukaryotes? As discussed in Chapter 4, eukaryotes appear to be more closely related to domain Archaea than either of these two domains is to domain Bacteria. It is generally thought that eukaryotes initially evolved from an archaean about 2 BYA. Who was that archaeal ancestor? The answer to this question has remained one of the most contentious puzzles in modern biology.

In 2015 European scientists announced the discovery of Lokiarchaeota, recovered from water samples collected deep in the Arctic Ocean. An analysis of these archaea indicated that they are likely to be more closely related to eukaryotes than any other archaea studied to date. In other words, Lokiarchaeota and eukaryotes share a common ancestor more recently with each other than they do with other lineages of prokaryotes (**Figure 9.14**). Indeed, Lokiarchaeota have been called the missing link between prokaryotes and eukaryotes.

More specifically, an analysis of Lokiarchaeota DNA showed that many of their genes were more similar to those of eukaryotes than any previously studied prokaryote. Not only that, some of the genes found in Lokiarchaeota code for proteins that are found in eukaryotes but not in other prokaryotes. These include certain proteins necessary for phagocytosis, which would have been necessary to carry out endosymbiosis. The Lokiarchaeota themselves cannot carry out phagocytosis, but the presence of genes related to this process strongly suggests that the common ancestral archaea that ultimately gave rise to both eukaryotes and Lokiarchaeota possessed these genes and passed them on to both of these lineages. Consequently, the first eukaryotes began with a rich genetic starter kit, which helped them gain the complexity that we see in modern eukaryotes.

Multicellular life arose from colonies of unicellular eukaryotes

Around 1.5 BYA, the first multicellular organisms evolved (**Figure 9.15**). To this day, some unicellular eukaryotes form colonies of cells that resemble a transition stage between unicellular and multicellular modes of life (**Figure 9.16**). As cells within such colonies became increasingly specialized, the capacity for independent life as a single cell was lost, and true multicellular organisms first appeared. As cells continued to specialize, they consequently became increasingly diverse, giving rise to the many types of cells currently seen in modern multicellular animals, fungi, and plants.

Multicellularity confers several advantages. First of all, with more cells, organisms become larger and therefore less likely to be consumed by

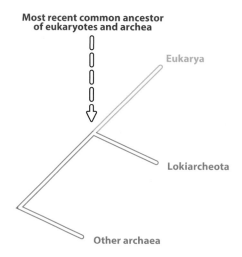

Figure 9.14 Our closest prokaryotic relatives. In 2015 the Lokiarchaeota were discovered deep in the Arctic Ocean, growing near a hydrothermal vent called Loki's Castle, for which they were named. Based on genetic analysis, the Lokiarchaeota and eukaryotes share a common ancestor more recently with each other than either group does with other prokaryotic lineages. The divergence into Lokiarcheota and eukaryotes is thought to have occurred about 2 BYA. Both Lokiarchaeota and eukaryotes have genes related to certain cellular processes such as phagocytosis that occur in eukaryotes but not in prokaryotes. Because of such similarities, Lokiarchaeota have been described as a transitional form between prokaryotes and eukaryotes.

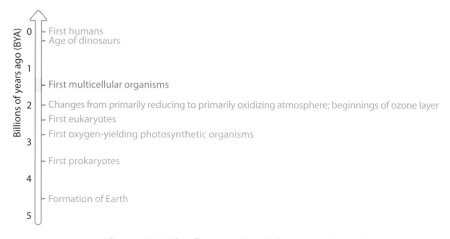

Figure 9.15 The first multicellular organisms. Eukaryotic organisms consisting of many cells first appear in the fossil record about 1.5 BYA.

another organism. Larger organisms are also better able to maintain suitable internal conditions. A single cell, for example, has little ability to maintain a particular temperature. With size, however, heat gain or heat loss is reduced, and a multicellular organism is to some degree buffered against the vagaries of the environment. Furthermore, multicellular organisms have less to fear from genetic mutations. A single serious mutation can spell doom for a unicellular organism. In a large multicellular organism, a mutation that causes the death of any single cell will probably go unnoticed.

Unicellular microorganisms must be doing something right, though, or they would no longer be so abundant. We have already reviewed several ways in which prokaryotes have the edge on eukaryotes. In Chapter 3 we discussed the advantage that small size alone confers, while in Chapter 7 we observed that protein synthesis occurs more rapidly in prokaryotic cells. These advantages mean that unicellular prokaryotes can reproduce at an accelerated rate, and when environmental conditions are to their liking, microbial populations can skyrocket. The take-home message is that no one mode of existence, whether it's prokaryotic or eukaryotic, single-celled or multicellular, is inherently superior. All come with advantages and disadvantages, and the bottom line is always the ability to reproduce and pass on genetic information. Because of variable environmental conditions, a successful evolutionary strategy in one environment may be considerably less successful in a different environment. Thus, the process of evolution relies on the environment to act as a sort of filter, permitting the survival of the best adapted, while poorly adapted organisms ultimately face extinction. The result is the almost limitless variety of life that surrounds us.

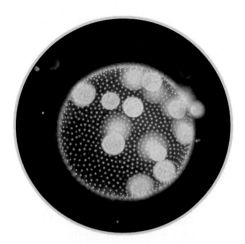

Figure 9.16 *Volvox*: a colonial green alga. A *Volvox* colony consists of a gelatinous matrix forming a hollow sphere, in which many individual cells are embedded. Such colonies, common in pond water, frequently reach sizes of up to 1 mm, making them visible to the naked eye. Colonial organisms such as *Volvox* may represent a transition state between unicellular and multicellular organisms.

The origin of viruses is obscure

So far our discussion has focused on cells. We know less about how viruses first evolved. Most likely, the earliest viruses were bacteriophages—viruses of bacteria that emerged during the time when prokaryotes were the only living things on the planet. As eukaryotes evolved, they also became targets for these parasites. But precisely how viruses originated is still open to considerable speculation. Currently, there are three principal hypotheses—called regressive, escaped-gene, and coevolved—to explain viral origins.

The **regressive hypothesis** postulates that viruses began as prokaryotes that gradually lost most of their genetic information and ultimately became dependent on host cells for most of their metabolic needs. According to this idea, an intracellular parasite could have become more and more dependent on its host cell, losing the ability to synthesize various biological molecules until it retained little except those genes needed for replication and transmission to new host cells. Eventually, this now-obligate parasite evolved into nothing more than a few genes within a protein coat. The regressive hypothesis therefore proposes an origin for viruses something like endosymbiosis, except that instead of evolving into intracellular organelles, viral ancestors evolved into parasitic life forms. If this hypothesis is correct, it means that with a bacterium like *Rickettsia* we could be witnessing viral evolution before our very eyes. *Rickettsia* cannot survive outside of a host cell, and they have very simple cell walls, lacking peptidoglycan (**Figure 9.17**). Have they lost the capacity for independent existence en route to becoming a new virus? Will they one day dispense with their cell walls altogether? We will have to return in a few million years to find out.

According to the **escaped-gene hypothesis**, fragments of nucleic acid from a host cell escaped and developed the ability to duplicate independently of the cell. These rogue DNA or RNA fragments would have needed to encode

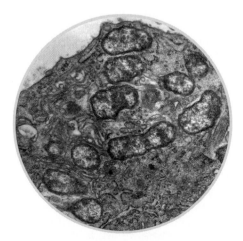

Figure 9.17 *Rickettsia*: an intracellular bacterium. The elongated, darker-staining bodies are *Rickettsia* in the cytoplasm of a eukaryotic cell. These small, intracellular parasites are unusual in that they have lost their capacity for independent life. According to the regressive hypothesis, this bacterial species may be in a transition stage on the road to becoming a virus.

the necessary information to replicate themselves, utilizing resources provided by the host cells. The original source of this independently replicating nucleic acid may have been plasmid DNA or transposons. Recall from Chapter 7 that transposons are short sequences of DNA that can move from one site on a chromosome to another site. They are found in both prokaryotes and eukaryotes. Evidence for the escaped-gene hypothesis has come from studying a specific viral transposon in retroviruses (see Chapter 5, p. 157). The transposon was found to contain base sequences similar to sequences found in the host, suggesting that it originally was part of the host's genetic material.

The third hypothesis proposes that viruses evolved alongside cellular forms of life and ultimately developed into obligate parasites of these early cells. While some forms of primitive life developed the cell membranes and metabolic capabilities previously discussed, others merely surrounded their genome with a protein coat, giving rise to the first viruses. As these primordial viruses coevolved with cells, they became increasingly dependent on them. Thus, the **coevolved hypothesis** suggests that life initially headed off in two separate directions. One resulted in increasingly complex cells while the other maintained the simple, streamlined structure that we now recognize as viruses.

As yet, there is no way to determine which of these three hypotheses is the most plausible. It is quite possible that viruses evolved more than once and that no one hypothesis explains the origin of all viruses.

Prions may have arisen from abnormal host proteins

As for prions, their origins are even murkier than those of viruses. Recall from Chapter 5 that prions are infectious proteins that have adopted an aberrant shape and that they can force normal forms of the protein to take on the altered conformation.

A clue to the origin of prions may stem from the fact that all known prion diseases in animals affect protein folding in the brain only. Recent research has suggested that certain proteins in the brain ordinarily change shape naturally as part of memory storage. That means that if the prion protein (PrP^{Sc}; see Chapter 5, p. 106) shows up, either due to gene mutation or infection, the brain already has many shape-shifting proteins that are vulnerable to attack. Because this sort of protein flip-flopping does not occur naturally outside the brain, proteins in other parts of the body are not susceptible to PrP^{Sc}, thus explaining why prion diseases are exclusively neurological. In other words, it is possible that prions are occasionally formed by normal brain proteins that are already prone to adopt different shapes, because of their function in memory formation. If they take on the shape of PrP^{Sc}, the result is Creutzfeldt–Jakob or another related prion disease.

Evolution explains life's diversity

"Nothing in biology makes sense, except in the light of evolution." This statement by the geneticist Theodosius Dobzhansky highlights the preeminence of evolutionary theory in our understanding of modern biology. Indeed, evolution is more than just a core biological principle; it is *the* core principle. Here we will investigate why evolution is central to biology and what exactly Dobzhansky meant.

We have already had ample opportunity in this book to acknowledge the great diversity of life on our planet. But within this bewildering variety there is a unity. Whether we are discussing *Streptococcus* bacteria, single-celled yeast, a eucalyptus tree, or a hippopotamus, we should remember that all life ultimately shares a common ancestor; we are all related. The same evolutionary forces are at work on all living things, and an understanding of how organisms evolve allows us to comprehend the living world in all its manifestations and complexity. Without evolution, the tree of life is nothing more than a thicket of unconnected branches. With it, the tree becomes a comprehensible road map, detailing the twists, turns, dead ends, and about-faces that have occurred over life's 3.5 billion year history.

The theory of evolution also provides us with a tool to explain new biological developments, many of which have direct relevance for the microbiologist. Why are antibiotics that used to eliminate bacterial infections now less effective or even useless? Why do some infectious organisms become less deadly over time, while others do not? With an understanding of how living things evolve, phenomena such as these not only become intelligible, they become predictable.

Natural selection is the driving force of evolution

Although not the first to propose that life evolved over time, Charles Darwin (**Figure 9.18**) was the first to lay out a detailed mechanism for evolution, supported by voluminous evidence. Darwin based his theory of evolution on two fundamental principles. First, he proposed that the organisms alive today descended from ancestral species. He called this view of life "descent with modification." Second, he suggested a mechanism by which such descent could occur. He called this mechanism **natural selection** and today it is understood that natural selection is the primary driving force behind evolution.

Natural selection can be broken down into a series of steps or observations. The outcome of these steps is evolution—changes in the genetically programmed characteristics of a population over time.

1. All living things have an enormous potential to reproduce. Under ideal conditions, all populations can grow exponentially if all individuals survive and reproduce.

2. Environmental resources will eventually become limited and because reproduction will produce more individuals than the environment can support, not all individuals will survive and reproduce.

3. Individuals within any population are not identical in terms of their characteristics and many of the differences between individuals are genetically based.

4. Those individuals most suited to their environment are the most likely to survive and reproduce. They therefore leave more offspring to the next generation, passing on those genetic traits that are best adapted to the environment.

5. This unequal capacity of individuals to survive and reproduce leads to a gradual change in the population, as those with favorable characteristics make up an increasingly larger proportion of the population over many generations. Less well-adapted individuals may eventually die out. The environment has thus "selected" the best-adapted genetic traits for survival.

It should be pointed out that natural selection works only on genetic variation, causing some genotypes to become more or less common (**Figure 9.19**).

Figure 9.18 Charles Darwin at age 51. In the mid-1800s Charles Darwin laid the groundwork for modern evolutionary theory by developing the concept of natural selection, which he published in *The Origin of Species*.

Cell undergoing a random unfavorable mutation is
removed by natural selection

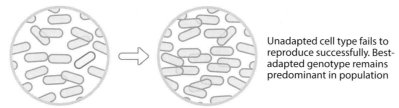

Unadapted cell type fails to
reproduce successfully. Best-
adapted genotype remains
predominant in population

Cell undergoing a random favorable mutation survives
and reproduces better than predominant genotype

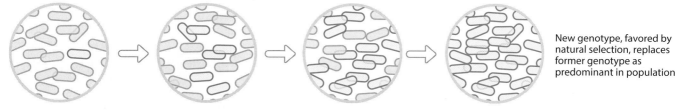

New genotype, favored by
natural selection, replaces
former genotype as
predominant in population

Figure 9.19 Natural selection in bacteria. When all members of a population are identical, natural selection cannot occur. When genetic variation is introduced, perhaps as a consequence of a random mutation, natural selection may select against or in favor of a mutated individual, depending on how the mutation affects the individual's adaptation to the environment.

The sources of genetic variation (mutation and genetic recombination) were discussed in Chapter 7. Characteristics such as blood type or the ability to make a particular enzyme have a genetic basis and can consequently be selected for or against by natural selection.

Traits that develop during an organism's life, on the other hand, are not subject to natural selection, because if a trait is not genetically based it is not transmitted during reproduction to the next generation. Animals that gain weight because of an abundant food supply do not pass genes for greater weight to their offspring. A tree that grows low to the ground because of exposure to strong winds will not transmit a low-growing characteristic to the next generation in its seeds.

A few other points about natural selection should be highlighted. First of all, natural selection is limited. It can work only on existing genetic variation. Such variation can arise by mutation at any time, but the environment cannot cause a specific favorable mutation to occur. Mutations simply occur, either spontaneously or in response to mutagens (see Chapter 7, p. 161). Once they do, natural selection is free to act upon them.

Furthermore, natural selection cannot result in organisms that are perfectly adapted to their environment. All organisms must be able to do many different things, and a characteristic that is valuable for one task may be less so for another. The same thick, waxy, acid-fast cell wall that protects *Mycobacterium tuberculosis* from damaging chemicals or antibiotics also means that it absorbs nutrients and oxygen very slowly, resulting in a very slow (for bacteria) growth rate (**Figure 9.20a**). Some species of *Staphylococcus* bacteria produce an enzyme called coagulase that causes host proteins to clump up around them, protecting the bacterial cells from host immune system cells. This protective armor, however, comes with a cost. Such bacteria are less able to spread easily to new host tissues (**Figure 9.20b**). An organism's various characteristics thus represent a compromise between the competing demands that any living thing must cope with.

Humans have guided the evolution of domestic plants and animals for thousands of years, and such **artificial selection** provides powerful evidence for natural selection. Consider, for example, the many breeds of domestic dogs. Over the centuries, humans have directed the evolution of dogs, by selecting those dogs with specific favorable traits and allowing only those individuals to reproduce. As a result, we now have breeds as diverse as pointers, selected for hunting ability; boxers, bred as guard dogs; and sheepdogs, selected for herding skills.

As microbiologists, however, we will now zero in on those organisms that are our primary focus and consider some examples of microbial evolution.

Microorganisms are subject to the laws of natural selection

CASE: VANCOMYCIN RESISTANCE: THE SEQUEL

In Chapter 7 (p. 159) we encountered the case of vancomycin resistance and how some bacteria are now unaffected by this once-powerful antibiotic. Recall that the antibiotic works by interfering with peptidoglycan synthesis by binding to the amino acids in peptidoglycan. Peptidoglycan synthesis is necessary for cell wall formation, and without a properly formed cell wall, bacteria are unlikely to survive. Due to a random mutation, some bacteria use different amino acids in their peptidoglycan. Vancomycin cannot bind to these altered amino acids, rendering bacteria with the mutation impervious to the effects of the antibiotic. Since vancomycin resistance first appeared in the 1980s, the number of resistant bacterial strains has increased, and today vancomycin is useless against certain infections.

1. **How was the initial mutation for vancomycin resistance affected by natural selection?**
2. **How might resistance to vancomycin be reversed?**

Drug resistance in bacteria provides an ideal example of how natural selection is involved in the evolution of specific characteristics. Like all other living things, bacteria that are better able to survive in a given environment tend to produce more progeny. Those that are less well adapted tend to perish over time. The arrival of widely available antibiotics on the scene, starting in the late 1940s, constituted a large shift in a microorganism's environment. Those individual microbes that were genetically able to survive in the presence of such antibiotics suddenly, and quite by chance, gained an enormous advantage over less resistant cells. Bacteria that were able to survive the introduction of antibiotics by whatever means reproduced more successfully and passed on characteristics that made resistance possible to their offspring.

Vancomycin resistance is an example; the altered peptidoglycan (**Figure 9.21**) did not originally evolve millions of years ago to prevent damage by an antibiotic that would not be used medically for eons. If on occasion a random mutation in a bacterium resulted in the production of the aberrant peptidoglycan, this organism would gain no advantage and would not increase in number relative to bacteria with normal cell wall structure. A single mutated bacterial cell carrying this mutation, without a selective advantage, would not be expected to become more numerous. In time any cells with the mutated cell wall might have died out entirely.

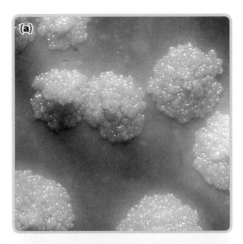

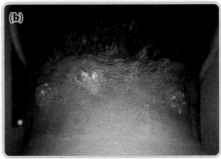

Figure 9.20 Natural selection is frequently a trade-off. All living things are subject to competing demands, and the results of natural selection are a compromise. (a) Colonies of *Mycobacterium tuberculosis* growing on an agar plate. Note the waxy appearance. Although *M. tuberculosis* is well protected from environmental hazards, its thick, waxy cell wall means that it absorbs nutrients and oxygen slowly and consequently has a relatively slow growth rate. (b) *Staphylococcus aureus* is a common cause of skin lesions such as boils. The bacteria secrete coagulase, which causes the coagulation of surrounding host proteins. Although this protects the bacteria from host immune system attacks, it limits their ability to spread throughout the host tissues.

(a)

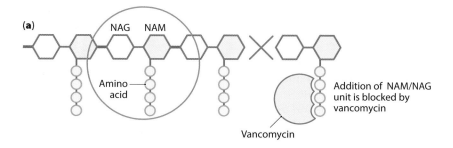

(b)

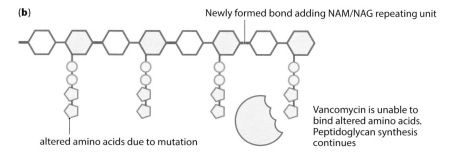

Figure 9.21 **Vancomycin action and resistance.** (a) During peptidoglycan synthesis, *N*-acetylmuramic acid (NAM) and *N*-acetylglucosamine (NAG) are assembled into the peptidoglycan repeating unit. The repeating unit is then added to the growing peptidoglycan molecule (see Figure 3.13 and Chapter 3, pp. 50–51, for a more complete description of peptidoglycan structure). Vancomycin can bind to the amino acid chain, preventing addition of new peptidoglycan repeating units and thus interfering with cell wall synthesis. (b) In vancomycin-resistant bacteria, a mutation changes some of the amino acids attached to NAM. Vancomycin cannot bind to these altered amino acids and therefore cannot interfere with cell wall synthesis.

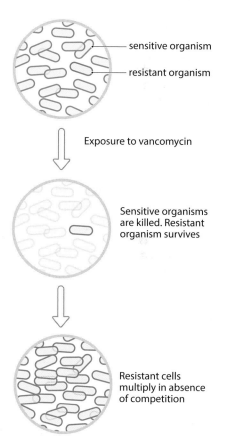

Figure 9.22 **Natural selection for vancomycin resistance.** In the presence of vancomycin, any cell that experiences a mutation that renders it resistant has a selective advantage. Sensitive cells are killed, but the resistant cell survives and reproduces, ultimately giving rise to a completely resistant strain. This selective advantage disappears if vancomycin use is discontinued. Any remaining sensitive cells might then again increase in abundance.

Now let's fast-forward a few million years to the age of antibiotics. Cells may now become exposed to vancomycin, and strictly by chance, the occasional mutation changing the cell wall structure still occurs (**Figure 9.22**). This means that if a cell happens to experience this mutation and is subjected to the antibiotic, this bacterium suddenly has an enormous advantage. It consequently survives and reproduces, passing on the now-favorable mutation to its progeny. Other cells, those with normal cell wall structure, are not so lucky. In the face of vancomycin exposure, they die and fail to reproduce. Over time, most members of the population will have descended from those cells that were initially able to resist the antibiotic. Cells bearing the mutation will continue to enjoy an advantage over normal cells as long as vancomycin exposure continues. This process, continuing over an extended period, might eventually result in a completely vancomycin-resistant strain of the bacteria.

The key phrase here is "as long as vancomycin exposure continues." Should vancomycin use stop, the advantage possessed by the mutated bacteria would disappear. Because they are no longer at a disadvantage, the proportion of normal bacteria in the population might once again begin to rise if, under antibiotic-free conditions, the normal cell wall is in any way superior to the mutated form. Consequently, by discontinuing the use of vancomycin or by using it judiciously, only in those cases where it is absolutely required, the reversal of resistance is possible.

Natural selection can influence the virulence of disease-causing organisms

CASE: A TALE OF TWO COUNTRIES— AUSTRALIA, ENGLAND, RABBITS, AND THE MYXOMA VIRUS

Rabbits are not native to Australia, but European rabbits were introduced there by settlers in the mid-nineteenth century. This seemingly harmless event triggered an ecological nightmare. With few natural enemies, the rabbit population exploded at an astonishing rate, alarming farmers as rabbits completely obliterated millions of acres of farming and grazing lands (**Figure 9.23**). In 1950, a bold solution was attempted. The myxoma virus, common in South American rabbits, was introduced into the Australian rabbit population. This virus has little effect on South American rabbits, but its impact on Australia's European rabbits was immediate and severe. The virus spread rapidly, and within only a few years, over 99% of all rabbits were dead. But just a few years later, rabbit populations once again began to increase. Research confirmed that not only had the rabbits begun to evolve resistance to myxoma virus, but the virus itself had become less deadly. Interestingly, this was not the case when the myxoma virus was introduced into England. In England, unlike Australia, the virulence of the virus has remained high, and exposed rabbits still usually succumb to infection.

1. **Why were South American rabbits unaffected by myxoma virus?**

2. **Why did the myxoma virus, so deadly initially for rabbits in Australia, become progressively less pathogenic?**

3. **Why didn't this reduction in pathogenicity occur in England, too? What is different about the situations in Australia and England?**

Figure 9.23 Rabbits gone wild. Following introduction into Australia, the population of European rabbits exploded, devastating millions of acres of farm and grazing land. Myxoma virus was introduced to control rabbits, but after initial stunning success, rabbit numbers began to increase once again. In England, the virulence of the virus has persisted.

The capacity of an organism to cause disease is an indication of that organism's **virulence** and any virulent, disease-causing microorganism is a **pathogen**. The organism the pathogen infects is called a **host**. Genetics play an important role in the host–pathogen interaction, and like other genetically based traits, the host–pathogen relationship is subject to natural selection. It consequently evolves with time.

It is often argued that over time, pathogenic microbes should become less virulent. After all, if you are a pathogen, you might think it would be a mistake to kill your host. Doing so sounds counterproductive in the extreme, because a pathogen that kills quickly has little chance to be transmitted to new hosts and suddenly faces extinction itself. Less deadly pathogens enjoy a selective advantage over more lethal strains of the same species, because the imperative to be transmitted is not so urgent. A more benign pathogen may remain in its host longer and consequently will have more time to reach a new host. Furthermore, the hosts are also evolving and often evolve resistance over time. Individual hosts who have a more vigorous immune response will also be more likely to have offspring and will pass on resistance to offspring. As a result, goes the argument, both pathogens and hosts evolve to the point where they reach a sort of evolutionary truce, allowing the survival of both.

We alluded to this principle previously in Chapter 6, when we considered the role of infectious diseases in the conquest of the Americas and in the exploration of Africa. In our case study we observe very much the same thing with myxoma virus in Australia. The virus evolved in South America, and South American rabbits evolved in response to it. After a long coexistence, the rabbit population was relatively unaffected by the virus, as those rabbits that developed resistance reproduced more successfully than nonresistant rabbits. Meanwhile, the virus was also evolving, with less-virulent strains of the virus better able to persist. Ultimately, both the virus and the rabbits in South America reached a point in their evolution where mutual survival was possible.

But the myxoma virus was poorly adapted to European rabbits in Australia when it was first introduced, and consequently the virulence of the virus was high. At the same time, rabbits in Australia had never been exposed to myxoma virus, so there had never been any selection for resistance. The result was the initial sky-high death rate for rabbits in Australia. This situation benefited neither the rabbits nor the virus. Once again, because of random genetic variability in both the rabbit and viral populations, natural selection went to work. Those few rabbits fortunate to have some resistance to myxoma virus were the only ones to survive and reproduce. Those rare viral particles that were genetically less virulent in European rabbits also fared better than more-virulent strains, because they allowed an infected rabbit to live long enough for the virus to get to a new rabbit.

It may be comforting to think that natural selection always promotes reduced virulence over time. Yet the English experience with myxoma virus demonstrates that it does not always work that way. Remember, when it comes to natural selection, the environment is all-important. No trait, including lower virulence, is necessarily always beneficial or always detrimental; environmental conditions set these ground rules. So, what is the difference between Australia and England? In Australia, myxoma virus is transmitted between rabbits by mosquitoes (**Figure 9.24a**). When a mosquito feeds on an infected rabbit, it picks up viral particles, which can then be transmitted to a different rabbit the next time the mosquito feeds. In England, the virus is transmitted by fleas (**Figure 9.24b**). And while most people probably consider both mosquitoes and fleas to be a similar sort of pest, there is actually a big difference between them. Adult mosquitoes die when the weather turns cold, but fleas survive.

During the Australian winter, there are no mosquitoes and consequently no viral transmission. The selection pressure on the virus for reduced virulence was therefore enormous. Virulent strains that killed their rabbit hosts quickly died out themselves, because once their host died, they were essentially stranded with no way to get to a new rabbit. Killing their host before they were transmitted meant these virulent viral strains effectively cut their own throat. Less-virulent strains that permitted rabbit survival, at least until the next mosquito season in the spring, had a large selective edge and ultimately came to predominate. The situation was quite different in England, where myxoma virus was transmitted year-round, because fleas were always available. There was never much difficulty getting to the next rabbit, so for a virus, killing its host was less of a problem. Less-virulent viral strains did not have a reproductive advantage. With the selection pressure for reduced virulence lessened, the virus retained its ability to kill most infected rabbits.

Biologists now understand that virulence, like other traits, is subject to natural selection and that pathogens become less virulent over time only if a reduction in virulence comes with a selective advantage. In other situations, becoming less virulent carries no such advantage. Often, as we saw in the case of myxoma virus, the determining factor is ease of transmission. When it is easy to get from one host to the next, less-vicious pathogens gain no real edge on their nastier cousins. When, alternatively, transmission becomes difficult, selection suddenly favors more benign disease organisms, allowing them to persist in their host until they can finally reach a new host.

(a)

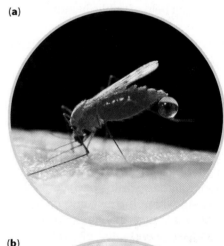

(b)

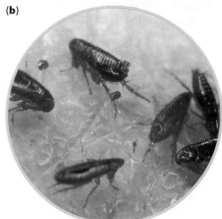

Figure 9.24 Myxoma virus transmission. (a) In Australia, myxoma virus is transmitted by mosquitoes, which do not survive the winter. Consequently, viral transmission ceases during the winter. (b) In England, the virus is transmitted by fleas. Since fleas survive throughout the year, virus transmission likewise is year-round. How does this fact explain the difference in viral virulence in Australia and England?

The differing virulence of cholera in Ecuador and Chile, used to introduce this chapter, illustrates this same concept in humans. Similarly, typhoid is another bacteria disease transmitted via contaminated water. In parts of the world where sanitation is especially poor (**Figure 9.25a**), typhoid bacteria are particularly virulent. Poor sanitary conditions ensure that the pathogen will have ample opportunity to achieve transmission. Routine sanitation, however, can change this. With safe drinking water (**Figure 9.25b**) and proper waste disposal, transmission is much less assured. Selective pressure on the bacteria now favors reduced virulence. The now more benign strains can survive indefinitely, until they get a relatively rare chance to reach a new host.

Accordingly, improved sanitation provides a double bonus. Not only do far fewer people contract water- or food-borne diseases in the first place, but when they do, they are likely to encounter a less-virulent strain of the disease-causing organism. Scientists are now beginning to consider the possibility of directing the evolution of disease organisms by changing the environment in ways that select for reduced lethality. Their thinking is that if we comprehend the underlying principles that control virulence, perhaps we can manipulate the host–pathogen relationship in ways that cause virulence to decrease. This emerging field of public health is called **Darwinian medicine**.

Microorganisms often influence the evolution of their hosts

In this chapter we have emphasized the evolution of microorganisms. We have paid little attention to how microorganisms influence the evolution of their hosts. Yet microbes are an important part of our environment and therefore act as a selective force on host evolution. Hosts that had certain adaptations, which best allowed them to fend off microbial attack, had higher rates of survival and contributed more offspring to the next generation. Microorganisms have certainly spurred the evolution of the immune system, which has become increasingly complex over evolutionary time. Many other host adaptations, less obviously related to immune defense, may also have been honed by microbial selection pressure. To make this point, let's consider just a few surprising examples.

Morning sickness may have evolved to protect the developing fetus

Any woman who has been through it understands that pregnancy can have all sorts of unusual effects on her body. One of the oddest is morning sickness. Just when a woman might require good nutrition the most, even the smell of certain foods can make her ill. More specifically, studies have shown that more women feel ill after smelling or tasting meat than any other food type (**Figure 9.26**). Many have suggested that morning sickness evolved as a way to protect a developing fetus from food-borne pathogens, and meat is especially likely to harbor organisms that might damage a developing baby. *Toxoplasma*, for instance, was discussed in Chapter 4. This protozoan is usually harmless for an adult, but an exposed fetus can suffer various birth defects, including deafness and mental retardation. Morning sickness might be realistically viewed as a sensible, if not particularly pleasant, evolutionary strategy to increase the likelihood of a healthy baby. Interestingly, morning sickness generally subsides after the first trimester, when the fetus is much less vulnerable and the benefits of extra nutrition outweigh the risk. In any event, perhaps a mother-to-be can take some solace from the fact that her

Figure 9.25 Transmission of water-borne pathogens and the evolution of virulence. (a) A slum in Latin America. When sanitation is poor, transmission is assured and there is little selection pressure on pathogens to reduce virulence. (b) A low-cost filtration device provides safe drinking water for these Bangladeshi villagers. Improved sanitation results in far fewer transmission opportunities. Less virulent pathogens now gain a selective advantage, because they can persist in their host without killing for a longer period of time and thus have more time to achieve transmission.

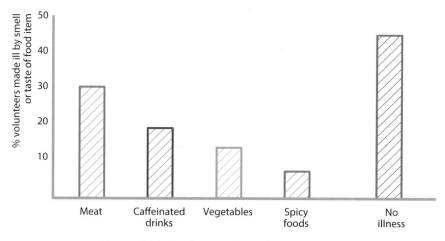

Figure 9.26 Mothers-to-be and morning misery. Researchers surveyed 79,000 pregnant women and found that meat was more likely to induce morning sickness than any other type of food. Is this an evolved, adaptive response to protect the fetus from potentially dangerous meat-borne pathogens?

discomfort is actually a sacrifice, maybe the first sacrifice, on behalf of her baby. Experienced parents know it will not be the last.

Prions may have affected evolution of the brain

In Chapter 5 we first learned about kuru, the prion disease once transmitted by cannibalism in New Guinea. Recall that if a person died of kuru, relatives who consumed portions of the deceased's brain became infected with the prion themselves and subsequently developed the disease. But not everyone who partook in this ritualized cannibalism developed kuru, and recent research has discovered why. Some people had a mutation that changed the shape of certain normal brain proteins, and in this mutated shape, the kuru protein could not bind and convert the normal protein to the kuru protein.

In societies where cannibalism was practiced, having the protective mutation would provide a large selective advantage. Individuals who had the mutation would be expected to have more children and to pass on the mutation to these children. Over time, a larger and larger proportion of the population would carry the protective version of the protein. In other societies where cannibalism was unknown, this mutation would provide no selective advantage. Consequently, if the mutation did occur, it would not necessarily increase in frequency and might ultimately disappear altogether. The investigators found that among New Guinea's Fore people, for whom cannibalism was once a regular part of ritual life, 55% of the population still carried the mutation. In Japan, with no history of cannibalism, the mutation was found in only 6% of individuals. Other sampled populations had rates of mutation similarly correlated with the historical occurrence of cannibalism.

Bacteria have influenced the formation of insect species

Sometimes infection can result in something as dramatic as the development of new species. In animals, a species is a group often defined by its members' ability to mate only with other members. Mating with a different species is almost always unsuccessful.

There are many reasons why mating between different species rarely succeeds. Eggs and sperm from different species are usually incompatible, and even if hybrid offspring are produced, in most cases they will not develop normally. Or, in the case of wasps within the genus *Nasonia*, the inability of different species within the genus to mate successfully is enforced by bacteria. The wasps and bacteria in the genus *Wolbachia* have one of nature's most bizarre partnerships. The bacteria live inside the wasp's reproductive cells, but they cause little if any damage (**Figure 9.27**). They do, however, have a huge impact on wasp reproduction.

Two closely related wasp species, *Nasonia giraulti* and *Nasonia longicornis*, are infected with different *Wolbachia* species. Investigators have found that the two wasp species were normally unable to mate successfully. Interestingly, if the wasps are treated with antibiotics, thus wiping out all bacteria in both wasp species, mating between the two species is successful. In other words, in *Nasonia*, the interspecies barrier to reproduction was completely due to the bacteria. The two types of wasps may actually provide us with a glimpse of new species in the process of forming and it is the bacteria that started the wasps on separate evolutionary roads. Should the now reproductively isolated wasps develop other genetic differences between them, the final species barrier may one day be complete.

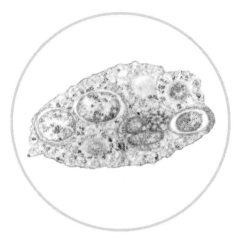

Figure 9.27 Wasps and *Wolbachia*. A wasp cell infected with *Wolbachia* bacteria. The bacteria are the oval-shaped structures within the cytoplasm of the larger, wasp reproductive cell. In some species of wasps, different species of these bacteria are the only things preventing successful mating.

Sex may have evolved in response to selection pressure imposed by pathogens

> ### CASE: ALICE IN WONDERLAND
>
> In Lewis Carroll's *Through the Looking Glass*, Alice passes through a mirror into a magical world where everything seems to be backward. Alice enters an enchanted garden where she meets the Red Queen, who tells her about the Red Queen's Race. Alice is mystified to learn that in this race the runners never actually move, and she observes, "In our country, when you run, you generally get somewhere." The Red Queen responds "Now, here, you see, it takes all the running you can do, to keep in the same place."
>
> 1. **How does the Red Queen's statement reflect the way hosts evolve in response to infectious organisms?**

Why is there sex? This question may sound surprising or even ridiculous at first, but consider the fact that most organisms on Earth get along just fine without it. Prokaryotes do not have males and females, and even many eukaryotes dispense with separate sexes. Many plants and animals reproduce primarily or exclusively asexually. Some lizard species, for instance, are all females. Offspring are produced when eggs, unfertilized by sperm, simply start developing into new lizards on their own. And compared with asexual reproduction, sexual reproduction (see Chapter 7, pp. 153–154) is slow and inefficient. After all, 10 asexually reproducing females can produce many more babies than five males and five females can.

One large advantage of sexual reproduction, however, is that it might permit host populations to survive in the face of a never-ending onslaught by pathogenic microorganisms. According to the **Red Queen hypothesis**, the continual selection pressure imposed by pathogens on their hosts favors genetic diversity in the host. And one prominent way to obtain genetic diversity is through sexual reproduction.

In asexual reproduction, all offspring may have the same genetic makeup as their parent. They are essentially clones of each other. In sexually reproducing organisms, each offspring represents a unique combination of genes from both the mother and the father. The genetic deck is shuffled with each reproductive event, so to speak. And such diversity might help prevent extinction at the hands of microorganisms, because microorganisms cannot infect and colonize all members of a host population equally. Some individuals, purely by luck, have genetically determined characteristics that make them a less appealing target for a specific pathogen. Individuals with these characteristics may in time come to be especially numerous in the population, because they are less susceptible to the pathogen in question.

Pathogens, however, are also subject to natural selection. There is intense selection pressure on the pathogens to colonize the most abundant host genotypes. Once pathogens colonize these hosts, however, the selective advantage previously enjoyed by these hosts is gone. Their abundance in the population begins to decline, setting the stage for a decline in the abundance of those pathogens that specialize on them. Meanwhile, however, in a genetically diverse, sexually reproducing population of hosts, a different host genotype is now the most resistant and begins to increase in abundance.

In this way, sexually reproducing hosts are genetic moving targets for pathogens. There is always a portion of the host population that is relatively unaffected by the pathogen. Both the host and pathogen undergo repeated cycles of selection, with increasing and decreasing abundance of different genotypes, ensuring the survival of both host and pathogen (**Figure 9.28**). In asexually reproducing organisms, this is far less likely. The largely clonal nature of such populations means that all members are very similar genetically. If a pathogen becomes able to colonize a particular clonal population, the entire population becomes an easy target for infection and may be exterminated.

But like Alice on her trip to Wonderland, in spite of constant natural selection and evolution, the best that sexual organisms can hope for is to stay in place. The eternal assault and counterassault between hosts and pathogens results in a never-ending race, with neither party ultimately winning, just as the Red Queen indicated. No one would argue that there are not other pressures besides pathogens that might favor genetic diversity and thus sexual reproduction, yet it is worth remembering that though unseen, microorganisms are an enormous part of our environment and in many ways have helped to shape us into what we are today.

Looking back and looking forward

In previous chapters we have described the diversity seen in the microbial world. Now we know how that diversity came to be. Starting with the evolution of life itself, and owing to the process of evolution via natural selection, life has diversified into the almost incomprehensible variety of living things with which we share our planet. Each of these living things has been honed by the interaction between their genomes and the environment, until each is best suited to a particular set of environmental conditions.

In Chapter 10 we will turn our attention to the consequences of this microbial diversity in term of ecological processes. In other words, we will investigate microbial ecology and the fundamental and critical role that microbes play in sustaining our environment. Put another way, in Chapter 4 we stated that life itself would not be possible on Earth without microorganisms. In Chapter 10, we will find out why.

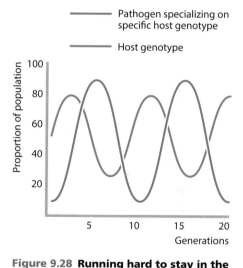

Figure 9.28 Running hard to stay in the same place: the Red Queen hypothesis. Pathogens that specialize on the most common host genotype have an advantage, simply because their host is most abundant. This gives a selective advantage to relatively rare host genotypes, which are relatively unaffected by pathogens. As numbers of the more abundant hosts decline because of high infection rates, the previously rare host type may come to predominate. There is now selection pressure on the pathogen to take advantage of this newly abundant host type. The result is a never-ending cycle of population peaks and crashes for hosts with a particular genotype and pathogens that specialize on that genotype. Sex may have evolved, at least in part, to introduce genetic variability into host populations, so that a pathogen cannot drive a specific host type to extinction.

Garland Science Learning System

- http://garlandscience.rocketmix.com/students/

- Discover why the outbreak of cholera was worse in Chile than in Ecuador

- Test your knowledge of this chapter by taking the quiz

- Familiarize yourself with the terminology used in this chapter by using the vocabulary review

- Get help with the answers to the Concept questions

Concept questions

These questions are designed to help you start thinking like a microbiologist. The answers are not always simply found in the text. Instead, you will need to take the concepts about which you have learned and apply them to new situations. Some of the questions may not even have just a single correct answer. Help is provided as part of the GSLS resources, which can be accessed through http://garlandscience.rocketmix.com/students/.

1. In this chapter we have reviewed the sequence of events that are thought to have occurred in the history of living things. Could they have occurred in a different order? Could animals, for example, or other multicellular eukaryotes have evolved first, with prokaryotes coming later?

2. The mutations that lead to antibiotic resistance have no doubt been occurring in bacteria for millions of years. Yet until the mid-twentieth century, bacteria that experienced these mutations did not increase in numbers. Over the last half century, however, such bacteria often thrive and come to predominate. Why?

3. The entire theory of evolution by natural selection can in some sense be summed up in the phrase "mutation proposes, selection disposes." What exactly does this mean?

4. How would the history of life on Earth be different if photosynthesis had never evolved? What would life be like if only glycolysis and photosynthesis had evolved, but aerobic respiration had not?

5. Explain how the discovery of ribozymes helps to resolve the issue of which came first, DNA or proteins.

6. After reviewing material in this chapter about the origin of the first cells, discuss why it appears crucial that these cells evolved in water.

7. Imagine a hypothetical and highly unrealistic situation involving two isolated human populations of the same size. Initially, rates of HIV infection are the same in both populations. In population A, however, a well-planned educational and prevention program to reduce HIV transmission is implemented. Almost all individuals are tested and subsequently given information as to how to best protect themselves and others. In population B, on the other hand, no such programs are implemented, and many people remain ignorant about HIV and how to protect themselves. If you compared these populations after 20 years, how might the virulence of the virus have changed in the two populations?

8. So far, nobody has ever discovered a type of bacterial cell that makes anything similar to the sort of protein coat that surrounds viruses. This fact would seem to argue against one of the three hypotheses put forward to explain the origin of viruses. Which one and why?

9. We concluded this chapter with a few examples of how microbes can affect host evolution. One well-known example we did not discuss was that of sickle-cell anemia, a debilitating condition in which red blood cells are malformed and cannot function properly. Some people, while not showing signs of disease are carriers of the gene for sickle-cell anemia. These individuals are much less vulnerable to certain types of malaria, because the protozoan causing malaria is less able to infect the red blood cells of these carriers. The gene for sickle-cell anemia is highest in parts of Africa where this form of malaria is common. Many people of African descent have moved over the years to countries outside of Africa where malaria is rare or absent. Over a long period of time, would you expect that in these people, the gene for sickle-cell anemia would become more or less common?

A MICROBIOLOGIST'S GUIDE TO ECOLOGY

When it comes to odd lifestyles, they don't get much more bizarre than that of *Heterorhabditis marelatus*, a nematode parasite of insects. These worms enter an insect as larvae through a natural body opening or across the insect's cuticle. But they don't come alone: They carry bacteria called *Photorhabdus* in their gut, and once inside the insect, they regurgitate these bacteria into the insect host's body cavity. The bacteria thrive in the nutrient-rich environment, and as they reproduce, they release chemicals that supress the insect's immune system, allowing the nematode to survive. Without this bacterial assistance, the nematode would be destroyed in the insect's body before it could mature to adulthood. As the parasites grow, they feed on the tissues of the insect, and produce their own new generation of larvae. As the larvae pass out of their mother's body, they acquire their own supply of *Photorhabdus* from special cells in the female's rectum. These larvae are now released into the environment where they set off in search of their own insects to infect (**Figure 10.1**).

For this nematode, these bacteria are crucial to their survival. The role that microorganisms play in other life processes is not always so apparent. When we think about interactions between living things, most of us tend to focus on the organisms that we can see. We observe rabbits and deer eating plants and we know that meat eaters like coyotes and wildcats consume the plant eaters. We notice insects such as bees as they pollinate plants, enabling the plants to reproduce. We may witness vultures feeding on dead animals. All the while,

Figure 10.1 It takes teamwork. A dead moth surrounded by numerous *Heterorhabditis bacteriophora*, which have just been released from the insect's body. The survival of the nematode parasite inside the insect requires the assistance of bacteria, released by the parasite, which supress the insect's immune system.

All case studies have a few questions at the end, the answers to which will become apparent as you read the sections following the case.

however, microbes, unseen and perhaps underappreciated, are quietly at work breaking down the waste products of other living things and increasing soil fertility. Microorganisms form the base of much of our food supply, and they produce much of the oxygen on which aerobes such as ourselves depend. In fact, all ecosystems rely on microorganisms if they are to survive and remain healthy.

Here in Chapter 10 we will focus on exactly what microorganisms do that is so critical to a healthy environment. After reviewing some basic ecological concepts, we will explore the roles of microorganisms in various habitats, including soil and aquatic environments. Next, we will learn how important elements such as carbon and nitrogen are cycled through the environment and how microorganisms drive this recycling. Finally, we will consider some of the interesting ways that microorganisms compete, prey upon, or otherwise directly interact with each other and with non-microorganisms. We will see that all life is indeed interlinked. No part of a healthy ecosystem can be altered without affecting every other part. We will now consider some of the smallest, but in certain ways some of the most important, cogs in the highly integrated system that makes up the interdependent web of living things.

Ecology is the study of how living things interact with each other and the environment

People often use the word "ecology" when they are discussing environmental issues like pollution or habitat preservation. To the scientist, **ecology** is the study of the relationships of organisms to each other and to their environment. The ecologist may therefore be concerned with who eats whom, which organisms compete for the same resources, or which ones interact in ways that benefit both organisms. Such biological interactions, known as **biotic** interactions, are influenced by numerous physical factors, such as soil chemistry, temperature, or availability of water. These physical factors, called **abiotic factors**, are important in determining which organisms are found in any given environment, and in many subtle and not so subtle ways they influence the nature of the biotic interactions. Biotic–abiotic interplay, however, is not a one-way street. Living things also alter abiotic conditions. To cite two examples, consider the difference plants can make on the air temperature and levels of sunlight in their immediate vicinity or how a herd of cattle alters the soil chemistry in a pasture. It is no coincidence that mushrooms, which require high levels of soil nitrogen to thrive, are so abundant where there is a lot of cow manure (**Figure 10.2**).

All the members of a particular species that live in a defined geographic area comprise a **population** of that species. The various populations within a defined region make up the **ecological community**. A coral reef community, for example, includes not only the coral, but the many other populations of various animals, plants, algae, and microbes that inhabit the reef. When we discuss both the community and the abiotic characteristics of a specific area, we are referring to an **ecosystem**. A freshwater lake or a large desert are examples of ecosystems.

Energy and nutrients are passed between organisms in an ecosystem

The manner in which energy and nutrients flow through any ecosystem is a key feature of that system. Often, this energy and nutrient flow is depicted as a **food chain**, which is the path of food consumption between the various members of a community (**Figure 10.3a**). At the bottom of the chain are

Figure 10.2 Biotic factors can affect abiotic conditions. Fungi require high amounts of soil nitrogen. Levels of this crucial nutrient may be especially high in cow pastures, where nitrogen continually enters the environment in cow manure.

(a)

(b)

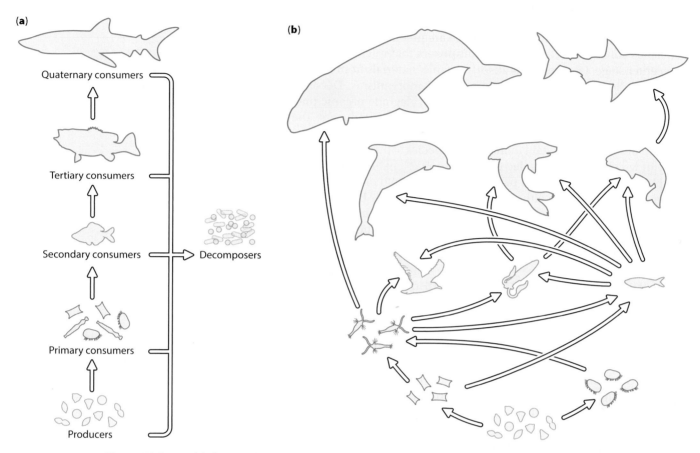

Figure 10.3 Trophic interactions. (a) Typical marine food chain. A food chain is a simple way of depicting feeding interactions between organisms. (b) Marine food web. A food web is a more accurate way to present trophic interactions in most ecological communities. Decomposers feed at all levels in a food web.

producers—those crucial organisms that first convert inorganic carbon into biological molecules through the process of carbon fixation. Many producers are photosynthetic organisms, but as discussed in Chapter 8 (p. 183), others rely on chemical energy to synthesize biological molecules. Producers are the only organisms able to introduce biological molecules into a food chain, and consequently, all life depends on them. Producers are consumed by **primary consumers**, also called **herbivores** (plant eaters), which in turn are consumed by **secondary consumers**, known as **carnivores** (meat eaters). Many food chains include tertiary- or quaternary-level consumers that feed on those lower in the feeding hierarchy (see Figure 10.3a). The position of an organism in a food chain is called that organism's **trophic level**. Producers occupy the first trophic level, while herbivores occupy the second level. Secondary and tertiary consumers occupy higher trophic levels in the food chain.

In all but the most basic ecosystems, however, the food-chain model is probably a gross oversimplification because many organisms are not restricted to a single location on the food chain. Many animals, for example, are **omnivores**, which eat both plants and other animals. Depending on what an omnivore is eating, it may be a primary, secondary, or even higher-level consumer. Accordingly, a more complex pathway, illustrating how different species in an ecosystem may feed at different trophic levels, is a more accurate way to depict feeding relationships. Such a depiction is called a **food web** (**Figure 10.3b**).

Some microorganisms are producers, and some, especially fungi, act as consumers. A few bacteria and many protozoa are predatory and consequently also act as consumers. But it is as **decomposers** that microorganisms truly take center stage. Decomposers absorb organic materials from dead organisms, including animals, plants, and other microorganisms. Decomposers play a vital role in all ecosystems, because they degrade organic materials into inorganic substances, which are then available to producers. They are absolutely essential to a healthy ecosystem. As discussed in Chapter 4, a single gram of soil might contain thousands of species of prokaryotes and fungi, many of which are active as decomposers. Aquatic habitats also teem with microbial recyclers. In certain places, such as very cold or very dry environments, decomposers are either scarce or slow-growing. Part of the reason that plants grow slowly in such environments is that with reduced decomposer activity, there are fewer nutrients available in the soil to support plant growth.

Nutrients are recycled in an ecosystem, whereas energy is not

Energy and nutrients do not move through ecosystems in the same way. Nutrients such as iron or phosphorus can be recycled through the environment indefinitely, but energy cannot. Recall from Chapter 8 (pp. 168–169) that for living things, energy initially resides in the carbon–hydrogen bonds of biological molecules. The conversion of energy from these molecules into a different chemical form such as adenosine triphosphate (ATP) is far from 100% efficient. Much of the energy in a food molecule is lost in the form of heat. Consequently, as energy moves up through a food web, there is less and less of it available at each higher level. When a herbivore eats a plant, as much as 90% of the energy in the plant material is lost as heat. When a carnivore eats the herbivore, most of the energy is once again lost. Put another way, it might take 100 kilograms of grass to produce a kilogram of antelope and 100 kilograms of antelope to produce a kilogram of lion. This explains why, if you could determine the mass of all organisms in an ecosystem, the collective mass of producers would far outweigh that of consumers. The collective mass of any population or community in a particular location is known as the **biomass**. In any ecosystem, the biomass of producers is greater than that of primary consumers, which is correspondingly greater than that of secondary consumers. By the time energy is transferred to secondary consumers, so much of that energy has been lost that very little is available for higher-level consumers. Only the most productive ecosystems, those where producers are especially prolific, can support small numbers of tertiary or quaternary consumers (**Figure 10.4**).

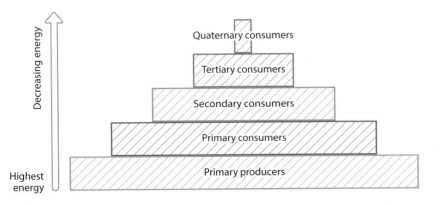

Figure 10.4 Amount of available energy at the various levels of a food chain. Because there is less and less energy available at progressively higher trophic levels, the numbers and/or biomass of organisms at each trophic level likewise decline. This relationship between biomass and trophic level is depicted here as an energy pyramid.

Microbial ecology investigates the role of microorganisms in their environment

Let's now focus our attention more specifically on the ecology of microorganisms. In many ways, microbial ecology is no different than the ecology of other organisms, and most of the same basic principles apply. But microorganisms are unique in that they live in all ecosystems. Not only are they found in places like oceans and soil that we commonly associate with plants and animals, but they also live in extreme environments, where few if any larger organisms are found. They colonize the digestive tracts of animals, the roots of plants, and essentially every other living surface. Wherever non-microbial life is found, microorganisms are there as well. As previously mentioned, their fundamental role in nutrient recycling and their phenomenal growth rate (see Chapter 8, pp. 184–185) set them apart from other living things. As we saw in Chapter 9, microorganisms did just fine without non-microorganisms for more than 1.5 billion years. The reverse could not occur. Without microbes, we would not be here.

Microorganisms live in microenvironments

Because microbes are so small, the environment in which they live can be described as a **microenvironment**—the area immediately surrounding a microorganism. Biochemical reactions carried out by microorganisms have a profound impact on the abiotic conditions of the microenvironment. As waste products are released, or as nearby nutrients are absorbed, chemical parameters such as pH or the concentration of ions can be significantly altered.

At first, it is difficult to fathom how small these microenvironments really are. As an example, consider a typical bacterial cell with a length of about 4 μm. If you measured environmental conditions 4 mm (4000 μm) away from the cell, it would be similar to measuring the conditions more than several kilometers distant from a large mammal. Over the seemingly small distance of 4 mm, conditions can vary dramatically. One spot, for instance, might be highly acidic, while just a millimeter or two away, basic conditions may prevail. To the casual observer, it might appear that acidophiles (acid loving) and non-acidophiles are living side by side, but in reality they occupy vastly different microenvironments. In a single soil particle one might find aerobes living on the particle surface, where oxygen levels are high, while only a short distance away, within the soil particle, a lack of oxygen allows strict anaerobes to grow (**Figure 10.5**).

Environmental conditions affect the growth rate of microorganisms

In Chapter 8 we learned that although microorganisms grow under various conditions, they tend to grow best under one specific set of conditions. For example, typical bacteria that live in the human intestine might divide about twice per day. If these same cells are cultured in the laboratory under optimal conditions, however, they may replicate in 30 minutes or less. Similarly, soil bacteria usually grow at a rate that is far below what might be observed under controlled laboratory conditions.

Why is this? First, in the laboratory microorganisms are often grown in pure culture (a culture of just one species). This is almost never the case in their natural environment. A single gram of soil might contain thousands of species and competition between these species can slow growth rate. Second,

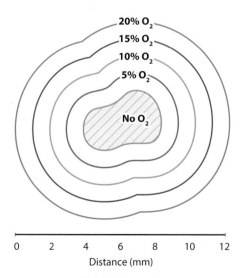

Figure 10.5 Microenvironments in a soil particle. Even a single soil particle that measures only several millimeters across may contain many microenvironments. This diagram depicts how just one abiotic factor, oxygen (O_2), can vary in different parts of the soil particle, creating different microenvironments for different microorganisms.

as discussed in Chapter 8, less-than-ideal abiotic conditions can result in submaximal growth. In many cases, growth rate reflects nutrient availability. Prolonged exponential growth (Chapter 8, p. 185) is actually quite rare in the environment, because microorganisms rapidly exhaust their supply of nutrients. Microbial growth often surges when nutrients are abundant and stalls when those nutrients are depleted. Accordingly, many microorganisms are adapted to a life of boom or bust. When resources are abundant, growth rate accelerates and may approach a maximum. When these resources are used up, growth slows or even stops as cells survive on stored biological compounds. To better visualize this concept, as microbial ecologists we must think tiny. A single dead insect or a rotting grass blade might represent a food bonanza for microorganisms, resulting in a rapid growth spurt. Once this bounty is used up, it is time to hunker down and wait for the next nutritional windfall.

Many microorganisms live in biofilms

Speaking microbiologically, what do your tongue and a corroded water pipe have in common? They are both likely to be coated with a slimy layer of microorganisms called a **biofilm**—a microbial community in which the cells are encased in a layer of polysaccharide. Although many bacteria live freely, separate from other individuals, they more commonly live as biofilms. Fungi are also frequently found in biofilms. The dental plaque that forms on teeth, as discussed in Chapter 3, is an example of a biofilm. So is the slippery layer found on rocks in a stream bed (**Figure 10.6**). Less-than-fastidious housekeepers are likely to have biofilms growing as scum that gradually builds up in toilet bowls or as a slimy coating on kitchen drains.

Bacterial biofilm formation begins when the microbes stick to a surface with their glycocalyx—the loose polysaccharide-based outer coating discussed in Chapter 3. Other unrelated cells may subsequently attach and begin to reproduce, until the entire surface is covered with a slippery layer of bacteria and polysaccharide. Biofilm growth, however, is not as haphazard as it sounds. Clumps of bacteria tend to grow in mushroomlike structures,

Figure 10.6 Bacterial growth in biofilms. Bacterial biofilms are found on many surfaces. (a) Biofilms are common on rocks in streams. The sticky sugars in the biofilm allow bacteria to adhere tightly to the rocks. (b) A biofilm is clearly visible in this toilet bowl, as the discolored area below or at the waterline. (c) Electron micrograph of streptococci growing as a biofilm on human teeth. The biofilm makes it less likely that the bacteria will be swept away by saliva or by tooth-brushing.

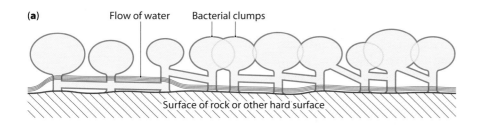

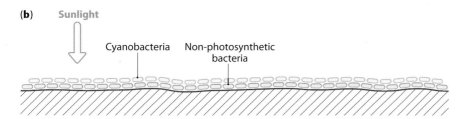

Figure 10.7 Structure of a biofilm. (a) Clumps of bacteria adhere to the surface via short stalks. Water can flow between the stalks, permitting efficient nutrient supply and removal of wastes. (b) Different microorganisms may be found in different layers of a biofilm. In this example, photosynthetic cyanobacteria occupy the top layer where they can more readily absorb sunlight. Non-photosynthetic bacteria form the lower layer.

supported by short stalks (**Figure 10.7a**). Openings between the stalks allow the circulation of water currents that carry resources like oxygen or nutrients to the cells. Some species of bacteria cannot attach to a surface on their own, but can sometimes bind other cells that have already attached. This colonization is facilitated by quorum sensing, as discussed in Chapter 7 (pp. 151–152). As cells join the biofilm, the regulation of various genes related to biofilm formation is altered and the positive feedback that occurs due to quorum sensing allows for continued biofilm growth.

Depending on the environment, biofilms can become increasingly complex, with different types of organisms forming separate layers (**Figure 10.7b**). On a rock in a steam, for example, there may be photosynthetic organisms on the surface, with a layer of non-photosynthetic bacteria just below.

Biofilms are certainly unsightly, but their importance goes far beyond the aesthetic. Persistent infections that seem to resist all antibiotic therapy are often caused by bacteria living in biofilms, since many antibiotics have difficulty penetrating the biofilm. Medical implants, such as artificial joints, are often sites of biofilm development. As many as 10 million infections per year in the United States are caused by biofilms on medical implants or because of invasive procedures with biofilm-tainted devices such as catheters.

Microbial biofilms also cause other, nonmedical problems. When they grow on a ship's hull, for example, they not only cause corrosion but also increase drag on the vessel, leading to higher fuel costs. Biofilms in pipes and drains cause extensive corrosion and block fluid transport. Submerged objects such as docks and oil rigs are often damaged by biofilm growth.

As unsavory as biofilms are, they have their proponents. As we will see in Chapter 17, **bioremediation**, the use of bacteria to degrade harmful chemicals, often relies on biofilms.

As we have already emphasized, microbes grow in almost any conceivable habitat. We will next review the role of microorganisms in some of their most important habitats, including soil, freshwater, and marine environments.

Soil often harbors rich microbial communities

Soil may appear fairly lifeless to the casual observer, but it actually constitutes a complex ecosystem that is constantly changing due to geological, meteorological, and biological processes. As rocks become weathered over long periods of time, they begin to fragment and crack. Microorganisms that colonize these cracks continue to break down the rock by producing acids and other chemicals. Once a thin layer of particles forms, the first plants may take hold, and as these plants die they contribute their organic material to the developing soil. As the soil continues to form, a top layer, called the topsoil, develops; it contains material rich in organic substances called **humus**. Below the topsoil, in the subsoil, the soil becomes increasingly inorganic; less humus is present, and less microbial activity takes place. Below the subsoil is a layer of parent material that includes the unweathered bedrock; little organic material or microbial activity is found here (**Figure 10.8**). Oxygen content of the soil also generally decreases with increased depth.

Although microbes thrive in soil, their distribution is not uniform. Soil microorganisms typically reproduce only when there is sufficient moisture, and soil moisture is a good indicator of microbial activity. Many soil microorganisms possess adaptations that allow them to persist during periods of dryness, but reproduction is slow or nonexistent during these times. The endospores produced by some bacteria, as discussed in Chapter 3, are an example of such an adaptation. Soil can be too wet, however. Certain fungi, including molds, are obligate aerobes and consequently are typically found

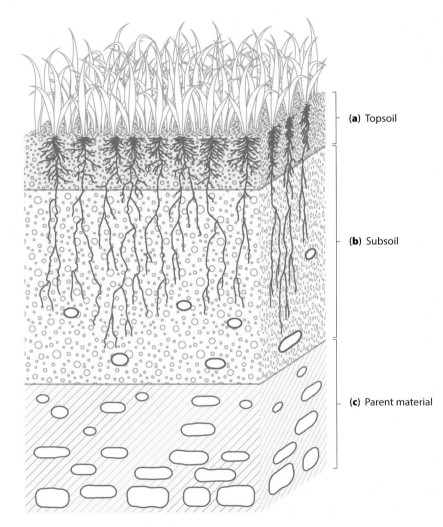

(a) Topsoil

(b) Subsoil

(c) Parent material

Figure 10.8 The soil environment. A soil profile reveals various layers differing in the amount of organic material and levels of microbial activity. (a) Topsoil contains living producers, undecomposed plant material, and high levels of decomposed organic material (humus). Levels of microbial activity are high. (b) Below the topsoil is the subsoil, where lower levels of both humus and microbial activity are found. (c) Parent material is mostly composed of unformed soil and bedrock. There is little microbial activity here.

in the uppermost layers of soil. If soil becomes waterlogged, oxygen levels drop to the point where these fungi cannot grow well. To persist during these less-than-optimal conditions, fungi typically produce reproductive structures called spores as part of their normal life cycle. These fungal spores can survive for long periods without growing, until the soil dries and oxygen levels increase.

Temperature, pH, and the availability of inorganic nutrients are other crucial abiotic factors that determine the microbial composition of the soil environment. Most soil bacteria are mesophiles and will consequently grow best between about 20 and 50°C (i.e. between 68 and 122°F). Very high or low pH inhibits the growth of most soil bacteria, but many fungi can tolerate wide fluctuations in soil acidity. In fact, some fungi grow especially well in highly acidic soils because of reduced competition from bacteria. In many soil ecosystems, low levels of one or more inorganic nutrients, like iron, place severe restrictions on the growth of some microorganisms.

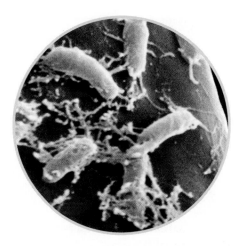

Figure 10.9 The rhizosphere. Plant roots release nutrients that allow growth of soil microorganisms on or near the root surface. In this electron micrograph, bacteria adhere to the roots of a cabbage plant.

Although photosynthetic microorganisms such as cyanobacteria and single-celled algae may be found on the surface of soils, most producers in this environment are green plants. As the plants die, microbial decomposers stand ready to return organic molecules to the environment. Plants also contribute to the formation of the **rhizosphere**, the soil environment immediately surrounding the roots (**Figure 10.9**). The microorganisms found in the rhizosphere grow in a biofilm on root hairs, absorbing amino acids and sugars from the plant. This is not a purely one-way relationship, however. The microorganisms return the favor by converting minerals into forms that can be absorbed by the plant. Fungi, for instance, often grow on the surfaces of roots in a mutually beneficial relationship with plants, called **mycorrhizal associations**. Recall from Chapter 4 (p. 83) that fungi in mycorrhizal associations often supply their plant partners with phosphorus or nitrogen in exchange for sugars. Some bacteria and fungi also produce substances that actually protect the plants from disease-causing soil nematodes or other plant pests.

Microorganisms are found deep below Earth's surface and even inside solid rock

Sometimes microbes are found in habitats that seem to stretch the meaning of the word habitat—far below Earth's surface, for example. Microorganisms have reportedly been recovered from drilling sites in northern Texas reaching as deep as 9000 m (30,000 ft), as well as from almost equally impressive depths in Alaska and South Africa.

We know little about what is living in this mysterious subterranean world, in part because it is so difficult to study. It is believed that primary production (the conversion of inorganic molecules into organic molecules) is based on chemical energy released by thermal geological sources and that many of the organisms are no doubt archaea. Other microorganisms have been found actually living inside solid rock. In Antarctica, for example, dry valleys are extremely cold and dry, and soil samples are almost microorganism-free (**Figure 10.10**). But this lunar landscape is not as lifeless as one might suppose. Lichens, symbiotic organisms consisting of fungi and algae, live inside porous sandstone. During the brief Antarctic summer, lichens produce organic molecules by photosynthesis, using minute amounts of melted snowfall as a water source. During the rest of the year, they exist in an essentially freeze-dried state. What this bizarre lifestyle lacks in excitement is certainly made up for in lifespan. Researchers have estimated that some of these lichens are thousands of years old, placing them among the oldest of all Earth's inhabitants.

Figure 10.10 Life at the bottom of the world. (a) The dry valleys of Antarctica are among the most extreme environments in the world. (b) Life, however, in the form of lichens persists inside porous sandstone rocks, just below the surface on the side of the rock facing the sun. Lichens are symbiotic organisms consisting of fungi and algae. Chlorophyll in the photosynthetic algae gives the lichen its green coloration, seen here as a streak just below the rock surface.

Many microorganisms are adapted to life in freshwater

Microorganisms abound in freshwater habitats. Even the most pristine stream is likely to support abundant microbial life. There are, however, substantial ecological differences between different freshwater habitats, such as a large lake and a quickly flowing river, and it should come as no surprise that the microbial community found in different freshwater habitats varies accordingly. Temperature, sunlight, oxygen, and nutrient levels all affect microbial abundance and distribution in freshwater environments.

In terrestrial ecosystems, multicellular green plants are the most important primary producers, but in freshwater, microorganisms take over the leading role. These photosynthetic organisms, mainly algae and cyanobacteria, make up the **phytoplankton**, which provide food for protozoa, small crustaceans, and other small primary consumers, which collectively make up the **zooplankton**. These small consumers are eaten, in turn, by larger invertebrates, small fish, and other secondary consumers, which provide food for higher-level consumers like large fish and birds. Meanwhile, bacterial decomposers actively convert dead or excess organic matter back into a form that can be utilized by producers.

Many lakes have several distinct ecological zones

CASE: WATER BIRDS AND BOTULISM

Avian botulism occurs worldwide, killing over a million water birds a year. It is especially problematic in the Great Lakes of North America. A recent 4-year study along the shores of Lake Michigan, for example, recovered a total of 4500 dead waterfowl, approximately 60% of which had died from botulism. The affected species included ducks, loons, mergansers, gulls, and cormorants. Most of the dead birds were recovered in late summer and fall, with mortality peaking in October. Because most dying birds are never found, wildlife experts believed that the reported deaths were just the tip of the iceberg.

1. Why is botulism such a problem for aquatic birds?
2. Are certain types of water bodies more prone to botulism outbreaks?
3. What is the relationship between the time of year and avian botulism?

Avian botulism is perhaps the most important disease threatening waterfowl in certain aquatic habitats. In certain years it can be especially deadly, as it was in 2002, when more than 5500 dead birds were found in a 2-week period along beaches near Buffalo, New York. The consequences for endangered birds, whose population is already low, can be particularly severe. Also in 2002, for instance, botulism killed 71 black-faced spoonbills in Taiwan out of a total species population of fewer than 1000 birds. Botulism is caused by the consumption of a toxin produced by *Clostridium botulinum*. To understand the link between lakes and this serious wildlife disease, we must first consider some basic principles of lake ecology.

Lakes can be divided into several zones, in which one expects to find different environmental conditions and therefore different microorganisms (**Figure 10.11**). The **photic zone** extends from the lake surface to the deepest point of sunlight penetration. Below the photic zone, photosynthesis cannot occur. Dissolved oxygen released as a photosynthetic waste is also highest near the surface, where the water can also be aerated by wind and wave action. In open water, phytoplankton are the only producers in the photic zone; close to shore in shallower water, multicellular aquatic plants are often abundant and they also serve as producers. The **littoral zone** (from the Latin word for "shore") is the portion of the photic zone through which light penetrates to the bottom.

Yet even in the photic zone, oxygen can be the limiting factor for microorganisms and other living things. Oxygen does not dissolve well in water, which means that water can hold only a limited amount of oxygen. Below the photic zone is the **profundal zone**, where oxygen is even more limited. Furthermore, because photosynthesis cannot occur in the profundal zone, consumers depend largely on organic material that has sunk from above. Many of the microorganisms living in the profundal zone are facultative anaerobes, with obligate anaerobes taking over in still deeper and even more oxygen-deficient water. The deepest region, called the **benthic zone**, includes the bottom sediments, composed of soil and organic material. Methanogens are often common here, as are other obligate anaerobes, including bacteria in the genus *Clostridium*.

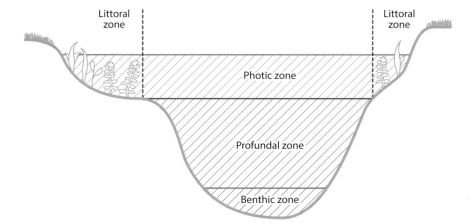

Figure 10.11 Ecological zones in a lake. Each zone differs in terms of abiotic factors such as light, temperature, and oxygen. These differences are reflected in the different microbial communities found in each zone. The photic zone is the topmost layer of the lake, through which light penetrates. If the water is shallow enough, the light may strike the lake bottom, a region of the photic zone called the littoral zone. Light does not reach the underlying profundal zone. The lake bottom beneath the profundal zone is called the benthic zone.

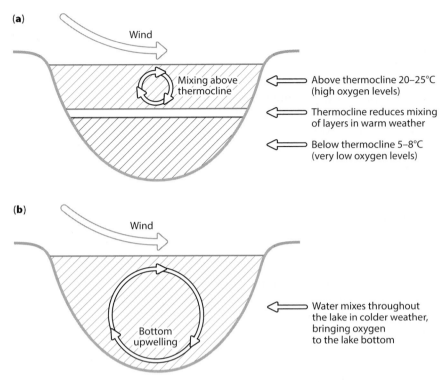

Figure 10.12 Seasonal change in a temperate lake. (a) During the summer, the development of the thermocline greatly reduces the mixing of surface and bottom layers. Oxygen levels are very low beneath the thermocline, because the oxygen consumed by organisms beneath the thermocline cannot be replenished. (b) In the autumn, the thermocline breaks down, allowing mixing between these different layers. Upwelling distributes oxygen levels throughout the lake.

If a temperate (nontropical) lake is deep enough, dissolved oxygen may vary considerably at different depths over the course of the year. During the spring and summer, surface water warms up quickly. Warmer water is less dense and it tends to remain near the surface. Colder water is denser and it therefore sinks. The result is a sharp demarcation in temperature known as a **thermocline** between the upper, warmer layers and deeper, colder layers (**Figure 10.12a**). Because there is little mixing between the two layers, very little oxygen is available below the thermocline. However, during the autumn, the surface water cools quickly and, as it sinks, it mixes with the deeper water, providing the deep-water habitats with increased oxygen (**Figure 10.12b**). Unless the lake freezes, such mixing occurs all winter long. Consequently, lake-bottom environments often alternate between being anaerobic in the summer and aerobic in the winter.

The lake's microbial community changes accordingly. This explains the relationship between avian botulism in lakes and time of year. During the warm summer months, a lack of oxygen in bottom sediments provides ideal growth conditions for obligate anaerobes like *C. botulinum*. Once the weather begins to cool and the thermocline breaks down, two things happen. First, increased oxygen in the benthic zone inhibits *C. botulinum* growth and induces spore formation. Second, mixing of the surface and bottom waters helps to move the bacterial spores toward the surface, where they may be ingested by birds feeding in the lake. The result can be a botulism outbreak. Deep lakes in temperate regions are especially prone to this problem, because they are most likely to develop a seasonal thermocline. The problem can likewise occur in other water bodies, like rice paddies, in which anaerobic conditions are likely.

Most of the bacteria that live in lakes can grow on relatively low levels of nutrients. Such organisms, which can prosper where others would starve, are called **oligotrophs**. They stand in contrast to the **eutrophs**, such as those living in an animal's intestine, which require relatively high nutrient levels. Because oligotrophs can survive on low levels of organic molecules and inorganic nutrients such as nitrogen and phosphorus, they have slow growth rates relative to eutrophs. Yet in most lakes and ponds, fast-growing eutrophs cannot compete with the oligotrophs that are adapted to starvation rations. As we will shortly see, however, things can change dramatically when lakes receive a large increase in nutrient material, either artificially or naturally.

Rivers and streams have their own unique ecological conditions

Flowing water habitats are quite different from lakes and ponds, where the water is stationary. First, rivers and streams are generally much shallower, and light can often penetrate to the bottom. Rapid water flow and turbulence ensure adequate oxygenation, at least in nonpolluted rivers. Many of the microorganisms living in flowing water are found in biofilms attached to rocks and other hard surfaces, where they come in contact with nutrients that flow past.

In larger rivers, photosynthetic organisms found in the river itself may serve as primary producers. In small streams, on the other hand, most nutrients may first enter the environment as runoff from nearby terrestrial habitats or in the form of leaves or other organic material that directly falls into the water.

Water pollution can lead to severe oxygen depletion

CASE: LAKE ERIE: A NEAR-DEATH EXPERIENCE

In the 1960s Lake Erie was dying. Oxygen levels in the water were declining, causing massive die-offs of fish and other aquatic vertebrates. Excessive blooms of algae gave the water a sickly green color. Beaches and shorelines, fouled by masses of rotting algae, reeked. Many native fish species disappeared from Lake Erie, replaced by species more resistant to the new conditions. Canadian scientists investigated the problem and found that the addition of inorganic compounds, especially phosphorus, was the primary culprit. Phosphorus entered the lake as runoff from agricultural fields, golf courses, or other places that were heavily fertilized. Untreated or partially treated domestic sewage was another major source, particularly because of phosphates used in household detergents. Following the implementation of laws to control water pollution as well as the removal of phosphates from detergents, Lake Erie began to recover. Today, in an ecological success story, the lake is considered to be largely rehabilitated. Nevertheless, such nutrient pollution is still perhaps the number one water quality problem worldwide.

1. What was the exact relationship of the agricultural and domestic runoff to the problems in Lake Erie?
2. How are microorganisms involved in this example of water pollution?
3. How have microbial processes caused the death of fish and the unpleasant odor of the lake?

We have previously mentioned that many of the microorganisms found in freshwater environments are oligotrophs that are well adapted to low levels of nutrients. In some circumstances, however, water bodies become inundated with organic material and nutrients. Sometimes this is due to a natural event such as a flood that washes large amounts of organic matter into the water. More often, human activity is to blame. When sewage enters a lake or river, it often carries a heavy load of organic material. Soils in agricultural areas often contain large amounts of phosphate and nitrate fertilizers, which can run off into lakes and streams following rains or irrigation. Phosphates in particular, especially those added to detergents to improve cleaning action prior to the mid-1970s, were an especially important part of the problem in Lake Erie.

The addition of these extra nutrients can wreak havoc on aquatic communities. Warm summer temperatures, combined with the added nutrients, stimulate an excessive growth of algae called an **algal bloom** (Figure 10.13). Phosphates are naturally low in such habitats and therefore are often a limiting factor in algal growth. With the addition of phosphate pollution, algae populations in Lake Erie skyrocketed.

Photosynthetic bacteria may run rampant as well. As populations of cyanobacteria explode, the water takes on an unhealthy green color. Populations of aerobic bacteria also soar, as they respond to the oxygen released by photosynthetic plants and microorganisms. As they deplete the oxygen in the surface water, the environment becomes increasingly anaerobic. Ultimately oxygen levels fall too low to support aquatic animals, including fish and large invertebrates, which may then die in large numbers. Anaerobic bacteria, on the other hand, thrive in the now oxygen-depleted environment. The hydrogen sulfide and other metabolic wastes that they release give the water a decidedly foul smell. This addition of excess nutrients to aquatic ecosystems, along with the subsequent ecological problems, is called **eutrophication**. While stringent water pollution control measures and the removal of phosphates from detergents have resuscitated Lake Erie, the problem of eutrophication is far from solved. As reliance on fertilizers in agriculture continues to increase, this environmental headache is likely to plague us for the foreseeable future.

Figure 10.13 Lake eutrophication. Water pollution in the form of inorganic nutrients is often due to runoff containing agricultural fertilizer. Algae, cyanobacteria, and aquatic weeds grow in response to this pollution. Although these photosynthetic organisms produce oxygen, their growth causes a surge in the growth of aerobic bacteria, which ultimately deplete the water of oxygen. The now oxygen-depleted water cannot support the survival of large invertebrates, fish, and other aquatic animals.

Although similar in many ways, the marine environment is distinct from freshwater

Earth is frequently called "the water planet" because most of its surface is covered with water. And the vast majority of water is found in the ocean. As in freshwater lakes, almost all primary production in the marine environment is due to phytoplankton, consisting of cyanobacteria and algae. These microbial organisms represent the base of the food chain, and they are eaten directly by zooplankton, which serve as primary consumers. As in freshwater habitats, higher levels of consumers include larger invertebrates, fish, and birds. Larger fish, sharks, and marine mammals are found at the top of the marine food chain.

Like lakes, the oceans can be divided into a photic zone, a profundal zone, and a benthic zone. Within the photic zone, we can distinguish between the near-shore littoral zone and the open-ocean **pelagic zone**. Along the shoreline there is also an **intertidal zone**, which is exposed to air at low tide and is covered by the sea at high tide.

While there are similarities, in many respects the marine and freshwater environments are quite different. Compared with lakes and rivers, oceans are far less variable in temperature and pH, and they have a much greater concentration of dissolved salts. Sea water is about 3.5% dissolved salts, whereas in freshwater, a typical value is around 0.5%.

The depth to which sunlight can penetrate varies considerably in different parts of the ocean. Depending on the amount of organic and inorganic material in the water, the photic zone may be as shallow as 50 m or as deep as 300 m. Photosynthetic organisms, wave action, and surface turbulence are sources of oxygen. Like lakes, deep sea environments lack both light and oxygen.

Nutrient levels may also range from very low to very high, and often the determining factor is proximity to shore. Near land, runoff from rivers and streams enriches coastal waters, and underwater currents frequently collide with land masses. As these currents are deflected upward from the sea floor, they carry organic material with them. The pelagic zone, alternatively, has neither coastal runoff nor current-driven upwelling, and consequently it is relatively nutrient-poor.

Because environmental conditions in the ocean are so different from freshwater conditions, it stands to reason that living things in these two ecosystems, including microorganisms, often display very different adaptations. For one thing, marine bacteria are somewhat **halophilic** (salt-loving) and grow best at salt concentrations of approximately 3.5%. Freshwater bacteria have a much lower optimal salt concentration. Furthermore, marine bacteria, unlike those found in freshwater or on land, generally require sodium for growth. Other adaptations vary, according to exactly where in the ocean a particular microorganism is found.

The littoral zone is typically richer in nutrients than the pelagic zone

If you have ever spent time on a boat, far from shore, you have no doubt been struck by the deep blue color of the water. By comparison, close to shore, the sea often takes on a greenish hue (**Figure 10.14**). Much of the reason for the color difference is microbial. Bereft of the upwelling and runoff found along the coast, nutrient-poor pelagic water contains little organic matter and relatively few microorganisms. In many marine environments, nitrogen is the primary factor limiting biological activity, while in certain areas, such as the

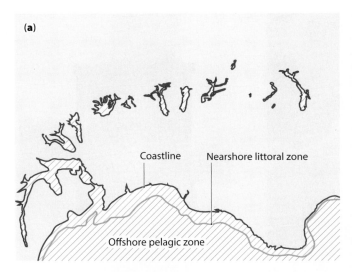

Figure 10.14 Littoral and pelagic waters. (a) Runoff from land introduces relatively large amounts of organic material and inorganic nutrients into the nearshore littoral zone. The green color is due to the multitude of tiny marine plants and photosynthetic microbes that grow well in the nutrient-rich water. The pelagic zone of the open ocean contains relatively little organic and inorganic material. Consequently, with fewer photosynthetic organisms, the water is blue in color. (b) Satellite image of the southern South American coastline. Note the greenish hue close to shore, indicating the presence of many photosynthetic organisms. Further offshore, deeper water reduces upwelling. With fewer nutrients available, photosynthetic organisms are far less abundant.

North Pacific Ocean, iron is the limiting factor. Understandably, most microorganisms found in pelagic waters are oligotrophs, adapted to low nutrient levels.

The green color typical of more inshore waters reflects the fact that compared with the pelagic zone, this littoral zone is far richer in nutrients and consequently supports more microorganisms, specifically chlorophyll-containing phytoplankton. It is no coincidence that the world's most productive fisheries are generally littoral or relatively close to shore; the rich and constant supply of nutrients provides ample resources for phytoplankton, which in turn support many primary and higher-level consumers. However, one does not have to look too hard to be reminded that too much of a good thing can be disastrous. With increased human activity along the coast come increased pollution, unwanted agricultural runoff, and microbial contamination of coastal waters. In many coastal areas, after a rainstorm, the harvest of mussels, oysters, and other shellfish must be temporarily suspended until the bacteria that have washed into the ocean have had time to die.

The area of the US Gulf Coast near the mouth of the Mississippi River provides a particularly grim example of how excess nutrients can negatively impact coastal ecosystems. Excess industrial and agricultural pollutants entering the Gulf of Mexico from the Mississippi result in eutrophication, characterized by explosive microbial growth and oxygen depletion. Without oxygen, fish and other animals cannot survive, creating a dead zone about the size of Massachusetts (**Figure 10.15**).

Red tide is another microbial problem familiar to many coastal dwellers. When excessive nutrients enter nearshore waters due to pollution runoff or increased upwelling, a rapid increase, or bloom, in dinoflagellate numbers turns the water red. Dinoflagellates are single-celled algae whose reddish pigments are responsible for the discoloration of the sea water (**Figure 10.16a**). Some dinoflagellates produce a powerful neurotoxin that can cause paralysis in many animals. During a bloom, levels of the poisonous substance increase in the bodies of many marine invertebrates. The shellfish are not harmed by the toxin, but as toxin levels in their bodies rise, animals consuming contaminated shellfish are at risk (**Figure 10.16b**).

Gulf of Mexico

Figure 10.15 The dead zone. Runoff of inorganic pollutants such as agricultural wastes, often at river mouths, can create large, eutrophic dead zones, such as this one on the US Gulf Coast at the mouth of the Mississippi River.

Figure 10.16 Red tides. (a) Increased nutrients can cause excessive growth, or blooms, of aquatic algae called dinoflagellates. The reddish color of the water is caused by the red pigments in the algal cells. (b) These algae can produce toxins that accumulate in shellfish. If animals, including people, consume contaminated shellfish, there is a serious risk of paralysis caused by shellfish poisoning.

Prokaryote numbers are highest near the surface and decline with increasing depth. In the upper 150 m or so, most prokaryotes are bacteria, whereas below that depth, numbers of archaea begin to increase, until in the profundal zone, the numbers of bacteria and archaea are about equal. This latter fact came as a surprise. Originally it was thought that archaea were found almost exclusively in extreme environments, but researchers now know that the sea literally teems with archaea.

Microbes in the deep sea and benthic zone require specialized adaptations

For organisms living below the photic zone, there are several major problems that must be overcome—namely, it is dark, it is cold, there is not much to eat, and the pressure is intense. Photosynthesis is not possible, and any primary nutrient production must rely on mechanisms that are not light-dependent (see Chapter 8 for a discussion of chemosynthesis). Organisms that live in the profundal and benthic zones are thus largely dependent on organic material and nutrients that fall from the overlying photic zone. For microorganisms living below 300 m, it can be slim pickings indeed. By the time organic material falls to this depth, most of it has already decomposed, and only about 1% of photosynthetically produced material actually makes it to the sea floor in an undigested form. Because there are such low levels of organic input, most microorganisms living at great depths are oligotrophs, able to grow at low nutrient levels.

At depths of more than about 100 m, the water temperature remains at approximately 2–3°C, and microorganisms found here are psychrophilic (cold-loving). Furthermore, for every 10 m of water, the pressure increases by 1 atm. (At sea level we experience 1 atm of pressure. One atmosphere equals 14.7 pounds per square inch.) This means that microorganisms living at 1000 m must be able to survive 100 atm of pressure. Such pressure-adapted organisms are called **barotolerant**.

One of the more remarkable discoveries in marine microbiology in recent years reveals a previously unknown ecosystem of astonishing proportions in the rocks beneath the sea bottom. Methanogens (methane-producing archaea) live in inconceivable numbers hundreds of meters beneath the sea floor. The methane that they produce has been estimated to have a greater mass than all the known reserves of oil, coal, and natural gas combined. Methane is a greenhouse gas linked to global warming, and it has been suggested that sudden bursts of methane, released from the sea bottom, have altered Earth's climate throughout its history and perhaps even ended ice ages.

So why is this superabundance of methane not being released into our atmosphere now, making current concerns about global warming seem like child's play? Other microorganisms living in more shallow sediments oxidize the methane as an energy source. Thanks to these organisms that break down the methane, our atmosphere is spared the input of an estimated additional 300 million tons of the gas a year.

The combined biomass of all these methane producers and methane oxidizers, living unobtrusively beneath the ocean floor, has been estimated at as much as a third of all life on the planet! Recall from Chapter 8 that when the atmosphere ultimately became oxygenated due to photosynthesis, living things had to either avoid oxygen or evolve mechanisms such as electron transport to detoxify it. The finding that a sizable proportion of Earth's life continues to pursue a strictly anaerobic existence, far removed from the aerobic world with which we are familiar, suggests that the former option was a popular one.

Cloud-dwelling bacteria may be important in promoting rainfall

Clouds scarcely seem like a promising habitat for living things, but some microorganisms are at home even here. Even more surprising is the role these microbes have as rainmakers. Before clouds can produce rain or snow, raindrops or ice particles must form. Tiny particles serve as the necessary nuclei around which the raindrops or snowflakes form. Most such particles are of mineral origin, but in 2008, researchers announced that in many cases bacteria, fungi, or tiny algae can do the job just as well. In some precipitation samples, over 70% of the nuclei were biological rather than geological. It is thought that when raindrops or snowflakes form around microbial cells, the airborne microbes can then return to earth. The researchers suggest that atmospheric scientists and meteorologists need to consider the implications of these microorganisms as they work to better understand the forces that drive our climate.

Exobiology is the search for extraterrestrial life

If life exists even in Earth's most challenging habitats, could it also exist on other planets? This is the compelling question asked by exobiologists—scientists who search for clues of life beyond our planet.

Currently, only Earth is known to support life. Yet microorganisms live in practically every conceivable habitat on Earth, including deep sea hydro-thermal vents where the water is above the boiling point, acidic environments where the pH is comparable to battery acid, or even inside rocks buried beneath Earth's surface. If life can exist in these extreme environments, exobiologists think that it might also exist on other planets, where similar conditions are found. These scientists often study extremophiles on Earth for insight into what extraterrestrial life might be like.

Because microbes thrive in even Earth's most extreme habitats, it is likely that if life is found on other worlds, it will also be microbial, rather than the sorts of large menacing creatures popular in science fiction movies. Europa, a moon orbiting Jupiter, is of special interest to exobiologists (**Figure 10.17**). And because water is essential for all life as we understand it, the 2015 announcement that there is water on Mars made this planet a prime candidate as a place where extraterrestrial life is possible.

In 1996 a meteorite found in Antarctica that bore apparent microbial fossils caused an immediate sensation (**Figure 10.18**). Analysis of the meteorite indicated that it came from Mars; upon heating, it emitted gases unique to the Martian atmosphere. What looked like fossils on the meteorite certainly resembled small microorganisms. Subsequent analysis, however, suggested that they were formed by geological rather than biological processes.

Microorganisms are primary driving forces behind biogeochemical cycles

The carbon atoms in the toast you may have had for breakfast this morning were once part of CO_2 molecules in the atmosphere, which were fixed into organic molecules during photosynthesis. Likewise, an iron atom that carries oxygen in your blood may have once done the same for a dinosaur, while in between it may have been in the soil, where it helped fertilize a cottonwood tree. Nature is indeed the great recycler, and although the term "biogeochemical cycle" may sound imposing, it simply refers to the movement

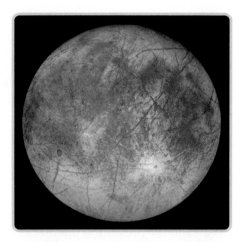

Figure 10.17 Life on other planets? One candidate location for extraterrestrial life is Europa, a moon of Jupiter. Europa is covered with ice, and examination of the moon indicates that large blocks of ice have moved, suggesting an underlying ocean. Conditions on cold, anaerobic Europa are thought to be similar to certain environments on Earth where prokaryotic life is abundant.

Figure 10.18 Alien life found? In 1996 a meteorite from Mars caused a sensation when it was reported that it bore fossils of Martian bacteria. The apparent bacterial fossils are now believed to be due to natural geological processes.

of chemical elements back and forth between living things and the nonliving environment.

All elements involved in life processes move through ecosystems in cyclical fashion, and in all these cycles, microorganisms play crucial roles. As microbial decomposers in soil and water break down organic matter, they release the inorganic building blocks back into the environment, which can once again enter food chains. In this section we examine a few of the most important of these cycles.

Carbon moves between living things and the environment

Carbon is truly the stuff of life, abundant in all biological molecules. In addition to these organic forms, carbon is also found in the atmosphere as carbon dioxide (CO_2) or in mineral form as calcium carbonate ($CaCO_3$).

Photosynthesis by green plants and cyanobacteria is the gateway by which CO_2 most typically enters the living world. Photosynthetic organisms fix atmospheric CO_2 in the Calvin cycle, incorporating it into biological molecules (see Chapter 8, p. 183). When animals or other consumers feed on plants or photosynthetic microorganisms, carbon begins its trip through the food chain (**Figure 10.19**). Some of the consumed carbon is released back into the atmosphere as a waste product of cell respiration in the form of CO_2. Other carbon atoms are incorporated into carbohydrates, proteins, and other biological molecules that will remain in a consumer until they are excreted as waste, or until the consumer is eaten by a higher-level consumer. Many consumers die before they are eaten, and when they do, bacteria and fungi, the decomposers, step forward to break down organic material, returning CO_2 to the environment. As discussed in Chapter 8, chemoautrophic microorganisms fix CO_2 using chemical energy instead of sunlight. Although less important to the overall carbon cycle than photosynthetic carbon fixation, these organisms also contribute to the overall ebb and flow of carbon between the living and nonliving worlds. Fermenting organisms also release CO_2 as a waste product.

Carbon is also stored as calcium carbonate ($CaCO_3$) in rocks such as limestone. Limestone itself, although inorganic, is the result of biological activity; it is secreted by marine organisms such as snails and clams to form their hardened shells and by corals during reef formation. Many microorganisms, including single-celled marine algae and protozoans, also secrete $CaCO_3$-containing shells. When these organisms die, their shells accumulate as marine sediments, and when limestone deposits that were originally formed in the ocean are eventually exposed to the atmosphere, they slowly decompose, releasing CO_2 back into the environment.

As organic carbon accumulates in anaerobic environments, the combined forces of heat and pressure can eventually convert it into fossil fuels like petroleum and natural gas. As discussed in Chapter 8, when these hydrocarbon fuels are burned, they are oxidized, releasing CO_2 as a waste. Because of the ever-increasing demand for fossil fuels, CO_2 is being released into the atmosphere faster than it can be fixed by photosynthesis. Atmospheric concentrations of CO_2 therefore continue to increase, upsetting the carbon cycle. This problem is only made worse as forests are cut down, especially in the tropics, reducing rates of photosynthesis and consequently carbon fixation. Because CO_2 is a greenhouse gas that traps heat in the atmosphere, most researchers believe that this imbalance in the carbon cycle is an important piece of the global warming problem.

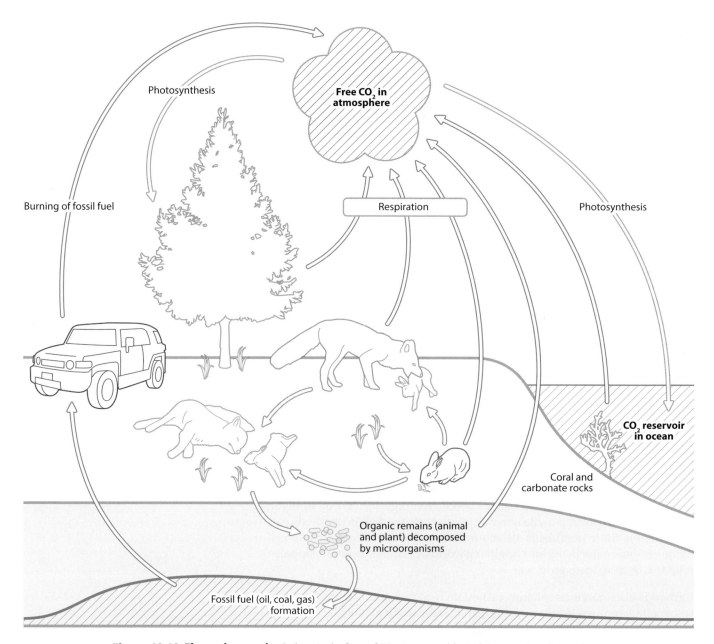

Figure 10.19 The carbon cycle. Carbon, in the form of CO_2, is removed from the atmosphere by producers, which convert it into biological molecules through photosynthesis. Consumers obtain carbon compounds in their diet and release CO_2 as a respiration waste product. Microorganisms likewise return carbon to the environment as they decompose organic matter. Some CO_2 is used by marine organisms to form calcium carbonate ($CaCO_3$) skeletons. When eventually exposed to the atmosphere, these skeletons can release CO_2 back into the environment. Combustion of fossil fuels adds to the atmospheric CO_2 pool.

"Iron fertilization" has been suggested as a way to decrease atmospheric CO_2 levels

The ocean can absorb some excess CO_2. Some have suggested that photosynthetic marine microorganisms could be coaxed into accelerating this process by spreading large amounts of iron-laced powder over the ocean surface. The logic behind this idea is that in many marine environments, low levels of iron limit microbial growth. By fertilizing the ocean with iron, the growth of marine phytoplankton would be stimulated. Removal of CO_2 from the atmosphere would be accelerated due to increased carbon fixation. This CO_2 might then remain in the marine environment as the phytoplankton die or are eaten by consumers. As atmospheric CO_2 decreases, the reduction in levels of this greenhouse gas might reduce the rate of global warming.

As an added bonus, such iron fertilization might enhance fishing. Recall that as energy, in the form of carbon–hydrogen bonds, moves up a food chain, much of the energy is lost in the form of heat. Consequently, at each level of a food chain, there is less and less biomass (see Figure 10.4). If the biomass of primary producers (in this case phytoplankton) could be increased, the biomass of consumers, including commercially valuable fish, may also increase. This strategy would work only in those environments where the availability of iron limits phytoplankton growth. The North Pacific, for instance, is noteworthy for its low levels of iron. In the Atlantic Ocean, on the other hand, nitrogen rather than iron limits phytoplankton growth. The impact of added iron would be less because nitrogen would still limit phytoplankton growth.

Over a dozen iron fertilization field trials have been attempted since the mid-1990s. Although atmospheric CO_2 levels fall in the test areas, controversy remains over whether or not it could be used on a practical basis. Many experts believe that the amount of iron required and the area of ocean that would need fertilization are far too large to result in a meaningful reduction in atmospheric warming.

Even if it did work, nobody knows what the consequence of introducing so much extra CO_2 into the ocean would be. We also know very little about the rate at which the ocean absorbs CO_2 and releases it back into the atmosphere or how increased levels of CO_2 in the ocean would affect other parts of the marine ecosystem. But the one thing we know for certain is that ecological interactions are complex. When one part is altered, the ramifications spread out in many and often unforeseen directions.

Bacteria convert nitrogen into forms that plants can absorb

CASE: THE KEY TO A BOUNTIFUL HARVEST

Veteran backyard gardeners know that different plant species require different amounts of fertilizer. Garden peas, beans, and other legumes provide a good example. An inexperienced gardener who fertilizes such plants might be surprised to find that although the plants appear healthy and continue to send out extensive leaves and vines, they produce very few pea pods (**Figure 10.20**). Usually the principal problem in over-fertilization is too much nitrogen. Legumes need very little nitrogen fertilizer or even none at all. With excess nitrogen, they funnel their energy into leaf and stem growth at the expense of flowers that develop into edible pods. Most garden supply stores, however, sell a product called "legume booster," which can be spread on pea seeds prior to planting or simply mixed with the soil. While not fertilizer, legume booster results in increased pea and bean pod production. Furthermore, many other cool-weather vegetables, including broccoli and cabbage, grow especially well when planted in proximity to pea plants.

1. **What is a legume, and why is over-fertilization a problem when growing them?**
2. **What is legume booster, and why should it be used with pea or bean seeds?**
3. **How does the presence of legumes stimulate the growth of other vegetables?**

Figure 10.20 Effects of too much nitrogen on plant growth. Legumes require little nitrogen; over-fertilization supplies them with too much nitrogen, which leads to excess leaf and stem growth and few edible pods.

Nitrogen is a major component of biological molecules, especially nucleic acids and proteins. Although pure nitrogen is extremely abundant in the environment, making up 79% of the atmosphere, this vast resource is unavailable to all but a few organisms. Atmospheric nitrogen is in the form of nitrogen gas (N_2), which animals and plants cannot use directly. These multicellular organisms, along with most microorganisms, need to obtain their nitrogen in other forms. Plants absorb nitrates (NO_3^-) or ammonium (NH_4^+), while animals obtain their nitrogen as part of proteins or other biological molecules in their food. How then does atmospheric nitrogen enter food chains? That's where microbes again take over, acting as the nitrogen conduit between the nonliving and the living worlds, via their participation in the nitrogen cycle (**Figure 10.21**).

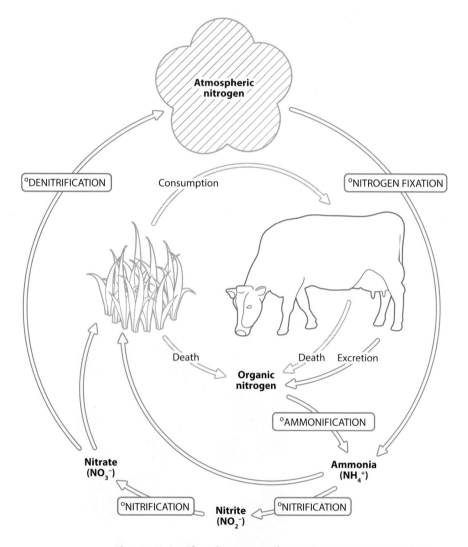

Figure 10.21 The nitrogen cycle. Small circles indicate microbial processes. Nitrogen from the atmosphere (N_2) is converted into ammonia by nitrogen-fixing bacteria. Some ammonia is then converted by soil bacteria into nitrite and nitrate, which, along with ammonia, are used by plants to produce nitrogen-containing organic molecules such as amino acids and nucleic acids. Organic molecules are then consumed by animals and other consumers. As plants and animals die, organic molecules release ammonia to the environment by the process of ammonification. This ammonia can be recycled by bacteria into nitrites and nitrates, thus making it available to plants again. Certain bacteria convert some nitrate back to atmospheric nitrogen gas in a process called denitrification.

The major way that nitrogen enters biological systems is through the process of **nitrogen fixation**—the conversion of atmospheric nitrogen into a form that plants can absorb (see Figure 10.21). Nitrogen fixation is carried out by certain nitrogen-fixing bacteria that live in soil or water. These bacteria have a unique enzyme that can break the bonds between two nitrogen atoms in atmospheric N_2. They then reduce the individual nitrogen atoms, producing the ammonium ion NH_4^+. Members of the genera *Azotobacter* and *Azospirillum* are examples of nitrogen-fixing bacteria.

Dead organic material also contains a large amount of nitrogen, and much of this is also released into the environment as ammonium ions through **ammonification** (see Figure 10.21). This additional NH_4^+ is formed as bacteria such as *Proteus* and *Clostridium* degrade proteins, nucleic acids, and other nitrogen-containing compounds. Ammonium, released as a waste, is then available to plants.

Not all ammonium, however, is directly absorbed by plants. Much of it is converted to nitrite and nitrate ions, through another microbial process called **nitrification** (see Figure 10.21). First, soil or aquatic bacteria including *Nitrosomonas* oxidize NH_4^+ as an energy source. Nitrite is released as a waste. Then, a second group of bacteria such as *Nitrobacter* converts NO_2^- (nitrite) to NO_3^- (nitrate), which can be utilized by other bacteria as well as by fungi and plants. To complete the nitrogen cycle, many common bacteria convert nitrate back to atmospheric nitrogen (N_2).

The role of bacteria in the overall nitrogen cycle is so essential to plants that some plants go to extremes to make life comfortable for nitrogen-fixing microorganisms. Snow peas and other legumes, including beans and alfalfa, develop symbiotic relationships with these nitrogen-fixers. When the roots of such plants are colonized by these bacteria, the roots start to form special structures called **root nodules** (Figure 10.22). These small aggregations of root cells provide not only housing for the bacteria but also nutrients and energy. The bacteria, in turn, supply their hosts with a constant supply of ammonium (NH_4^+), which is used for the synthesis of nitrogen-requiring biological molecules.

Because legumes can utilize nitrogen from the air, they are essentially self-fertilizing. If colonized with the proper nitrogen-fixing bacteria, they can thrive even in nitrogen-poor soils. Extra fertilizer is unnecessary, and if provided, legumes utilize it for extra growth rather than for flower and pea pod development. If soil does not contain the necessary nodule-stimulating bacteria, the farmer or gardener can actually inoculate the seeds or the soil with them as a means of increasing crop yield. The "legume booster" mentioned in our case is simply a powdered formula containing nitrogen-fixing bacteria.

Cabbage, broccoli, and other plants benefit by close proximity to legumes. Some of the excess ammonia is released into the soil, where it becomes available to plants that cannot fix nitrogen on their own. Sometimes, entire fields are sown with legumes simply to increase soil fertility. After the legumes are harvested, other crops are better able to grow in the nitrogen-enriched soil.

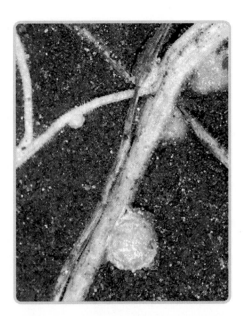

Figure 10.22 Root nodules in a soybean plant. Nitrogen-fixing bacteria such as *Rhizobium* form mutually beneficial relationships with legumes. The bacteria are housed in special structures called root nodules where they fix nitrogen for the plant. The root nodules are seen here as swollen areas along the length of the root. In return, the bacteria are supplied with nutrients by the plant.

Like other cycles, the phosphorus cycle relies on microbial activity

At one time, the tiny Pacific island nation of Nauru was among the richest nations in the world, all thanks to bird droppings. When seabirds eat phosphorus-containing fish, they deposit phosphorus in the form of guano

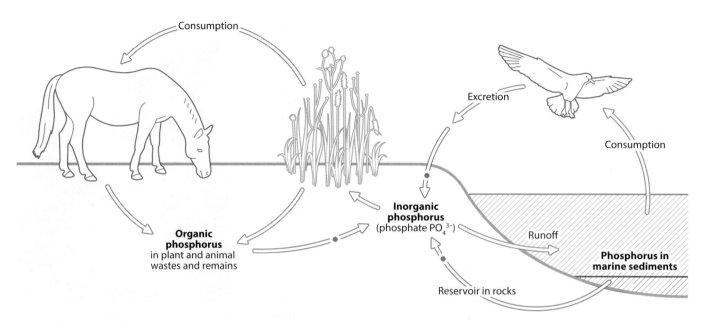

Figure 10.23 The phosphorus cycle. The small circles indicate microbial processes. Phosphorus in rocks or guano is released into the environment as phosphate (PO_4^{3-}), either by erosion or by the dissolving power of bacterial acids. Phosphate then enters the food chain, where it is incorporated into organic molecules such as nucleic acids, ATP, or phospholipids. These organic molecules are returned to the environment by microbial decomposers.

on islands where they nest. These phosphorus deposits are mined for use as fertilizer, and in this regard Nauru was particularly blessed. The phosphorus mines on the island are now exhausted but they serve as a good illustration of the value placed on this essential nutrient, and provide insight into yet another biogeochemical cycle (**Figure 10.23**).

Phosphorus, essential for DNA, RNA, ATP, and many other biological molecules, cannot enter the atmosphere and ends up accumulating in the ocean, which explains why marine fish and the seabirds that eat them acquire so much of the nutrient. Except for seabirds, the only other ways for phosphorus to return to the land are through the geological uplifting of ocean floors or the decomposition of organic material. Because the supply of phosphorus is limited in many soils, the demand for phosphorus-enriched fertilizers is high.

Phosphorus in rocks and guano deposits is released as the phosphate ion (PO_4^{3-}) by the dissolving power of bacterial acids. Sulfuric acid, secreted by bacteria in the genus *Acidithiobacillus*, is an example. Once in the soil or water, phosphate can be absorbed by primary producers including plants or photosynthetic microorganisms. Here it is incorporated into organic molecules, where it can be passed up the food chain to higher trophic levels. Organic phosphate in molecules such as DNA or ATP is returned to the cycle by decomposers. Eventually it returns to the inorganic reservoir in the sea.

Phosphorus is certainly essential for living things, but as in so many cases, too much of it has negative consequences. When humans interfere in the phosphorus cycle, by allowing phosphate-containing fertilizers to run off into lakes and streams, the enriched water bodies can quickly become eutrophic, causing algal blooms, oxygen depletion, and the death of animals. As previously discussed, until phosphate-containing laundry detergents were banned in many places, they were a serious part of this pollution problem.

No living thing, microorganisms included, operates in a vacuum. All life is interconnected, and organisms depend upon and interact with other organisms in a myriad of complex and indirect ways. But not all interactions are so roundabout and subtle. In the remainder of this chapter we take a closer look at some of the ways microorganisms interact directly with those around them.

Both organisms in a mutualism benefit from the relationship

CASE: CORAL REEFS AND DARWIN'S PARADOX

There's a reason scuba divers love coral reefs. Not only are the corals themselves beautiful, but they support a staggering amount of biological diversity (**Figure 10.24**). Coral reefs cover less than 0.1% of the ocean surface, yet they support over 25% of all marine species. Furthermore, corals do best in very clear water with excellent visibility, making for a spectacular diving experience.

Which leads to the paradox first noted by Darwin in 1842; we have recently learned that in nutrient-rich sea water, there is reduced water clarity due to the abundance of photosynthetic plankton. These producers give the water its greenish color and support consumers at higher tropic levels. The crystal clear water in areas where coral reefs are found reflects the fact that nutrients are so low in the area that phytoplankton are scarce. Yet a coral reef is like an oasis in a biological desert, supporting exceedingly complex food webs with multiple trophic levels, all in the absence of the photosynthetic, planktonic microbes we would normally expect to see in such a productive ecosystem.

1. **What explains the seemingly paradoxical distribution of coral reefs?**
2. **In addition to nutrient-poor water, coral reefs are also restricted to the littoral zone. Why are they unable to grow elsewhere?**

Figure 10.24 Healthy coral reefs. Coral reefs are perhaps the most diverse ecosystem on Earth. A single acre of healthy reef may support more fish species than there are bird species in all of North America. The color of the coral comes from the pigments in the photosynthetic zooxanthellae living in the tissue of the coral polyps.

A coral reef is actually a massive colony of small, individual animals called polyps, each of which is embedded within a calcium carbonate shell. Like other animals, these polyps, closely related to sea anemones and jellyfish, are consumers. But coral polyps actually consume only a very small (often less than 10%) portion of their nutrients. For most of their nutrient requirements, they rely on single-celled photosynthetic algae called zooxanthellae, which live in the polyp tissue. These microbes supply the polyp with up to 90% of its biological molecules. The polyp, in turn, releases CO_2 as a waste product of cell respiration, providing the algae with a carbon source for carbon fixation.

Like the relationship between the nematodes and bacteria with which we opened this chapter, coral polyps and zooxanthellae are involved in a **mutualism** (**Figure 10.25**): a biological interaction in which both partners benefit. And once we understand the interdependence between the polyps and the algae, Darwin's paradox is no longer such a puzzle. In areas of high productivity, there is so much plankton in the water that the zooxanthellae in coral do not get adequate light for photosynthesis. The clearer the water, the better both the coral polyps and their algal partners do. Furthermore, the reliance on photosynthesis for this mutualistic duo explains why coral reefs are restricted to the littoral zone—that portion of the ocean where sunlight can reach the bottom.

Mutualism

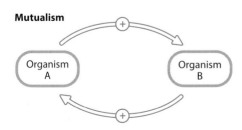

Figure 10.25 Mutualism. Both partners in this microbial interaction benefit.

Mutualism involving microorganisms is common. Recall, for instance, the case of the sick horse in Chapter 2 that was given "beneficial bacteria" as part of his treatment.

Another fascinating mutualism involves termites. Termites cannot digest wood on their own—they do not produce the necessary enzymes to break down the cellulose and other complex carbohydrates in wood. Without their associated gut microorganisms, termites would pose little threat to homes and other wooden structures.

Some termites do not eat wood. Instead, they depend on fungi, which the termites actually cultivate in their nests. When termites move to a new nest, they carry fungal spores with them to start fungal gardens in their new homes. These termites farm the fungi, which they tend and provide with organic matter. As the fungi grow, the termites feed upon them.

Returning to coral reefs, these magnificent ecosystems are exceedingly fragile and under enormous threat. One of the most worrisome dangers is **coral bleaching**—the loss of zooxanthellae, leaving only the white, bleached polyp behind (**Figure 10.26**). When polyps are stressed for any reason, they become less able to support their algal mutualists. The algae may be lost, putting the polyp at risk of starvation. During this time, polyp growth and reproduction may slow or stop altogether. If the stress on the polyps is reduced, they may reacquire zooxanthellae and recover. But all too often, the polyps die before this happens. Many factors can stress the polyps, but perhaps the most important is elevated water temperature. And with no end in sight to the current problem of global warming, the danger to coral reefs is likely to increase. It is estimated that 10% of the world's coral reefs are already lost, with another 60% of those remaining at risk. Unless global warming abates, that figure may be 100% of reefs at risk by 2050: a vivid and worrisome reminder of how important mutualistic interactions involving microbes can be.

Figure 10.26 Bleached coral. Environmental stress can cause the coral polyps to lose their zooxanthellae. The remaining coral polyp lacks color and is consequently referred to as bleached.

In a commensal relationship one organism is benefited while the other is unaffected

Commensalism differs from mutualism in that while one organism benefits, the other is neither helped nor harmed (**Figure 10.27**). In some commensal relationships, one microorganism makes a living from the waste product of another. For example, in our previous discussion of nitrification, recall that bacteria like *Nitrosomonas* release nitrite as a waste. Other bacteria such as *Nitrobacter* then use the nitrite, converting it to nitrate. While *Nitrobacter* is benefited by this relationship, *Nitrosomonas* is unaffected one way or the other. Another example of a bacterial commensal relationship is when facultative anaerobes living in an animal's intestine use up most of the available oxygen. Under these conditions, obligate anaerobes can now grow, without obvious cost or benefit to the facultative anaerobe.

Many of the microorganisms growing on human body surfaces are commensals. For example, bacteria such as *Corynebacterium* on your skin obtain nutrients and you are neither hurt nor assisted by your microbial guests. The same is true of bacteria dwelling in your armpits or moist regions of the body. Many of these bacteria use secreted lipids as an energy source, releasing fatty acids as a waste. Some of these fatty acids readily dissolve in the air and produce a strong, foul smell; they cause body odor. You might argue that this is not necessarily a commensal relationship, due to the cost the human host can pay in social ostracism. Hence, the popularity of underarm deodorants, many of which contain antimicrobial compounds.

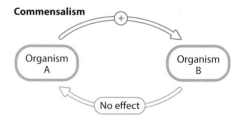

Figure 10.27 Commensalism. In commensal relationships, one organism benefits while the other is neither helped nor harmed.

Microorganisms in the same environment may compete for certain resources

CASE: A MICROSCOPIC DOG-EAT-DOG WORLD

As the grand finale of her microbiology course, Dr Brant expects her students to identify unknown species of bacteria, using techniques that they have learned during the semester. Each student is provided with a liquid broth, into which Dr Brant has placed one Gram-positive and one Gram-negative species. The student then streaks this mixed broth onto two media. One is a special selective medium on which only Gram-positive bacteria can grow (Figure 10.28), whereas the other is selective for Gram-negative growth only. With the two species now isolated, students can perform various tests on each one to make their identifications. When mixing her samples, Dr Brant adds 0.9 ml of the Gram-positive species but only 0.1 ml of the Gram-negative species. When mixing one of the unknown samples, however, she is distracted and adds 0.9 ml of both species. This sample is given to Steve, a student in the class. Steve streaks his sample onto the two selective media as directed, but following a standard 24-hour incubation period, he is disappointed to find that although his Gram-negative species grew well, the Gram-positive species did not grow at all. No other students have this problem, and Steve seeks advice from his instructor about how to proceed.

1. Why didn't Steve's Gram-positive species grow?

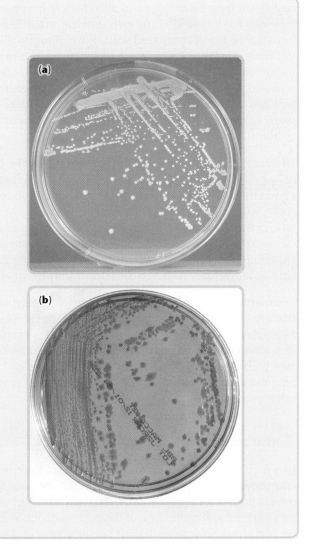

Figure 10.28 Bacterial isolation by use of selective media. (a) Gram-positive bacteria, growing here as whitish colonies, can be isolated from Gram-negative bacteria in a mixed sample by streaking the sample on a colistin–nalidixic acid agar plate. Acids in the agar inhibit the growth of Gram-negative bacteria without affecting Gram-positive species. (b) Similarly, Gram-negative bacteria can be isolated by streaking a mixed sample on a MacConkey plate. Compounds in this selective medium permit Gram-negative growth (pinkish colonies) while inhibiting the growth of Gram-positive organisms.

When two species within the same habitat try to acquire the same resources, **competition** is the result. Such an interaction has a negative impact on both participants (Figure 10.29). Often, one of the two competing organisms, the one that is better adapted to the habitat in question, will drive the less adapted organisms to extinction. This principle of **competitive exclusion** was first demonstrated in 1934 with two different protozoan populations, forced to compete for the same resources. No matter how the environmental conditions were modified, one species "won" while the other "lost" and disappeared.

If you examine a typical soil or water sample, brimming with many microbial species, you might conclude that competition among microorganisms is rare, since so many different organisms seem to be peacefully coexisting. Remember, however, the concept of microhabitats from earlier in this chapter and the need to think small in microbial ecology. Although you might find dozens of species in a single soil sample, closer inspection reveals that each species occupies its own microenvironment, where it outcompetes all comers.

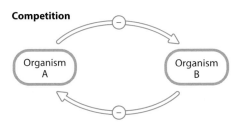

Figure 10.29 Competition. Competition is the microbial interaction in which two species within the same habitat vie to acquire the same resources.

The reality is that competition in the microbial world is especially fierce, and while competitive exclusion between animal or plant competitors may take weeks, years, or even decades, microbial competition is often resolved within hours. The ability of a microorganism to compete depends largely on its growth rate, which in turn is based on its metabolism, and how well it is adapted to any given environment. Because microbial growth rates are so rapid compared with those of multicellular organisms, we do not have to wait long to learn which species reproduces faster. Another important factor influencing microbial interactions is the chemicals released by some organisms that actually interfere with the growth of competitors. As we will see in Chapter 13, humans have taken advantage of many of these antagonistic chemicals, using them in the development of antibiotic drugs.

Dr Brant knew that under laboratory conditions, the Gram-negative bacteria in her unknown samples would grow faster and therefore outcompete and eliminate the Gram-positive species, unless the slower Gram-positive bacteria were given a numerical head start. This is why she mixed her samples at a ratio of 9 to 1. But when she mistakenly added equal amounts of both bacteria in Steve's sample, the Gram-positive organisms didn't stand a chance. Outcompeted by the rapidly growing Gram-negative species, they were eliminated before Steve had the opportunity to isolate them.

Competition between bacteria probably explains, at least in part, why the bacterial community in the human intestine stays so constant over time. Every day we introduce many new bacterial species into our digestive system along with the food and liquids we consume. Yet the newcomers are rarely able to colonize the intestine, because of competition with well-adapted species that are already colonizing their preferred microhabitats in large numbers.

Some microorganisms are predators

In a predatory interaction, one organism (the predator) consumes another (the prey) (**Figure 10.30**). This benefits the predator, while the impact on the prey is negative, to put it mildly. Bacteria are often prey, but only a few species are true predators. One of the best studied is *Bdellovibrio*. These bacteria first attach to the cell of their bacterial prey and penetrate into the space between the cell wall and the plasma membrane (**Figure 10.31**). Here *Bdellovibrio* replicates, absorbing proteins, nucleotides, and other biological molecules directly from its host cell. Eventually, the host cell lyses and dies, releasing progeny *Bdellovibrio*. Interestingly, *Bdellovibrio* can prey only on Gram-negative species. Gram-positive bacteria are safe from attack. Another predatory bacterial species with the colorful name *Vampirococcus* kills its bacterial prey in a different manner. Following attachment, it releases an enzyme that bursts the host cell, releasing its contents.

Many protozoa, especially ciliates and amebas, are predatory, and bacteria are frequently the victims. Although fungi are generally saprophytic or parasitic (see Chapter 4, p. 83), there are some predators as well. Fungi in the genus *Arthrobotrys*, for example, actually trap nematode worms in the soil, using their hyphae as nooses (**Figure 10.32a**). Once the nematode is stuck in the hyphae, new hyphae grow directly into the worm, where they absorb nutrients directly from the nematode's body cavity.

And having opened this chapter with a description of the nematode–bacterial team that targets insects, let's conclude with another reason to be glad you're not an insect: the fungi that prey on insect victims. Fungal spores

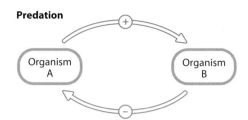

Predation

Figure 10.30 Predation. In a predatory interaction, one organism consumes the other; the predator benefits at the expense of the prey.

Figure 10.31 *Bdellovibrio*, a bacterial predator. Replication of *Bdellovibrio* occurs after these bacteria (in red) gain access to the space between the cell wall and the plasma membrane of their Gram-negative prey (in blue).

stick to the insect's body as it forages for food. Hyphae produced by the spore penetrate the insect's body, and once inside, the fungus grows, consuming the insect from within. After a few days, the now seriously ill insect crawls up to a high spot, where it dies. Shortly thereafter, the fungus literally pushes its way out of the insect's abdomen, spewing new spores into the air (**Figure 10.32b**). Any nearby insect unlucky enough to be caught in this spore shower will now become infected. Several species of so-called entomopathogenic (causing disease in insects) fungi go through this gruesome life cycle. Fortunately for us, vertebrates are unaffected.

Looking back and looking forward

Although the basic principles of ecology apply to all ecosystems and organisms, microbial ecology is distinctive in that microorganisms are found in virtually any environment. Furthermore, because microorganisms are so small, we must alter our perspective when we consider their ecological interactions, thinking in terms of the microhabitat, rather than the larger habitats with which we are more familiar. Microorganisms form the base of many food chains, and they serve as the principal decomposers in most ecosystems. Their role in driving biogeochemical cycles underscores their critical importance in maintaining a healthy, properly functioning ecosystem.

However, our description of ecological interactions involving microorganisms is not yet complete. We have yet to fully explore parasitism, a relationship in which one organism (the parasite) benefits, while the other (the host) is harmed, but often in a way that does not cause death. Many microbes parasitize humans and other animals and we will devote Chapters 11 and 12 to investigating the relationship between microbial parasites and their hosts. More specifically, we are exposed to parasitic microorganisms every day. Usually we remain healthy, but sometimes we get sick or even die. We will now find out why.

Garland Science Learning System

- http://garlandscience.rocketmix.com/students/

- Discover what have microbes ever done for you

- Test your knowledge of this chapter by taking the quiz

- Familiarize yourself with the terminology used in this chapter by using the vocabulary review

- Get help with the answers to the Concept questions

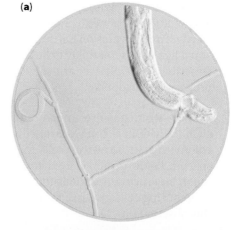

Figure 10.32 Predatory fungi. (a) Fungi in the genus *Arthrobotrys* live in the soil, where they trap nematode worms. Here, a nematode is trapped in the constricting ring of the fungus. Note the unsprung ring to the left. (b) An entomopathogenic fungus that has burst through the body wall of an infected ant.

Concept questions

These questions are designed to help you start thinking like a microbiologist. The answers are not always simply found in the text. Instead, you will need to take the concepts about which you have learned and apply them to new situations. Some of the questions may not even have just a single correct answer. Help is provided as part of the GSLS resources, which can be accessed through http://garlandscience.rocketmix.com/students/.

1. Select any habitat and construct a hypothetical food web for that habitat. Indicate which organisms are producers, primary consumers, secondary consumers, tertiary consumers, and decomposers.

2. There is often a striking difference in the rate at which plants grow in tropical versus cold, temperate regions. Explain the role played by microorganisms in this difference.

3. You are hired to write a screenplay for a new movie in which, due to a genetic engineering experiment gone awry, all prokaryotes on Earth are rapidly going extinct. You'd like your screenplay to be as realistic as possible. Describe a possible scenario for what would happen as prokaryotes disappear.

4. If microorganisms are spread out over an agar plate, each individual cell may start to divide, giving rise to a colony of cells. However, cells that were initially well spaced, some distance from other cells, give rise to much larger colonies. Cells that were initially close to neighboring cells give rise to smaller colonies. Explain this observation.

5. Imagine that a large earthquake struck a large city and that sanitation facilities in the city were badly disrupted. All of a sudden large amounts of raw sewage are pouring into a nearby waterway. What do you think would happen over the next several weeks in the waterway in terms of the population of photosynthetic microorganisms? Of oxygen content in the water? Of fish populations?

6. Although we did not discuss it, oxygen, like other elements crucial to life, cycles between the living and nonliving worlds. On the basis of what you learned about biogeochemical cycles in this chapter and about photosynthesis and aerobic respiration in Chapter 9, draw a simple diagram of how you think oxygen cycles in the environment.

7. Some people have argued that a vegetarian lifestyle is more sustainable because larger numbers of people can be fed on plant material than can be fed on animals that ate the plant material. Is there any truth to this? If so, what is the logic?

8. As discussed in the section on exobiology, the 2015 announcement that water was found on Mars generated great excitement regarding the possibility of extraterrestrial life. Why?

9. Tsetse flies, which transmit African sleeping sickness, feed exclusively on blood. They rely on bacteria in their gut to synthesize other nutrients that are not found in a diet limited to blood. What type of biotic interaction is this?

THE NATURE OF DISEASE PART I: A PATHOGEN'S PERSPECTIVE

11

Even Nobel Prize winners make mistakes. In 1960, the Australian Macfarlane Burnet won the award in Physiology or Medicine for his work in immunology. In 1962 he concluded that infectious disease was on the way out, and that "the major practical problems of dealing with infectious disease had been solved." Who could blame him for feeling that way? Certainly, after millennia as the primary scourge of civilization, pathogens were in full retreat. New antibiotics, breakthroughs in vaccine development, and steadily improving sanitation all pointed to a future free of infectious disease.

Burnet wasn't the only optimist. In 1967, US Surgeon General William Stewart declared that "the time has come to close the books on infectious disease." Meanwhile, the World Health Organization was laying out a time line for disease eradication. Tuberculosis, for instance, would be gone by the year 2000, with measles, polio, and diphtheria close behind.

Things didn't work out exactly as planned. A half century later, infectious disease is still with us, and in some ways it is a bigger problem than ever. Indeed, it is difficult to pick up a magazine or watch the news without hearing about yet another infectious disease that threatens us. The more spectacular diseases such as AIDS and Ebola usually grab most of the headlines, but other, perhaps less newsworthy maladies like bacterial dysentery still inflict more than their share of misery. Diarrhea alone still accounts for well over 700,000 deaths a year in children worldwide. Tuberculosis still kills over a million.

In Chapter 9 we began to see why infectious microorganisms are such intractable foes. Evolution via natural selection ensures that pathogens remain moving targets and our efforts to eradicate them are rarely completely successful. Moreover, examples such as the evolution of drug resistance remind us that when control efforts are poorly planned, the situation can become even worse in the long run.

Yet in spite of the continuing threat, most of us are usually healthy and when we do contract an infectious illness we generally recover. But not always. What determines how and when we develop infectious disease? When we do, how serious will it be? Will we recover and if so, will that recovery be rapid or prolonged? To answer these questions, we must understand how pathogens and hosts interact with each other. In this chapter, we will concentrate on what pathogens must do if they are to cause disease, and how and why they actually make us ill. In other words, we will view the disease process from the pathogen's point of view. Disease, however, is a two-way street. How we as hosts respond to microbial attack is often just as important, or even more important than the attack itself. To fully comprehend the overall disease process, we must also consider the host response, which we will do in Chapter 12.

> All case studies have a few questions at the end, the answers to which will become apparent as you read the sections following the case.

Contact with microorganisms only rarely results in disease

The simple presence of pathogens in or on a host does not mean that illness will follow. Rather, disease is the consequence of a series of increasingly important effects on the host that begins with **contamination**—the mere presence of microorganisms. Contamination does not imply that the microorganisms are causing any problem or that they are reproducing. It simply means that the contaminated host or inanimate object is not sterile. Indeed, almost all surfaces, living or not, are contaminated and in the vast majority of cases the microorganisms involved cause no particular difficulty. In fact, many such organisms provide valuable services to the host. If microorganisms survive and reproduce in or on a host, we say that an **infection** has occurred. Although we are conditioned to view infections as bad things, this is not necessarily true. Many infections have few if any negative consequences. However, if the infection interferes with normal host functions, we are now faced with a **disease**. Depending on the tissues affected and the degree of interference, disease may range from mild to severe and the changes in the host caused by the pathogen may be temporary or permanent. For example, once a person recovers from a bout of flu, there are ordinarily no lingering side effects. Recovery is complete. If the eyes of an infant are infected at birth with *Chlamydia trachomatis*, however, permanent scaring of the cornea can occur (**Figure 11.1**). Most *Streptococcus* infections cause only temporary disease, but occasionally there is permanent damage to the heart valves or to kidney tissue. This condition, called rheumatic fever, most often occurs when *Streptococcus pyogenes* infections are not properly treated.

Parasitic microorganisms able to cause disease are known as **pathogens**. The capacity to cause disease does not imply that disease will occur every time the pathogen encounters a host. Some pathogens almost always cause disease following infection, while others do so only rarely. *Mycobacterium tuberculosis*, the causative agent of tuberculosis, almost always interferes with normal host function to some degree. *Staphylococcus epidermidis*, on the other hand, generally is not considered to be a pathogen, but under certain unusual circumstances it may cause disease. Most infectious agents fall somewhere between these two extreme examples.

Figure 11.1 Eye damage caused by *Chlamydia trachomatis*. Colonization of the eye's conjunctiva leads to tissue inflammation. Repeated inflammation due to re-infection can ultimately lead to scarring. If scar tissue accumulates over the cornea, blindness results. This condition, called trachoma, is an important form of blindness. Rare in the developed world, trachoma is most common in Africa, Asia, and South America, where an estimated 230 million people are at risk.

Hosts are colonized with normal microbiota that is usually harmless

CASE: SAYONARA SALMONELLA

Improperly cooked chicken is a frequent source of exposure to the *Salmonella* bacteria that cause food poisoning. Today, however, chicken farmers can ensure that their birds are almost *Salmonella*-free by spraying newly hatched chicks with a product called Preempt®. This spray consists of 29 live bacterial species that are part of the chick's normal microbiota. As the chicks preen they ingest these bacteria, which take up residence in their intestines. A single treatment can prevent *Salmonella* infection in almost 99% of treated chicks. Infections with other pathogens, including *Listeria* and *Campylobacter*, are similarly prevented.

1. **What is meant by normal microbiota?**
2. **How does spraying the chicks with Preempt® prevent later infection with *Salmonella*?**

As previously mentioned, all animals, humans included, are populated by microorganisms that generally cause no health concerns. These commonly encountered, largely benign microbes are called the **normal microbiota**. We begin to acquire them at birth (**Figure 11.2**) and they are typically found on all exposed body surfaces (**Figure 11.3**). Hundreds of bacterial species and smaller numbers of other microorganisms call the human body home and their individual numbers can truly be astronomical. It is estimated, for instance, that a typical human body consists of 10^{13} cells. However, approximately 10^{14} prokaryotes live on your skin, in your mouth, in your intestine, and elsewhere. Even within our own bodies, human cells are outnumbered by microbial cells by approximately 10 to 1! Fortunately, under normal circumstances, these denizens of our bodies do not cause disease.

The normal microbiota is different in different host species. The organisms found in Preempt® are normal microbiota for chickens. Humans likewise have their own typical species, although many of them would also be found in other animals, including chickens. Even within humans there is considerable variation. The normal microbiota differs between an adult and an infant, as it does between people living in Calgary and Cairo. Even individuals of the same age and in the same geographic area do not have an identical normal microbiota. Each of us possesses a unique combination of species and strains.

The microorganisms composing the normal microbiota vary from one body region to another. On the skin, microorganisms are especially abundant in moister areas such as the groin or armpits. Most of the bacteria found here are Gram-positive; the dryness of the skin prevents its colonization by most Gram-negative species. In the female genital tract, most resident bacteria

Figure 11.2 Acquisition of normal microbiota. A baby giraffe enters the world at the Dallas Zoo. In giraffes and other mammals, the acquisition of microorganisms begins as the baby exits the birth canal, picking up bacteria present in the mother's reproductive tract.

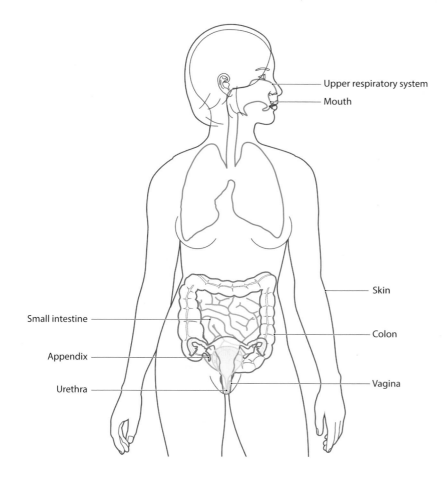

Figure 11.3 Body surfaces colonized by normal microbial microbiota. All of these surfaces are contiguous with the external environment. Other areas of the body not in direct contact with the exterior should remain sterile.

are Gram-negative. The upper respiratory tract and lower urinary tract, and especially the digestive system, also harbor an abundant normal microbiota. In a typical fecal sample, bacteria make up as much as a third of the weight. Other parts of the body, however, including the blood, cerebrospinal fluid, and deep tissues are normally sterile. The presence of any microorganisms in these locations is suggestive of disease. Even typically harmless organisms can cause disease if they gain access to such sites. This explains, for example, why dentists often prescribe a course of antibiotics following invasive dental work. The mouth supports a rich array of bacteria, which generally are not involved in disease. Serious problems may arise, however, if these microorganisms gain access to the blood due to a dental procedure. When members of the normal microbiota cause disease due to unusual circumstances, we say that the host has an **opportunistic infection**.

In the above dental example, the "unusual circumstance" is sudden access to an unaccustomed site. Now, normally harmless organisms have the opportunity to cause disease. Gram-negative bacteria such as *Escherichia coli* and *Klebsiella*, to provide another example, usually cause few problems in the large intestine. But it is not too unusual for severe alcoholics to develop pneumonia caused by these species. How do these usually harmless residents of the bowels gain access to the lungs? People suffering from severe alcoholism often pass out and when they do, they sometimes vomit. If vomit, which may contain microorganisms from the digestive system, is aspirated back into the lungs by an unconscious individual, a lung infection can result.

Normal microbiota may also become opportunists when the host's immune system is in some way suppressed. This explains why the likelihood of infection increases in patients undergoing cancer chemotherapy. Such therapy often has immunosuppressive side effects, making patients vulnerable to a variety of opportunistic infections. The same is true of patients infected with the human immunodeficiency virus (HIV). For example, almost everyone has a small number of the fungus *Candida albicans* present in their mouth. In those with a normally functioning immune system, they are unable to grow well, and typically cause no problems. In HIV-infected individuals, however, the decreased immune protection provides an opportunity for the fungus to proliferate, causing a condition known as thrush (**Figure 11.4**). This example is illustrative in an era when HIV continues to represent a serious health issue. You cannot catch an opportunistic infection such as *C. albicans* from an HIV-positive individual. You are already infected with this organism yourself, but if your immune system is functioning properly, you stand no risk of developing disease.

While it is true that the normal microbiota can cause problems as a source of opportunistic infections, by and large these microbial guests are relatively well behaved, and in many instances they prevent rather than cause disease. One of the characteristics of the normal microbiota is that they are particularly adept at attaching to the surfaces on which they live. Structures like fimbriae and glycocalyces (see Chapter 3, pp. 47–48) ensure that these species are unlikely to be swept away by intestinal fluid in the digestive system, sweat on the skin, or saliva in the mouth. Many normal microbiota organisms are so well adapted to particular regions of the host's body that they literally cover every square micrometer of space. There simply is no room for latecomers that may be pathogenic but are not necessarily so skilled at attaching to surfaces. Furthermore, many members of the normal microbiota secrete compounds that actually inhibit the growth of other, competing microorganisms.

Because they are outcompeted by the normal microbiota, many pathogens, even if they gain access to a host, cannot colonize. This is exactly how Preempt® keeps chickens free of *Salmonella* and other pathogenic bacteria.

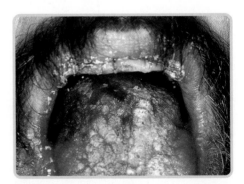

Figure 11.4 An opportunistic infection. A severe case of thrush, caused by an overgrowth of *Candida albicans* on the tongue of an HIV-infected patient.

By establishing a rich normal microbiota in young chicks, it leaves simply no room if and when *Salmonella* shows up.

We are sometimes vividly reminded of the protective value of our normal microbiota when we take an antibiotic to treat an infection with pathogenic bacteria. As we will see in Chapter 13, many antibiotics do not discriminate between pathogens and normal microbiota. Such a drug, in addition to destroying the pathogen, can also disrupt the normal microbiota. Such a disruption sets the stage for an opportunistic infection. For example, we recently learned that almost all people have a small number of *C. albicans* in the mouth. In women, this single-celled fungus is also present in the reproductive tract. Normally the growth of this potential pathogen is kept in check by the normal bacterial microbiota, but if an individual takes antibiotics, the microbiota may also be negatively impacted. This provides an opportunity for *C. albicans*, which is now able to proliferate. The result can be thrush in the mouth or vaginitis in the female reproductive tract.

The microbiota may impact many other host processes

CASE: SHY MOUSE VS COURAGEOUS MOUSE

A recent study has shown that even your tendency to walk down a dark alley at night, or how likely you are to strike up a conversation with a stranger at a party, isn't entirely your own decision: at least not if you're a mouse. Researchers first identified mice that they categorized as either "timid" (unwilling to explore new surroundings) or "bold" (those that readily explored new surroundings). Next, the makeup of their intestinal microbiota was determined and it was found that significant differences existed in the bacterial species found in the two groups of mice. The mice were then treated with antibiotics to kill off their intestinal bacteria. Finally, once the intestinal bacteria were destroyed, each group of mice was inoculated with the normal microbiota from the other group. Timid mice got their new microbiota from the bold mice and bold mice were colonized by bacteria from the timid mice. Remarkably, their personalities changed; previously timid mice became more exploratory, while the former bold mice became more timid.

1. **How can intestinal bacteria affect the behavior of mice?**
2. **Might the same phenomenon occur in humans?**

One of the most exciting new fields of inquiry in microbiology is the role played by the organisms of the microbiota in many diverse and surprising aspects of our wellbeing. Indeed, the microbiota is now frequently referred to as a newly discovered or virtual organ: one with impacts far beyond its ability to outcompete potential invaders.

The proper development of the immune system, for example, is now understood to depend on early exposure to microorganisms. Exposure to such microbes at a young age can lessen the severity of immune problems such as allergies later in life. This may explain the fact that in more developed parts of the world with better sanitation and hygiene, certain allergies such as asthma have become much more common (**Table 11.1**). In less developed countries, in contrast, children are more reliably exposed to a wide range of microorganisms. Allergies in such places have not increased in frequency. Likewise, the likelihood of certain autoimmune diseases, including type 1 diabetes and multiple sclerosis, in which there is a misdirected attack on the host's own tissues, may be influenced by the microbiota. Comparisons of children with and without diabetes, for example, have shown that children with diabetes have a significantly different microbiota in their digestive systems than

Table 11.1 Allergies and autoimmunity in the twentieth century. Data are for the United States between 1950 and 2000. For each disease listed, the number 1 represents the baseline number of cases in the year indicated. The numbers in later years indicate the fold-increase in the number of cases over baseline. Thus, for multiple sclerosis, by 1960, the number of cases had increased by roughly 1/3 over the 1950 value. By 1990 the number of cases had increased by approximately 3.3 times over the baseline number. Could the sharp rise in these examples of immune disorders be due to decreased exposure to microorganisms?

	YEAR					
	1950	1960	1970	1980	1990	2000
Asthma	—	—	—	1	× 2.3	× 3.0
Multiple sclerosis	1	× 1.3	× 2.0	× 2.3	× 3.3	—
Crohn's disease	1	× 1.6	× 1.6	× 2.3	× 3.0	× 4.0
Type 1 diabetes	—	—	—	× 1.5	× 2.3	× 2.75

healthy children. The microbiota of diabetic children is less stable over time and has fewer and different bacterial species. Even some forms of cancer seem to be at least in part attributable to the makeup of the microbiota.

And the list doesn't stop there. Studies with twins, in which one is obese and one is lean, highlight the importance of the microbiota in body weight maintenance. Obesity seems to be associated with a less diverse microbiota, but one in which intestinal bacteria produce more of the enzymes that help to digest nutrients in the intestine. This enhanced digestion, in turn, leads to an increased nutrient absorption, higher calorie intake, and weight gain. And because various members of the microbiota can influence the release of certain chemicals in the brain, the microbiota has been linked to a laundry list of neurological conditions, ranging from depression to anxiety to even the severity of jet lag!

This capacity to influence the chemistry of the brain is no doubt behind the results described in our case. It is well known that certain chemical compounds called neurotransmitters, used by nerve cells to communicate with each other, can have a profound effect on animal behavior. In the last few years it has become clear that some bacteria do more than influence neurotransmitter release—they actually produce chemicals with structures similar to neurotransmitters that mimic their activity. It remains unclear if humans are affected precisely the same way mice are in terms of shyness. It is increasingly clear, however, that microbes exert all sorts of influences over us that we are just beginning to comprehend.

A lack of complete understanding, however, has not discouraged some from suggesting that various conditions might be treated by altering a patient's microbiota. In fact, although not yet routine, and perhaps not for the squeamish, fecal transplant continues to gain acceptance as a possible treatment for certain conditions. In this procedure, fecal bacteria from a healthy individual are introduced into the colon of a recipient in order to restore a healthy intestinal microbiota. Although still experimental, fecal transplant has already proven to be an effective treatment for certain types of chronic diarrhea and other digestive system disorders.

To cause disease, a pathogen must achieve several objectives

Having reviewed some of the basic concepts of relevance to infectious disease, let's now take a closer look at exactly how and why disease occurs.

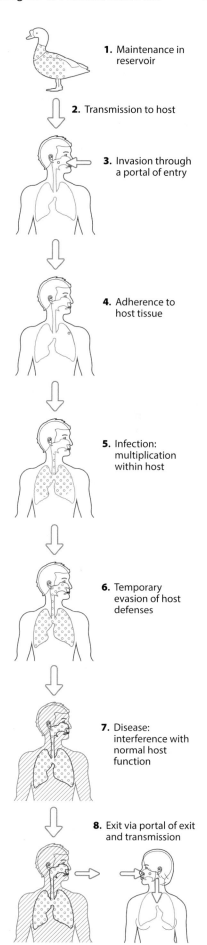

Figure 11.5 Requirements for infectious disease. Even normal microbiota must meet most of these requirements. Whereas pathogens interfere in some way with normal host function, the normal microbiota ordinarily does not. In this example a bird is serving as the reservoir, and the pathogen is being transmitted via the respiratory route.

1. Maintenance in reservoir

2. Transmission to host

3. Invasion through a portal of entry

4. Adherence to host tissue

5. Infection: multiplication within host

6. Temporary evasion of host defenses

7. Disease: interference with normal host function

8. Exit via portal of exit and transmission

In general, when pathogen meets host, the likelihood of disease depends on variables including the number of pathogens to which a host is exposed, the virulence of the pathogen, and the status of the host's defenses. These factors are all closely interrelated. For example, if a particular pathogen is highly virulent, fewer pathogens are needed to cause disease. If we know something about these factors in a particular host–pathogen interaction, we can make a reasonable prediction about the likely outcome of that interaction. Yet even the most-virulent pathogen cannot make you ill unless it accomplishes several things. We will now consider these requirements individually (**Figure 11.5**). To cause disease, a pathogen must:

1. have a place to survive before and after infection (termed a reservoir)

2. be transmitted to the host

3. invade the host's body

4. adhere to the host in an appropriate site

5. multiply in or on the host

6. evade the host's defenses, at least initially

7. interfere with normal host functions

8. leave the host and reach either a new host or its reservoir.

Reservoirs provide a place for pathogens to persist before and after an infection

For many pathogens, following the infection of a susceptible host, there are really only two possible outcomes: The host either eventually recovers or dies. Host survival usually results in at least temporary immunity to further infection by the same microorganism. Either way, this means that if the pathogen is to persist over the long term, it must leave the host at some point and find a new host. Eventually, however, the pathogen is faced with the prospect of a host shortage if most of the hosts are either dead or immune. In that case, the pathogen would also cease to exist, because it has no place to survive. Such pathogen extinction is rare, however, because even without a limitless supply of susceptible hosts, pathogens can persist in their **reservoir** until new hosts become available. A reservoir is a place where the pathogen can survive indefinitely without causing serious disease and from which it can infect new susceptible hosts if and when they are available. New hosts typically become available through new births or the waning of immunity in previously infected hosts. Reservoirs therefore permit the long-term survival of pathogens, and all pathogens have at least one.

Many pathogens use animal reservoirs

A human disease for which certain animals serve as reservoirs is called a **zoonosis**. In general, the pathogen does not cause disease in the animal reservoir because the microorganism and the particular animal host are well

Table 11.2 Representative infectious diseases. Reservoirs, typical mode of transmission, and portals of entry are indicated.

DISEASE	MICROORGANISM	RESERVOIR	USUAL MODE OF TRANSMISSION	USUAL PORTAL OF ENTRY
Bacterial diseases				
Typhoid	*Salmonella typhi*	Human	Water- or food-borne	Digestive system
Legionnaire's disease	*Legionella pneumophila*	Environmental	Airborne	Respiratory system
Tuberculosis	*Mycobacterium tuberculosis*	Human, animal	Airborne	Respiratory system
Whooping cough	*Bordetella pertussis*	Human	Airborne	Respiratory system
Syphilis	*Treponema pallidum*	Human	Contact	Reproductive system
Plague	*Yersinia pestis*	Animal	Vector	Flea bite through skin
Viral diseases				
Hantavirus pulmonary syndrome	Hantavirus	Animal	Airborne	Respiratory system
Measles	Measles virus	Human	Airborne	Respiratory system
Common cold	Many viruses	Human	Airborne	Respiratory system
Rabies	Rabies virus	Animal	Contact	Animal bite through skin
Yellow fever	Yellow fever virus	Animal	Vector	Mosquito bite through skin
Protozoan diseases				
Chagas' disease	*Trypanosoma cruzi*	Animal	Vector	Kissing bug bite through skin
Malaria	*Plasmodium* species	Human, animal	Vector	Mosquito bite through skin
Giardiasis	*Giardia intestinalis*	Animal	Food- or water-borne	Digestive system
Fungal diseases				
Valley fever	*Coccidioides immitis*	Environmental	Airborne	Respiratory system
Athlete's foot	Various fungi	Environmental	Contact	Colonization of skin

adapted to each other. Should a human become infected, disease may result, because there is no similar adaptation between humans and the pathogen. Because no disease is caused in the reservoir, there is no danger of these reservoirs all dying off. Humans, although susceptible, are not essential for the pathogen's long-term survival.

Examples of zoonotic disease are numerous (**Table 11.2**). Reptiles, for instance, often serve as reservoirs for *Salmonella*, and reptile owners sometimes develop *Salmonella* infections after handling their pets. Respiratory disease caused by hantavirus was first described in 1993. The virus, however, is thought to have been present in the Southwest United States for eons, surviving quietly in deer mice, in which it causes no illness whatsoever (**Figure 11.6**).

Some pathogens rely at least in part on human reservoirs

Other pathogens utilize humans as their reservoirs (see Table 11.2). An example is provided by *Streptococcus pyogenes*, the causative agent of strep throat. Some people carry this potential pathogen in their throats for decades without becoming ill. They do, however, serve as a source of infection for susceptible people.

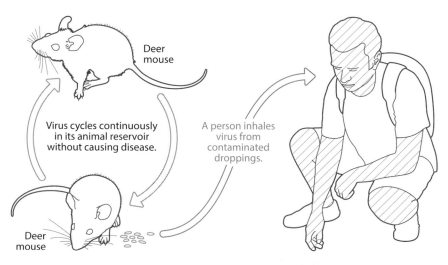

Figure 11.6 Hantavirus: a zoonotic disease. The virus cycles continuously in its animal reservoir, the deer mouse, without causing disease. Virus shed in the mouse's feces or urine can be respired into the lungs of a human, who may then develop a severe, often lethal respiratory illness called hantavirus pulmonary syndrome (HPS). Outbreaks of HPS often seem to follow wet years, which increase vegetation and food for mice. Mice populations then rise, increasing the likelihood of contact between humans and mice.

The infamous case of Typhoid Mary is a classic example of a human reservoir from the early 1900s (**Figure 11.7**). Infected with *Salmonella typhi*, Mary herself remained healthy. But because she worked as a cook, and because typhoid can be transmitted through tainted food, people around her often developed this disease. When it became apparent that in some poorly understood manner, Mary was making those she worked for ill, she was institutionalized for 26 years on a small island in New York's East River, always proclaiming as outrageous the idea that she could make anyone sick. It is now known that when most people contract *S. typhi* they develop the disease but eventually kill off the microorganism and survive. Approximately 3% of humans can act as human reservoirs. They do not develop symptoms of typhoid and carry the bacterium indefinitely, potentially infecting others.

There is no guarantee that a human reservoir will never develop illness. Indeed, in many cases, eventually the reservoir also succumbs to disease. This is true of the human immunodeficiency virus (HIV), which frequently does not cause serious disease until years after infection. During this long symptom-free period, an HIV-positive individual serves as a reservoir for the virus, fully capable of infecting others.

Some pathogens can survive indefinitely in soil, water, or other environmental reservoirs

In Chapter 3 we discussed spore-forming organisms such as *Clostridium tetani*, the causative agent of tetanus. This bacterium can survive for decades as a spore in the soil, until it reaches the warm, moist, and anaerobic environment of an animal or human where it can germinate. Similarly, in Chapter 4 we discussed *Coccidioides immitis*, a fungus that lives in the soil of the desert Southwest (see Table 11.2). When inhaled, it sometimes causes the respiratory ailment valley fever. Water can serve as a reservoir for many pathogens, including *Vibrio cholerae*, which causes cholera.

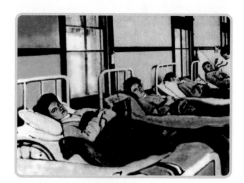

Figure 11.7 Something about Mary. Mary Mallon (in the closest bed) on North Brother Island in New York's East River. She was originally sent to the island when it was realized that she was responsible for cases of typhoid in those she worked for. She was later released on the condition that she did not work as a cook, but when a typhoid outbreak began in a hospital, Mallon was found working there under an assumed name. She was shipped back to the island, where she remained until she died in 1938.

The type of reservoir used by a pathogen has implications for disease control

It is no coincidence that to date the natural transmission of only one important human pathogen, smallpox, has been completely eliminated. Smallpox relies exclusively on human reservoirs; it cannot persist in either the environment or in an animal. Furthermore, a potent anti-smallpox vaccine offers long-term protection against this scourge, and vaccinated individuals cannot serve as reservoirs because any virus is immediately destroyed by the immune system. In the 1950s, international health authorities reasoned that if enough people were vaccinated, smallpox virus, without a reservoir, would go extinct. The World Health Organization coordinated an intensive international vaccine effort, and the last victim of naturally occurring smallpox is believed to have been a young man in Somalia, who contracted and later recovered from smallpox in 1977 (**Figure 11.8**). Other diseases that are targeted for eventual eradication, such as polio, likewise depend exclusively on human reservoirs.

When human pathogens utilize animal reservoirs, the problem of control is considerably more problematic. A case in point is provided by efforts to control yellow fever in Cuba and Panama in the early 1900s. In Cuba there were no significant animal reservoirs, and yellow fever was effectively eradicated through mosquito control (see Chapter 6, pp. 119–120). This strategy was never entirely successful in Panama, however, where monkeys serve as reservoirs and as an ongoing source of infection.

If all animal reservoirs could be vaccinated, the problem would be solved. Vaccinating wild animals, however, is tricky at best. A more commonly used, if brutal, approach is simply to slaughter animals believed to serve as reservoirs. When avian influenza (bird flu) first appeared in 1997, the response of officials in Hong Kong was to kill 1.5 million chickens thought to be potential reservoirs. Nevertheless, the virus was able to elude this control effort and has now spread to many countries in Asia and beyond.

Figure 11.8 End of the line for smallpox. (a) Villagers in West Africa line up for smallpox vaccines. Under the supervision of the World Health Organization, smallpox was eradicated thanks to a worldwide vaccination campaign starting in the 1950s. An effective vaccine and the lack of reservoirs other than humans were crucial factors in the success of this effort. (b) This Somali man, who contracted smallpox in 1977, was the last victim of naturally occurring smallpox.

Pathogens must reach a new host via one or more modes of transmission

---

CASE: LAID LOW ON THE HIGH SEAS

Ah, the joys of a tropical cruise! The dancing! The exotic ports of call! The vomiting! The diarrhea! Yes, a cruise ship can be fun, but danger lurks in those all-you-can-eat buffets. Cruise ships bring together many people from different communities into a crowded, semi-enclosed environment that facilitates transmission of certain diseases. Contaminated food increases the likelihood of food- and water-borne transmission. Norovirus, in particular, causes over 90% of food-borne illness. Prior to 2010 there were 27 outbreaks of norovirus reported annually on cruise ships. There are still about 10–12 annual norovirus outbreaks, which may affect up to 5–6% of passengers. On large cruise ships carrying 3000 passengers, this translates into over 150 people with less-than-fond memories of their vacation at sea. In addition to contaminated food, norovirus can be spread via aerosol and contact transmission. The disease caused by norovirus (commonly called stomach flu), while certainly unpleasant and able to spoil your holiday, is usually not dangerous and resolves on its own within 2–3 days.

1. **What exactly is meant by food-borne, aerosol, and contact transmission?**
2. **Why is norovirus in particular such a problem?**
3. **Are there other environments in which norovirus is especially worrisome?**

All pathogens need at least one **mode of transmission**—a means of traveling between hosts or between host and reservoir. Modes of transmission can conveniently be categorized into several basic types.

Contact transmission may be direct or indirect

Numerous microorganisms can move from an infected individual to an uninfected, susceptible host via physical contact (see Table 11.2). Such contact can be direct or indirect. Hand shaking, kissing, sexual contact, and other forms of physical contact that allow pathogen transmission are examples of **direct contact transmission**. In **indirect contact transmission**, an inanimate object acts as an intermediary between an infected and a susceptible individual. Such is the case if you become infected with a cold virus, for example, after using a hand towel or drinking from a glass previously contaminated by an infected individual. The inanimate object that served as the intermediary is called a **fomite**. Other examples of fomites are improperly sterilized catheters in a hospital or needles shared between drug users. Many pathogens, including HIV and hepatitis C, can be transmitted either sexually or via shared needles and are therefore transmitted by both direct and indirect contact transmission. Norovirus, as mentioned in our case, can also be spread by direct and indirect contact transmission. If viral particles contained in vomit contaminate skin, clothing, or other objects, for example, viral transmission to new hosts is possible. Norovirus is particularly problematic because it is so hardy; routine cleaning measures do not destroy it, and viral particles can remain infectious for several weeks on hard surfaces.

Many pathogens can move through the air in an aerosol released by an infected individual

Aerosol transmission or **airborne transmission** (see Table 11.2) is also utilized by norovirus. Aerosol transmission simply implies that the pathogen

Figure 11.9 Airborne transmission. This can occur via the aerosol spray expelled during a sneeze.

leaves a host via a cough, sneeze, or even microscopic droplets released while talking (**Figure 11.9**). Pathogens residing in the upper respiratory system, including cold viruses and influenza virus, are transmitted in this fashion. Norovirus infects the digestive tract, but viral particles present in vomit may be aerosolized and breathed in by susceptible individuals. If some of these viral particles are swallowed, the virus arrives at a suitable site for replication in the new host.

Intestinal pathogens most commonly rely on food- or water-borne transmission

Although norovirus is transmitted by both contact and aerosol transmission, transmission also regularly occurs via **food-** or **water-borne transmission**. Microorganisms transmitted in this manner are released through the feces of an infected individual. This contaminated waste must then contact either food or water, whereupon it is ingested by another susceptible individual. Another reason norovirus is such a problem, not only on cruise ships, but also in other crowded, enclosed places like college dorms, prisons, and daycare centers, is that this virus has multiple options when it comes to transmission.

Other pathogens using food- or water-borne transmission include *Salmonella*, poliovirus, and parasitic amebas, just to name a few. Because of their reliance on transmission by contaminated food or water, these microorganisms are readily controlled through improved sanitation. For this reason, diseases transmitted this way are often associated with undeveloped parts of the world where sanitation is poor. Great strides have been made in many places to control such illness, but even the most developed countries remain susceptible. When cows, chickens, and other animals are slaughtered for meat, the meat can become tainted with microorganisms that were living in the animal's digestive system. If such meat is improperly handled during preparation, food-borne illness is a real possibility. We are often advised, for example, not to use the same knife to cut up both chicken and salad vegetables. If the knife used on raw chicken is then used to dice vegetables, contamination of the vegetables and subsequent food-borne transmission may occur, even if the chicken itself is well cooked. Occasional outbreaks of hepatitis A are often traced to a single infected food worker who failed to properly wash up following a trip to the restroom (**Figure 11.10**).

Figure 11.10 Hand washing. Even in places with excellent sanitation, food- and water-borne transmission can occur when simple hygienic measures are not followed.

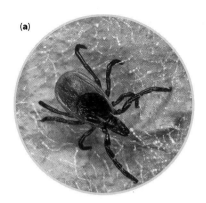

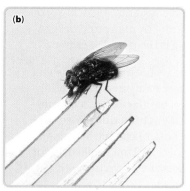

Figure 11.11 Vector transmission. (a) The black-legged tick, *Ixodes scapularis*, is the biological vector of Lyme disease. The tick itself is infected with the pathogen and is an essential part of the pathogen's life cycle. (b) Mechanical transmission can occur when pathogens stick to the body of a vector such as this housefly. The pathogen does not develop or reproduce in or on the vector.

Vector transmission relies on insects or other invertebrates to transmit infectious organisms

Mosquitoes and other invertebrates are responsible for the transmission of many pathogens (see Table 11.2). Vector transmission is generally subdivided into one of two types—biological or mechanical. **Biological vector transmission** means that the invertebrate vector is essential to the life cycle of the microorganism, which replicates and/or undergoes necessary developmental changes within the vector. In biological transmission, the vector itself is infected with the disease agent. *Plasmodium*, the protozoan responsible for malaria, relies on mosquitoes as a biological vector, as do West Nile virus and many others. *Borrelia burgdorferi*, the bacterium responsible for Lyme disease (**Figure 11.11a**), is biologically transmitted by ticks, while the protozoan that causes sleeping sickness utilizes tsetse flies.

When an insect serves as an incidental vector in which no pathogen replication or development has occurred, we say that **mechanical vector transmission** has taken place (**Figure 11.11b**). Unlike a biological vector, the mechanical vector is not actually infected with the pathogen. The microorganism merely sticks to its body or otherwise hitches a ride on the vector. For example, we have already noted that norovirus is transmitted in several ways. If virus is present in feces, however, a fly might land on this contaminated waste to feed. If it does, viral particles may stick to hairs on its body. Later, if this same fly lands on uncovered potato salad at a picnic, viral particles may contaminate the food item, posing a risk to anyone who consumes it. Other typically food-borne pathogens can likewise be transmitted in this manner by mechanical vectors.

Some pathogens can pass from a female to her offspring

Some infectious agents can pass from a pregnant female to a fetus across the placenta, while others are transmissible through the breast milk. Transmission of pathogens in this manner is called **vertical transmission**. German measles (rubella), HIV, and syphilis can all be contracted vertically.

Pathogens gain access to the host through a portal of entry

Following successful transmission, the pathogen must invade the host through an appropriate **portal of entry** (**Figure 11.12**). Major portals of entry include

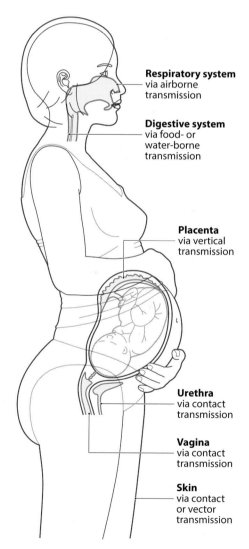

Respiratory system
via airborne transmission

Digestive system
via food- or water-borne transmission

Placenta
via vertical transmission

Urethra
via contact transmission

Vagina
via contact transmission

Skin
via contact or vector transmission

Figure 11.12 The principal portals of entry into a human host. All pathogens require a portal of entry into a host. Major portals of entry are illustrated here. Pathogens transmitted by biological vectors are usually introduced directly into the circulatory system as the vector feeds (not shown).

surfaces of the respiratory, digestive, and reproductive systems. A cut, burn, or other break in the skin can be utilized as a portal of entry by many microorganisms. In vertical transmission to a fetus, the placenta is the portal of entry.

Whether or not the pathogens that invade the host will actually establish an infection depends in part on the number of invading organisms. The relationship between pathogen numbers and the likelihood of infection can be studied experimentally in animals and expressed as the **ID$_{50}$ (infectious dose-50)**: the number of microorganisms that must enter a host to establish an infection in 50% of exposed hosts. The value may be as low as 10 microbial cells for *M. tuberculosis* or as high as 100 million for *V. cholerae*. Norovirus, with an ID$_{50}$ of fewer than 20 viral particles, is highly infectious. This fact, along with its toughness in the environment and its numerous routes of transmission, help to explain why norovirus in particular is such a problem on cruise ships.

A related value is the **LD$_{50}$ (lethal dose-50)**. The LD$_{50}$ is the number of organisms that must enter a host to kill 50% of the hosts. Typically, we expect more-virulent organisms to have lower LD$_{50}$ values.

Once they have entered, pathogens must adhere to the host

Like the normal microbiota, successful pathogens must resist being swept away by host defenses or body fluids. Many pathogens adhere to host surfaces by means of fimbriae or capsules. Others have various proteins on their cell surfaces that can bind to other proteins normally found on host cell membranes (**Figure 11.13**). The specificity of certain pathogens for particular regions in the host's body is often determined by specific binding between microbial adherence proteins and host proteins present only in certain cell types. For instance, *Neisseria gonorrhoeae* can colonize genital areas as well as the throat and the conjunctiva of the eye, because the host proteins recognized and bound by *Neisseria* adherence proteins are found on those host tissues. In Chapter 5 we discussed how viruses attach to cells via their glycoproteins or other surface proteins and how the interaction between these viral proteins and host receptor proteins determines which cell types a specific virus can infect. Norovirus, for example, recognizes receptors located on the surfaces of cells lining the intestine.

Most pathogens must increase in number before they cause disease

The number of organisms to which a host is exposed is usually too small to cause disease. Before their presence is felt by the host, the pathogens must multiply; in other words, they must establish an infection. The time between the initial exposure and the onset of symptoms is called the **incubation period** (**Figure 11.14**). During the incubation period the number of pathogens is growing, and there is no disease until a given population threshold is reached. Beyond this threshold, the number of pathogens is sufficiently high to cause disease symptoms. If the threshold is not achieved by the pathogen, no illness is seen. Once again using norovirus as an example, the incubation

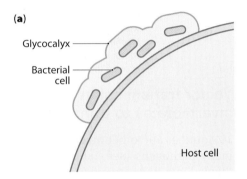

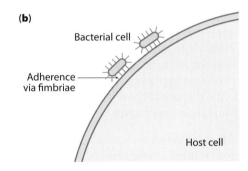

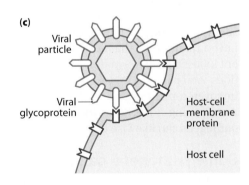

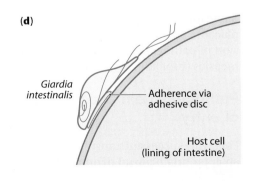

Figure 11.13 Adherence by pathogens. Bacterial pathogens may adhere to host cells in several different ways, including (a) a glycocalyx or (b) fimbriae (see Chapter 3, pp. 47–48, for a description of these adherence structures). (c) Viral pathogens often use their envelope glycoproteins (Chapter 5, pp. 95–96) to adhere to membrane proteins on host cells. (d) The protozoan parasite *Giardia intestinalis* uses its ventral adhesive disc to adhere tightly to cells lining the host's intestine.

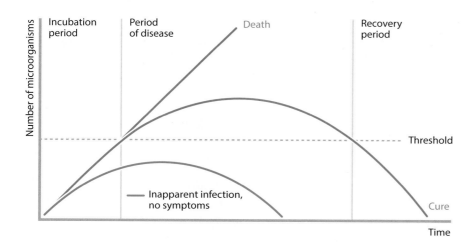

Figure 11.14 Microbial multiplication and disease. Before a patient suffers disease symptoms, a specific threshold of microbial numbers must be reached. The time between infection and the onset of symptoms, a time of microbial multiplication, is the incubation period. As the host fights off the infection, and the number of microorganisms drops below threshold, the host enters the recovery period. If the host cannot mount an effective response, microbial numbers may increase until the host dies. If the number of microorganisms never reaches the threshold, the infection will be asymptomatic and unapparent.

period is usually between 12 and 48 hours; symptoms begin somewhere between 12 and 48 hours post-infection.

Hosts can be infected with multiple pathogens at the same time

A simultaneous infection with more than one species is called a **co-infection.** The hepatitis D virus (HDV), for instance, cannot infect liver cells unless a hepatitis B virus (HBV) has already done so. HDV relies on HBV, which carries genetic information encoding certain crucial proteins that HDV cannot itself produce. In other words, HDV is a parasite of HBV! Another example is provided by the ticks that transmit *Borrelia burgdorferi*, the bacterium causing Lyme disease. These ticks are commonly co-infected with a protozoan called *Theileria microti*. A human bitten by a co-infected tick is now not only at risk for Lyme disease, but also for a malaria-like disease called theileriosis. The AIDS virus, HIV, is notorious for opening up the door to co-infections, because it supresses host immunity. Co-infection with HIV and tuberculosis is an important example. In some places, up to 80% of tuberculosis patients are co-infected with HIV.

Successful pathogens must at least initially evade host defenses

Fortunately for us, the threshold of disease is often not reached. Many infections are destroyed by the host's immune system well before there are clinical indications of disease. To cause disease, therefore, the pathogen must evade these defenses, at least until its population rises above threshold. In most cases, even when disease is evident, the immune system eventually gains the upper hand, pathogen numbers start to fall, and we enter the recovery phase, ultimately ending in the complete destruction of the pathogen. At this point the patient is cured (see Figure 11.14). In some cases the immune response is not vigorous or specific enough to completely destroy the pathogen, and the patient may undergo a long, **chronic** illness. The topic of how hosts defend themselves against pathogens is discussed in Chapter 12.

Pathology can be the result of the host's immune response

In many cases, an overly vigorous or inappropriate immune response is itself to blame for much of the tissue damage observed during an infection. The primary symptom in a gonorrheal infection is the production of a large amount of pus in the urethra. Pus is a semisolid accumulation of dead white blood cells, the material they have engulfed by phagocytosis, and tissue debris. It is a natural consequence, along with swelling, pain, and redness, of inflammation. Inflammation is a natural if sometimes unpleasant component of a normal immune response to infection. Inflammation, however, when it occurs inappropriately, too strongly, or for prolonged periods, can itself cause damage to host tissues. Although *N. gonorrhoeae*, the causative agent of gonorrhea, has stimulated the inflammatory response, this microorganism is not directly responsible for symptoms; the host response is. Similarly, the fever, fatigue, and aches associated with many viral infections are the result of our immune response to the virus, rather than the virus itself.

The symptoms of tuberculosis (**Figure 11.15**) and food poisoning from *Salmonella* are also mainly a consequence of host response as opposed to a direct effect of the pathogen. In fact, the very term food poisoning is something of a misnomer. Poisons (or toxins, to be discussed shortly) are only part of the reason for the illness caused by *Salmonella*.

Bacterial pathogens cause disease in several different ways

So far in our exploration of the disease process, we have discussed several things that a pathogen must do for disease to occur. But by and large, the microorganisms of our normal microbiota do these same things. What, then, is the fundamental difference between a pathogen and a member of our normal microbiota? The difference is that pathogens have the capacity to interfere with normal host activities. This interference is actually how we define infectious disease. Exactly how a bacterial pathogen interferes varies from species to species. In general, however, bacterial capacity to cause disease, known as bacterial **pathogenesis**, can be categorized among several basic mechanisms.

Some pathogenic bacteria produce toxins

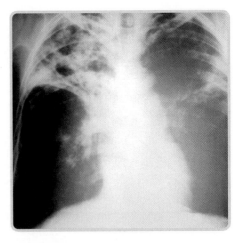

Figure 11.15 Tuberculosis. In this chest X-ray, dark areas in the lungs are unaffected areas. White areas reveal damage, primarily the result of the host response to infection with *M. tuberculosis*. The inflammation that occurs in response to infection, while limiting the spread of the bacteria, forms fibrous or calcified lesions. These tubercles persist throughout life and are seen as nodules on X-rays.

CASE: TROUBLE IN PARADISE

Howard and Heidi take a trip to the Galápagos Islands to view the exotic wildlife unique to the archipelago. During the vacation, Howard, who suffers from acid reflux (the release of stomach acid into the esophagus) and is especially vulnerable to a nervous stomach while traveling, regularly takes antacid tablets to relieve his discomfort. Heidi takes no such medication. Following their visit to the islands, they spend a few days in the Ecuadorian coastal resort of Salinas. On their last day they dine at an outdoor café that features ceviche, a marinated raw seafood dish. Although they were warned about raw seafood, they can't resist the local delicacy. They fly home the next day, and within 24 hours, Howard is suffering from severe watery diarrhea, muscle cramps, and dizziness. Alarmed, Heidi takes him to the emergency room, where the attending physician finds that Howard also has a weak, rapid pulse and very low blood pressure. When Heidi mentions that they just returned from South America, the doctor suspects cholera. Howard is immediately admitted to the hospital, given intravenous fluids, and started on antibiotics. A stool sample is tested and the diagnosis of cholera is confirmed when large numbers of *Vibrio cholerae* are found. Once treatment begins, Howard's diarrhea decreases, and after 2 days he is released from the hospital.

1. **How exactly does *V. cholerae* cause the symptoms observed in Howard?**
2. **Why didn't Heidi get sick?**

Although poisons are of limited or no importance in some diseases, for plenty of bacteria, poisons, generally called **toxins**, are the principal factor responsible for pathogenicity. All toxins act by altering the normal metabolism of host cells in some way, and because they contribute to disease, they are important examples of virulence factors. There are two basic types of toxins: **exotoxins** and **endotoxins**.

Exotoxins are bacterial proteins that interfere with cellular processes in specific ways

Exotoxins are bacterial proteins that are usually secreted into the surrounding environment. They vary considerably in specificity—some act only on certain cell types, while others affect a wide range of tissues or cells. A specific bacterial pathogen may produce one or more exotoxins. Other species produce none. Endotoxins consist of components found only in the outer membrane of Gram-negative bacteria. As we will see, they act as important toxins only under certain circumstances.

The *V. cholerae* contracted by Howard is a Gram-negative bacterial species, transmitted through contaminated food or water. Because ceviche is served raw, and because *V. cholerae* uses sea water as a reservoir, the seafood was the likely source of the infection. Once arriving in the intestine, the bacteria produced a powerful exotoxin that caused Howard's symptoms. Fortunately for him, he received treatment rapidly. Untreated cholera is often lethal; the explosive diarrhea can cause severe dehydration (**Figure 11.16**). Some victims lose up to 25% of their body weight, simply in fluid, within 24 hours.

Like many exotoxins, cholera toxin consists of an **A subunit** and a **B subunit** (**Figure 11.17**). The B (binding) subunit gives cholera toxin its specificity for cells lining the intestine. Like a lock and key, the B subunit protein binds to a specific receptor found only on these host cells. Next, the host cell brings the toxin molecule into the cell by endocytosis. Once inside, the A and B subunits separate. The A (active) subunit now exerts its toxic effect. Specifically

Figure 11.16 A cholera bed. Such beds are sometimes seen in hospitals in countries where cholera is present. The hole allows the patient to remain in bed while experiencing the explosive diarrhea that accompanies a cholera infection.

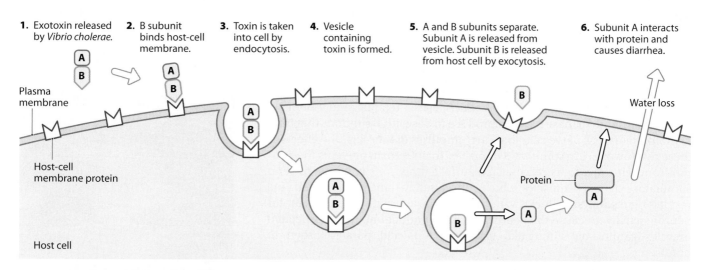

1. Exotoxin released by *Vibrio cholerae*.
2. B subunit binds host-cell membrane.
3. Toxin is taken into cell by endocytosis.
4. Vesicle containing toxin is formed.
5. A and B subunits separate. Subunit A is released from vesicle. Subunit B is released from host cell by exocytosis.
6. Subunit A interacts with protein and causes diarrhea.

Figure 11.17 Mechanism of cholera toxin. The B subunit binds specifically to receptors on host cells lining the intestine. Following binding, the toxin is taken into the cell by endocytosis. The A and B subunits separate and the A subunit interacts with cell proteins, initiating a series of events that ends with the host-cell membrane becoming leaky. The water that flows out of intestinal cells and into the intestine accounts for the diarrhea that occurs during a bout of cholera. Other exotoxins have different B subunits, explaining the specificity that exotoxins usually have for certain host-cell types. The activity of the A subunit also is different in different exotoxins.

it acts as an enzyme and alters a specific host-cell protein called a G protein. The now enzymatically altered G protein then starts a series of events that causes cells lining the intestine to leak water. It is this loss of water by intestinal cells that explains the massive diarrhea and fluid loss seen in cholera victims like Howard. Note that the bacteria themselves never enter the host cells. Only the secreted toxin does.

The other symptoms that Howard endured were also the result of this loss of water from intestinal cells. As he lost fluid, Howard's blood volume decreased, causing the low blood pressure and weak pulse. *V. cholerae* infections generally respond well to a combination of rehydration and antibiotics if they are treated quickly.

Why didn't Heidi get sick? Recall that *V. cholerae* has a relatively high ID_{50}. Typically, consumption of 100 million or more bacterial cells is required to ensure an infection in half the people that consume them, since the vast majority of bacteria are destroyed by stomach acid and never reach the intestine. Howard was taking antacid tablets to combat his nervous stomach, so the pH of his stomach was higher (less acidic) than normal, making it easier for the bacteria to survive and reach his intestine. Heidi's stomach remained acidic enough to prevent infection.

Other types of exotoxins have different B subunits and therefore bind to different target cells. They also have different A subunits, which means they will interfere with metabolism in a different way. Diphtheria, for instance, is caused by the Gram-positive bacterium *Corynebacterium diphtheriae*. The disease usually starts with sore throat, fever, and fatigue, which if untreated can eventually lead to heart and kidney failure. The B subunit on the diphtheria toxin binds to specific receptors on the membrane of target cells. The specificity of the B subunit explains why certain cells, like those in the heart and kidney, are affected while others are not. Once inside the cell, the A subunit binds and inactivates a protein involved in moving ribosomes along mRNA during translation. In effect, the toxin blocks protein synthesis.

Diphtheria, fortunately, is rare in Europe and North America because of a highly effective vaccine routinely administered to children. However, the breakup of the former Soviet Union in the early 1990s reminds us what can happen when public health is neglected. Due to political upheaval, vaccine programs broke down and an epidemic of diphtheria resulted. By the end of the outbreak, 125,000 cases and 4000 deaths were reported from the onetime Soviet republics.

Other well-understood exotoxins include those produced by *Clostridium tetani* and *Clostridium botulinum*, which cause tetanus and botulism, respectively. Both of these bacteria produce toxins with B subunits that bind to motor neurons—nerve cells that carry impulses to muscles. The effects of these two toxins are in some ways opposite. Tetanus toxin prevents the neuron from returning to its relaxed state. Affected neurons continue to send stimulatory signals to muscles, which consequently remain contracted. The result is rigid paralysis or lockjaw. Botulism toxin, alternatively, blocks motor neuron activity. Because the neurons cannot conduct impulses, associated muscles cannot be stimulated, causing the flaccid paralysis seen in botulism.

Botulism toxin is perhaps the most potent of all biological poisons—a few milligrams would be enough to kill thousands of people. Yet recently botulism toxin has found a more benign niche—as a cosmetic or beauty aid or to treat certain medical conditions. Used under the name BOTOX® Cosmetic, minute amounts of this toxin are injected directly into wrinkles (**Figure 11.18**). The

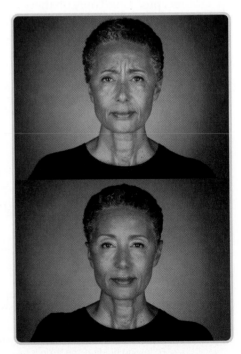

Figure 11.18 Cosmetic use of botulism toxin. A person photographed (top) before and (bottom) after treatment with the BOTOX® Cosmetic.

muscles causing the undesired facial lines cannot contract, and as they relax, the wrinkles disappear. BOTOX® Cosmetic also can be used to treat conditions such as vocal cord paralysis, in which muscles surrounding the larynx contract involuntarily, making speech difficult. The effect of the BOTOX® Cosmetic is temporary and treatments need to be repeated every 4–6 months.

Endotoxins are components of the Gram-negative outer membrane

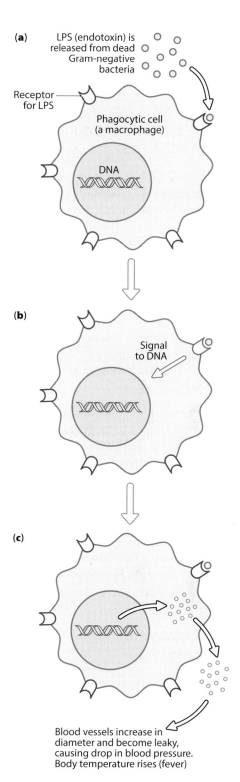

Figure 11.19 **Mechanism of endotoxin action.** (a) When Gram-negative bacteria die, they release LPS (in red). Certain phagocytic cells called macrophages have LPS receptors on their surfaces. (b) When LPS binds its receptor, a signal is sent that causes certain genes in the DNA to be transcribed and translated into proteins. (c) These proteins (blue circles) are ultimately released from the cell. Once released, they can travel through the blood, causing the symptoms associated with endotoxin.

> ### CASE: A RUPTURED APPENDIX
>
> When Robyn felt severe pain in her lower abdomen, her husband Gary became concerned and took her to the hospital. The doctor who performed the initial examination found her abdomen to be extremely hard and tender. An elevated temperature and low blood pressure suggested impending shock. On the basis of these symptoms, the doctor diagnosed Robyn's condition as a ruptured appendix and immediately began supportive treatment, consisting of IV fluids and large doses of antibiotics. Ten hours later, once her condition had stabilized, Robyn was taken to surgery, where her ruptured appendix was removed. Robyn made a slow but uneventful recovery.
>
> 1. **What type of bacterial infection is due to a ruptured appendix?**
> 2. **How do the involved microorganisms cause Robyn's symptoms?**
> 3. **In infections of this type, symptoms often temporarily worsen once antibiotic treatment has begun. Why?**

In Chapter 3 (see pp. 51–52), we learned that Gram-negative bacteria have an outer membrane composed in part of lipopolysaccharide (LPS). When Gram-negative bacteria die, their outer membrane degrades and LPS is released. In its released form, LPS acts as an **endotoxin**. Unlike an exotoxin, where each toxic protein has specific and unique effects, all LPS impacts the host in the same way, regardless of the bacterial species that releases it.

The mechanism of endotoxin activity is illustrated in **Figure 11.19**. Certain phagocytic cells like macrophages have receptors for lipopolysaccharide on their surface. When LPS, released when Gram-negative bacteria die, binds these receptors, a signal is sent, which causes certain genes in the macrophage's DNA to be transcribed and translated. The result is the production of specific proteins. These proteins are released into the blood or tissue fluid, where they cause blood vessels to increase in diameter and become leaky. These effects can cause blood pressure to drop, resulting in weakness and malaise. Other proteins released by macrophages induce a rise in body temperature. In other words, they induce fever. The macrophage receptor for LPS is an example of what are called Toll-like receptors. We explore their role in the host–pathogen interaction more fully in Chapter 12.

The symptoms of endotoxin are not too serious when LPS is released by dead bacteria in small amounts. As we have mentioned previously, however, the large intestine is inhabited by trillions of bacteria, most of them Gram-negative. When the appendix ruptures, these bacteria stream into the normally sterile peritoneal cavity. This opportunistic infection is called peritonitis. As these organisms die, they release endotoxin in large amounts—large enough to cause a severe drop in blood pressure and to cause the other symptoms experienced by Robyn. If blood pressure falls too far, shock is a real possibility. In such a situation, it is crucial to treat the bacterial infection rapidly with large doses of antibiotics, but the antibiotics themselves often initially make symptoms even worse. As the antibiotics kill large numbers of bacteria, levels of endotoxin in the peritoneal cavity can increase rapidly. During such treatment, therefore, supportive care is especially crucial, and as soon as the patient is stabilized, surgery is required to seal the rupture and prevent further bacterial invasion.

Some bacteria produce enzymes with pathogenic properties

In Chapter 2 we mentioned streptokinase, an enzyme produced by certain *Streptococcus* species. Streptokinase degrades the protein fibrin, which is common in host connective tissue. By degrading fibrin, the bacteria are better able to spread through the connective tissue, reaching new sites in the host. Some *Streptococcus* species, including *S. pyogenes*, the causative agent of strep throat, produce hemolysins, which lyse red blood cells and release hemoglobin (**Figure 11.20a**). The bacteria thus gain access to the iron in the hemoglobin, which they require as a nutrient. In Chapter 9 (p. 211) we described the action of coagulase, produced by some *Staphylococcus* bacteria. Coagulase is in some ways the opposite of streptokinase; instead of degrading fibrin, it causes fibrin to form from a precursor protein called fibrinogen. Fibrin plays a role in clot formation, and the clotting that forms around the *Staphylococcus* bacteria protects them from host immune response.

Other tissue-damaging enzymes produced by different pathogens include hyaluronidase, which degrades the carbohydrate hyaluronic acid that binds many host cells together, and collagenase, which digests collagen, an important structural protein in connective tissue. DNase, as the name implies, degrades host DNA, making the nucleotides available to the bacteria. Flesh-eating bacteria are usually a strain of *S. pyogenes* that comes equipped with an impressive arsenal of tissue-degrading enzymes, including streptokinase, hyaluronidase, collagenase, and DNase. An infection with this strain may start with a tiny cut or puncture wound, which quickly spreads over a large area, "eating" flesh in the process (**Figure 11.20b**). The less colorful, more technical name for this condition is necrotizing fasciitis.

Bacterial virulence is sometimes regulated by quorum sensing

Pseudomonas aeruginosa is usually a fairly peaceful bacterium, causing few if any problems. But on occasion, it turns vicious. It is especially worrisome in patients who have suffered burns, where it can colonize the burn site and cause serious complications. It is also an important cause of pneumonia in immunocompromised patients. The secret of how this species becomes virulent lies in quorum sensing, discussed in Chapter 7 (pp. 151–152). Recall that in this process, bacteria do not express certain genes until the number of cells present reaches a certain threshold level. In other words, the bacteria are able to sense the presence of other bacterial cells, and do not respond until there are enough bacteria present to make gene expression beneficial. *P. aeruginosa* typically cannot reproduce sufficiently to reach this threshold. Certain genes are therefore generally not expressed. But this opportunistic pathogen can take advantage of unusual conditions, such as badly burned skin or an immunocompromised host, to reach its threshold. Once it has done so, there is a high enough concentration of the specific protein used for intercellular

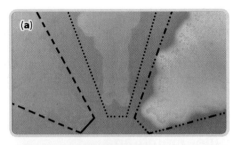

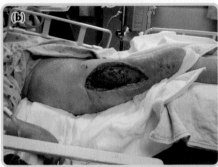

Figure 11.20 Damage caused by bacterial enzymes. (a) Blood agar medium contains sheep red blood cells. When *Enterococcus faecalis* (left) is inoculated onto a blood agar plate, no hemolysis of red blood cells is seen, because this species does not produce the necessary hemolysins. *Streptococcus mitis* (center) causes partial hemolysis, while *S. pyogenes* (right) causes complete hemolysis of red blood cells. The pale areas indicate where red blood cells have been destroyed. (b) Necrotizing fasciitis caused by "flesh-eating bacteria." The term "necrotizing" comes from the Greek "necro," meaning "dead." Fascia is the connective tissue that binds together internal organs or parts of the body. Fasciitis is an inflammation of the fascia.

communication (the autoinducer) in the immediate environment to stimulate gene expression in nearby bacteria. And for *P. aeruginosa*, many of these activated genes code for virulence factors. These include genes for a specific toxin that blocks host protein synthesis and tissue-degrading enzymes. Other species, such as the *Salmonella* responsible for food poisoning, also activate certain virulence factor genes only when their population has risen to a certain point.

Viruses cause disease by interfering with the normal activities of the cells they infect

Viruses must enter an appropriate host cell to replicate. Once inside the host cell, most viruses have specific **cytopathic effects (CPE)** that interfere with normal host function. A cytopathic effect is the specific impact that a virus has on a host cell, which may result in host-cell damage or death. Depending on the specific virus, the CPE may or may not actually kill the host cell. Many viruses exert lethal effects when they divert host-cell resources and synthetic processes to viral replication. The virus effectively commandeers the host cell, utilizing the cell's protein synthesis machinery for its own replication.

In some viral infections, the host cell is destroyed when cellular lysosomes release their digestive enzymes, and in many others the plasma membrane becomes leaky due to the effect of viral proteins. In other cases, viral proteins found in association with the host-cell plasma membrane target the cell for destruction by the host immune system. In some viral infections, large accumulations of viral proteins and nucleic acids build up in the cytoplasm of an infected cell. These **inclusion bodies** are essentially concentrations of viral garbage, which will never be assembled into new viral particles, but they may simply get in the way of normal cell processes. Some specific viral infections can be diagnosed from the presence of inclusion bodies. Rabies is a notable example (**Figure 11.21**).

It is now well understood that some forms of cancer are caused by viral infection. Cervical cancer, caused by the human papillomavirus, and liver cancer, caused by the hepatitis B virus, are two examples. Cancer is basically a situation in which cells divide in an uncontrolled manner. Some viruses carry genes that, when inserted into the host DNA, disrupt the normal regulation of cell division. Alternatively, host cells have their own genes that tightly regulate cell division; if these genes are disabled by a virus, the cell may begin unregulated division, ultimately leading to tumor formation. Cancer-causing genes, whether of host or viral origin, are called **oncogenes**.

The point at which the CPE of a given virus becomes noticeable varies considerably (**Figure 11.22**). As we learned in Chapter 5, viral infections can be acute, chronic, or latent. In acute infections, numerous host cells are rapidly destroyed as the virus undergoes repeated rounds of replication. Symptoms may be severe, but they are generally of relatively short duration. Once the immune system reins in viral reproduction, the infection is eradicated and damaged host cells are replaced by new cells. Influenza virus and norovirus, discussed earlier in this chapter, cause acute infections of this sort. Some viruses, such as hepatitis B and hepatitis C, replicate slowly over long periods of time, resulting in chronic infections. Other viruses can enter latent stages where all viral reproduction is halted. The period of latency may last years, until factors such as a weakened immune system, other infections, or stress allow the virus to begin replicating once again. The varicella zoster

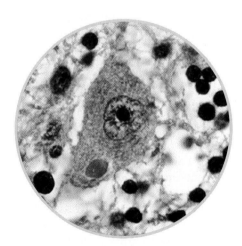

Figure 11.21 Inclusion bodies. A cell from the human brain, taken from a patient who died of rabies. The dark-staining structures in the cell cytoplasm are inclusion bodies. The presence of these inclusion bodies, called Negri bodies, is used to confirm a case of rabies.

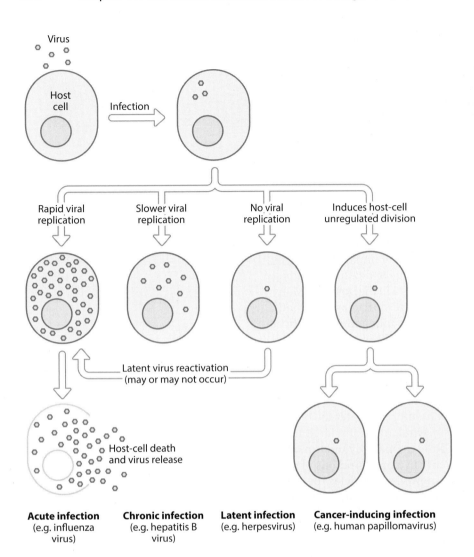

Figure 11.22 Some possible outcomes of a viral infection. Depending on the type of virus and the type of cell infected, there can be several possible outcomes when a virus infects a cell. In acute viral infections, the virus replicates rapidly and may kill the cell. Slower viral replication may result in a chronic infection. In some cases, a virus may not replicate at all, yet it remains viable within the host cell. Such an infection is called a latent infection. Latency may or may not eventually result in a new acute infection. Some viruses, able to disrupt the normal control of cell division, can cause cancer.

virus provides a useful example. When it infects a child, it causes the disease typically known as chickenpox. As the infection is brought under control by the host immune system, the virus retreats to nerve cells, where it can remain latent for years. Later in life it may re-emerge as shingles.

Different eukaryotic parasites affect their hosts in diverse ways

Protozoan parasites are a varied lot, and the means by which they cause disease are equally variable. The intestinal parasite *Giardia intestinalis* so thoroughly covers the surface of the intestinal wall that it prevents absorption of nutrients and the reabsorption of water, causing diarrhea (see Figure 11.13d). The *Plasmodium* parasites that cause malaria destroy red blood cells and cells in the liver, by entering them and replicating intracellularly (see Figure 4.15d). The trypanosomes that cause sleeping sickness (see Figure 3.20b) repeatedly change the proteins on their cell surface. A different host immune response is generated against each of these changing proteins. This immune response soon becomes of limited value in controlling the parasite, but massive immune response itself damages host tissues, contributing to the disease. One type of protozoan parasite that you are likely to have heard of is the ameba. *Entamoeba histolytica*, in particular, is a souvenir of foreign travel that you do not want to pick up (see Figure 4.15b). The parasite is transmitted via the food- or water-borne route and is therefore a risk wherever sanitation is poor. After attaching to cells in the intestine, *E. histolytica* releases proteins that form pores in the target cell membrane. It also releases enzymes that destroy host cells, and in severe cases it may actually digest its way through the intestinal wall. When this occurs, amebas may spread to other organs of the body, such as the liver and lungs, where they continue to cause tissue damage.

Recall from our discussion of basic fungal biology in Chapter 4 that fungi digest nutrients extracellularly and then absorb the digested material. This property forms the basis of much fungal pathology. Parasitic fungi often exert their harmful effects by releasing digestive enzymes that degrade host tissue. As these tissues are digested, the fungi absorb required nutrients and subsequently invade adjacent tissue.

Figure 11.23 The Salem witch trials. In this painting a young woman is accused of practicing witchcraft. More than 200 individuals were similarly accused in 1692 and 1693, 20 of whom were executed. Some have suggested that witch hysteria was due to ergot poisoning in local residents.

Some fungi also either produce toxins or provoke an allergic response in the host. As an example, *Claviceps purpurea* is a parasite of rye. If infected grain is used to produce bread, anyone consuming the bread may also ingest a fungal toxin called **ergot**. Ergot causes blood vessels to constrict, and if circulation to fingers or toes is cut off, gangrene may result. Ergot also causes mental disturbances, including hallucinations. Before the advent of modern milling techniques, a contaminated batch of rye would occasionally cause mass hysteria among those consuming it. Historians even suspect that the Salem witch trials, which occurred in Salem, Massachusetts, in 1692, were instigated by mass hysteria induced by ergot poisoning (**Figure 11.23**).

Pathogens leave the host through a portal of exit

Although it is not necessary for a pathogen to leave its host in order to cause disease, it certainly is necessary for the long-term survival of the pathogen. Pathogens leave their host through a **portal of exit**.

For pathogens of the intestinal tract, the usual portal of exit is the anus, although the case involving norovirus demonstrates that at least occasionally, alternative portals of exit are possible for an intestinal pathogen. For airborne pathogens, sexually transmitted pathogens, and those pathogens relying on vectors, the portal of exit is generally the same as the portal of entry. Airborne pathogens, after multiplying in the respiratory system, exit the respiratory system in droplets when the host coughs or sneezes. Sexually transmitted pathogens use genital contact for both entry and exit, and vectored pathogens both enter and leave the blood through the bite of an insect or other arthropod.

In some cases, no portal of exit exists, and the host can be considered a dead end from the pathogen's point of view. This is often the case in zoonotic infections, where humans are not the normal or appropriate host. *Bartonella henselae* or *B. quintana*, for example, cause cat scratch disease in humans. These bacteria are typically found in cats, where they cause no illness. The bacteria are usually transferred between cats by fleas. Humans may become infected if scratched or bitten by a cat harboring the bacteria. Symptoms in an infected human may include swollen lymph nodes, headache, chills, and malaise. Humans, however, provide no portal of exit. Some pathogens that utilize environmental reservoirs also only incidentally infect humans. When they do, although they may cause illness, they cannot maintain a cycle of transmission because they cannot leave the host. *Clostridium tetani*, the causative agent of tetanus discussed earlier in this chapter, is an example. As far as this pathogen is concerned, infecting a human is a mistake—it will never get out alive.

Symptoms of disease often assist the pathogen in its transmission to a new host

Remember that here in Chapter 11 we are considering the disease process from the pathogen's perspective. In Chapter 12, we concentrate on how hosts respond to invasion by pathogens, but before we change our point of view, we ask what may sound like an odd question—why do pathogens cause disease? In other words, what do they get out of making us ill?

For many pathogens the answer is often not much. Many pathogens infect humans only accidentally, and disease is an unfortunate consequence of this host–pathogen interaction that benefits neither party. In such cases, symptoms may sometimes best be viewed as an attempt by the host to rid itself of the pathogen. An example is provided by the bacteria in the genus *Shigella*, one of the primary causes of human diarrhea. For these bacteria, the diarrhea might be described as an efficient, if not particularly pleasant, way to get intestinal pathogens out as quickly as possible. If diarrhea is stopped with medication, the bacteria actually do better and the infection lasts longer. Things are different, however, with *Vibrio cholerae*, the species causing cholera. Available data suggest that diarrhea occurs in spite of what is best for the host—the massive loss of fluid, containing millions of bacteria, is actually induced by the microorganism to facilitate its transmission.

Other examples in which pathogens may actually be inducing certain symptoms for their own purposes include those pathogens transmitted by biological vectors. Such pathogens might gain a real advantage if they make their host as sick and lethargic as possible; thus, when an appropriate mosquito or other vector lands to take a blood meal, the host is in no condition to swat it away (**Figure 11.24**).

Figure 11.24 Sleeping sickness. A young woman watches over her comatose husband, who is suffering from the disease. Is the extreme lethargy induced by the parasite a strategy to make it easier for its tsetse fly vector to take a blood meal?

Looking back and looking forward

The likelihood of disease in any host–parasite interaction depends on both the effect of the parasite on the host and the response of the host. To best understand the first part of this equation—how they affect us—we have tried to view the world from the pathogen's point of view, and think about what they must accomplish if they are to cause disease.

In Chapter 12, we will complete this story by considering how we respond to pathogens. Only by understanding both sides of this dynamic interaction does a complete picture of disease emerge. We will therefore next investigate immune defense. Sometimes these defenses work and sometimes they don't. Either way, once we understand these defenses, we will be well-positioned to answer the question we posed earlier in Chapter 11: Why do we sometimes get sick, and when we do, why do we usually (but not always) recover?

Garland Science Learning System

- http://garlandscience.rocketmix.com/students/

- Investigate bacterial weapons and discover how horseshoe crabs prevent hospital infections

- Watch a cell infected with *Listeria Monocytogenes*

- Test your knowledge of this chapter by taking the quiz

- Familiarize yourself with the terminology used in this chapter by using the vocabulary review

- Get help with the answers to the Concept questions

Concept questions

These questions are designed to help you start thinking like a microbiologist. The answers are not always simply found in the text. Instead, you will need to take the concepts about which you have learned and apply them to new situations. Help is provided as part of the GSLS resources, which can be accessed through http://garlandscience.rocketmix.com/students/.

1. You've just been hired by the World Health Organization, where your job is to develop a list of diseases that might realistically be targeted for global eradication. You must also develop a list of diseases for which eradication is most unlikely. What might be the characteristics of disease agents on your two lists?

2. Bubonic plague is caused by the Gram-negative bacterium *Yersinia pestis*. These bacteria can persist in certain rodent populations for long periods of time without causing severe disease. Animals, including humans, generally become infected when a flea feeds first on an infected individual and later on a susceptible individual. Occasionally, if the bacteria spread to the lungs, the patient develops pneumonic plague, which is spread via airborne transmission. Although cats can develop plague similar to that of humans, dogs are generally resistant to infection.

   With the above information, what advice would you give somebody who lives in an area where plague is known to occur on how to best protect themselves and their pets?

3. Patients admitted to hospitals often acquire new infections during their hospital stays. Careful hand washing by hospital personnel is one of the easiest and most important ways to reduce the numbers of such infections. What do you think the link between hand washing and reduced numbers of hospital-acquired infections is?

4. Imagine two bacterial species, A and B. Species A causes disease by releasing endotoxin, whereas species B produces no endotoxin but produces a disease-causing exotoxin. If a doctor has two patients, one infected with species A and one with species B, for which patient will it be easier to diagnose the exact species causing illness, based solely on the symptoms of the patient?

5. Those pathogens that are most virulent often have the most efficient means of transmission. Pathogens with less efficient modes of transmission are often less virulent. What do you think explains this relationship?

6. In Chapter 8 we learned that *Corynebacterium diptheriae* produces a powerful toxin when the environment is low in iron. Would the LD_{50} of this species be higher or lower when iron is abundant in the environment?

7. You are told that the ID_{50} for a particular bacterial species is 1000 cells. Do you think the LD_{50} is greater than or less than 1000? Explain.

8. Recent discoveries about our normal microbiota have led some to propose that we should start "stool banking" as a routine measure. In other words, stool samples would be taken when we are healthy and stored indefinitely, to be used in case of certain illnesses. What do you think the rationale is for this seemingly odd suggestion?

9. Suppose a patient is infected with a respiratory virus that causes severe coughing. The patient takes a cough suppressant. Although there is less coughing, the infection actually takes longer to resolve than it would without the cough suppressant. Does this suggest that the coughing is a host response to clear the pathogens from the respiratory system, or a pathogen-induced symptom to facilitate the transmission of the pathogen?

THE NATURE OF DISEASE PART II: HOST DEFENSE

It's a dangerous world out there. Almost anything you touch is covered with microorganisms. Every breath you take and every bite of food you eat is laden with microbes, some of which are potential pathogens. If you dwell excessively on the unseen microbial world seething around you, it can seem as if death and disease are lurking around every corner.

Nevertheless, even though our exposure to potentially deadly microbes is relentless and never-ending, it should be comforting to remember that usually most of us feel fine. Infectious disease is the exception rather than the rule, and our encounters with pathogens are frequently not much more serious than an occasional cold or bout of diarrhea. Our ability to fend off most pathogens is based on the fact that we do not go through life unprotected. Indeed, animals come equipped with a formidable arsenal of defensive weapons, constantly at the ready to repel microbial invaders.

Infectious disease can still threaten our well-being. In Chapter 11, as we investigated the host–pathogen interaction from the pathogen's point of view, we began to understand why. However, in any host–pathogen encounter, the likelihood of disease depends on more than just what pathogens do to us. Disease is a two-way street; how we defend ourselves against microbial attack is equally important. We will now investigate that defense. The scientific study of these defensive mechanisms is called **immunology**.

In other words, having considered in the previous chapter how pathogens can impact us, we will now examine the disease process from our perspective—how we respond to them. After we have done so, we will have the complete answer to the questions posed at the beginning of Chapter 11—namely, why, when we are exposed to pathogens, do we sometimes, but not always, get sick? And why, if we do get sick, do we usually, but not always, recover?

When infection cannot be blocked, first innate immunity and then adaptive immunity is activated

Before we get too far, we need to introduce a few basic concepts. Host defenses, in their simplest form, take a sort of plan A–plan B approach (**Figure 12.1**). Plan A is to block invading organisms from infecting us in the first place. As we learned in Chapter 11, infection is one of the crucial steps on the path to disease. To prevent infection, animals have several types of barriers to entry, which stop most potential invaders before they can colonize us. Skin is an example.

All case studies have a few questions at the end, the answers to which will become apparent as you read the sections following the case.

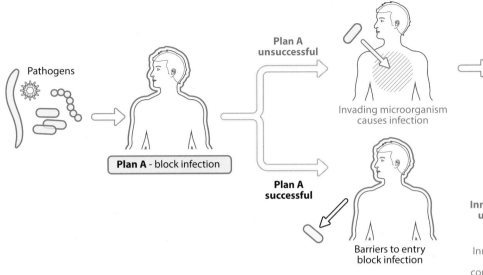

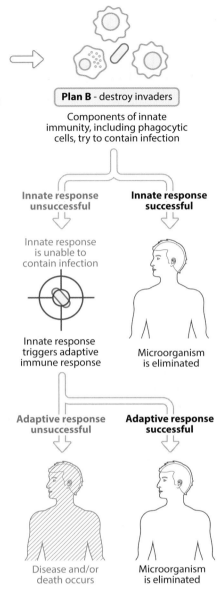

If microorganisms circumvent initial barriers and successfully infect the host, the host switches to plan B—namely, destroy the invading microbes. This immune response to microbial invasion consists of two components. First, elements of our **innate immune system** engage the invader (see Figure 12.1). Innate immunity refers to those components of our defensive repertoire that are inborn. Always in place and always at the ready, they represent our first line of defense. They are not directed specifically against any particular pathogen, and they are not enhanced in response to prior exposure. The activation of phagocytic cells, which engulf and destroy invading pathogens, is an example of an innate immune process.

If infection is not rapidly contained, certain elements of innate defense alert and activate the **adaptive immune response**. The adaptive response is a highly directed immune counterattack to specific pathogens. When activated, a variety of cells and molecules are generated in response to infection and the adaptive immune system can recognize and eliminate an essentially limitless variety of foreign invaders. The production of antimicrobial proteins called antibodies, which bind to and help destroy specific microorganisms, is an example of an adaptive immune response.

Many microorganisms are cleared by innate processes before adaptive immunity is even activated. But if the innate response cannot contain an infection on its own, it activates the adaptive response (see Figure 12.1). Unlike the innate immune response, the adaptive immune response to a particular organism grows stronger with each exposure. This key feature of the adaptive immune response is called **immunological memory**.

Figure 12.1 The strategy of immune defense. Initially, the host attempts to prevent infection by utilizing various barriers to entry (plan A). Should these barriers fail, plan B goes into action: Elements of innate immunity try to eliminate the invading microorganisms. If innate mechanisms alone are insufficient, the innate response triggers the activation of adaptive immunity, which mobilizes a highly specific, targeted attack on the infecting microorganism.

CASE: INFLUENZA: EXPOSED!

Jeff is an American journalist based in Hong Kong. In early March, Jeff makes plans to fly to Los Angeles for an important meeting. His flight is scheduled for a midday departure, and that morning he awakens with the vague feeling that something is wrong. He has a headache, feels unusually tired, and cannot escape the feeling that he is coming down with something. He briefly considers postponing his trip, but this meeting is crucial. He simply has to be there, so he tries to put his symptoms out of his mind and leaves for the airport. As the morning progresses, Jeff's symptoms worsen. Yet he boards the aircraft as planned, hoping that he can relax and will start to feel better.

Shortly after takeoff, however, he can no longer deny it; he has the flu. Jeff now has a fever, and he begins to experience the shivering, aching muscles, and malaise that typify an influenza infection. Unfortunately for his fellow passengers, Jeff is still highly infectious to others, and on a crowded airplane there is simply nowhere to hide. As Jeff coughs, infectious influenza viral particles are wafted through the cabin, where they are inhaled by others on the plane. By the time the plane reaches Los Angeles, almost everyone on board has been exposed.

1. **Although most people on the flight were exposed, the outcome of this exposure is different in different individuals. Why?**

We will spend the remainder of this chapter answering this single question. To do so, we will see what happens to four representative exposed passengers, named Rob, Laura, Bonnie, and Oscar. The outcome of their exposure will be somewhat different for each of them. The reasons for these differences are based on how each person responds immunologically. By following these various responses, we will come to understand the highly sophisticated and carefully orchestrated defensive strategies that constitute our immune system.

Plan A: Barriers to entry may prevent infection before it occurs

CASE: ROB

Rob is a 22-year-old exchange student, currently studying in China. He is returning home to the United States for a brief visit. Although he heard Jeff coughing and was somewhat worried about catching whatever his fellow passenger might have, he seems to get off lucky. After his long flight he still feels fine and after several days he is pleased to realize that he has no symptoms of illness whatsoever. He soon forgets all about the sick passenger on his flight.

1. **Even though Rob was exposed to influenza virus, he did not become ill. Furthermore, he was not even infected by the virus. How did he avoid infection entirely?**

Influenza virus is transmitted via the respiratory route (see Chapter 11, p. 262). Upon reaching the nose or sinuses of a susceptible individual, it may initiate an infection by entering cells of the upper respiratory tract, where it begins to replicate. But for Rob, plan A worked! His barriers to entry prevented an infection from ever occurring (**Figure 12.2**).

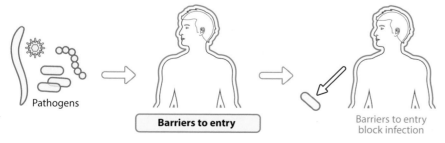

Figure 12.2 Crossing barriers to entry. Before a potential pathogen can cause an infection, it must enter the body of a host. To do so it must cross barriers to entry such as the skin or mucous membranes. Microbes often cannot cross such barriers and this prevents many infections from occurring.

In this example, the primary barrier to entry was the thick layer of mucus coating the membranes of the respiratory system. Any airborne pathogen must cross this mucous layer before it can reach the underlying membranes and subsequently enter otherwise sterile regions of the body. In many cases, including Rob's, pathogens become trapped in this mucus and are thereby prevented from reaching the underlying mucous membranes. Rob was also aided by the cilia lining the respiratory tract (**Figure 12.3**). These cilia move any invading microorganisms back toward the mouth, where they are caught in saliva and either spit out or swallowed. Respiratory infections in general are more common in places where air pollution is a problem, because pollutants can destroy cilia, making it easier for microbes to reach the lungs. Smokers are more prone to respiratory infections for similar reasons. Rob, who neither smokes nor lives in an area with serious air pollution, had the full benefit of these cilia.

In at least one other respect, Rob was lucky. In addition to the cramped quarters and recirculated air on an airplane that enhance airborne transmission, low humidity on airplanes tends to dry out mucous membranes. As mucus dries, it becomes less effective at blocking access to the underlying membranes, and infection becomes more likely. Perhaps Rob was assisted by drinking lots of water on the flight and thereby avoiding dehydration. Avoiding alcohol may have also helped, because alcohol hastens the drying of mucous membranes.

Other membranes lining the digestive and urogenital tract are likewise protected by mucus (**Figure 12.4**). In addition to mucus, these surfaces have other barriers that help to thwart invading microbes. Acid produced by the

Figure 12.3 Cilia. Cilia are hairlike extensions of the cell, composed of protein. They help protect the respiratory tract from microbial invasion and are found on many cell types, including those lining the respiratory tract. As respiratory tract cilia beat back and forth, they move fluid away from the lungs and back toward the mouth or nose. This movement prevents many invading microorganisms from reaching the lower respiratory tract.

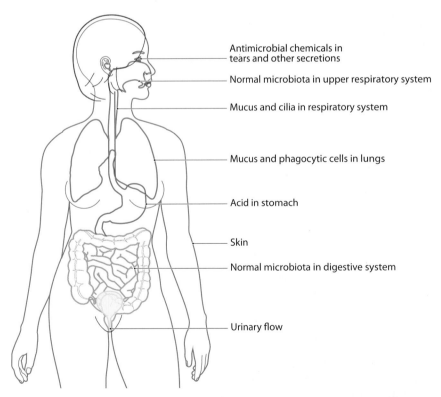

Antimicrobial chemicals in tears and other secretions

Normal microbiota in upper respiratory system

Mucus and cilia in respiratory system

Mucus and phagocytic cells in lungs

Acid in stomach

Skin

Normal microbiota in digestive system

Urinary flow

Figure 12.4 Important barriers to entry. All body surfaces are lined with skin or mucous membranes, which impede microbial colonization. These surfaces are often associated with additional chemical or physical barriers to further hamper microbial entry. Some of the most important barriers are depicted here.

stomach lining destroys many of the microorganisms that might otherwise reach the intestine. In the urogenital tract, regular urine flow impedes microorganisms from advancing to the bladder.

Other body surfaces are protected by the skin—one of our most formidable barriers to infection. The skin surface is composed of tightly packed epithelial cells, with an outermost layer of dead cells filled with protein. A thicker region containing blood vessels as well as oil and sweat glands, which bathe the skin with antimicrobial substances such as fatty acids, lies beneath the epithelial layer. Very few microorganisms can penetrate intact skin.

Mucous membranes and skin, which cover and protect all exposed body surfaces, are examples of **anatomical barriers**—barriers to entry by physical and chemical means. We are also protected by **microbial barriers**. In Chapter 11 (pp. 252–254), we discussed the normal microbiota and its importance in preventing infection. These well-adapted organisms provide another layer of defense, by producing substances that inhibit the growth of competitors, by monopolizing available nutrients, or by simply taking up all available space. With an intact normal microbiota, less-well-adapted pathogens are much less likely to colonize the host.

Certain other barriers to entry can be termed **genetic barriers**. Although you, for example, might contract hepatitis B virus from an infected person, you cannot infect your cat with this virus. Likewise, your cat cannot infect you with feline leukemia virus. The genetic differences between hosts are simply too great for these pathogens. Even within a particular species, some individuals have greater or lesser genetic resistance to a specific pathogen. Some humans, for instance, are relatively resistant to malaria because their red blood cells lack certain molecules required by the protozoan parasite to enter these cells.

Barriers to entry do not always work

Although our barriers to entry are formidable, they are not foolproof. Recall, for instance, the case in Chapter 11, in which a person who took antacid tablets, reducing his stomach's acidity, became more vulnerable to cholera. Anyone with injuries to the skin, including burns or cuts, becomes much more susceptible to infection of underlying tissues. And many pathogens have adaptations to circumvent such barriers. *Entamoeba histolytica*, to cite just one example, is an ameba that on occasion causes severe and bloody diarrhea. This problem only arises if the amebae can get past the mucous layer in the intestine and reach underlying cells. *E. histolytica* has a complex bag of tricks which it uses to get past the mucus, including enzymes that degrade components of the mucous layer.

CASE: LAURA

Laura is in her mid-fifties. She is returning home after a three-week vacation in East Asia. She never really noticed on her return from Hong Kong that one of the other passengers was coughing throughout the flight. Forty-eight hours after arriving home, however, she has developed all the symptoms associated with influenza, including headache, chills, fever, and achy muscles, and she spends most of the next 3 days in bed. Shortly thereafter the symptoms subside, and although she still feels somewhat weak, Laura makes a full recovery.

1. **Laura has suffered an acute infection of influenza. Although she was ill for several days, the incident ended with the elimination of the pathogen and a complete restoration of health. What occurred immunologically in Laura to account for this pattern of disease?**

If barriers to entry fail, the pathogen is confronted by the cells and molecules of innate immunity

In Laura's case, at least some viral particles reached the epithelium in her upper respiratory tract and then entered these epithelial cells to initiate an infection. Unlike Rob, who got off easily, Laura must now switch to plan B—destroy the invaders. This backup plan begins by confronting the pathogen with various innate defenses.

Think of these innate defenses as built-in security systems that are always on alert for invaders and can respond to infection quickly. Within a few hours these cellular and molecular first responders are on the scene, attempting to eliminate invading microbes. If successful, they will terminate their defensive activities. If they need assistance, they will sound the alarm, activating adaptive immunity.

An innate response often begins when cells at the site of invasion release certain chemicals that summon white blood cells, called **leukocytes**, to the site of infection. In Laura's case, epithelial cells lining her upper respiratory tract released these chemicals as they suffered damage due to viral replication. There are several different types of leukocytes and the first to arrive are phagocytic cells called **neutrophils** (the commonest type of immune cell in the blood) and **macrophages** (**Figure 12.5**). The details of phagocytosis were described previously (see Chapter 3, p. 64). Upon arrival at the site of infection, these phagocytic cells begin to engulf invading microorganisms, generally destroying them in large numbers. In addition to their innate phagocytic activity, macrophages also help to activate an adaptive immune response. **Dendritic cells** are another type of leukocyte capable of engulfing pathogens. They are vital to a successful immune response and are discussed in detail later in this chapter. As we will see, macrophages and dendritic cells form the critical link between innate and adaptive immunity.

Phagocytic cells need a way to recognize various microbes that pose a threat. One of the ways they do so is through **pattern recognition**—the ability to recognize certain types of molecules that are unique to microbes. The capacity to recognize such molecules, which are not normally found in humans and other animals, is a powerful feature of innate immunity. Pattern recognition means that certain microbial molecules act as danger signals, indicating that an infection is under way. Macrophages, for example, have special receptors on their surfaces that allow them to recognize microbe-associated molecules. These receptors are called **Toll-like receptors** (**Figure 12.6**), or **TLRs** for short, and they are found on several different types of immune system cells, including dendritic cells.

At least 10 different types of Toll-like receptors have been discovered in humans. Each type of TLR binds a different microbial product, allowing the immune system cells to identify different invading microbes as threats. For example, there is a TLR that binds specifically to bacterial peptidoglycan. Other TLRs bind molecules typically found only on acid-fast bacteria or fungi. Each type of TLR interacts with microbial molecules that are typically not found in animals. When Laura was infected by influenza virus, the TLRs on the arriving macrophages latched onto the viral RNA, indicating to the macrophage that a viral infection was in progress. The macrophages knew they were dealing with a virus because of the binding of viral RNA to the specific TLRs. This allows the macrophages to respond appropriately.

When TLRs and similar receptors are activated on the surface of macrophages and dendritic cells, these cells respond by releasing chemical

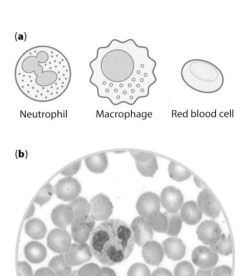

(a)

Neutrophil   Macrophage   Red blood cell

(b)

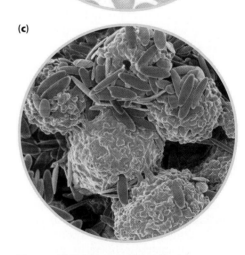

(c)

Figure 12.5 Phagocytic cells. (a) Neutrophils and macrophages are among the most important phagocytic cells. A red blood cell is shown for comparison. (b) A human blood smear containing a neutrophil with its characteristic multilobed nucleus. Note the preponderance of the smaller red blood cells. (c) Macrophages phagocytosing small polystyrene beads (in blue).

messengers called **cytokines** into the surrounding environment (see Figure 12.6). Activation of different TLRs can cause different types of cytokines to be released.

Inflammation and fever are key components of innate immunity

In response to many infections, cytokines are released that promote **inflammation**. Inflammation refers to a series of events that take place in reaction to tissue damage. The site of an injury becomes swollen, red, hot, and painful. While unpleasant, this process is part of the normal response to tissue damage and is an important defense against possible microbial invasion. The symptoms we observe are caused when the blood vessels in the affected area increase in diameter, a process called **vasodilation**. They also become leakier than usual, allowing blood serum to leave the blood vessels and enter the tissues. This increased leakiness is referred to as increased **permeability**. Although these changes contribute to the symptoms we experience during an infection, they also allow immune system cells and molecules to leave the blood to confront invading microbes. In Laura, inflammation in the respiratory tract helps to explain symptoms such as her persistent cough.

Laura's high fever was also due to cytokine release. Following her infection, a specific cytokine called interleukin-1 (IL-1) acted on Laura's brain to raise body temperature. Fever is actually a beneficial aspect of the innate immune response, at least to some extent. At elevated body temperature, phagocytic cells are more efficient, and some pathogens, including influenza virus, are less able to reproduce. So even though Laura's fever made her uncomfortable, it helped to limit the spread of the virus in her respiratory system. You may be surprised to learn that even cold-blooded animals, which cannot physiologically regulate their body temperature, will find a way to heat themselves up if they suffer from an infection. Grasshoppers infected with bacteria, for instance, seek out warmer than normal locations. If prevented from doing so, they are less likely to survive. Tropical fish hobbyists know to raise the temperature of their aquariums slightly if their fish are suffering from an infection. If the fever gets too high, though, it can be dangerous or even life-threatening. Fever, like inflammation, is therefore a good example of an important point—namely, many disease symptoms are not directly due to damage caused by the pathogen. Rather, they are often the unpleasant side effects of an immune response. The immune system is very much a double-edged sword. The powerful immunological weapons that keep us safe from most microbial assaults can sometimes cause discomfort, or even serious damage.

Special chemical messengers help combat viral infections

The **interferons** are especially important cytokines in viral infections like Laura's. Their name is based on the fact that they interfere with viral replication. These chemical messengers are produced by any cell that becomes infected with a virus. When cells in Laura's respiratory tract became infected with influenza virus, they released interferons, which alerted uninfected cells nearby that a virus was on the loose. The uninfected cells could then take protective measures. In other words, interferons induce an antiviral state in neighboring cells, greatly reducing the ability of the virus to spread (**Figure 12.7**).

When an antiviral state is induced in an uninfected cell, the cell produces special proteins that will shut down further protein synthesis if a viral

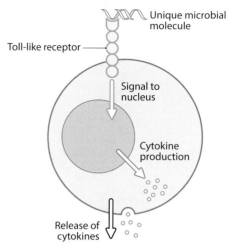

Figure 12.6 Pattern recognition by Toll-like receptors. When a pathogen molecule binds to a particular TLR, a signal is sent to the nucleus. This signal causes the cell to start producing specific chemical messengers called cytokines. Leukocytes produce many different cytokines, which regulate and coordinate various immune responses. Different pathogen molecules bind different TLRs, resulting in the release of different cytokines. In this way, the immune response is tailored to the general class of pathogen infecting the host.

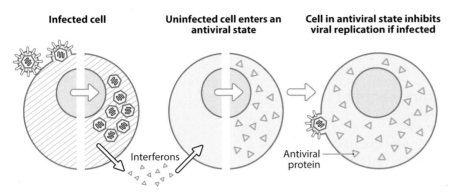

Figure 12.7 Interferons induce an antiviral state in uninfected cells. After a virus has infected a host cell and replicated within it, the infected cell produces interferons. Interferons allow an infected cell to warn its uninfected neighbors that a viral infection is in progress. The uninfected cells respond by taking protective measures, which include the production of antiviral proteins that inhibit protein synthesis by invading viruses. If a virus infects a cell that has produced these antiviral proteins, the virus is less likely to replicate.

invasion occurs. Not only is translation of mRNA to proteins inhibited but also the mRNA itself is degraded. Consequently, if a virus infects a cell in which the antiviral state has been induced by interferons, the virus cannot hijack the cell's protein-making machinery to produce the proteins it needs for viral replication. Once the viral infection is cleared and interferon stimulation ceases, surviving cells can restart their protein synthesis machinery. Interferons act very quickly and help "hold down the fort" until an adaptive response can be generated. However, like inflammation and fever, they contribute to the symptoms we suffer during an infection. Specifically, many of the flu-like symptoms that Laura suffered, including fatigue, headache, and chills, are provoked by interferons.

When they were first discovered, many scientists believed that interferons might be the magic bullet with which they could treat any viral infection. These hopes largely turned to disappointment when it was learned that interferons can cause the side effects mentioned above. Additionally, interferons are active in the body only for a very short time. Their short lifespan makes it difficult to deliver adequate doses of interferons to appropriate parts of the body. Nevertheless, interferons have shown some benefit in the treatment of certain viral infections such as hepatitis C.

When innate mechanisms fail to eliminate an infection, the adaptive immune response is activated

Laura's innate immune defenses could not contain the infection on their own (see Figure 12.1). Consequently, certain leukocytes, specifically macrophages and dendritic cells, sounded the alarm, rousting the cells of the adaptive immune system to action.

Why do innate mechanisms sometimes, but certainly not always, contain an infection on their own? Recall from Chapter 11 that the likelihood of disease in any encounter between pathogen and host depends upon the number of pathogens to which a host is exposed, the overall virulence of the pathogens, and the status of host defenses. If we know something about these factors in a particular host–pathogen interaction, we can make a reasonable prediction about the likely outcome of that interaction. In Laura's case, where innate mechanisms were insufficient to prevent illness, it may be that she was exposed to a particularly large dose of infectious virus that simply overwhelmed the capacity of innate defenses. Alternatively, for various reasons, Laura may have been especially vulnerable to viral infection at the time of her flight. Factors such as stress, lack of sleep, or inadequate diet can compromise immune defenses. It is possible that in Laura's case, a combination of these factors was involved.

Many pathogens are notorious for the adaptations they have evolved to avoid innate immune destruction. In Chapter 3, for instance, we reviewed a case of parrot fever, in which we learned that the involved bacteria were able to avoid destruction in a phagocytic cell by preventing fusion between the vesicle containing the bacteria and the lysosome, which contains enzymes that would otherwise digest and kill the bacteria. Other microbes are similarly sneaky. New parents, for instance, who have been kept awake all night by a baby with an ear infection, can usually blame *Streptococcus pneumoniae*. The bacterial strains responsible for infection have cells surrounded by a thick carbohydrate capsule that interferes with phagocytosis by phagocytic cells.

Antigen-presenting cells activate those cells responsible for adaptive immunity

When the Toll-like receptors of macrophages and dendritic cells (**Figure 12.8**) are bound by pathogen molecules, these important cells begin not only to release certain chemical messengers (cytokines) but also to actively ingest invading pathogens. Once ingested, by either phagocytosis or other similar processes, the pathogen is digested within the endomembrane system of the cell (see Chapter 3, p. 64). Following digestion, small protein fragments that were part of the pathogen bind onto cellular proteins called **major histocompatibility complex** (**MHC**) proteins. These MHC proteins, bound to pieces of pathogen protein, are then transported to the plasma membrane of the macrophage or dendritic cell and displayed on the surface (**Figure 12.9**). It is as if the cell is saying, "We've been invaded by a pathogen and here's a piece of that pathogen." These tiny pathogen fragments are called **antigens**. Antigens are pieces of foreign molecules that the adaptive immune system recognizes as part of a specific microorganism. The macrophages and dendritic cells, with these antigens displayed prominently on their surfaces, are serving as **antigen-presenting cells**.

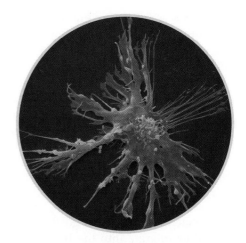

Figure 12.8 A dendritic cell. Dendritic cells, components of the innate immune response, play an essential role in stimulating the initiation of an adaptive response by acting as antigen-presenting cells. The long, fingerlike extensions characteristic of dendritic cells, called dendrites, give these cells their name.

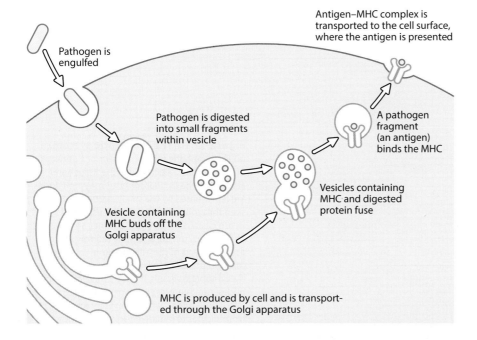

Figure 12.9 Foreign antigens are presented with MHC. When a macrophage or a dendritic cell (an antigen-presenting cell) engulfs a pathogen (in red), the pathogen is degraded by digestive enzymes. Meanwhile, MHC proteins (in green) are produced by the antigen-presenting cell. The MHC proteins move through the cell's endomembrane system, and eventually bud off the Golgi apparatus in a membrane-enclosed vesicle. The vesicles containing the pathogen and the MHC fuse, and small digested fragments of the pathogen bind to the MHC. These small pathogen fragments, recognizable as foreign by the immune system, are called antigens. The antigen–MHC complex is then transported in a vesicle to the cell membrane, where it is prominently presented.

In the case of influenza, the most important antigens are parts of the proteins embedded in the viral envelope (see Chapter 5, p. 94, to review the structure of the viral envelope). As we will see, once an antigen-presenting cell has these antigens on its surface, it will present them to certain cells responsible for adaptive immunity, thus letting them know what sort of highly targeted response is required. A different invading pathogen, *Staphylococcus* bacteria for instance, would have its own unique antigens, permitting a different adaptive response.

Antigen-presenting cells migrate to lymphatic organs to activate adaptive immunity

Once macrophages and dendritic cells are presenting antigen on their surface, they migrate into the **lymphatic system** (**Figure 12.10a**). This network of vessels returns any fluid that has leaked out of the blood back to the

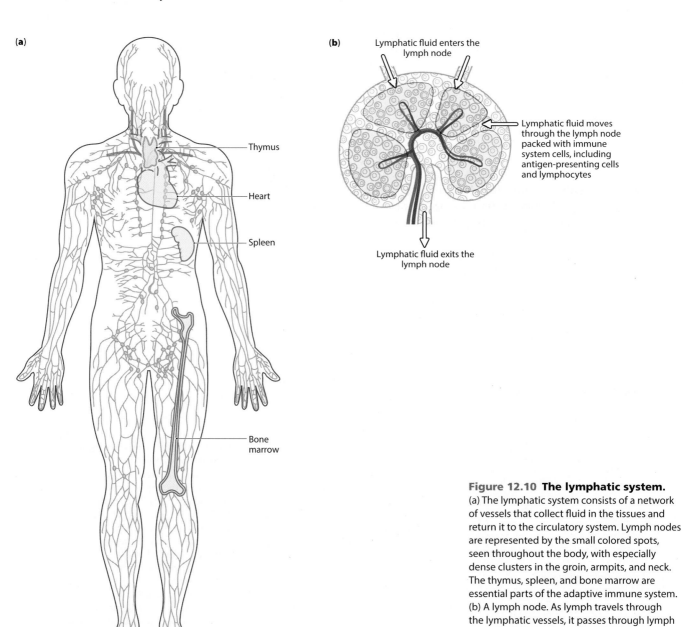

(a)

Thymus

Heart

Spleen

Bone marrow

(b) Lymphatic fluid enters the lymph node

Lymphatic fluid moves through the lymph node packed with immune system cells, including antigen-presenting cells and lymphocytes

Lymphatic fluid exits the lymph node

Figure 12.10 The lymphatic system.
(a) The lymphatic system consists of a network of vessels that collect fluid in the tissues and return it to the circulatory system. Lymph nodes are represented by the small colored spots, seen throughout the body, with especially dense clusters in the groin, armpits, and neck. The thymus, spleen, and bone marrow are essential parts of the adaptive immune system. (b) A lymph node. As lymph travels through the lymphatic vessels, it passes through lymph nodes. Antigen-presenting cells migrate into lymph nodes, where they interact with lymphocytes to initiate an adaptive response.

circulatory system. The fluid in the lymphatic system (the **lymph**) passes though increasingly larger lymphatic vessels until it is returned to the blood through a duct, which empties into a large vein near the heart. As the lymph progresses, it passes through lymphatic organs called **lymph nodes** (**Figure 12.10b**). These small, bean-shaped structures are generally found in places where several smaller lymphatic vessels come together to form larger vessels.

Lymph nodes also contain numerous **lymphocytes**. Lymphocytes are the specific white blood cells that are responsible for adaptive immunity. Lymphocytes, as well as all other blood cells, are initially formed in the red marrow of the bone. As we will see, there are two principal types of lymphocytes, called the **T cells** and the **B cells**. Each cell type plays a specific role in an adaptive response.

Lymphocytes circulate through the blood, but they can also leave the blood and enter the lymphatic system by adhering to capillaries and squeezing between the cells of the capillary walls. Once they have entered the lymphatic system, they may be eventually returned to the blood. But before they get there, they must pass through lymph nodes. If there are no antigen-presenting cells in a lymph node, they will pass through the node and continue circulating. If, however, antigen-presenting cells (macrophages and dendritic cells) are also in the lymph node, the lymphocytes may be activated to initiate an adaptive response.

The lymph nodes are strategically placed throughout the body, where they concentrate all the components necessary for an adaptive immune response: antigen, antigen-presenting cells, and lymphocytes. In Laura's case, since influenza virus has invaded the epithelium of her upper respiratory system, we would expect the adaptive response to be generated in the lymph nodes closest to the site of infection: in the neck and throat. Once activated, lymphocytes within the node begin to rapidly divide, greatly increasing in number. This rapid rise in lymphocyte number causes the swelling of lymph nodes during infection.

Lymph nodes are not the only lymphatic organ. When microbes reach the blood, they are filtered by the **spleen** (**Figure 12.11**), instead of lymph nodes. The spleen is an organ about the size of a bar of soap, found in the upper left portion of the abdomen (see Figure 12.10). It basically works like a lymph node for blood-borne pathogens, providing a site for the initiation of an adaptive response. Other areas in the body that are particularly prone to invasion and infection, such as the intestine, are equipped with their own lymphatic organs.

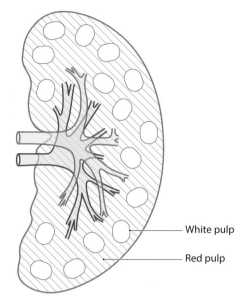

White pulp

Red pulp

Figure 12.11 The spleen. The spleen consists of red pulp, where old red blood cells are destroyed, and white pulp, where adaptive immune responses to blood-borne pathogens are generated.

Antigen-presenting cells activate helper T cells to initiate an adaptive response

When antigen-presenting cells arrived in the lymph nodes of Laura's neck and throat, they prominently displayed pieces of the influenza virus on their surface, bound to MHC molecules. They also released certain cytokines that attracted lymphocytes called **helper T cells**. These lymphocytes are just one type of T cell, and like all T cells they migrate from the bone marrow, where they are formed, to the **thymus**, where they complete their development. The thymus is a lobed organ, found just above the heart (**Figure 12.12**). Once they mature, T cells leave the thymus and begin to circulate. As we will see, helper T cells (abbreviated as **Th cells**) are especially crucial in adaptive immunity, because once they are activated by antigen-presenting cells, they coordinate the overall adaptive response.

(a)

(b)

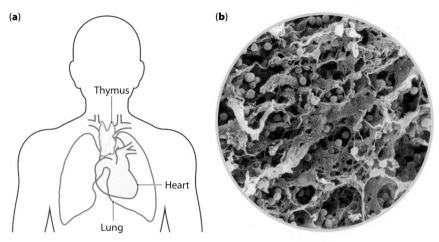

Figure 12.12 **The thymus.** (a) The lobed thymus is found immediately above the heart. Lymphocytes that leave the bone marrow and migrate to the thymus develop into T cells. (b) In this image taken with an electron microscope, developing T cells (the small spherical cells) can be seen among the extensive epithelial tissue of the thymus.

Helper T cells all have protein complexes on their surface called **T-cell receptors**. The T-cell receptor allows the Th cell to recognize a specific antigen bound to MHC on the surface of an antigen-presenting cell (**Figure 12.13**). Each helper T cell has 100,000 or so of these receptors, and they are all identical. But different Th cells have different T-cell receptors, all of which recognize different antigens. One Th cell may have receptors that recognize a certain part of an influenza envelope protein. A different Th cell may have receptors that recognize a different piece of the same protein, while a third Th cell may recognize a piece of a different viral protein. All of these Th cells will be activated by antigen-presenting cells (macrophages and dendritic cells) in Laura's lymph nodes. Other Th cells have receptors that might recognize antigens from other viruses or from various bacteria or fungi. These Th cells will not be activated if Laura is infected only with influenza, because they will not come in contact with their specific antigen. The Th cells that do recognize their influenza antigens, on the other hand, begin to replicate furiously. The newly replicated Th cells all have the same T-cell receptor that was found on the Th cells that were originally activated. In other words, only a relatively small number of Th cells initially recognized their antigens. Now that they have replicated, however, their numbers have increased enormously.

The number of different kinds of Th cells, all with different T-cell receptors on their surfaces, is almost limitless. We even have Th cells with receptors for antigens to which we will probably never be exposed; if you take a trip to Antarctica and turn over a rock exposing yourself to an unusual bacterium of some sort, you probably have T cells that would recognize its antigens. This stupendous diversity is made possible because T-cell receptors are coded for by a collection of genes. Although these genes are finite in number, each receptor is encoded by a different combination of these genes. And even with a limited number of genes, different gene combinations allow the assembly of millions of different receptors. The genetics behind this process is complex and beyond the scope of this text. However, it is safe to say that, thanks to this remarkable process of gene recombination, almost any invading pathogen will be recognized as a foe to be combated.

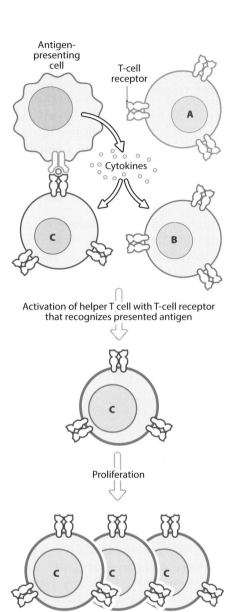

Activation of helper T cell with T-cell receptor that recognizes presented antigen

Proliferation

Activated, effector helper T cells all with same T-cell receptor

Figure 12.13 **Activation of helper T cells by antigen-presenting cells.** Once an antigen-presenting cell (a macrophage or dendritic cell) is presenting antigen on its surface, it migrates into a nearby lymph node (or the spleen if the pathogen is blood-borne), releasing cytokines that attract helper T cells. In this figure, three representative helper T cells (labeled A–C) are illustrated. They are all attracted to the antigen-presenting cell, but in this example only helper T cell C has a T-cell receptor that recognizes the antigen being presented. This helper T cell is therefore activated and begins to divide. Once division is complete, there are now many activated helper T cells, all bearing the T-cell receptor for the same antigen. These now-activated helper T cells are called effector Th cells.

Some activated helper T cells activate cytotoxic T cells to initiate a cell-mediated response

The activation of the Th cells that we have described takes a few days. Meanwhile, in spite of innate defenses, the influenza virus that infected Laura has been multiplying and as viral levels cross the threshold of disease (see Chapter 11, p. 265), she starts to feel ill. Soon, however, the appropriate Th cells are ready for action and they begin to activate a second type of T cell called a **cytotoxic T cell** (Figure 12.14). Like Th cells, cytotoxic T cells (Tc cells) develop in the thymus, and like Th cells, different Tc cells have specific T-cell receptors on their surface. Furthermore, like Th cells, once they are activated, they begin to divide and increase in number.

Cytotoxic T cells are activated when Th cells stimulate them with certain cytokines. Interleukin-2 (IL-2) is an example. The now-active cytotoxic T cells then leave the lymph nodes, in search of infected cells. They can recognize these infected cells because the infected cells display foreign antigens on their surface, bound to MHC (see Figure 12.14). By displaying these antigens on their surface, infected cells are essentially broadcasting a message saying, "I'm infected. Kill me before the pathogen can replicate and infect other cells." If a now-activated, patrolling Tc cell interacts with this cell, and if its T-cell receptor matches the displayed antigen, it releases toxic chemicals that cause the death of the infected cells (Figure 12.15). Uninfected cells do not display foreign antigens on their surface and are unharmed. This targeted cell killing, called the **cell-mediated response**, is similar to performing microscopic amputations—infected host cells are sacrificed to prevent uninfected cells from becoming infected.

The Th or Tc cells that have been activated during a cell-mediated response are collectively referred to as **effector T cells**. Before we go on, it is important to point out that when either Th or Tc cells are first stimulated, the dividing T cells do not all develop into active, effector T cells. Some remain as relatively inactive and long-lived **memory T cells**. As we will soon see, these memory cells are a big part of the reason why, if Laura is infected with influenza again in the near future, she will probably remain healthy and blissfully unaware of her second encounter with the same virus.

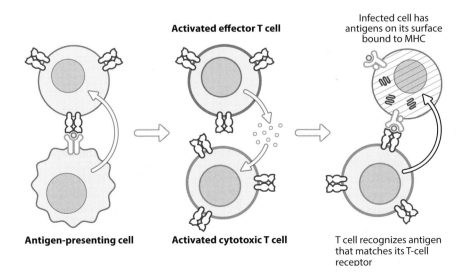

Activated effector T cell

Infected cell has antigens on its surface bound to MHC

Antigen-presenting cell   **Activated cytotoxic T cell**

T cell recognizes antigen that matches its T-cell receptor

Figure 12.14 The cell-mediated response. Once helper T cells are activated by antigen-presenting cells, they may release specific cytokines that activate cytotoxic T cells (Tc cells). Cytotoxic T cells also have T-cell receptors on their surface; they only recognize antigens for which their T-cell receptor is specific. The now-activated effector Tc cells seek their antigen, bound to MHC, on the surface of host cells.

Infected host cell   Infected cell is killed

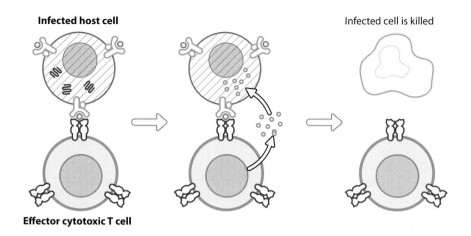

Effector cytotoxic T cell

Figure 12.15 Elimination of infected cells by cytotoxic T cells. If Tc cells recognize their antigen, bound to MHC on the surface of an infected cell, they release toxic chemicals that kill the infected cell.

Helper T cells also activate B cells to initiate a humoral immune response

While some Th cells are busy activating Tc cells, others release different cytokines, including IL-4 and IL-5, to activate a different type of lymphocyte called a B cell. As opposed to T cells, which migrate to the thymus, B cells remain in the red marrow of the bone, where their development is completed. B cells were so named because their site of development was first discovered in birds, where they complete their development not in the bone marrow but in a structure known as the bursa.

Recall that both Th and Tc cells have T-cell receptors on their surface. These receptors recognize specific antigens when the antigen is bound to MHC on the surface of either an antigen-presenting cell (in the case of Th cells) or an infected cell (in the case of Tc cells). B cells, on the other hand, have **antibodies** in their plasma membranes (**Figure 12.16**). Like T-cell receptors, antibodies are composed of protein. Antibodies also recognize specific antigens. But unlike T-cell receptors, they bind antigen that is floating freely in the body's fluids. Similar to what we described for T-cell receptors, each B cell has many identical antibodies on its surface, all of which recognize the same antigen. Furthermore, we produce literally billions of different kinds of B cells, all with different surface antibodies. This practically limitless variety of antibodies, all recognizing different antigens, ensures that whatever pathogen happens to come along, we will have antibodies that can recognize it. As with T-cell receptors, this diversity is based on the ability to recombine a limited number of antibody genes in an astronomical number of different ways.

After they mature in the bone marrow, the B cells begin to circulate through the blood and lymph. As they circulate, they pass through lymphatic organs, including lymph nodes or the spleen. Foreign antigens also accumulate in these lymphatic organs, much as a drain in a kitchen sink might trap bits of solid material. If a B cell has a surface antibody that matches antigens in the lymph node, the B cell binds this antigen. This antigen–antibody binding and the stimulatory cytokines released by helper T cells activate the B cell and provide it with the two signals that it needs to start dividing (**Figure 12.17**). After a series of divisions, the cells have developed into one of two types: **memory B cells** and **plasma cells**. We will discuss memory B cells shortly. Like memory T cells, memory B cells ensure that Laura won't have to worry about influenza again, a least for a while.

Plasma cells are antibody-secreting cells. Instead of retaining their antibody on their surface, plasma cells release antibody into the blood or lymph. Soon

(a)

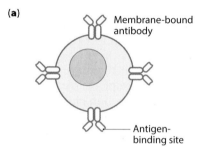

Membrane-bound antibody

Antigen-binding site

(b)

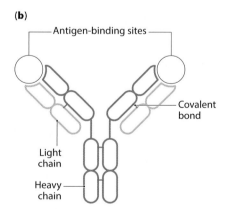

Antigen-binding sites

Covalent bond

Light chain

Heavy chain

Figure 12.16 B cells and antibodies. (a) Mature B cell expresses antibodies on its membrane. Each antibody has two antigen-binding sites, and all antibodies on a particular B cell are specific for the same antigen. (b) Antibody structure. Each individual antibody is actually composed of four proteins, held together by covalent bonds. The two smaller proteins (called light chains) are identical, as are the two larger proteins (called heavy chains). Each antibody molecule has two antigen-binding sites.

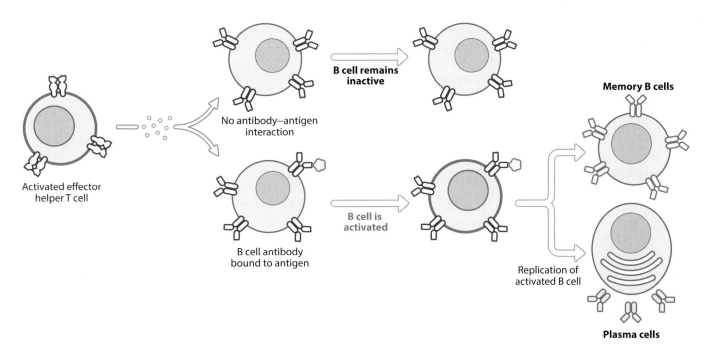

Figure 12.17 B-cell activation. To be activated, a B cell must be bound to its antigen and be stimulated by activating cytokines from an activated effector T cell. Once activated, it starts to divide rapidly and develops into one of two cell types: a short-lived plasma cell, which secretes its antibodies into the lymph, or a long-lived memory cell that retains its antibodies on the cell surface.

the body fluids are flooded with antibodies, all of which recognize and bind the same antigen. The production of antibody by B cells is called the **humoral response**. This is the portion of an adaptive response against antigen that is free in the body's fluids, once called humors.

In a humoral response to influenza, the recognized antigens will again be pieces of the virus, such as pieces of envelope proteins. Once antibodies are locked into place on a viral particle, that particle can no longer bind to appropriate host cells and therefore cannot enter these cells (see Chapter 5, pp. 95–96, to review viral binding to host cells). This capacity to lock out viruses from the host cells in which they replicate is called **neutralization** (**Figure 12.18a**). Bacteria or other pathogens can likewise be neutralized by antibody, if the antibody bound to their surface prevents them from reaching a particular target site in the body. Furthermore, note in Figure 12.18 that antibodies have two binding sites for antigen (the two tips of the Y). Because they can bind two antigens at once, antibodies can form large masses together with antigen. Such clumping is called **agglutination** (**Figure 12.18b**). Any pathogens trapped in these large agglutinated clumps will never reach their target tissue. They are therefore neutralized and readily available to phagocytes.

Antibodies are also useful in other ways. Observe in **Figure 12.18c** that, in addition to their antigen-binding sites, antibodies also have a binding site (at the base of the Y) for phagocytic cells like macrophages and neutrophils. Because they can bind antigen at one end and phagocytic cells at the other end, antibodies form a physical link between the phagocytic cell and the antigen. If the antigen is part of a pathogen, the phagocyte can simply engulf both the antibody and the pathogen. Antibodies functioning in this way are acting as **opsonins**. An opsonin is any molecule that forms a physical connection between a phagocyte and an antigen. Although phagocytes do not require opsonins to phagocytose foreign material, opsonins make their job easier and increase phagocytic efficiency.

Before we conclude our discussion of antibody function, it is worth noting that when antibodies act as opsonins, they are promoting an innate immune process—phagocytosis. This is a good example of an important point—namely, the separation of immunity into innate and adaptive components is in some ways arbitrary. In a properly functioning immune system, innate and adaptive processes overlap and very much depend on each other.

In a humoral response, several different classes of antibodies may be produced

Secreted antibodies are categorized as one of four basic classes: IgM, IgG, IgA, and IgE. The prefix Ig stands for immunoglobulin, a term that is interchangeable with antibody. Our discussion here focuses on IgM, IgG, and IgA, the three most important antibodies in most humoral responses (**Figure 12.19**).

The membrane-bound antibodies on B cells are mostly of the IgM class. A fifth antibody type, IgD, is also present on the membrane. Its overall role in humoral immunity is still not completely clear, and we will not consider it further. Once a B cell is activated and matures into a plasma cell, it first produces a secreted form of IgM that is released into the body fluids. Later, plasma cells start to produce other antibody types instead, primarily IgG. The antigen specificity of the antibody, however, does not change. In other words, if plasma cells first secrete IgM and later IgG in response to a particular antigen, both the IgM and the IgG will react with the same antigen.

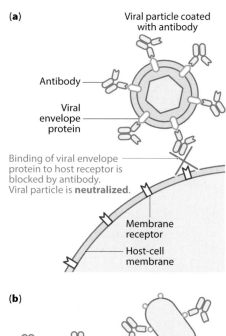

(a)

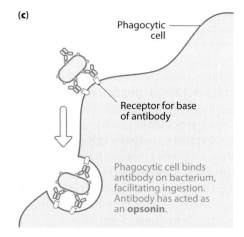

(b)

(c)

Figure 12.18 Antibody functions.
(a) Antibody can coat the surface of viral particles, preventing them from attaching to their cell receptors. This process is called neutralization. (b) Because each antibody has two antigen-binding sites, large agglutinated clumps of antigen and antibody often form. These clumps are readily available to phagocytic cells. In this example the antigens (in green) are on the surface of bacterial cells. (c) Antibody can also enhance phagocytosis by acting as an opsonin for phagocytic cells, which have receptors on their surface for the base of Y-shaped antibodies.

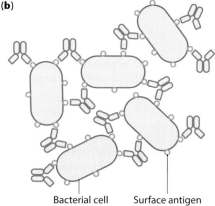

	Structure	Location	% of total antibody made	Function
IgM	Pentamer	Surface of B cells (as monomer) Blood Lymph	~10%	First secreted antibody in adaptive response Agglutination
IgG	Monomer	Blood Lymph Breast milk Fetal circulation	~80%	Predominant antibody in most humoral responses Neutralization Opsonin Agglutination Newborn immunity
IgA	Dimer	Secretions such as intestinal fluid, saliva, tears, and breast milk	~10%	Protection of mucous membranes

Figure 12.19 Structure and function of the three most important antibody classes.

The reason for the switching from IgM to other antibody classes is that the different classes of antibody have somewhat different functions. Therefore, depending on the type of infection, different types of antibodies might be more beneficial. As stated, the first type of antibody secreted in a humoral response is IgM. When secreted, five IgM antibodies join together to form covalently bound pentamers. With 10 antigen-binding sites, such pentamers are especially effective at agglutinating antigen. IgM is produced only during the first few days of a humoral response. As levels of IgM begin to decline, levels of other antibodies, mostly IgG, start to rise.

IgG is by far the most abundant antibody type produced in a humoral response. IgG can agglutinate antigen, and it enhances phagocytosis as an opsonin. Because it can cross the placenta and enter breast milk, IgG is maternally transferred to a fetus or newborn, providing an important source of immunity in infants. In Laura's immune response to influenza, IgG is important because of its ability to bind influenza envelope proteins, neutralizing the virus before it can attach to respiratory cells in the upper respiratory tract.

Individual IgA antibodies bind to each other in pairs, forming dimers. Once secreted, IgA can cross from the blood to body fluids such as saliva and intestinal fluid. It is also found in tears and mucous secretions in the respiratory and urogenital tracts. IgA thus plays a special role in immunity on mucous membranes and is particularly valuable in combating pathogens found in places like the digestive tract. Like IgG, IgA can enter the breast milk and consequently helps to provide immunity for newborns. IgA is also produced during an influenza infection, but it is probably less important in viral neutralization than IgG, because it is produced in substantially lower amounts.

A successful adaptive response culminates in the elimination of the pathogen

Once her bout of influenza began, Laura felt awful for several days. The virus, while slowed down by innate defenses, was still able to replicate prior to the activation of a full-fledged adaptive response. The time between initial infection and the start of the adaptive response is called the **inductive period** (**Figure 12.20**). The inductive period thus corresponds to the time that is necessary for T cells and B cells to be activated, to proliferate, and to begin their antimicrobial activity. During the inductive period, with no adaptive response to contend with, the pathogen has something of a free rein. In Laura's case, as the amount of virus rose above the threshold of disease, symptoms commenced, and she was forced into bed. But after several days, with antibodies now limiting viral spread and cytotoxic T cells now on their "seek-and-destroy mission," levels of virus in Laura's body began to fall, and she started to recover. Gradually, activated lymphocytes died off, levels of antibody in the body fluids waned, and the adaptive response came to an end. A third group of T cells, called the **regulatory T (T_R) cells**, is also involved in bringing an adaptive response to an end. Although their exact mode of action is still not precisely understood, it is believed that T_R cells play a role in suppressing the activity of other T cells.

While the activated lymphocytes do not persist, both T cells and B cells have formed populations of long-lived memory cells. Memory cells can last for years, decades, or even a lifetime, and it is these memory cells that form the basis of immunological memory—the ability of the adaptive immune response to a particular organism to grow stronger with each exposure. We will shortly investigate immunological memory in greater depth.

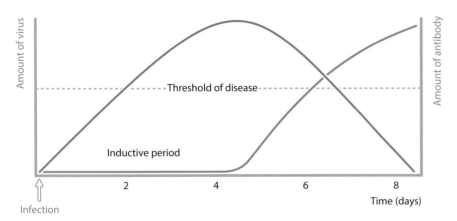

Figure 12.20 The course of infection and immune response. The amount of virus in an infected person is shown in red. Once an individual is infected, the amount of virus in the body begins to rise. For influenza virus, it usually takes about 2 days before viral levels are high enough to cause symptoms. This is the point at which the virus passes the threshold of disease. Other pathogens may take more or less time to reach their thresholds. The pathogen replicates relatively well for the first few days of the infection because there is no adaptive response to contain it. The amount of time necessary to generate an adaptive response is called the inductive period. For simplicity, only the humoral response, measured as the amount of antibody, is shown here (in blue). Activation of the cell-mediated response is similar. Once antibody levels begin to rise, viral levels begin to fall, and the patient begins to recover. Symptoms wane as viral levels, contained by adaptive immunity, fall below the threshold of disease.

An adaptive response is not always successful

Unfortunately, not everyone is as lucky as Laura was. Influenza still kills between 250,000 and 300,000 people worldwide during a typical year. During an epidemic, this number can rise considerably. Most fatalities occur among the elderly, in whom the immune response is less vigorous, or in those with underlying respiratory problems, which lessen the effectiveness of barriers to entry and may also impede an effective immune response. In many fatal cases, damage to the respiratory epithelium by influenza virus opens the door to opportunistic bacterial infections, which may result in pneumonia.

Other agents of infectious disease still take their toll in terms of fatalities. As mentioned earlier in this chapter, the likelihood of disease in a host–pathogen encounter depends primarily on three interrelated factors: the number of pathogens to which the host was exposed, the virulence of the pathogens, and/or the status of host defenses. When hosts are exposed to too many pathogens, or if those pathogens are too virulent, even the most robust immune response may not be enough. Alternatively, if the host's immune system is in some way compromised, the adaptive response may be insufficient to protect against even less-virulent microorganisms.

Many pathogens have all sorts of tricks with which they avoid immune destruction. We have discussed a few of these in relation to getting past barriers to entry and the innate response. Adaptive immunity, too, is vulnerable to microbial evasion. To make this point, we will provide a single example. The Epstein–Barr virus (EBV) is responsible for mononucleosis (also known as "the kissing disease"). When it is replicating inside of host cells, although invisible to antibodies, you would think it would still be vulnerable to cell-mediated destruction. But EBV uses a remarkable ploy to manipulate the immune system. It actually encodes a specific viral protein that looks and even acts like the host cytokine, IL-10. This cytokine normally supresses cell-mediated immunity and promotes humoral immunity. The viral dummy molecule does exactly the same. The promoted humoral response has little effect against the intracellular virus, while the supressed cell-mediated response insures viral survival.

CASE: BONNIE

Bonnie is a 30-year-old flight attendant based in Hong Kong. She makes the flight from Hong Kong to Los Angeles approximately every 10 days. She missed one of her flight assignments earlier in the year, in late January, when she was ill with the flu. As Bonnie lay miserably in bed at that time, she became angry with herself for not getting a flu vaccine, which was offered free of charge to all employees of her airline. On this most recent flight, in early March, she interacts with Jeff, the sick passenger, several times, and she can see that he is seriously ill. Bonnie dreads the thought of getting sick again herself, but to her relief, she remains healthy.

1. **Bonnie was infected with influenza virus when Jeff coughed repeatedly in very close proximity to her. In spite of this infection, she developed no signs of disease. Why not?**
2. **How exactly would a vaccine have prevented Bonnie's earlier illness?**

Subsequent exposure to the same pathogen results in a stronger and faster adaptive response

Although Laura and Bonnie were both infected with influenza on their flight to Los Angeles, they had very different experiences; while Laura became quite ill, Bonnie was not even aware that she had been infected. The difference is the immunological memory established in Bonnie when she was sick 2 months earlier. Immunological memory is one of the most powerful and distinctive features of the adaptive immune response. It refers to the fact that an adaptive response to a particular pathogen is faster and stronger with each exposure.

When Bonnie became ill in January, the sequence of events was much the same as what we saw in Laura. Innate defenses could not contain the infection on their own, and consequently, antigen-presenting cells activated helper T cells in lymphatic organs. Once they were activated, helper T cells activated both Tc cells, to initiate a cell-mediated response, and B cells, to stimulate a humoral response. As we have seen, this series of events takes time. The time between the initial infection and the generation of a full-fledged adaptive response (the inductive period) is usually about 5–7 days. During the inductive period, with neither antibodies nor cytotoxic T cells to impede it, the virus is replicating successfully, and as it increases in numbers, it crosses the threshold of disease and symptoms of illness develop (see Figure 12.20). Once adaptive immunity is in full swing, viral replication slows and as viral numbers fall back below the disease threshold, the patient starts to recover. This pattern of disease and recovery is typical when the infected individual generates a **primary immune response**—the adaptive response that occurs the first time an individual is infected with a particular pathogen.

Recall that when T cells and B cells are activated and undergo proliferation, some develop into active effector T cells and plasma cells. As the pathogen is destroyed, these relatively short-lived cells die off and the adaptive immune response comes to an end. However, as previously noted, not all the proliferating lymphocytes became effector T cells or plasma cells. Many became long-lived memory T and B cells. These memory cells, some of which may survive for months, years, or even decades, circulate until the same pathogen shows up again, sometime in the future. When it does, memory cells stand ready to quickly differentiate into effector T cells and plasma cells. Rather than a 5–7 day inductive period, within only a day or two, effector Tc cells are destroying virally infected cells, and antibody is present in the blood at high levels (**Figure 12.21**). Although the pathogen might replicate for a short time, it is quickly confronted by an adaptive response before it passes the disease threshold. Thus, although the pathogen has established an infection, it is stopped dead in its tracks by the presence of memory T and B cells before it can replicate enough to cause symptoms. This is what occurred in Bonnie. During her primary response in January, she became ill and subsequently generated large numbers of memory T and B cells to the antigens on the influenza virus. When she was exposed a second time on the flight to Los Angeles, these memory cells responded so quickly and so strongly that the infection remained **subclinical**—at a population below the threshold of disease. Consequently, she suffered no symptoms. This second, much stronger and faster adaptive response is called a **secondary response**. If Laura, who responded with a primary response to her initial infection, is infected a second time, she too will respond with a secondary response and will probably remain healthy. At worst, she will suffer only mild symptoms. During a secondary response, even greater numbers of new memory cells are generated. This means that if the same pathogen infects for a third time, there will be an even more vigorous adaptive response.

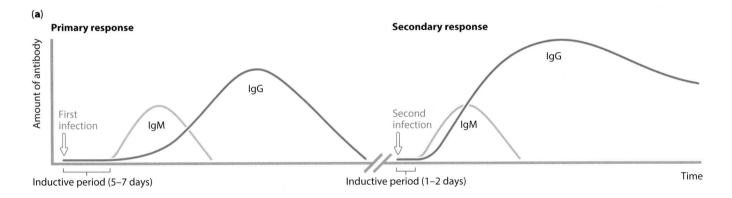

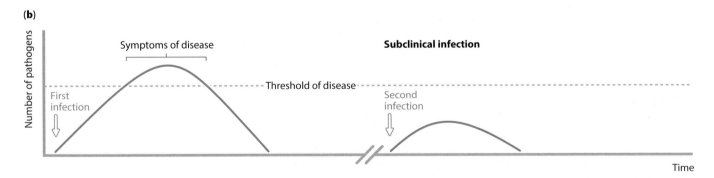

Figure 12.21 Primary and secondary humoral responses. (a) Only the humoral response is shown, for simplicity. The activity of the cell-mediated response (not shown) is similar. During the primary response, the inductive period is about 5–7 days. Initially a small amount of secreted IgM is produced. A larger amount of IgG follows. Later, during a secondary response, memory cells that have persisted since the initial primary response quickly differentiate into plasma cells, resulting in a shorter inductive period and heightened antibody production. (b) The course of infection over this same time period. During the first infection, pathogen numbers cross the disease threshold, and symptoms develop in the infected individual. During the second infection, pathogen numbers stay below the threshold of disease, because the secondary response is so much faster and stronger. Symptoms of disease do not develop. In other words, the infection remains subclinical.

Vaccines induce immunological memory without causing disease

The ability to mount ever-stronger adaptive responses with each exposure is the basis of vaccine therapy. **Table 12.1** provides a list of some currently available vaccines**.** In most vaccines, a primary response is stimulated by exposing the vaccinated individual to either a dead or a weakened form of the pathogen that cannot cause illness. Both sorts of vaccine exist for influenza. If you have been vaccinated against influenza with an injection, you were being exposed to killed virus. Each year these vaccines are reformulated to protect against whichever strains of influenza are most common at the time. Recently, a second type of influenza vaccine has been introduced. Rather than using killed virus (generally called **inactivated virus**), this new vaccine relies on weakened but still viable virus. Although the weakened virus can cause an infection, it replicates poorly and is very unlikely to cause illness. Such a virus is called an **attenuated virus**. The new, attenuated influenza vaccine is inhaled as a nasal mist.

Regardless of how vaccines are made, they all work in a similar way. When a vaccinated person is later exposed to a live and virulent form of the pathogen, he or she responds with a secondary response, because memory cells for that pathogen were generated by the exposure to the vaccine.

Sometimes, even a secondary response will not reliably prevent disease. In these cases, more than one vaccine is necessary. Hepatitis B is a good

Table 12.1 Representative vaccines. A partial list of some currently used vaccines. Not all of these vaccines are available in all countries.

DISEASE	PATHOGEN
Anthrax	Bacterium
Cervical cancer	Virus
Diphtheria	Bacterium
Hepatitis A	Virus
Hepatitis B	Virus
Human papillomavirus	Virus
Influenza (seasonal flu)	Virus
Lyme disease	Bacterium
Measles	Virus
Meningococcal pneumonia	Bacterial
Mumps	Virus
Pertussis (whooping cough)	Bacterium
Pneumococcal pneumonia	Bacterium
Poliomyelitis (polio)	Virus
Rabies	Virus
Rotavirus	Virus
Rubella (German measles)	Virus
Shingles	Virus
Smallpox	Virus
Tetanus	Bacterium
Tuberculosis	Bacterium
Typhoid fever	Bacterium
Varicella	Virus
Yellow fever	Virus

example. Vaccination against the hepatitis B virus requires three injections. With the first injection, a primary response is mounted. The second injection some time later is termed the first booster because it results in a secondary response, boosting the number of memory cells to the viral antigens. The third injection (the second booster) causes a third response (also called a tertiary response), generating even more memory cells. Therefore, if a fully vaccinated individual is infected with the actual hepatitis B virus, the immune system responds for the fourth time with a quaternary response. So many memory cells would now be ready to respond that the chances of actually developing the disease are extremely slight.

The hepatitis B vaccine does not rely on either inactivated or killed virus; it represents a third type of vaccine called a **subunit vaccine**. A subunit vaccine does not even contain an entire pathogen. Rather, the vaccine contains only those pieces of protein known to be strong antigens for the pathogen and therefore powerful stimulators of adaptive immunity. In the hepatitis B vaccine, all three injections contain these same viral proteins.

Generally, attenuated vaccines result in better immunological memory than killed vaccines; although the pathogen is weakened, it does cause an actual infection. Consequently, an attenuated vaccine is a closer mimic to what occurs during an infection with the pathogenic microbe. Because the attenuated pathogen replicates, the adaptive immune response is stimulated strongly, resulting in the development of strong immunological memory. A killed pathogen may not persist in the body long enough to fully stimulate adaptive immunity in this way.

Why do we use inactivated vaccines at all, if attenuated vaccines result in stronger memory and therefore better protection? In some cases, killed vaccines are safer. With an attenuated vaccine, because the pathogen is alive, there is always a slight chance that it will mutate back to its virulent state. In such a case, a vaccine can, on rare occasions, cause the disease it was designed to prevent. This problem, however, has not been observed with the attenuated influenza vaccine.

The attenuated polio vaccine does occasionally revert to virulence. Assuming you have been vaccinated against polio, depending on when and where you were born, you received either the injected Salk vaccine or the liquid Sabin vaccine that you drank in a sugary solution. The Salk vaccine is an inactivated vaccine; it contains killed poliovirus. The Sabin vaccine is an attenuated vaccine, composed of weakened strains of the virus that elicit immunity without causing disease.

Because the virus in the Sabin vaccine is alive, it replicates slowly in the body, and it mimics a natural poliovirus infection, inducing a strong primary response. The Salk vaccine does not replicate and its persistence in the body is reduced. The memory that is generated in response to the Salk vaccine is consequently somewhat less, which explains why the Salk vaccine is fully protective only after several boosters.

Occasionally, however, the virus in the Sabin vaccine mutates from its weakened form back into the virulent form. This possibility is slight, but about one in every 2.6 million people who get the Sabin vaccine actually develops polio for just this reason. This may be an acceptable risk if the likelihood of a natural poliovirus infection is great, which accounts for the preferential use of the Sabin vaccine when polio was more common. In recent years, however, the chances of naturally contracting polio in Europe, North America or other developed regions are exceedingly slight. Consequently, health authorities opt to use the safer, albeit less protective, Salk vaccine.

HIV is an ongoing public health emergency

CASE: OSCAR

Oscar has been HIV-positive for 10 years. He originally sought medical attention when he developed severe respiratory illness, later diagnosed as pneumonia caused by *Pneumocystis jirovecii*. This opportunistic fungal pathogen does not ordinarily cause disease in healthy individuals. When he was found to be HIV-positive, his physician prescribed appropriate antiviral drugs, and Oscar's health improved dramatically. For the past several years, he has had no serious health problems, and except for the drugs he must take daily, Oscar has resumed a normal life. Lately, however, he has become lax about taking his medications. On this recent trip to Asia, which lasted well over a month, he became especially negligent, often going several days at a time without taking the antiviral drugs. He rationalizes this by thinking to himself that he will get serious again about his medications once he returns home.

On the flight to Los Angeles he becomes infected with influenza virus and develops a very severe case of the flu, which requires hospitalization. In addition to serious symptoms of influenza, Oscar develops a case of bacterial pneumonia and he is placed on a respirator in intensive care. With careful monitoring and supportive care, he recovers, but he receives a stern warning from his doctor to take his anti-HIV medications regularly and without fail.

1. **Why was Oscar's bout of influenza so severe? Why did he suffer from a bacterial infection in addition to the influenza?**
2. **Could an influenza vaccine have prevented illness in Oscar?**

By any measure, the human immunodeficiency virus (HIV) represents a public health crisis of catastrophic proportions. Over 35 million people are believed to be infected worldwide, and in some parts of the world, notably sub-Saharan Africa and Southern and Southeastern Asia, new infection rates remain disturbingly high (**Figure 12.22**).

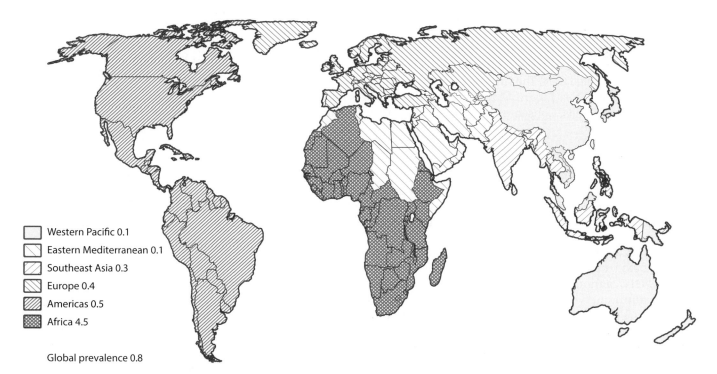

Western Pacific 0.1
Eastern Mediterranean 0.1
Southeast Asia 0.3
Europe 0.4
Americas 0.5
Africa 4.5

Global prevalence 0.8

Figure 12.22 The global AIDS epidemic. Estimated worldwide distribution of AIDS cases as of 2013. Values represent the prevalence (% infected of total population) of HIV infection.

Because acquired immune deficiency syndrome (AIDS) is essentially a disease of the immune system, it is appropriate to consider Oscar's experience and how HIV interferes with an effective immune response. Such a discussion will not only reinforce recently reviewed principles but will also allow us to better understand one of the most important global health issues of our time.

Recall that Oscar's first indication that he had a serious health condition was the pneumonia caused by *Pneumocystis jirovecii*. Almost all cases of pneumocystis pneumonia occur in individuals with severely suppressed immune systems. In fact, before the AIDS epidemic began, the illness was so rare that if a physician in the United States needed to treat it, he or she had to request the drug of choice, pentamidine, from the Centers for Disease Control (CDC) in Atlanta. In a normal year there might be a few requests for the medication. A sudden spike in requests for pentamidine in 1981 was the red flag that alerted health officials that something unusual was going on. Investigations into the mystery revealed that a strange new disease had reared its head, and in June 1981, the CDC's *Morbidity and Mortality Weekly Report* (MMWR) described five previously healthy, homosexual men in Los Angeles who had been diagnosed with pneumocystis pneumonia (**Figure 12.23**). The report raised the possibility of a link between lifestyle and disease, and on the basis of clinical examination of the five patients, it speculated that the ailment might be caused by reduced immune function.

It was soon realized that the condition was not limited to the homosexual population and the new disease was given its current name of acquired immune deficiency syndrome or AIDS. In 1983 the responsible pathogen was discovered, a previously unknown human retrovirus (see Chapter 5, pp. 99–100) named the human immunodeficiency virus, or HIV.

HIV causes immune system suppression

HIV is typically transmitted via sexual intercourse, shared blood products, or passage from mothers to infants. As we learned in Chapter 5, viruses recognize their target cells when proteins on the viral surface are able to bind to specific molecules found on the surface of host cells. In the case of HIV, that host molecule is called CD4. The virus recognizes and binds to CD4 by virtue of its envelope proteins (**Figure 12.24**). And crucially, CD4 is found on the surface of helper T cells. Consequently, helper T cells are targeted for infection by this virus.

Once the virus's envelope proteins bind CD4, a series of events begins that allows the virus to enter the host cell's cytoplasm. The virus then releases its genetic material, and viral replication begins. Newly produced progeny virus leaves the infected cell and the cell is killed. Normally T cells survive for years, but after HIV infection, most infected cells die within a day or two.

Notice in **Figure 12.25** how HIV hamstrings an effective immune response. Neither cytotoxic T cells nor B cells have CD4 on their surface. Consequently, HIV cannot infect them. But as helper T cells decrease in number due to continuing viral replication, it becomes progressively more difficult to activate either B cells or cytotoxic T cells. Although healthy B- and Tc-cell populations persist, eventually neither an effective humoral nor cell-mediated response can be mounted. At this point infected individuals become susceptible to the opportunistic infections that are the hallmark of AIDS (see Chapter 11, p. 254, for a discussion of opportunistic infections).

Morbidity and Mortality Weekly Report (MMWR) 1981;**30**:250-2 (June 5, 1981)

Pneumocystis Pneumonia Los Angeles

Center for Disease Control

In the period October 1980-May 1981, 5 young men, all active homosexuals, were treated for biopsy-confirmed Pneumocystis carinii pneumonia at 3 different hospitals in Los Angeles, California. Two of the patients died. All 5 patients had laboratory-confirmed previous or current cytomegalovirus (CMV) infection and candidal mucosal infection. Case reports of these patients follow.

Figure 12.23 June 5, 1981. The first report of the new disease that would come to be called AIDS.

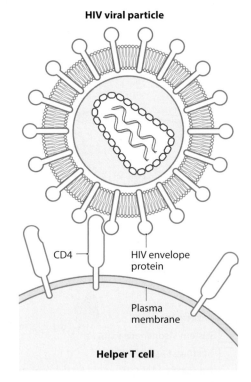

Figure 12.24 Recognition of a helper T cell by HIV. The CD4 protein is bound by the HIV envelope protein, allowing the viral particle to recognize a helper T cell. Following binding, the virion enters the host cell's cytoplasm, where it releases its genetic material and begins replicating (see Chapter 5, pp. 99–100, for a review of the retroviral replication cycle).

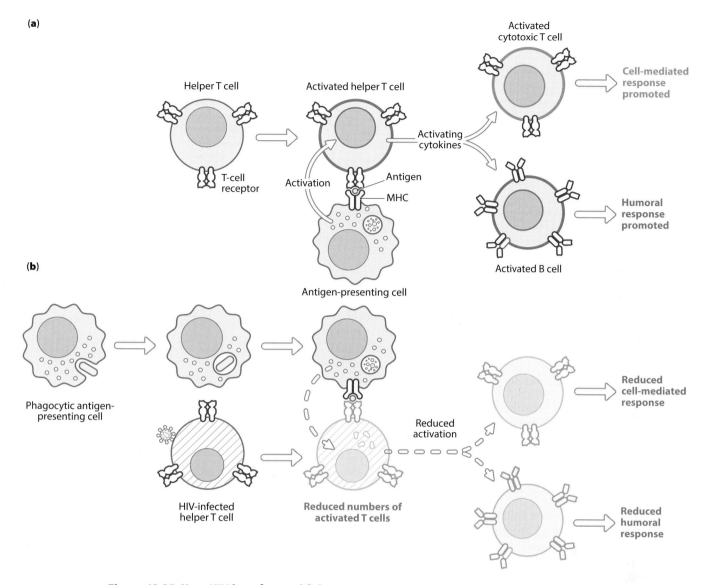

Figure 12.25 How HIV interferes with immune response. (a) In a normal adaptive immune response, helper T cells are activated by antigen-presenting cells. The now-activated, effector helper T cells then activate both cytotoxic T cells and B cells, promoting cell-mediated and humoral responses, respectively. (b) In an HIV-infected individual, helper T cells are reduced in number. This reduction in Th cells results in reduced activation of both cell-mediated and humoral immunity.

The good news is that if HIV-positive individuals take their medications regularly, viral replication is inhibited and the number of helper T cells can rebound. In Oscar's experience, as is often the case, helper T cells recover sufficiently to the point where opportunistic infections become far less common. As we explain in Chapter 13, however, these anti-HIV drugs do not cure the infection; the virus persists at low levels. If the drug regime is not followed closely, viral replication resumes and helper T-cell numbers once again begin to fall. This is what happened to Oscar. Once he became sloppy with his medications, the virus resumed rapid replication and his helper T cells declined in number. When he was exposed to influenza, his adaptive response was therefore less vigorous than normal, and the influenza virus was better able to replicate. Furthermore, in his compromised immune state, and with damage being done to his respiratory system by the rapidly replicating influenza virus, Oscar became susceptible to a bacterial respiratory infection as well. He is lucky to have survived. When people die of influenza, it is often due to such secondary bacterial pneumonia.

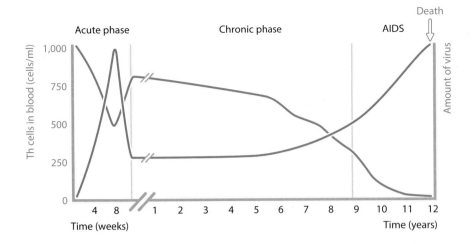

Figure 12.26 Progression of an HIV infection. A normal individual has approximately 1100 helper T cells per milliliter of blood serum. Over the course of an untreated HIV infection, the number of these cells (in blue) rapidly declines during the acute phase, but then rises early in the chronic phase. T cells then begin a long numerical decline. A Th cell count of approximately 200/ml marks the point where a protective adaptive immune response can no longer be initiated and serious opportunistic infections set in. At this point the patient is said to have clinical AIDS. Levels of virus in the blood (in red) rise sharply during the acute phase and then stabilize during the chronic phase. Viral levels slowly climb late in the chronic phase and then rise abruptly, signaling the onset of clinical AIDS. All of these time periods are approximate. The chronic phase can be extended, perhaps indefinitely, with proper treatment.

For HIV-infected individuals who are controlling their virus with proper medications, an influenza vaccine is certainly a good idea. Only the inactivated vaccine would typically be used on an immunocompromised individual. In such a person, even the slow growth of an attenuated virus might be enough to result in disease symptoms. If a person has full-blown AIDS, any vaccine would be of limited value, due to the patient's inability to generate protective immunological memory.

In untreated HIV-infected individuals, the time between infection and severe disease averages approximately 10 years. In the interim there may be few signs of disease, although sometimes there is a short, acute period immediately following infection where the patient may experience fever, rash, and swollen lymph nodes. Note in **Figure 12.26** that this acute phase corresponds to a rapid rise in viral levels and a sharp decline in Th cells.

Why is the period between initial infection and the onset of disease so lengthy in an HIV infection? The initial infection event stimulates a powerful immune response on the part of the host. This response, both humoral and cell-mediated, keeps the virus in check after its initial burst of replication during the acute phase, and the amount of virus in the blood levels off (see Figure 12.26). Although the patient usually shows few signs of illness during this time, the virus is continuing to replicate while the immune response continues to counterattack. As many as a billion new viral particles are produced daily, which infect and destroy helper T cells. New Th cells continue to be produced, replacing those that have been killed, but eventually the immune response cannot keep pace, and the virus gains the upper hand. Helper T cells begin a steady decline, while viral levels surge upwards. Opportunistic pathogens, including *P. jirovecii* and *Candida albicans*, the cause of a fungal overgrowth in the throat called thrush, now gain the opening they require to establish serious infections. A list of some of the more common and medically significant opportunists affecting AIDS patients is provided in **Table 12.2**.

Host versus pathogen: a summary

So, to rephrase the questions we asked earlier, why do we sometimes get sick, and when we do, why do we usually but not always recover? As we have learned in the last two chapters, the answer depends both on how pathogens affect us (our theme in Chapter 11) and how we respond to them (reviewed in this chapter). If a potential pathogen infects us, the likelihood of disease is influenced by interrelated pathogen factors, such as the number of infecting

Table 12.2 Some opportunistic infections seen due to HIV infection. A partial list of some of the more common opportunistic pathogens observed in AIDS patients.	
MICROORGANISM	**OPPORTUNISTIC DISEASE**
Fungal	
Pneumocystis jirovecii	Pneumonia
Candida albicans	Thrush
Bacterial	
Mycobacterium avium	Tuberculosis
Viral	
Varicella-zoster	Shingles
Human herpesvirus type 8	Kaposi's sarcoma
Cytomegalovirus	Inflammation of retina
Protozoan	
Cryptosporidium species	Chronic diarrhea

organisms, the rate at which the pathogen reproduces, and the number and type of virulence factors that the pathogen possesses.

For example, pathogens with powerful virulence factors generally have a lower threshold of disease (**Figure 12.27a**); fewer pathogens are needed before this threshold is crossed and symptoms of disease appear. If a pathogen replicates rapidly and/or the infective dose is large, the pathogen can approach the threshold before an immune response can be generated (**Figure 12.27b**). Alternatively, low virulence, slow growth, and/or a low infective dose reduce the likelihood of reaching the disease threshold.

Opposing the pathogen is the host's immune response. If the response is fast enough or strong enough to keep the pathogen below the disease threshold, the patient will remain healthy. This is what we expect to happen in a memory response. If the pathogens do cause disease, in most cases, once an adaptive immunity is generated, the immune response will generally gain the upper hand, pathogen numbers fall below the disease threshold, and health is restored. Occasionally, however, because of pathogen or host factors, the host cannot contain the infection, pathogen numbers continue to rise, and the host succumbs as a result.

Looking back and looking forward

The immune response relies on highly integrated innate and adaptive mechanisms that communicate extensively with each other to protect us

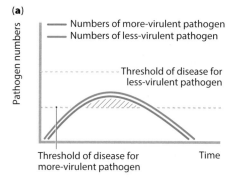

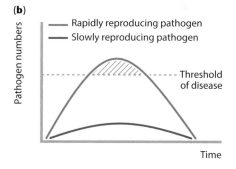

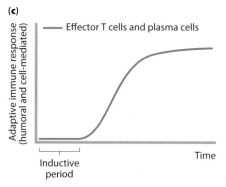

Figure 12.27 The likelihood of disease. All graphs are on the same timescale, and we are making the artificial assumption that the immune response (graph c) is the same for all four scenarios. (a) The more virulent pathogen has a lower threshold of disease. If the immune response to both of these pathogens is the same, the less virulent pathogen does not reach its threshold, because it is contained in time by the immune response. Only the more virulent pathogen crosses its threshold and causes illness. (b) Two pathogens have the same threshold but they reproduce at different rates. The more rapidly reproducing pathogen reaches the threshold before the inductive period is complete. The more slowly reproducing pathogen is contained by the immune response before it can reach the threshold, and therefore does not cause disease.

against invading pathogens. Repeated exposure to the same pathogen can result in even stronger protection. But our defenses are not foolproof. In any host–pathogen encounter three important factors—microbial virulence, microbe numbers and host defense—interact to determine the outcome. When we know something about these factors we can begin to make reasonable predictions about what will happen when pathogen and host meet.

Luckily for us, gone are the days when if someone was infected with a pathogen, the best we could hope for was a successful immune response. Microbial growth can be effectively controlled in many situations and in Chapter 13 we investigate how such control is achieved. This includes the use of antimicrobial medications. Also in Chapter 13, in addition to other means of microbial control, we investigate these miracles of modern medicine, and how they are used and, unfortunately, how they are misused.

Garland Science Learning System

- http://garlandscience.rocketmix.com/students/

- Meet the superheroes of the immune system

- See some immune superheroes in action

- Test your knowledge of this chapter by taking the quiz

- Familiarize yourself with the terminology used in this chapter by using the vocabulary review

- Get help with the answers to the Concept questions

Concept questions

These questions are designed to help you start thinking like a microbiologist. The answers are not always simply found in the text. Instead, you will need to take the concepts about which you have learned and apply them to new situations. Some of the questions may not even have just a single correct answer. Help is provided as part of the GSLS resources, which can be accessed through http://garlandscience.rocketmix.com/students/.

1. The bacterium *Neisseria meningitidis* is often found in the throat of healthy people, but it sometimes penetrates the epithelium in the throat and invades the nervous system, causing meningitis. In North Africa, there are regular meningitis epidemics. These outbreaks almost always occur during the dry season. During the rainy season, although the bacteria are still present in the throat, they are far less likely to penetrate the epithelium and cause disease. Explain this observation, based on what you learned about barriers to entry.

2. Although they make us feel better, medications to reduce inflammation during illness may in some cases actually impede immune function. How?

3. Innate and adaptive immunity are often discussed separately in textbooks as a matter of convenience. Yet this separation is somewhat artificial. Provide two examples of how innate immunity and adaptive immunity depend on each other.

4. People who have had their spleen removed can usually live normal lives, but some, especially children, are placed on prophylactic antibiotics to prevent infection. If they must undergo surgery or even dental work, the dose of the antibiotic is often increased. Based on what you learned about the spleen, what do you think is the rationale for the use of prophylactic drugs?

5. In a famous experiment in immunology that demonstrated the concept of antigen presentation, scientists found that they could stimulate antibody production under laboratory conditions in one of the following situations. In which of the following scenarios do you think antibodies would most likely be produced? Explain your reasoning.

   - Scenario A: Macrophages are exposed to antigen and are then mixed with B and T cells.

- Scenario B: B and T cells are exposed to antigen and are then mixed with macrophages.

- Scenario C: Macrophages are exposed to antigen and then mixed with B cells only.

- Scenario D: Macrophages are exposed to antigen and then mixed with T cells only.

6. Imagine that mice have their thymus removed at birth, before the immature T cells in the bone marrow that have migrated to the thymus can mature. What sorts of immune responses would these mice still be capable of? Which types of responses would no longer be possible?

7. Colds are caused by many different viruses. As we get older, we get fewer and fewer colds. Children get frequent colds, but older adults rarely get colds. Explain this phenomenon based on what you have learned about the immune system.

8. A common metaphor used to describe the immune response is that of fire fighters arriving to put out a fire. The fire might be extinguished, but the water used to put it out can also cause damage. Explain the biology behind this metaphor.

9. As discussed in Chapter 4 (p. 80), protozoa in the genus *Leishmania* are intracellular parasites that cause the disease leishmaniasis. Some forms of this disease cause disfiguring skin lesions. It has been suggested that leishmaniasis might be treated by injecting these lesions with IL-12, a cytokine released by antigen-presenting cells that inhibits humoral and promotes cell-mediated immunity. What is the logic behind this treatment option?

CONTROL OF MICROBIAL GROWTH

A hospital can be a scary place. In the United States, it is estimated that up to 10% of all hospital patients acquire a **nosocomial** (from the Greek for "hospital") infection—an infection that is the result of a hospital stay. An estimated 1.7 million patients a year become infected simply because of their hospitalization. Over 90,000 people a year probably die of nosocomial infections, making such infections a leading cause of death. The story is much the same in countries like France, Italy, and Great Britain.

Why are hospitals such hotbeds of infection? Hospitals house numerous people who are already sick, many of whom have weakened immune systems or damaged barriers to entry. Medical personnel move from patient to patient, providing an easy way for pathogens to spread. Even normal microbiota on the hands of a staff member can prove dangerous if mistakenly introduced into a wound or a surgical incision.

Many nosocomial infections are preventable with basic control measures. Some of these measures are so simple that they are easily overlooked or performed haphazardly. Proper hand washing is probably the single most important precaution that can reduce the transfer of microorganisms between individuals in hospitals and therefore decrease the likelihood of nosocomial infections. Many infections are caused when medical devices such as catheters are not sterilized properly prior to their use. Other problems, however, are more complex. Antibiotic resistance, for example, is an especially thorny issue in the hospital environment, and many infections that could once be easily cured simply no longer respond to the best available drugs.

Yet there is still much that can be done to control microorganisms in the hospital, the home, the laboratory, and elsewhere, and here in Chapter 13 we will investigate many of the strategies, as well as the problems, associated with such control. We will also see that the control of microbial growth does not only mean limiting their growth. In some situations, including the production of many foods and industrial products, microbial growth is something we wish to encourage, and we will also explore how the growth of useful microbes can be facilitated.

But as the example of nosocomial infections should make clear, we rarely if ever truly control microorganisms. The best we can really hope for in our relationship with microbes is coexistence—encouraging the growth of some, while trying to keep the dangerous ones at bay.

Killing microorganisms requires no great insight or cleverness; we do it every day without a moment's thought. We boil, incinerate, or dehydrate microorganisms and we douse them in lethal chemicals as we clean our

> All case studies have a few questions at the end, the answers to which will become apparent as you read the sections following the case.

Table 13.1 Some representative methods for controlling microbial growth.

METHOD	COMMON USES	EFFECTIVENESS
Physical methods		
Oven	Glassware	100%
Autoclave	Glassware, fabrics, liquids, medical instruments	100%
Boiling	Liquids	Kills bacteria and viruses; does not destroy all spores
Flash pasteurization	Beverages	Kills most pathogens
Filtration	Heat-sensitive liquids	Removes bacteria
UV light	Air and surfaces	Kills bacteria with direct exposure
Chemical methods		
Iodine	Antiseptic	Kills bacteria
Alcohol	Skin and surfaces	Removes bacteria and some viruses
Soap	Antiseptic	Removes bacteria
Bleach	Disinfectant	Kills most bacteria and viruses

houses, cook our food, or perform other ordinary tasks. The challenge is to select the right technique for the right situation. High temperature will kill bacteria, but if your goal is to sterilize plastic tubing on a piece of medical equipment, excessive heat may damage the plastic. If you want to be certain that milk is microbe-free, boiling the milk will certainly do the trick, but the resulting milk will not taste very good. When selecting a method to control microbial growth, we need to consider where and what we want to control. Different tasks require different strategies.

Physical control of microorganisms involves manipulation of environmental factors

We can control microorganisms with various physical methods, including heat, cold, filtration, and radiation (Table 13.1). Some of these techniques have been utilized for millennia. Ancient people used salt and drying to preserve food. Romans knew that burning corpses during epidemics helped limit the spread of disease, and the ancient Greeks learned to boil water before drinking it to avoid certain ailments. We will next describe some of these physical methods and consider when their use is appropriate.

Temperature is an effective means to control microbial growth

CASE: AN ARMY MARCHES ON ITS STOMACH

When he first uttered this now-famous quotation, the French emperor Napoleon knew that an army must be well fed to be successful. And in 1795, under his guidance, the French military offered a cash prize of 12,000 francs for a new method to preserve food for the troops. Nicholas Appert, a Parisian chef, finally claimed the award in 1810. Appert's simple technique involved placing food items into glass bottles, leaving room for an air space. He then sealed the bottles with a cork and placed the bottles into boiling water for a prolonged period. Using this technique, Appert successfully preserved soups, jellies, meat, eggs, and various other foods. He then went on to publish *L'Art de Conserver les Substances Animales et Vegetales* (*The Art of Preserving Animal and Vegetable Substances*), the first cookbook of its kind on modern food preservation.

1. **What problems did this method of food preparation have?**

2. **How did Appert's technique differ from pasteurization?**

In Chapter 8 we learned that all microorganisms have preferred temperatures at which they grow best. They also have minimum and maximum growth temperatures, below or above which they cannot grow. Temperature, accordingly, is one of the most direct and inexpensive ways to encourage or inhibit microbial growth. When the goal is to produce large numbers of microorganisms quickly, as it often is in the food industry or in a research laboratory, the optimum growth temperature for the species in question is selected. Recall from Chapter 8 (see Figure 8.38) that microorganisms differ in their temperature preferences and that the optimum temperature is quite different for psychrophiles, mesophiles, and thermophiles. Likewise, if high or low temperatures are to be used to limit or halt microbial growth, the particular temperature limits of the species of interest must be considered.

Heat is an especially effective means of **-cidal** control; the suffix "-cidal" indicates that microorganisms are killed. Bactericidal agents kill bacteria, and fungicidal agents kill fungi. The term "germicidal" (germ killing) is used to describe the general killing of microorganisms. Unlike UV radiation and certain other methods, which kill only those microorganisms living on a directly exposed surface, heat can penetrate an object and kill microorganisms living throughout. Heat kills microorganisms in large part because of its ability to denature proteins (see Chapter 2, pp. 35–36).

Heat is often used to **sterilize** certain objects or materials. Sterilization refers to the complete elimination of all organisms. Either dry or moist heat can be used in this way (see Table 13.1). Students in a microbiology laboratory quickly learn that passing a metal inoculating loop through a flame is a fast and effective way to eliminate any organisms on the loop (**Figure 13.1a**). Likewise, an oven can easily be used to sterilize glassware and other heat-tolerant materials. Dry-heat sterilization, however, usually requires considerable time, especially for heat-resistant organisms or endospores. Moist heat penetrates more quickly and is effective at lower temperatures. In laboratory or medical settings, an **autoclave** is often used to sterilize materials with moist heat (**Figure 13.1b**). An autoclave works on the same principle as a pressure cooker. It consists of a metal container that can be filled with pressurized steam. As the

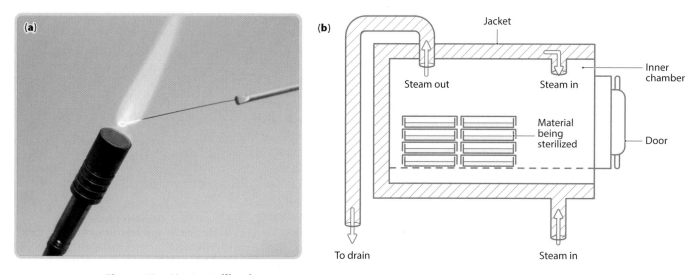

Figure 13.1 Heat sterilization. (a) Sterilization of an inoculating loop. Incineration destroys all microorganisms that may contaminate an inoculating loop. The loop is routinely used in microbiology laboratories to inoculate media. (b) An autoclave. Steam pumped into the inner chamber both heats the chamber and increases its pressure, usually to 121°C and 15 pounds per square inch of pressure. Proper autoclaving completely sterilizes any material in the inner chamber. Following autoclaving (usually after 15 minutes), steam is allowed to escape through an outlet valve.

steam penetrates the objects placed inside the autoclave, microorganisms are killed. Autoclaves are completely effective in sterilizing a wide range of materials, including less heat-resistant objects like towels, which would be destroyed by the higher temperatures necessary for dry-heat sterilization. Autoclaves may also be used to sterilize liquids. Although boiling can also eliminate microorganisms, endospores and some thermophiles can survive prolonged periods of time in boiling water. Autoclaves work better because at high pressure, the temperature in an autoclave rises considerably above the boiling point, even above the temperature necessary to kill endospores. However, when in doubt, boiling is a useful way to make sure certain foods and water are safe. A camper who takes water from a stream and boils it for 15–20 minutes has little to fear from drinking the water.

Some liquids, however, as well as other food products are damaged by exposure to prolonged or excessive heat. This was the principal problem with Appert's preservation methods. Because the boiling process killed most or all of the microbes in the food, they remained safe and edible for longer periods of time, but many of these foods simply didn't taste very good any more.

Orange juice, wine, and other beverages retain their flavor but are still safe to drink if they have been pasteurized. Pasteurized beverages, however, are not microbe-free. **Pasteurization**, the temporary heating of liquids (see Chapter 6, p. 126), destroys pathogens, although some highly heat-tolerant species survive. How does pasteurization target those organisms that are the most worrisome to us?

During flash pasteurization, liquids are generally exposed to a temperature of 72°C for 15 seconds (see Table 13.1). In batch pasteurization, the liquid is subjected to a temperature of 65°C for 30 minutes. Because it causes less flavor change, flash pasteurization is usually preferable. Both of these processes kill mesophiles, which are most likely to be pathogenic. Surviving thermophiles are far less likely to be involved in disease processes, simply because body temperature is too cool for them to thrive. Therefore, even pasteurized milk contains microorganisms, but they are not the ones that pose a threat to our health. They can cause milk to spoil, however, which explains why even milk in an unopened container eventually goes bad. When a longer shelf life for milk is required, ultrapasteurization might be used. In this process, the liquid is exposed to a temperature of 134°C for only 1–2 seconds. Not only does ultrapasteurized milk keep for up to 3 months, it can also be stored without refrigeration—an important consideration in parts of the world where a refrigerator is still considered a luxury.

Cold temperatures alone generally do not kill microorganisms. Rather, when the temperature dips below the minimum growth temperature, microbes cannot reproduce. Such control measures that inhibit microbial growth are indicated by the suffix **-static** (**Figure 13.2**). Food in a refrigerator remains fresh for longer periods of time because the growth of most microorganisms, including most pathogens, is held in check. Yet anyone who has hauled some long-forgotten casserole or piece of cheese out of a corner of the refrigerator, only to find it spoiled, has learned that cold does not eliminate all growth. The organisms able to grow at such low temperatures usually pose little disease risk, but *Listeria monocytogenes* is an exception. This Gram-positive bacterium is able to grow at refrigerator temperatures, especially in dairy products. Most infections are asymptomatic, but occasionally the pathogen can invade the central nervous system, where it can cause meningitis.

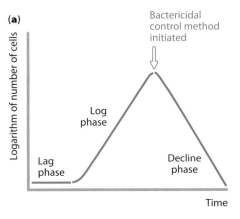

(a)

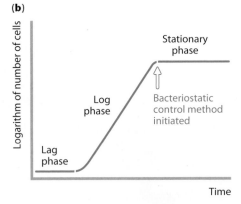

(b)

Figure 13.2 Growth curves for bactericidal and bacteriostatic control methods. (a) Bactericidal control methods kill bacteria and move the growth curve immediately into the decline phase. (b) Bacteriostatic control inhibits further cell replication, moving the curve into the stationary phase. For a review of the microbial growth curve, see Chapter 8, (p. 192).

Liquids that cannot tolerate elevated temperatures may be filtered

If even the sub-boiling temperatures of pasteurization cause problems, filtering may provide an alternative way to render liquids microbe-free (see Table 13.1). For example, a biologist studying a protein in solution might need to keep the solution cold to avoid any change in the protein's three-dimensional structure but may still need to eliminate any bacterial contaminants. The most common filters used in laboratories for heat-sensitive liquids are membranes made of nitrocellulose (a modified form of cellulose) that have a pore size of about 0.45 micrometer (μm). This is small enough to remove bacteria, although some viruses can pass through pores of this size. In fact, viruses were first discovered when it was found that some sort of infectious agent could pass through filters that retained bacteria. Filtering is still used by virologists today to ensure that viral cultures are bacteria-free.

Beverages can also be filtered when pasteurization is thought to adversely affect flavor. This is the rationale, for instance, behind cold-filtered beer. Most beer, like most commercially sold beverages, is pasteurized to ensure its safety.

But if you believe the television advertisements, cold-filtered (or the even more evocative, frost-filtered) beers are especially refreshing because bacteria are removed while keeping the beer at a low temperature. The result is supposed to be superior taste. We'll let the consumer judge whether or not this is really true.

Radiation can be used to kill microorganisms in some situations

In Chapter 7 (p. 162) we saw how radiation can damage genetic material, by causing thymine dimers that inhibit replication, and thus act as a mutagen. This ability to introduce mutations explains how radiation can be used to control microbial growth. Gamma radiation (ionizing radiation) causes water to split into highly reactive molecules such as OH^- and O_2^-. Ultraviolet radiation (non-ionizing radiation) causes the formation of thymine dimers that inhibit proper DNA replication. Both of these forms of radiation have applications in microbial control.

Gamma radiation can be used to kill microorganisms in food products, although this practice remains controversial. It is widely used to sterilize medical equipment, including plastic instruments, vaccine preparations, drugs, and other materials that cannot be filtered or subjected to heat. It may also be used to eliminate any microbial contaminants from tissue intended for transplantation, such as heart valves.

Although sterilization is rarely achieved with ultraviolet (UV) lamps, microbial numbers on surfaces or in the air can be dramatically reduced. Such lamps have value wherever bacterial contamination is a concern. For instance, UV radiation can be used in hospital operating rooms to reduce the likelihood that someone undergoing surgery will be exposed to airborne microorganisms that might otherwise settle into an open surgical incision. Because UV light can also damage our DNA, and because it poses a risk to eyesight, UV lights are used only between surgeries—not when people are actually in the operating room. Other similar situations where UV lamps may be used in this way include nursing homes, child care facilities, prisons, and food preparation areas. UV radiation is also commonly employed in water treatment facilities, where it is used as a means of water purification (**Figure 13.3**).

Figure 13.3 Controlling microorganisms with ultraviolet light. At this water treatment plant ultraviolet light is being used to kill microorganisms in drinking water.

Certain foods are preserved by drying

The next time you eat a piece of beef jerky or a slice of salted fish, you might consider that these products have been preserved with methods that date back to antiquity. A newer but similar process is freeze-drying, a technique that produces the Stroganoff you might take on a backpacking trip.

Native Americans have dried meats for hundreds of years as a way of preserving them (**Figure 13.4**). The practice of using salt to draw water out of foods has an equally ancient pedigree. For fruits, water may be removed with sugar instead of salt. Whether water is removed by simple evaporation or through salt or sugar preservation, the consequences for microorganisms are the same; they cannot reproduce in the absence of water. Few microorganisms are actually killed, however, and should water become available, growth will begin again.

Figure 13.4 Drying: an age-old way of preserving fish and other food products. In Arctic villages, fish is preserved in this manner to provide food throughout the winter.

Freeze-dried foods are likewise preserved by removing water. After being placed in a chamber under a partial vacuum, the food is quickly frozen at temperatures of about $-75°C$ while water is removed by vacuum pressure.

Chemical methods can control microorganisms on living and nonliving material

Many chemical compounds can also be used to control microbial growth (see Table 13.1). Like physical methods, different chemicals are appropriate in different situations. Although bleach, for instance, is bactericidal, you would not use it as a mouthwash or on an open wound to reduce the risk of infection. Agents such as bleach, appropriate for use on nonliving material, are termed **disinfectants**. To control microorganisms on living tissue, less damaging **antiseptics** like iodine and topical sprays such as Bactine® may be used. In this section we will briefly discuss some common disinfectants and antiseptics used to control microorganisms in the environment. In the next section, we will investigate how **chemotherapeutic agents** or anti-microbial drugs are used to combat microorganisms inside the body of a human or animal host.

Various chemicals have antiseptic and disinfectant properties

If you are ever fortunate enough to go skin diving on a tropical coral reef, don't forget to take some iodine along. Coral cuts are common, and because coral is covered with bacteria, an infection is almost certain if the cut is left untreated. Swabbing iodine onto the cut will prevent infection. Or in a pinch, if you forget the iodine, you might follow the lead of the ancient Greeks, who knew to wrap cuts in seaweed, which is rich in iodine. Likewise, campers know that iodine tablets can be added to water to ensure its safety. Iodine acts by binding the amino acids of enzymes and other proteins, inhibiting their activity. It also appears to alter cell membranes by binding fatty acids.

In Chapter 6, we learned how the English physician Joseph Lister first treated wounds and bandages with phenol to control infection in the late 1800s. Phenolic compounds kill microorganisms by interacting with and denaturing proteins. They are not commonly used as antiseptics today, but they remain one of the active ingredients in Listerine®, named in Lister's honor. When the makers of this product advertise that their mouthwash "kills the germs that cause bad breath," the phenolics are the reason why. Furthermore, every time you spray Lysol® on your toilet bowl you are using a similar phenolic compound called *para*-cresol to eliminate bacteria and other microorganisms.

Chlorine gas, compressed into a liquid form, is often used to protect drinking water and swimming pools from potentially harmful microorganisms. Chlorine exerts its germicidal activity by forming an acid when added to water (see Chapter 2, pp. 26–27). Chlorine, in the form of sodium hypochlorite ($NaOCl$), is also the active ingredient in bleach, which can be used as a household disinfectant, or, in small amounts (2–4 drops per liter), to ensure the safety of drinking water.

Alcohol is an effective disinfectant for surfaces, a thermometer, telephone mouthpieces, or other small objects. Alcohols kill microorganisms by denaturing proteins and disrupting lipid bilayers in plasma membranes. They are effective against bacteria, fungi, and enveloped viruses. Nonenveloped viruses and endospores are not destroyed. Alcohol is also used to remove microorganisms from the skin before an injection is given. In spite of what you have seen in movies, where the grizzled old cowboy soaks a bullet wound in whiskey to prevent infection, alcohol is a poor wound antiseptic because it causes host proteins to congeal. Beneath this coagulated layer of proteins, bacteria can continue to reproduce.

Soaps and detergents remove microorganisms from surfaces

CASE: HAND WASHING

In her microbiology laboratory class, Martina has heard her instructor say it over and over; before beginning work, wash your hands carefully to prevent contamination. Martina understands the reasons for this. Normal microbiota is common on the hands. Some of these bacteria, if you work without first removing them, are likely to come off on your cultures. Then, when you perform tests on these cultures, you cannot be sure whether the results you see are caused by the bacteria in your cultures, as you intended, or by contaminating organisms from your hands. Yet Martina wonders if this is really such a problem. To find out, she obtains three agar media plates and performs the following experiment; on plate 1, she presses her right index finger onto the agar, without washing her hands. She then gives her hands a routine washing—the way someone might before sitting down to eat. She inoculates plate 2 with her middle finger. To conclude, she next washes her hands vigorously for 12 minutes, using lots of soap, hot water, and a scrub brush. She then presses her ring finger onto plate 3.

After a 24-hour incubation at 35°C, she examines her plates. There must be some mistake! As expected, plate 3 has few growing bacterial colonies but plate 1 actually has less bacterial growth than plate 2 (**Figure 13.5**). What's going on here?

1. What explains Martina's results?

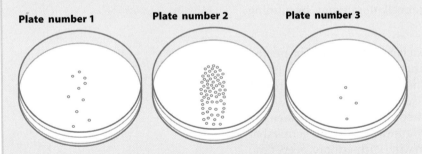

Plate number 1   **Plate number 2**   **Plate number 3**

Figure 13.5 Hand washing and bacterial numbers. Plate numbers correspond to plates described in the case. Each red dot represents one bacterial colony.

What do a bar of soap and laundry detergent have in common? They both get rid of dirt and they both reduce microbial numbers. Furthermore, they both do it in the same way. We have learned that microorganisms often have various structures, such as capsules and fimbriae, to help adhere to surfaces like skin and clothing. Soaps and detergents disrupt this adherence. They do so by inserting themselves between the microorganisms and the surface to which they are adhering. At this point, the microorganisms are washed away with water. Because molecules of the active ingredients of soaps and detergents have both hydrophobic and hydrophilic regions (see Chapter 2, p. 25), they also kill some microorganisms by penetrating their plasma membranes, causing leakage and ultimately cell death.

In light of this information, perhaps Martina's results make sense after all. The normal skin microbiota, adhering tightly to her skin, did not easily come off Martina's dry finger onto plate 1 (see Figure 13.5). The routine washing she performed before pressing her middle finger onto plate 2 was enough to loosen many of the bacteria on her skin, meaning lots of cells were freed up to inoculate the plate. The final, thorough scrubbing she gave her hands detached and washed away almost all of the cells before she inoculated plate 3.

So does this mean that when we briefly wash up before a meal, we are doing more harm than good? Normal washing with soap certainly removes dirt and some bacteria. And remember, these bacteria are part of the normal microbiota. You are already exposed to them all the time, so the risk from them is minimal. But Martina's results do show that a routine wash is not good enough before working in a microbiology laboratory, and it is definitely insufficient in a hospital setting. Especially before performing surgery, a brisk 15-minute scrub with soap containing bactericidal chemicals is required. Without such precautions, there is a serious risk that normal microbiota on the hands could be introduced into surgical wounds, causing an opportunistic infection.

Heavy metals are toxic to cells

Salts of heavy metals latch onto sulfhydryl groups (–SH) on certain amino acids, thereby interfering with proteins and killing microbial cells (**Figure 13.6**). Unfortunately, they do the same to eukaryotic cells, which explains why heavy metal compounds are rarely used today as antiseptics. Before the development of antibiotics, however, silver salts were commonly used to prevent infection, especially in the eyes of newborns where trachoma could result in blindness. Mercuric chloride was once widely used as an antiseptic, but it is no longer employed due to its toxicity.

Antimicrobial chemotherapy has led to major improvements in health

For a real-life horror story, read a little history about microbial disease before the mid-1940s. Pneumonia and bacterial sepsis are still causes for serious concern, but before World War II, infections such as these were downright terrifying. Likewise, there was no such thing as minor surgery. Procedures that are considered fairly routine today could be life-threatening in the pre-antibiotic era. During any invasive procedure, there is a reasonable chance that bacteria will find their way into what should be sterile body regions. Today, prophylactic antibiotics prior to surgery greatly reduce that risk, and if an infection should occur, doctors have an arsenal of powerful drugs to combat it. Before antibiotics became available, besides trying to relieve symptoms and hoping for a successful immune response, there was little that could be done to treat an infection.

There was and is, however, plenty that can be done to prevent infection in the first place. So before we delve into a discussion of antimicrobial drugs, consider **Figure 13.7**, which graphs changes in human survival since the mid-1800s. As we can see, the largest improvement in survivorship occurred prior to the mid-1930s—before antibiotics were even available. During that time, a better understanding of the germ theory of disease was accompanied by the introduction of aseptic techniques in medicine. Likewise, improved sanitation greatly reduced the transmission of food- and water-borne illness. The smaller, but by no means insignificant, increase up through the mid-1950s can be attributed in large part to the development of antibiotics. Note that the more recent breakthroughs in medicine, since about the mid-1980s, have had

Figure 13.6 Effect of heavy metals on bacterial growth. The early-twentieth-century cent and dime were composed mainly of copper and silver, respectively. The clear zones around these two coins are zones of inhibition where the bacteria cannot survive because these metals are present. The metal found in the nickel has little negative effect on the bacterial species in this culture.

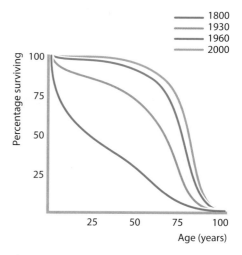

Figure 13.7 Human survival over the last 200 years. The graph depicts the approximate percentage of individuals surviving at any given age at four points in time. The large increase in survivorship between 1800 and 1930 is largely due to improved sanitation. The significant increase between 1930 and 1960 is largely due to the introduction of antibiotics. The increase in survivorship since 1960 has been relatively small.

only a modest impact on average human life expectancy. The contribution of antibiotics has been great, but that of good plumbing has been much greater.

Nevertheless, the development of antimicrobial drugs must be considered among the milestones of scientific achievement, and the history of their development is a fascinating tale of foresight, inquiry, persistence, and simple luck. Even dumb luck, though, is of little value unless the lucky individual has the presence of mind to recognize and act upon the luck when he or she sees it. As the golfer Lee Trevino once said, "The more I practice, the luckier I get." Like golf, the history of drug development is replete with people who got plenty of practice.

An ideal antimicrobial drug inhibits microorganisms without harming the host

In the late 1800s, Paul Ehrlich, a German physician, became intrigued with the way certain dyes used to prepare cells for observation under the microscope stained some cells but not others. When he observed that some of these dyes stained bacterial cells but not animal cells, he had one of those Eureka! moments that scientists live for; the differential staining properties of the cells must mean that the cells were fundamentally different in structure. Consequently, he thought, it might be possible to develop toxic dyes that would bind to and kill bacterial cells without adhering to and harming animal cells. Such a compound might therefore be able to zero in on and destroy bacterial infections, without undue side effects on the host. Ehrlich's search for what he called the "magic bullet" had begun.

Ehrlich's first target was *Treponema pallidum*, the bacterium that causes syphilis. He initially developed an arsenic-based compound that could easily kill *T. pallidum* but was also disappointingly toxic for human cells. He then began to systematically alter the structure of his arsenic compound until finally in 1910, after 605 attempts, he came up with a derivative that was highly effective in treating infected laboratory animals. The drug, which he called salvarsan (from the words "salvation" and "arsenic"), was subsequently used to treat human syphilis patients. Its use, however, was eventually discontinued because its **selective toxicity**—its ability to kill microorganisms without harming human cells—was far from absolute. Salvarsan was still toxic to humans and even occasionally killed patients. It did, however, offer the possibility of a cure in cases that were previously considered hopeless and it demonstrated that selective toxicity was a realistic goal. Pathogens could in theory be eliminated without causing irreversible harm to the human host.

Almost a quarter of a century passed before a truly selectively toxic magic bullet would become a reality. In 1932, another German, the chemist Gerhard Domagk, found that prontosil, a reddish dye, could be used to cure ordinarily fatal bacterial infections in mice. Oddly, prontosil had no effect *in vitro* (under laboratory conditions) when it was added to bacterial cultures. It was effective only *in vivo*, meaning when it was actually administered to living, infected animals. It was later determined that an enzyme in an animal's body breaks the prontosil down into a smaller molecule called **sulfanilamide**, and it is sulfanilamide, not prontosil itself, that has antibacterial activity. In a lab culture, this conversion from prontosil to sulfanilamide did not occur. Sulfanilamide was the first of the so-called sulfa drugs, many of which are still used today to treat a wide variety of bacterial infections.

Sulfa drugs work because they have structures very similar to a crucial bacterial nutrient, *para*-aminobenzoic acid, often abbreviated as PABA (**Figure 13.8**). Bacteria enzymatically convert PABA to folic acid, a

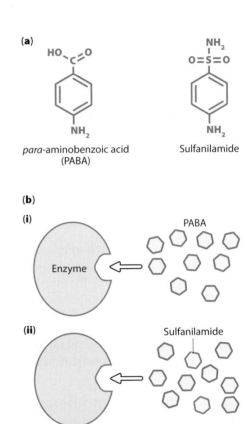

Figure 13.8 The structure and activity of sulfanilamide. (a) Bacteria require *para*-aminobenzoic acid (PABA) to produce folic acid, which is needed for DNA replication. Note the similarity between the structures of PABA and sulfanilamide. (b) As seen in diagram (i), a specific enzyme converts PABA to a different compound, which is ultimately converted into folic acid by a second enzyme. If sulfanilamide is present at high enough concentrations, as seen in diagram (iii), the enzyme may bind sulfanilamide instead of PABA. Because sulfanilamide cannot be converted into the next compound, the bacterial cell ultimately suffers from a lack of folic acid, preventing it from replicating. Although sulfanilamide is present in the middle diagram, (b) (ii), it is not available at high enough concentrations to block all enzyme interaction with PABA, highlighting the importance of proper dosage for this antibacterial drug.

substance essential for the replication of both eukaryotic and prokaryotic cells. But the enzyme that acts on PABA cannot distinguish between PABA and the sulfa compounds. Consequently, if sufficient sulfa is present, the enzyme binds sulfa instead of PABA, and because sulfa cannot be converted to folic acid, the bacteria cannot replicate. Side effects from sulfa drugs are generally limited, because humans lack the targeted enzyme. We cannot convert PABA to folic acid and must obtain all of our folic acid in our diet.

Antibiotics are antimicrobial compounds that were initially produced by microorganisms

Sulfa drugs are synthetic chemicals, and therefore they are not technically **antibiotics**, which are a category of specific antibacterial agents that originally came from natural microbial sources themselves. In Chapter 6, we described the discovery of the first true antibiotic when, in 1928, Alexander Fleming noticed that some of his bacterial cultures had become accidentally contaminated with mold. Recall that a chemical, which Fleming named "penicillin," was being produced by the mold, inhibiting the growth of bacteria in the cultures.

Fleming became frustrated by his inability to purify penicillin, and he did not study it for long. A decade later, however, other British scientists successfully purified penicillin, and in 1941 the antibiotic was first tested on human patients. Penicillin proved to be extremely useful in treating previously untreatable infections, and with World War II wreaking havoc on Europe, production facilities were transferred to North America. Penicillin changed medicine forever. The antibiotic was so effective that it must have seemed to be a true miracle (**Figure 13.9**). Gone were the days when physicians were more or less helpless when confronted with serious, potentially lethal infections.

Figure 13.9 Penicillin: the miracle drug. An advertisement from the mid-1940s for penicillin as a cure for gonorrhea.

Now that penicillin had been developed, the floodgate was opened. Next came streptomycin, an antibiotic isolated from a soil bacterium called *Streptomyces griseus*. With thousands of microorganisms now being screened for useful compounds, the list quickly expanded to include many of the antibiotics we know and use today. The era of antibiotics had begun. Doctors now had an arsenal of drugs from which to choose and could select the most appropriate drug for a given situation. But there can be too much of a good thing. A cavalier attitude developed, and misuse of antibiotics was common. It wasn't long before the first signs of drug resistance appeared, a topic we explore in more detail later in this chapter.

Many factors influence the selection of the most appropriate antibiotic in a given situation

With so many drugs to choose from, how does the physician decide which one to use? Selecting the right drug for the right situation is a complex enterprise, influenced by many factors. Does the patient have any known drug allergies? Has the infectious organism been specifically identified? In which organ or tissue is the infection? These and other factors must be considered before one can decide upon the best antibiotic. Sometimes the deciding factor is not even a medical one. Some drugs, for instance, are more expensive than others, and for someone without health insurance, cost may dictate drug choice. For doctors in undeveloped rural areas where electrical supply is uncertain, it may be necessary to opt for drugs that do not require refrigeration. However, here we will focus on some of the important purely medical criteria for drug selection.

Some drugs work faster than others

The sulfa drugs previously described, while effective, do not really begin to inhibit bacterial growth for up to a few hours. Most bacterial cells have at least some surplus supply of folic acid. As long as this supply remains, the bacteria can continue to divide. After a number of divisions, the supply of folic acid is exhausted. Sulfa drugs then exert their influence by preventing further folic acid from being made. Erythromycin, on the other hand, is an extremely fast-acting antibiotic. By binding to and inhibiting the activity of the bacterial ribosome, erythromycin quickly brings bacterial protein synthesis to a halt. This is an especially attractive feature if the bacteria are producing a protein toxin. Erythromycin stops such toxin production almost immediately. This does not imply, however, that the faster drug is always preferable. For example, the physician might consider whether the patient's condition is rapidly deteriorating. If so, speed is of the essence, and the rapidity with which the drug acts may be of overriding importance. In other situations, where the patient is relatively stable or for chronic infections, speed is much less of an issue. In such cases, doctors have the luxury to select a slower-acting drug if it is superior for other reasons.

Antibiotics are either bactericidal or bacteriostatic

Bactericidal antibiotics kill bacteria. Penicillin, for example, by preventing dividing bacteria from properly synthesizing new cell wall material, causes cells to rupture as they absorb surplus water from the environment. **Bacteriostatic** drugs are more subtle; they merely inhibit bacterial cell division, thereby holding the numbers of microorganisms in check. The host immune system can then gain the upper hand more easily and destroy the bacteria in question. Both sulfa drugs and erythromycin are bacteriostatic.

Generally speaking, a bactericidal drug is preferable because of its more direct impact. Activity of bacteriostatic drugs relies on the continual presence of the drug until the infection is cleared, because if medication is stopped, the bacteria may begin to reproduce again. With bactericidal drugs, an effect is seen even before an immune response begins. Right from the start, bacterial numbers begin to fall, and as the number of bacteria decreases, symptoms in the patient may begin to wane. In immunocompromised individuals, such as cancer patients undergoing chemotherapy or AIDS patients, bactericidal drugs are essential.

In some situations, however, a bacteriostatic drug may be preferable. When Gram-negative bacteria invade normally sterile body regions, for example, as they might after rupture of the appendix, it might be necessary to use bacteriostatic drugs exclusively. As you may remember from Chapter 11 (see pp. 269–270), when Gram-negative bacteria die, they release toxic components of their outer membrane called endotoxins. If such bacteria are killed quickly in large numbers, as might happen if a bactericidal antibiotic is used, large amounts of endotoxin are suddenly released, and the patient may go into shock. It is much safer in such situations to use a bacteriostatic compound. The bacteria are then held in check while the patient's immune response eliminates the bacteria slowly. Because endotoxin is then released slowly, and because it is toxic only in large amounts, the likelihood of shock and other serious complications is greatly reduced.

Drugs differ in their spectra of activity

CASE: A MYSTERY ILLNESS

A warning to English dog owners: You may want to avoid taking your dog for a romp in the woods in late summer and early fall. A new, potentially fatal disease is lurking in the Midlands, Nottinghamshire, and elsewhere. The disease, called seasonal canine illness (SCI), begins between 24 and 72 hours after a visit to forested areas. Symptoms include diarrhea, vomiting, lethargy, and tremors. The cause remains unknown, although certain possible explanations, including man-made poisons, contaminated water, or the consumption of toxic plants have been ruled out. Many affected dogs are infested with harvest mites, leading to speculation that SCI may be caused by a mite-transmitted pathogen of some sort. When it was first reported in 2009, the fatality rate for dogs was about 20%. Currently, with greater awareness among veterinarians and the public, the rate of survival has increased dramatically, even though there is no agreed-upon course of treatment or means of protection (**Figure 13.10**).

The public is advised to not let their dog roam freely in the woods between August and November and to protect their dogs against mite infestation with a topical spray. Treatment of affected dogs relies on rapid rehydration. Therapy with a broad-spectrum antibiotic may be effective.

1. **What is the rationale behind the preventative and treatment measures utilized for dogs with SCI?**
2. **What is a broad-spectrum antibiotic?**

Figure 13.10 Seasonal canine illness. A poster distributed by the Animal Health Trust in Great Britain to help raise awareness of this recently described serious veterinary disease.

With a potentially deadly and mysterious disease on the loose, the beautiful English woodlands of Sherwood Forest, Thetford Forest, and elsewhere are not so appealing these days. And although the cause of the disease remains unknown, the recommendations make good sense, at least until the cause of the disease and how dogs become infected are determined. Let's find out why.

Some antibiotics can be used to combat a wide variety of microorganisms. Such drugs are termed **broad-spectrum** because they are effective against a broad range of bacterial types. Tetracycline, for instance, can be used to treat both Gram-positive and Gram-negative bacterial infections (**Table 13.2**). It is even effective against intracellular bacteria. Penicillin, on the other hand,

Table 13.2 Some representative drugs with their spectrum of activity indicated.		
BROADEST SPECTRUM	Tetracycline	Effective against both Gram-positive and Gram-negative bacteria, including intracellular organisms such as *Chlamydia*
	Sulfa drugs	Effective against Gram-positive and Gram-negative species
	Streptomycin	Effective against Gram-negative and acid-fast bacteria
	Erythromycin	Effective against Gram-positive bacteria and bacteria without cell walls (*Mycoplasma*)
	Penicillins	Original penicillin G is mainly active against Gram-positive bacteria; newer penicillins have broader range and are active against many Gram-negative species
NARROWEST SPECTRUM	Isoniazid	Effective against acid-fast bacteria only

is relatively **narrow-spectrum**, because it is primarily active against Gram-positive bacteria only. In other words, broad-spectrum drugs take a shotgun approach, while narrow-spectrum drugs are more highly targeted to specific types of infections.

In most cases, doctors will opt for a narrow-spectrum drug if possible, because broad-spectrum drugs, by definition, will kill other species besides those causing illness. In Chapter 11, for instance, we learned about the important beneficial roles played by the normal microbiota. Broad-spectrum antibiotics inhibit normal microbiota as well as pathogens, which explains some of the side effects associated with broad-spectrum drugs. Those who have taken such an antibiotic may find that their tongues develop a coated appearance; as microbiota living on the tongue is destroyed, fungal opportunists have the chance to proliferate. Women on a broad-spectrum antibiotic are more likely to suffer from vaginal yeast infections for the same reason. Others taking a broad-spectrum drug may experience intestinal irregularities such as diarrhea because the gut microbiota is disrupted.

In some situations, on the other hand, a broad-spectrum drug is preferable. This is the case when disease is caused by a pathogen that has yet to be identified. If the pathogen responsible for an illness is unknown, a broad-spectrum antibiotic is more likely to prove effective, simply because such a drug is active against a wider range of potential bacterial causes. Consequently, veterinarians in Britain who are treating dogs with SCI frequently employ a broad-spectrum antibiotic, most commonly Baytril®, or a tetracycline-like drug. The mechanism of both of these antibiotics will be discussed shortly. Tetracycline in particular makes sense in light of the possible link to harvest mites. Although such mites are not known to transmit disease in Great Britain, they do transmit pathogens in other parts of the world, including the *Rickettsia* bacteria, for which a tetracycline-like drug is effective. If it is eventually revealed that SCI is *not* caused by a bacterium, then neither of these drugs will have helped. In the meantime, however, their use certainly sounds like a worthwhile intervention.

Broad-spectrum drugs may also be called for when antibiotics are used prophylactically to protect individuals against infections that may occur in the future. This is especially necessary in the case of immunocompromised individuals who may be exposed to any number of potentially opportunistic species.

Drug selection may be based on where in the body an infection has occurred

Some drugs work better in certain parts of the body. For example, streptomycin works well only in parts of the body where oxygen is present. Lung infections are therefore responsive to streptomycin, whereas intestinal infections, infections of the blood, or infections in other relatively anaerobic sites are not. A different drug, metronidazole, which inhibits ATP synthesis, is oxidized to an inactive form if oxygen levels are high. This drug, therefore, is used only to treat infections in relatively anaerobic parts of the body. Nitrofurantoin is used almost exclusively for urinary tract infections. Nitrofurantoin is rapidly filtered by the kidneys and quickly ends up in the urine in high amounts. Consequently, this drug cannot achieve sufficiently high levels in other body regions to be effective. It is ideal, however, for bladder or other urinary tract infections, because this is exactly where the drug becomes concentrated.

A drug's possible side effects are always an important consideration

Even the best antibiotic cannot be used if it causes unacceptable side effects in a patient. We have already mentioned how some antibiotics, especially those that are broad-spectrum, may cause problems by disrupting the normal microbiota. Some individuals are allergic to specific antibiotics, ruling out the use of the allergy-provoking drug, no matter how appropriate it may otherwise be. Finally, for some antibiotics, the goal of specific toxicity is only partly achieved. Such drugs must be used judiciously and only in specific situations. Streptomycin can cause hearing loss, kidney damage, and other problems. For this reason, streptomycin is no longer commonly used.

Drug resistance is a serious and growing problem

In Chapter 7 (see pp. 159–161), we discussed the genetic basis of microbial drug resistance. In Chapter 9 (see pp. 212–213), we considered the situations in which drug resistance is likely to become a significant problem. We will conclude this chapter with a more in-depth look at the problem of drug resistance and what, if anything, can be done about it. For now, it is sufficient to say that resistant microorganisms have greatly undermined our ability to treat many infections and that when resistance crops up, the options available to a physician become limited. In many cases, doctors are forced to use their second- or third-choice antibiotics, because the pathogen in question is resistant to what would ordinarily be the drug of choice. For example, doctors may be forced to use streptomycin (despite its serious side effects) to treat tuberculosis because the patient is infected with strains of bacteria that are resistant to other, more effective antibiotics that cause fewer side effects.

Antibiotics work by interfering with specific bacterial structures or enzymes

Having considered some of the factors involved in antibiotic selection, let's take a look at a few of the basic types of these drugs. A comprehensive description of all antibiotics is beyond the scope of this text. Instead, we will describe how antibiotics work in general and provide representative examples. We will discuss how these examples can be used in light of the criteria for antibiotic selection that we have just reviewed.

To meet the goal of selective toxicity, most antibiotics work by targeting a feature of the microbial cell that is fundamentally different from the host's eukaryotic cells. Sulfa drugs achieve this, for instance, by interfering with an enzyme that is unique to bacteria (see Figure 13.8b). Because humans lack the enzyme, suppressing the enzyme's activity poses no particular risk to our cells. Likewise, in Chapter 3 (pp. 50–51) we described how penicillin acts. Briefly, when bacterial cells divide, penicillin interferes with the synthesis of new pentaglycine bridges between different layers of peptidoglycan in the bacterial cell wall (see Figure 3.13). Without these bridges, the integrity of the cell wall is severely compromised, and the cell is likely to burst as it takes in water from its environment. Animal cells have no peptidoglycan to begin with and consequently are unaffected by penicillin. Bacterial cells offer up a variety of such targets (Figure 13.11).

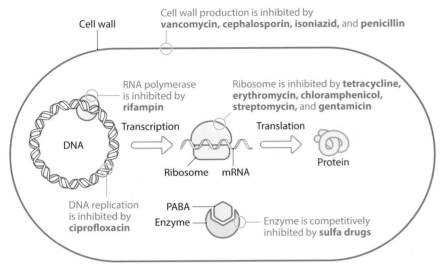

Figure 13.11 Representative targets for antibacterial drug activity. Most antibiotics work by interfering with some cellular process in bacteria that is fundamentally different or absent in eukaryotic cells. Cellular targets include cell wall material, bacterial ribosomes, transcription, DNA replication, and enzymes unique to bacteria.

Many antibiotics inhibit cell wall synthesis

"Penicillin" actually refers to a large class of antibiotics that may be collectively called the penicillins. The first commercially available penicillin, penicillin G, is still with us, but over the years a variety of other, modified penicillins have been developed. All members of this antibiotic class work the same way, but because of slight structural modifications they all have unique characteristics, which might make them more or less useful in specific situations. A few of these compounds are diagrammed in **Figure 13.12**. Notice, for example, that penicillin G is a narrow-spectrum drug, primarily useful against Gram-positive bacteria. It is also acid-sensitive, which means that it must be administered by injection. If taken orally, penicillin G would be destroyed by stomach acid. Bacteria that are resistant to penicillin produce an enzyme called penicillinase that can destroy penicillin. Penicillin G is especially sensitive to penicillinase and is therefore not recommended when drug resistance is a problem.

Compare the properties of penicillin G and those of ampicillin. Modifications in the structure of ampicillin make it more broad-spectrum than penicillin G, and because it is acid stable, it can be taken by mouth. Like penicillin G, it is vulnerable to penicillinase. If drug resistance is a concern but a penicillin-like drug is still called for, a physician may opt to use oxacillin instead. Structural modifications in oxacillin greatly reduce the ability of penicillinase to destroy it. Cephalosporins are a separate class of antibiotics, although their structure and mode of action make them similar to the penicillins. Like the penicillins, these antibiotics are available in many different forms. They are generally active against both Gram-positive and Gram-negative bacteria and are resistant to penicillinase. They are thus useful against bacteria that produce this enzyme, although some bacteria secrete a different enzyme, rendering them resistant to cephalosporin as well.

Although it is structurally unrelated to penicillin and cephalosporin, vancomycin also inhibits cell wall synthesis. Its mechanism of action has already been described (see Chapter 9, pp. 212–213, and Figure 9.21). Vancomycin is a narrow-spectrum antibiotic, highly active against certain Gram-positive

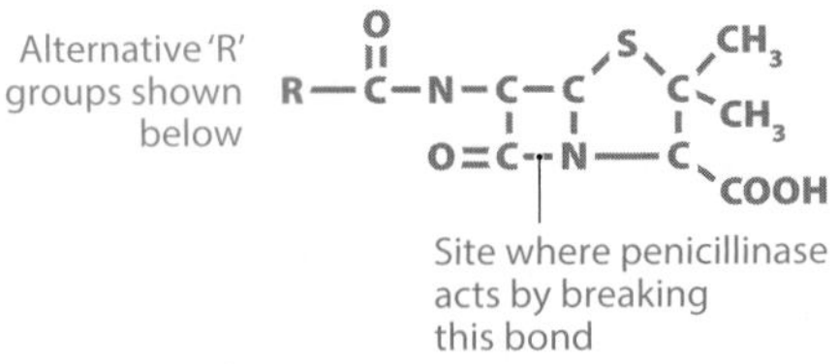

Alternative 'R' groups shown below

Site where penicillinase acts by breaking this bond

Penicillin G

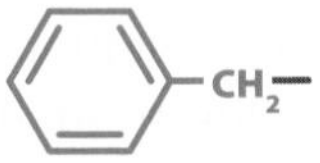

• Highly active against Gram-positive bacteria
• Sensitive to acid
• Sensititve to penicillinase

Ampicillin

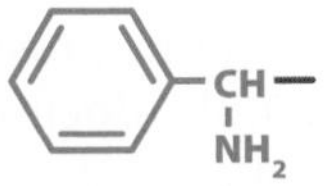

• More active than penicillin G against Gram-negative bacteria
• Acid stable
• Sensitive to penicillinase

Oxacillin

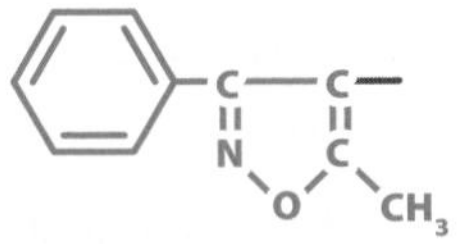

• Acid stable
• Resistant to penicillinase

Cephalosporin

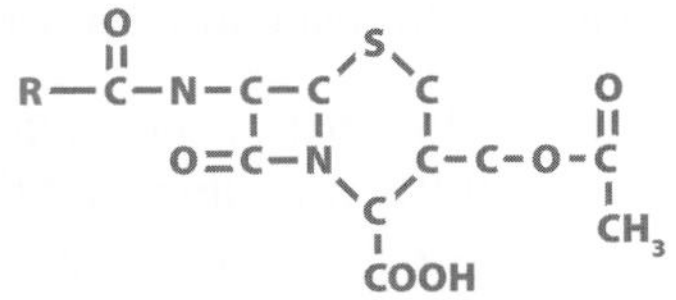

• Similar mode of action to penicillins, but a separate class of antibiotic

Figure 13.12 Representative cell wall inhibitors. Three examples of penicillins are shown. All have the same basic structure but differ in their R group (diagrammed in blue). Different R groups explain the different characteristics of each penicillin. Cephalosporin is an unrelated drug that has a mode of action similar to the penicillins. It too varies at the R group (in blue), and there are many cephalosporins, all with slightly different properties.

species. Because it has toxic properties, vancomycin must be used with care, but it is especially valuable against strains of *Staphylococcus aureus* that are resistant to most of the penicillins. In some cases, vancomycin has become the last option when patients are infected with certain strains that are resistant to all other antibiotics. As described in Chapter 9, this makes the appearance of vancomycin-resistant *Staphylococcus* especially worrisome.

Isoniazid is a very narrow-spectrum cell wall inhibitor that targets *Mycobacterium*, including *Mycobacterium tuberculosis*. As we learned in Chapter 3 (p. 53), *Mycobacterium* has an acid-fast cell wall, rich in a long-chain molecule called mycolic acid. Isoniazid inhibits mycolic acid synthesis. Because non-acid-fast bacteria do not include mycolic acid in their cell walls, this antibiotic has no effect against them.

Antiribosomal drugs inhibit bacterial protein synthesis

Recall from Chapter 3 (p. 57) that bacterial ribosomes are smaller and contain less protein and ribosomal RNA than eukaryotic ribosomes. Many antibiotics achieve selective toxicity because they can interfere with the smaller bacterial but not larger eukaryotic ribosomes. These drugs comprise a diverse group and are not all closely related. They are grouped together here simply as a convenience because they all attack the same bacterial target.

As mentioned earlier, erythromycin is an example of such an antibiotic (**Figure 13.13**). By binding to the bacterial ribosome, it prevents movement of the ribosome along mRNA and thereby blocks translation. Because it cannot penetrate the outer membrane of Gram-negative bacteria, erythromycin is mainly used against Gram-positive species. It is, however, highly effective against *Mycoplasma pneumoniae*, the species that causes atypical (or walking) pneumonia. Penetrating the cell wall is not an issue for this unusual species, because, as you may recall from Chapter 3, *Mycoplasma* have no cell wall. As previously mentioned, erythromycin is a fast-acting bacteriostatic drug, especially useful for quickly shutting down the production of protein toxins. If its use is discontinued prematurely, however, the bacteria may begin to reproduce again, resulting in a return of symptoms.

Tetracyclines, mentioned in our recent case describing SCI, are another group of bacteriostatic antiribosomal drugs that are active against a large array of bacterial species. They work by blocking the binding between transfer RNA and the mRNA–ribosomal complex (see Figure 13.13). Tetracycline is not recommended for children, since it can cause a discoloration of the

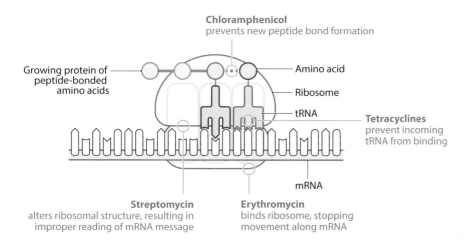

Figure 13.13 Representative antiribosomal drugs. Bacterial ribosomes, because they are smaller and of somewhat different structure than those of eukaryotes, make useful targets for antibiotics. Various antibiotics interfere with the ribosome's ability to translate messenger RNA into proteins, but the manner in which they interfere differs according to the antibiotic. The specific mode of action for several representative antiribosomal drugs is illustrated here.

teeth in younger people. Chloramphenicol is also a broad-spectrum bacteriostatic antiribosomal antibiotic. It works by blocking the formation of peptide bonds between amino acids during translation. Because it does not cause tooth discoloration, it may be used in children.

Streptomycin and gentamicin also inhibit translation, but they are bactericidal rather than bacteriostatic, because they cause irreversible damage to bacterial ribosomes. Specifically, they alter the structure of the ribosome in such a way that codons on the mRNA are misread, and translation occurs incorrectly (see Figure 13.13). These antibiotics are mainly active against Gram-negative bacteria, although streptomycin is sometimes used to treat tuberculosis when the strain of bacteria is resistant to less toxic, first-choice drugs such as isoniazid. Neither streptomycin nor gentamicin can be used to treat infections in anaerobic parts of the body, such as puncture wounds, because they are active only in their oxidized forms. Consequently, they are used only to treat infections that occur in well-oxygenated parts of the body. Because of toxic side effects that can result in hearing loss and kidney damage, these antibiotics must be used with caution.

Some antibiotics target bacterial DNA

The list of drugs in this class of antibiotics is relatively short because bacterial DNA is essentially the same in structure as eukaryotic DNA. Selective toxicity is therefore more difficult to achieve, because anything damaging bacterial DNA might harm host DNA as well. But that does not mean that such selectivity is impossible. Rifampin, for example, binds to and inhibits the activity of bacterial RNA polymerase and consequently blocks transcription. Since it does not bind eukaryotic RNA polymerase, it can be used safely. Its use can be disconcerting, however, because as it is secreted from the host, the urine and even sweat and tears take on a reddish hue. It is especially useful against tuberculosis, because it can penetrate macrophages in which *M. tuberculosis* often resides. Rifampin can also penetrate the connective tissue nodules called tubercles that form around infected macrophages in the lungs and elsewhere in tuberculosis patients.

Experienced travellers to parts of the world where sanitation is poor often know to obtain a prescription for ciprofloxacin before they embark. Ciprofloxacin is active against a wide range of Gram-positive and Gram-negative bacteria and is especially useful for respiratory, digestive tract, or urinary tract infections. Because it is so broad-spectrum, it is often the drug of choice when the organism causing the infection has not been identified. Baytril®, also considered for therapy in cases of seasonal canine illness, is very similar to ciprofloxacin, and is used by veterinarians in similar situations.

Both cyprofloxacin and Baytril® act by binding to and inhibiting a bacterial enzyme necessary for DNA replication (see Figure 13.11). Specifically, this enzyme reduces the coiling of the DNA double helix during DNA replication. When this enzyme is inhibited by the antibiotic, the DNA becomes so badly twisted that further DNA replication is not possible.

Selective toxicity, while possible, is harder to achieve against eukaryotic pathogens

Compared with the antibiotics used against bacteria, a physician in need of a drug to treat a protozoan or fungal infection has relatively limited options. Why is the list of drugs that fight eukaryotic pathogens so short? Simply because they are eukaryotic. Fungal and protozoan cells are so similar to ours that there are far fewer unique targets to attack. Selective toxicity is consequently harder to achieve. Ribosomes, for instance, are the same in human, fungal, and protozoan cells, meaning that using antiribosomal drugs against eukaryotic pathogens is not an option.

The hands of a doctor confronting such a pathogen are not completely tied. For certain fungal infections, for instance, a group of drugs called the polyenes provide a treatment option. Polyenes disrupt the fungal plasma membrane. Fungi incorporate a sterol called ergosterol in their plasma membrane to increase membrane stability and flexibility, much as animal cells use cholesterol (see Chapter 3, p. 61). Polyenes bind egosterol better than they bind cholesterol and therefore can generally be used safely to treat fungal infections, at least when those infections are on the skin. Systemic fungal infections are much harder to control. Polyenes can also damage host cells and therefore are considered toxic. Amphotericin B, for example, is used to treat systemic fungal infections, but it is used only when the situation is considered life-threatening. Because of its toxic properties, it must be administered at relatively low doses for a long period of time. Side effects can include fever, vomiting, headache, anorexia, and low blood pressure. As a result, this antifungal drug has the nickname "Ampho the Terrible." Recently, however, less toxic forms of amphotericin B have been developed. In these newer preparations, the drug is imbedded within lipids, which bind to the fungal cell wall. Because this drug–lipid preparation does not interact well with animal cell membranes, toxic affects to animal cells are reduced. Other drugs, the azoles, interfere with ergosterol synthesis by inactivating a fungal enzyme used to produce this sterol compound. Azoles are typically found in topical creams used to treat conditions such as athlete's foot.

Protozoa, too, while more difficult to treat than bacteria, have a few Achilles' heels that can be exploited in the event of an infection. In fact, the Chinese have known for centuries that malaria could be treated with certain plant extracts, while the Inca in Peru used the bark from cinchona trees for the same purpose. The active ingredient in cinchona turned out to be quinine (**Figure 13.14**). When you see old photos of British colonialists in British East Africa or India, lounging on a veranda with a gin and tonic, now you know part of the reason why this particular cocktail was so popular. The quinine in the tonic may have provided some protection against malaria.

World War II played an unusual role in spurring on the development of new antimalarial drugs. Early in the war, Japanese capture of cinchona plantations in Indonesia created a quinine shortage in the United States. This stimulated a flurry of research, culminating in the development of several new effective drugs, including chloroquine, which is still a workhorse in malaria treatment. Malaria parasites reproduce inside human red blood cells. The hemoglobin in these cells is a crucial nutrient for these parasites, but chloroquine prevents hemoglobin digestion. Unfortunately, resistance to chloroquine is now widespread (**Figure 13.15**). Newer antimalarial drugs like Malarone® target other features of the parasite, such as components of the electron transport chain or specific enzymes. Yet even for these newer drugs, resistance looms as a dangerous threat.

Metronidazole, also known by the trade name Flagyl, is a drug widely used for Amebae, *Giardia*, and other intestinal protozoan infections. Metronidazole inhibits ATP synthesis in anaerobic environments such as the gut. Because it is not well absorbed across the intestinal lining, it is relatively safe to use. If taken with alcohol, however, it can cause vomiting and headache, as well as possible dangerous low blood pressure.

Figure 13.14 Protection against malaria. A quinine tin from the 1860s. Extracted from the cinchona tree, quinine was widely used during the second half of the nineteenth century.

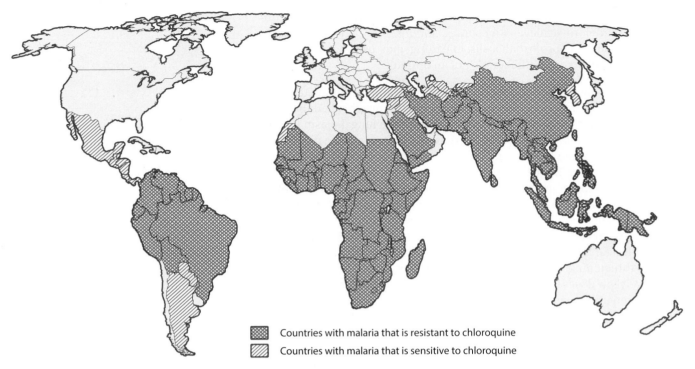

Countries with malaria that is resistant to chloroquine

Countries with malaria that is sensitive to chloroquine

Figure 13.15 Chloroquine resistance. Resistance to chloroquine was first detected in South America and Southeast Asia in the 1950s. Since then, resistance has spread widely across Africa and Southern Asia.

Antiviral drugs must interfere with a particular step in the viral replicative cycle

The development of drugs to combat viral infections has special problems all its own. As we learned in Chapter 5, viral replication is tied intimately to the host. Once an animal virus infects a host cell, the virus's genetic material is released, and the subsequent production of new viral particles depends largely or even exclusively on the machinery of the host cell. Attacking any of these host-cell components means attacking the host cell itself. Furthermore, when viruses are outside the host cell, they are metabolically inactive for the most part. There is little or nothing to attack because the virus is not undergoing replication. The fact that replication takes place only within host cells presents an additional complication; an antiviral drug must ordinarily be able to penetrate host cells without harming them.

Nevertheless, viruses do have unique features, meaning that, at least in theory, selectively toxic antiviral drugs are possible. In recent years, this possibility has begun to be realized. Although the number of antiviral medications is small compared with the number of antibiotics, that number continues to grow, and for at least some types of viral infections, antiviral therapy has become reasonably commonplace. Several steps in the viral replication cycle are vulnerable to attack (**Figure 13.16**).

Acyclovir acts by interfering with viral DNA replication

The first truly effective antiviral drug to become available was acyclovir, which is still commonly used to treat herpesvirus infections. To understand how acyclovir works, remember that a DNA nucleotide is composed of a phosphate, a sugar, and a base (see Chapter 7, pp. 135–136). After the sugar

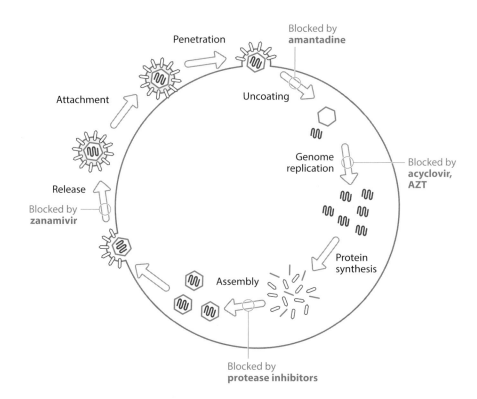

Figure 13.16 The viral replication cycle and some points of intervention. The stages in the cycle where representative antiviral drugs interfere are indicated. Amantadine, used against influenza virus, prevents viral particles from uncoating inside the host cell. Drugs such as acyclovir, used to combat herpesvirus outbreaks, and AZT, an anti-HIV drug, prevent the virus from making copies of its genetic material. Protease inhibitors interfere with HIV at a different point in its replicative cycle—the assembly of new viral particles. Zanamivir, another anti-influenza drug, prevents newly made viral particles from exiting the host cell.

and base are assembled, phosphates are added by an enzyme called thymidine kinase. The now-complete nucleotide can be used in DNA replication by DNA polymerase. When a herpesvirus infects a human cell, it uses sugar–base compounds of the host and converts them to nucleotides by adding phosphates to the combined sugar and base with its own thymidine kinase. The addition of phosphates to another molecule in this way is called **phosphorylation**. The virus then uses these stolen nucleotides to copy its DNA and produce new viral particles.

Now look at the structure of acyclovir (**Figure 13.17a**) and notice that the drug resembles a DNA sugar–base that is missing part of the sugar. The herpesvirus thymidine kinase, however, cannot tell the difference between a genuine sugar and the dummy acyclovir sugar, and it phosphorylates the acyclovir (**Figure 13.17b** and **c**). DNA polymerase now uses this false nucleotide to attempt replication of the viral DNA, but because part of the sugar is missing, DNA replication cannot continue, and the herpesvirus cannot make new copies of itself. Human thymidine kinase is not fooled in this way; it will not phosphorylate acyclovir. Consequently, there is no effect on host cells, and selective toxicity is achieved.

Other viral infections are now amenable to antiviral drug therapy

Flu patients now have antiviral options as well. Amantadine is thought to interfere with the ability of the influenza virus to shed its protein coat following entry into the host cell (see Figure 13.16). The anti-influenza activity of this drug was first noted in 1961, when it was being used to treat Parkinson's disease. Surprisingly, it was observed that these Parkinson's patients were less likely to get the flu. Zanamivir is another anti-influenza drug that works by inhibiting the release of newly replicated viral particles from the infected cell (see Figure 13.16).

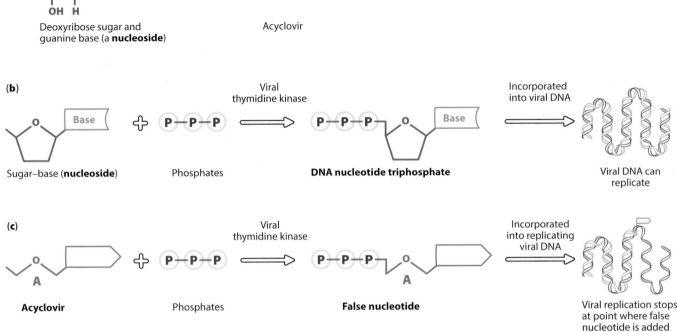

Figure 13.17 How acyclovir stops the herpesvirus. (a) The structure of acyclovir is similar to that of a DNA sugar–base complex with part of the sugar missing. (b) The sugar–base complex must be phosphorylated via the viral enzyme thymidine kinase before it can be incorporated into viral DNA. Initially three phosphates are added. Later, when the nucleotide is incorporated into DNA, two of these phosphates are removed. (c) If acyclovir is phosphorylated instead, this false nucleotide may inappropriately be incorporated into the DNA. At this point, viral DNA replication is terminated.

Although both amantadine and zanamivir can reduce flu symptoms considerably, speed is of the essence in both cases. To make a significant difference in the course of the illness, these drugs must be taken within 24–48 hours of the onset of symptoms. High-risk populations such as the elderly may take these drugs prophylactically to greatly reduce the likelihood of developing the flu in the first place.

Zanamivir is particularly interesting because, unlike most antimicrobials, it is an example of a **designer drug**. The vast majority of antimicrobial drugs used against bacteria, eukaryotes, or viruses were basically found by trial and error. Scientists test thousands of compounds for antimicrobial activity, and when they find something that works, they investigate further. Since the 1990s, however, scientists have started to construct specific molecules that they think will interfere with a microorganism in a precise manner. In the case of zanamivir, the exact three-dimensional structure of a viral protein crucial in the virus's release from an infected cell was first worked out. Scientists then synthesized the antiviral drug atom by atom, resulting in a molecule with the exact shape and characteristics needed to latch onto the viral protein. Not many designer drugs are currently available, but the success of zanamivir demonstrates that such drugs are possible and that other designer drugs are in our future.

The first antiviral drug used to treat HIV-positive patients was zidovudine, also known as AZT. Similar to acyclovir, AZT looks something like, but not exactly the same as, a DNA nucleotide. When the viral enzyme reverse transcriptase converts viral RNA to DNA (see Chapter 5, pp. 99–100), the enzyme may incorporate AZT into the DNA instead of a real DNA nucleotide. If this happens, DNA synthesis cannot continue, and viral replication is halted. Many of the newer **reverse transcriptase inhibitors** work in the same way.

Although reverse transcriptase inhibitors can extend lifespan and reduce symptoms in HIV-infected individuals, the virus rapidly develops resistance to such drugs. Therefore, these drugs offer no long-term solution by themselves. But the treatment of HIV-positive patients changed dramatically in the mid-1990s with the introduction of **protease inhibitors**, the first designer drugs to be successfully developed. They attack the virus at still another step in its replicative cycle (see Figure 13.16). Specifically, protease inhibitors bind to and inhibit a viral enzyme called protease, which cleaves large proteins into the small proteins the virus needs for its assembly (see Chapter 5, p. 97). The use of combination drug cocktails, usually consisting of two reverse transcriptase inhibitors and one protease inhibitor, has literally given patients with HIV a new lease on life. Patients may remain symptom-free for prolonged periods of time, and the likelihood of resistance is considerably reduced. Some experts have speculated that many patients treated with these drug cocktails may survive and remain relatively healthy indefinitely. Certainly many patients have done so since the mid-1990s.

Yet these drug combinations do *not* represent a cure. If the medications are stopped, the virus quickly rebounds. They might mean, however, that HIV is now a manageable disease, somewhat akin to diabetes, heart disease, and other chronic diseases. Only time will tell how effective these and perhaps newer drugs will remain over the long run.

Other drug combinations, however, may sometimes offer cures to previously hard-to-treat viral infections. Until recently, hepatitis C infections were generally treated with a combination of interferon (see Chapter 12, pp. 284–285) and ribavirin, a drug that interferes with viral RNA polymerase. Although cure rates were 70–80%, the flu-like side effects were at best unpleasant for patients. In 2014 a new combination with the trade name of Harvoni® became available. This combination, with a cure rate of about 95%, consists of two drugs, each of which inhibits a specific viral protein crucial in protein synthesis or genome replication, all with greatly reduced side effects.

Newer, innovative antiviral drugs are on the horizon

In the future, some viral infections may be treated with **antisense molecules**. These are very short segments of DNA or RNA that have a base sequence complementary to a crucial viral gene. The antisense molecule binds this complementary nucleic acid and blocks its expression or replication. Experimental antisense molecules are currently being investigated for viral infections caused by West Nile virus, dengue virus, and others. For at least one viral infection, infection of the eye caused by cytomegalovirus, the future is already here. Formivirsen was the first, and is still the only antiviral, antisense drug currently available for clinical use. The 21 nucleotides in the drug bind the viral mRNA, blocking its translation (**Figure 13.18**). By putting the brakes on viral replication in this way, the likelihood of extensive eye damage and even blindness is considerably reduced.

Another exciting, very recent development is an experimental antiviral drug with a name that sounds like a villain in a James Bond movie: DRACO. Many viruses, as they replicate, briefly form double-stranded RNA (dsRNA). When RNA viruses replicate, for instance, the RNA in their genome is transcribed

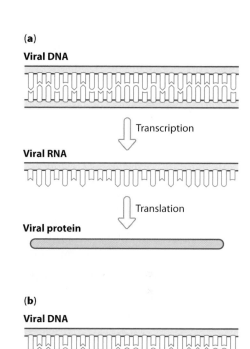

Figure 13.18 The action of formivirsen. (a) For DNA viruses such as cytomegalovirus, viral DNA is used to produce proteins in the same two-step process used by other organisms. (b) Formivirsen is a short segment of RNA that is complementary to the base sequence of the mRNA for a viral gene. This antisense RNA binds the viral mRNA, blocking translation, because the ribosome cannot move past the short section of double-stranded RNA.

into complementary RNA, and while this transcription is taking place, the RNA is briefly double-stranded (See Chapter 5, p. 99). When the dsRNA is formed in the presence of DRACO, the drug latches onto it, and then binds a host enzyme called caspase. This binding activates caspase, which then initiates a series of steps that ultimately cause the infected host cell to die. In other words, DRACO induces the infected cell to commit suicide, which simultaneously nips the viral infection in the bud; the virus cannot replicate in a dead cell. And because uninfected cells lack the dsRNA, DRACO poses no threat to them. DRACO has already been shown to be highly effective against the viruses that cause the flu, colds, polio, and others. This suggests that if it is eventually approved, DRACO might represent that holy grail of antiviral drug research—a broad-spectrum antiviral drug with very limited side effects. Or, as James Bond himself might put it, with a real antiviral "license to kill."

Antibiotic misuse has led to the problem of drug resistance

CASE: THE IMPORTANCE OF COMPLETING PRESCRIPTIONS

Frank, a healthy third-grader, complained to his school nurse of a sore throat. The nurse found that he had a fever and his mother was called. Frank's mother took him to his pediatrician, who saw that Frank had enlarged lymph nodes, a reddish throat, and swollen tonsils covered with grayish-white patches. The doctor took a throat swab and streaked it on a blood agar plate. Test results confirmed her suspicion of a strep throat infection caused by *Streptococcus pyogenes* (see Figure 11.20a). Frank's mother was given a 10-day prescription of ampicillin with instructions to complete the treatment, no matter how Frank felt. Within 72 hours, Frank's sore throat was gone and he felt much better. His mother decided not to give him his remaining medication, saving it to treat either Frank or one of his two sisters the next time one of the children became ill. When the pediatrician heard about this on a later office visit, she strongly voiced her anger.

1. **Why was the pediatrician angry?**
2. **How does the sort of self-medication considered by the mother with the remaining antibiotic contribute to drug resistance?**
3. **Suppose that Frank developed another sore throat soon after the first episode and so did one of his sisters. How might treatment need to be changed?**

Frank was lucky. *S. pyogenes* has so far shown only limited signs of drug resistance. Had he been infected with certain other bacteria, particular species of *Staphylococcus* or some strains of *E. coli* (**Figure 13.19**), for instance, he might have paid a heavy price for his mother's inability to follow the doctor's instructions.

Although it has been a major public health issue only in the last several decades, antibiotic resistance is not new; the first signs began to appear in the late 1940s, soon after the use of antibiotics became common. As an example, a sharp rise in penicillin-resistant *Staphylococcus aureus* strains began in the 1940s soon after the drug's introduction, and resistance has continued to rise since (**Figure 13.20**). We should have seen larger problems looming—after all, the mechanism of resistance was already well understood. But with most bacterial infections suddenly curable, a complacency developed over our ability to treat infectious disease. With the elimination of

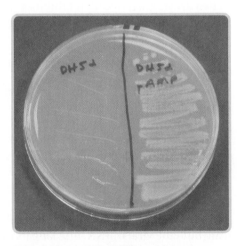

Figure 13.19 Ampicillin-resistant and -susceptible bacteria. The agar used to construct this plate contains ampicillin. A susceptible strain of *E. coli* was streaked on the plate to the left of the blue line. In the presence of the antibiotic, the bacteria failed to survive. To the right of the blue line, resistant *E. coli* display heavy growth, in spite of the ampicillin.

such disease seemingly on the horizon, the warning signs were largely ignored. Today, drug resistance is a frightening reality. The modern medical student must now become familiar with a veritable alphabet soup of acronyms to describe various resistant organisms. These include MRSA (methicillin-resistant *Staphylococcus aureus*, VRSA (vancomycin-resistant *Staphylococcus aureus*) and MDRTB (multi-drug-resistant tuberculosis). A 2014 study commissioned by the British government estimated that about 700,000 lives are lost each year worldwide due to drug resistance. The same study estimated that this number would balloon to 10 million by 2050.

Antibiotic resistance may work in one of several ways. Tetracycline, for instance, as described previously, inhibits bacterial protein synthesis. In resistant cells, however, although tetracycline enters the cell, it is rapidly excreted by a protein pump on the bacterial plasma membrane. Consequently, although the drug enters the cell, it is quickly expelled before it reaches its target. Bacteria resistant to erythromycin employ a different strategy; they have ribosomes with a slight structural modification. This modification prevents erythromycin from binding to the ribosome, meaning that protein synthesis in these cells can continue unimpeded. Resistance to penicillins, as previously stated, is due to the production of a bacterial enzyme called penicillinase that cleaves the drug, rendering it inactive (see Figure 13.12).

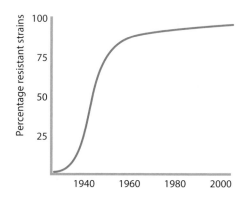

Figure 13.20 Increasing penicillin resistance in *Staphylococcus aureus*. The graph illustrates the proportion of *S. aureus* that developed resistance during the second half of the twentieth century.

Failure to complete a course of antibiotics contributes to resistance

Antibiotic resistance, like other genetically determined characteristics, evolves through the process of natural selection. In Chapter 9 (pp. 212–213) we considered the evolution of vancomycin resistance and learned precisely how such resistance might arise. The development of resistance to other antibiotics occurs in a similar manner.

So, why is it so crucial that patients like Frank complete antibiotic prescriptions? When Frank became infected, the bacteria causing the infection would be expected to vary genetically in terms of how antibiotic-sensitive they were (**Figure 13.21**). Some cells would be killed easily, while some would be somewhat resistant. The majority of the bacteria would be somewhere between these two extremes.

We saw in Chapter 11 (see p. 265) that the ability of a pathogen to cause disease is partly a function of its population within the host. An important reason that Frank felt so ill was that he was infected with a large number of *S. pyogenes* bacteria. When he started his treatment, many of these bacteria were quickly killed, so it wasn't long before he started feeling better (see Figure 13.21). But those that died first were most sensitive to ampicillin. The most resistant were the last to succumb. Within 72 hours the number of bacteria was so low that Frank felt well. By this time all highly sensitive bacteria, and perhaps most of those with intermediate resistance, had been eliminated. Frank was fortunate because this species, *S. pyogenes*, rarely shows high levels of resistance, and even the most resistant members of this species were still relatively easy to kill. However, had he been infected with a more highly resistant microorganism such as *S. aureus*, the most resistant individuals might still be viable after 72 hours, even if overall numbers had been reduced and Frank felt better. Had he stopped taking his medication early in this case, the most resistant, still-living cells might have recommenced reproducing, and very soon Frank might have been ill again. This time, however, all of the bacteria would have increased resistance. If he then infected one of his sisters, she would also have been infected with this

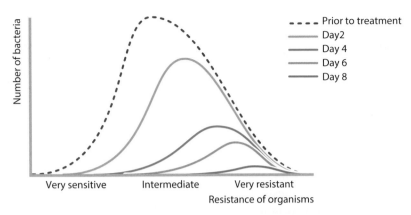

Figure 13.21 Hypothetical changes in bacterial numbers and resistance over a typical 10-day treatment course. Before treatment begins, bacterial numbers are high, causing numerous and possibly severe symptoms. Resistance is mostly intermediate, with only a few very sensitive and very resistant organisms. After 2 days, the patient generally feels much better because bacterial numbers are much lower, but those bacteria that do survive tend to be the more resistant ones. After 3–5 days, the patient may feel well because bacterial numbers continue to be reduced. Once again, however, the few remaining bacteria are increasingly those that have the highest resistance. If treatment is halted prematurely at that time, those now more resistant organisms may once again increase in numbers, causing renewed symptoms. If treatment is continued for the recommended 10 days, all organisms, even the most resistant ones, are destroyed.

more-resistant strain. The pediatrician may then have had to either switch antibiotics and/or increase the dosage of ampicillin to deal with these more-resistant bacteria.

Such misuse of antibiotics is an important reason for the drug resistance that currently confronts us. The 10-day prescription would ordinarily eliminate the entire infection, including the more-resistant bacteria. By stopping treatment prematurely, Frank's mother could have incrementally increased the resistance of the *S. pyogenes* by giving only the most-resistant cells the chance to reproduce. Multiply this by millions of patients not fully complying with their doctor's instructions, and the result is the evolution of highly resistant strains and ineffective antibiotics.

Other forms of antibiotic misuse contribute to resistance

Antibiotic resistance is also promoted when such drugs are used inappropriately to treat nonbacterial conditions. Although Frank's sore throat was caused by bacteria, most sore throats are due to viral infection. Antibiotics, though, have no effect on viruses, because the targets of these drugs are simply not present in viruses. Nevertheless, antibiotics continue to be misused in situations where they do nothing of benefit. A study by the Centers for Disease Control and Prevention (CDC) estimated that in the United States, 30% of all antibiotic prescriptions for ear infections and 50% of those for sore throats are of no value because the illness is nonbacterial. In these situations, although there is no bacterial pathogen, the normal microbiota is exposed to the antibiotic. These harmless bacteria may likewise develop resistance, which they can then pass on to a genuine pathogen by processes such as conjugation, as discussed in Chapter 7.

The use of antibiotics in animal feed also contributes to resistance. In the 1950s, farmers began to routinely add antibiotics to animal feed, because it

Figure 13.22 Antibiotic use in livestock. Cramped conditions promote transmission of infectious organisms. Prophylactic antibiotics in animal feed can protect the health of domestic animals and, incidentally, promote rapid weight gain. Unfortunately, they also promote antibiotic resistance.

was correctly assumed that conditions for transmission of disease were favorable in closely penned livestock (**Figure 13.22**). The use of prophylactic antibiotics would reduce disease, protecting both the animals and the farmer's investment. Unexpectedly, it was also found that antibiotics resulted in faster animal growth. It is thought the growth of such animals is somewhat slowed by intestinal bacteria that produce toxins or excessive gas. Because livestock fed antibiotics this way reached marketable size sooner, the use of antibiotics in animal feed soon became widespread.

Over time, the normal microbiota in these animals develops resistance as well. Humans can become exposed to resistant microorganisms by consuming contaminated dairy products, eggs, or meat. When such food products are undercooked or eaten raw, these bacteria may gain access to human hosts. They may then either cause disease directly, as in the case of *Salmonella*-contaminated chicken, or they may pass their resistance genes on to other bacteria in the human intestine via conjugation, transduction, or transformation (see Chapter 7, pp. 154–157).

Antibiotic resistance can result in superinfections

One of the thorny challenges that has become more prevalent in the age of antibiotic resistance is the problem of superinfections. A **superinfection** is a second infection by a different microorganism that is resistant to the antibiotic used against the first infection. A highly relevant example is provided by *Clostridium difficile*, a spore-forming, anaerobic Gram-positive species that is frequently involved in superinfections. *C. difficile* (often called C. diff) is common in the environment, and because it is a spore-former, it is notoriously difficult to control. Roughly 3–5% of people harbor this species in their intestines, but because it grows so slowly and is outcompeted by the normal microbiota it may reside there indefinitely without causing problems. In the hospital environment, however, there can be problems aplenty. Many patients in hospitals are being treated for various infections, with broad-spectrum antibiotics like ciprofloxacin. These antibiotics, because they are broad-spectrum, can disrupt the normal microbiota. Many strains of *C. difficile*, however, have developed resistance to various antibiotics, including ciprofloxacin. With the disruption of the normal microbiota, *C. difficile* gets the opportunity it needs to proliferate. The various toxins produced by *C. difficile* can then cause diarrhea and a serious, hard-to-treat inflammation of the intestine called colitis.

New strategies provide options for circumventing antibiotic resistance

With the emergence of multi-drug-resistant microorganisms a reality, what does the future of antimicrobial therapy hold? The simplest solution to the problem of resistance is more judicious and selective use of antibiotics. Consider the experience of Hungary, where resistance to penicillin in some *Streptococcus* species rose to 50% in the 1980s because of overuse of the drug. From 1983 to 1992 the use of penicillin in Hungary declined by 50% because doctors found that it was ineffective in too many cases. And as penicillin use went down, so did resistance. By the mid-1990s the proportion of resistant *Streptococcus* dropped to 34%.

When drug use is restricted, the microbial environment is altered. Resistant cells lose their selective advantage and no longer reproduce at an accelerated rate relative to sensitive cells. The example in Hungary demonstrates that at least some resistance can be reversed and previously ineffective drugs can become useful again. The use of antibiotics in animal feed might likewise be restricted, resulting in a reversal of resistance. In the United States, the use of Baytril was banned in poultry flocks in 2005, because of its potential to promote resistance in the closely related drug ciprofloxacin, used in humans. In 2006, the European Union instituted a total ban on the use of antibiotics in animal feed. Nevertheless, lacing animal feed with antibiotics is still a common practice in many places. Another way to decrease the likelihood of resistance is to reduce the duration of treatment when possible. For some infections, such as certain cases of bacterial pneumonia, for example, antibiotic treatments with a short course 5-day regimen have been developed. The shorter the treatment, the more likely patients will comply.

Other appealing strategies to fight resistance might involve looking for new antimicrobial compounds against which resistance is unlikely to develop. For example, although insects do not have a typical vertebrate immune system, they have been successfully dealing with microbial pathogens for more than 500 million years. As a result, insects and other invertebrates have compounds in their blood with powerful antimicrobial properties. Recent studies have found, for instance, that many insects are protected from bacterial infections by blood-borne proteins that perforate the bacterial plasma membrane. Because membrane structure is a fundamental property in all cells, any modifications to it are likely to be detrimental. Consequently, any bacteria that modify their own membrane structure in an attempt to avoid the action of antimicrobial proteins are unlikely to survive. Resistance under such circumstances would not develop.

Vertebrates, too, should be considered as potential sources of new antimicrobial compounds. For instance, a strain of bacteria living in the gut of turkeys, *Staphylococcus epidermidis* strain 115, has been found to produce a compound that serves as a powerful inhibitor of protein synthesis in other competing bacteria. Or, consider alligators. In spite of the fact that they frequently inflict wounds on each other, and that they live in environments teeming with microbes, their wounds rarely become infected. To find out why, researchers isolated 45 proteins from alligator blood and found that eight had powerful antibiotic properties. Finally, when a female koala is not reproducing, her pouch contains a rich bacterial microbiota. When she ovulates, bacterial numbers start to decline, reaching their lowest level just as a baby koala enters the pouch (**Figure 13.23a**). Australian researchers have

Figure 13.23 Possible alternatives to traditional antibiotics. (a) Other animals may produce novel antimicrobial peptides. This baby koala faces a reduced risk of infection, due to antimicrobial compounds secreted into the pouch by its mother. (b) Nobel Prize-winning American author Sinclair Lewis. In his novel *Arrowsmith*, a doctor stops a plague epidemic in the Caribbean using phage therapy.

found that female koalas secrete antimicrobial proteins into their pouch to prevent infection in their babies. Interestingly, these proteins are most effective against aerobic bacteria. This is especially vital for baby koalas, because they are born with an underdeveloped respiratory system and are unusually susceptible to infection by pathogenic aerobes. Such proteins might one day provide the basis for a new narrow-spectrum antibiotic against aerobes in humans.

An intriguing alternative to antibiotics is the use of **phage therapy**. As we learned in Chapter 5, bacteriophages, or phages, are viruses of bacteria. After replicating in a bacterial cell, they cause the cell to rupture as new progeny phages are released. Phages were described in 1917 by Felix d'Hérelle of the Pasteur Institute, who immediately touted their potential as antibacterial agents. Prior to the development of antibiotics, phage therapy was attempted against many diseases, including plague, typhoid, and cholera (see **Figure 13.23b**). Success, however, was spotty, because not enough was understood about the specificity of certain phages for certain bacterial species. With the development of antibiotics, phage therapy was largely forgotten. In light of increasing resistance and a more complete knowledge of phage biology, however, scientists are taking a fresh look at this strategy.

The evolutionary battle between microorganisms and humans is likely to continue for the foreseeable future. But through careful and selective use of antimicrobial drugs, as well as the use of alternative strategies where appropriate, it is a battle that can and must be won. Certainly the alternative, returning to an era where infectious disease was the number one cause of human mortality, is not an acceptable option. New antibiotics will no doubt buy us time, but in the long run our success depends on using drugs with a wisdom that matches our cleverness in creating them.

Looking back and looking forward

The manipulation of microbial growth for specific purposes requires that we understand the basic biology of the microorganisms in which we are interested. With this knowledge, we can attempt to limit the growth of harmful species or encourage growth when such growth is beneficial. Various means of controlling microbial growth are available, but different strategies are appropriate in different situations.

Antimicrobial drugs are especially important for the control of pathogens. Most of these drugs work by attacking microbial features that are missing or sufficiently different in the host. Like other control strategies, there are particular situations in which the use of a specific drug is or is not warranted. Drug misuse leads to the problem of resistance, which severely undermines our ability to treat or prevent certain infections.

The problem of resistance highlights a point made early in Chapter 13; complete microbial control is something of an illusion. We never completely control microorganisms and the best we can realistically hope for is to encourage the growth of useful microbes and to minimize the impact of the harmful ones. New disease outbreaks are a certainty. Predicting such outbreaks, responding to those that occur, or tracking down the source of a new microbial menace are the responsibility of epidemiologists. In Chapter 14 we ride along with these medical detectives as they sift the clues, interview witnesses, and otherwise try to identify the perpetrator of new microbial crimes.

Garland Science Learning System

- http://garlandscience.rocketmix.com/students/

- Learn how magic bullets work

- Test your knowledge of this chapter by taking the quiz

- Familiarize yourself with the terminology used in this chapter by using the vocabulary review

- Get help with the answers to the Concept questions

Concept questions

These questions are designed to help you start thinking like a microbiologist. The answers are not always simply found in the text. Instead, you will need to take the concepts about which you have learned and apply them to new situations. Some of the questions may not even have just a single correct answer. Help is provided as part of the GSLS resources, which can be accessed through http://garlandscience.rocketmix.com/students/.

1. Pick any microorganism you have learned about in this textbook. List one possible scenario in which you would wish to encourage its growth and one in which you would wish to suppress its growth.

2. You have a common household product, which advertises that it is effective in controlling microbial growth. Describe a simple experiment you could perform to determine whether the effects of this product were cidal or static in nature?

3. In 1994, when a mysterious, new illness erupted in the American Southwest, doctors were advised to treat patients with tetracycline or other broad-spectrum antibiotics, even though the cause of disease was unknown. The pathogen turned out to be a new type of hantavirus, meaning the recommended antibiotics were of no value. Nevertheless, the recommendation made sense at the time. Why?

4. A friend of yours who is taking microbiology becomes somewhat obsessive about hand washing after learning about all the microorganisms to which we are regularly exposed. Even before having lunch, he scrubs his hands for several minutes, as he was taught to do in microbiology laboratory. Explain to him why this is not necessary.

5. As we learned here in Chapter 13, many commonly used antimicrobial drugs were derived from compounds that certain microorganisms themselves produce. Why do you think microorganisms produce such compounds?

6. A new antibiotic is found to be very broad-spectrum and fast-acting. It shows a high degree of selective toxicity. Think of a medical situation in which you might choose to use this drug and a different situation in which you probably would not want to use this particular drug.

7. In Chapter 11 (p. 269), we reviewed the case of Robyn, who had a ruptured appendix. After reviewing her case, if you were a physician in charge of her treatment, which antibiotic might you select and why? Are there any that you would avoid?

8. If it were your job to develop a policy for your state regarding the use of antibiotics in animal feed, what issues might you consider if you were interested in reducing the likelihood of drug resistance, as well as the economic well being of farmers and ranchers?

9. You are hired by a large hospital to develop a plan for antibiotic use that will decrease the emergence of resistance. What are the components of your plan?

EPIDEMIOLOGY: WHO, WHAT, WHEN, WHERE, AND WHY?

14

On the morning of January 5, 2015, the California Department of Public Health was notified about a suspected measles case in an 11-year-old girl. Later that same day, four additional suspected cases were reported. Measles is highly infectious and even a single case elicits an immediate response. Usually there are only about 200 or so measles cases in the United States annually. So, a sudden cluster of cases had alarm bells ringing and suggested that all the cases were in some way related. Public health personnel then asked the routine questions. Did the patients know each other? Had they all engaged in the same activity or had they all visited a common location? And it wasn't long before investigators uncovered a vital clue: All five of these initial patients had visited Disneyland in the past several days (**Figure 14.1**). Disneyland, with over 20 million guests annually, is among the most visited tourist attractions in the world. And for an easily transmissible disease like measles, the dense throngs of people in the park provide optimal conditions for efficient transmission.

Things didn't stop there. The girl who was the initial case next flew to Washington State with her family for a few days, before returning to Southern California. As we learned in Chapter 12, air travel can easily promote disease spread, and it was not long before other measles cases started popping up across North America. Ultimately, at least 173 cases were confirmed before the outbreak was declared over in April 2015. Most of these people had not visited Disneyland; they were either secondary cases (they had interacted with someone who was infected at Disneyland) or tertiary cases (those who interacted with secondary cases). Neither the young girl who was initially diagnosed nor many of the other patients had been completely protected with vaccinations. In fact, only one patient had definitely been fully vaccinated with two doses, which typically protects 99% of vaccinated individuals from infection.

For epidemiologists, identifying the source of disease outbreaks such as this one, and tracking their progression, is all in a day's work. **Epidemiology** is the study of disease in populations. Epidemiologists try to understand the factors that influence a disease's frequency and distribution. They try to predict future problems, and they make recommendations that might prevent disease or limit its spread. When something new or unexplained crops up, epidemiologists try to make sense of it. Here in Chapter 14, we take a closer look at the work of these medical detectives in order to gain a better understanding of how disease behaves in human populations.

All case studies have a few questions at the end, the answers to which will become apparent as you read the sections following the case.

Figure 14.1 Measles in the magic kingdom. The so-called "happiest place on Earth" was a little less cheerful in early 2015, when a measles epidemic began in the amusement park. There were close to 200 confirmed cases, making this the largest US outbreak since the early 1990s. The initial confirmed case was presumably infected by a different, unidentified visitor, perhaps from a country where measles is still endemic.

The first modern epidemiological study identified cholera as a water-borne disease

In the mid-1800s, residents in one part of London obtained their drinking water from one of two sources. Some were served by the Southwark and Vauxhall Company, while others relied on water provided by the Lambeth Company. Both companies pumped their water from the River Thames. During this time, cholera was still a serious problem in London, and John Snow, a local physician, suspected that contaminated water could transmit the disease.

In 1853, London was struck by a cholera outbreak and Snow followed the **incidence** of the disease (number of new cases per specified time period) until 1855. When he compared a person's likelihood of contracting cholera with where that person obtained water, an interesting trend emerged. In the first month and a half of the epidemic, about 3% of Southwark and Vauxhall's customers died of cholera. For customers of Lambeth, the death toll was under 0.4%. Why the difference?

Snow investigated where exactly the two companies got their water. He discovered that Southwark and Vauxhall's water came from a pump near Broad Street—an area where untreated sewage entered the Thames (**Figure 14.2**). The Lambeth Company, on the other hand, drew its water from the river upstream of the city, before it could become contaminated with human waste. Snow concluded that water from the Broad Street pump was the source of the epidemic. The science of modern epidemiology was born.

Figure 14.2 The first epidemiological study. (a) John Snow's map of London, drawn about 1854, showing the area around the Broad Street pump and the incidence of cholera. Water pumps are indicated by Xs, and each black spot represents a cholera death. (b) An 1866 drawing depicting "King Cholera" pumping water in London. Thanks to Snow's pioneering work, the mode of transmission for cholera was recognized.

Florence Nightingale found that improved hygiene reduced the likelihood of typhus

Not much later, Florence Nightingale attempted to analyze the factors that put British soldiers at risk for typhus. This bacterial disease, caused by *Rickettsia typhi*, is transmitted by human body lice. Historically, military personnel have been especially at risk, because the lack of regular opportunities for bathing or washing their clothes permits the lice to thrive.

Nightingale carefully compared cases of the disease in the civilian and military populations, looking for factors that seemed to link typhus victims in both groups. Unsanitary conditions seemed to be the principal factor that both civilian and military patients had in common. In 1858 she published her report, and based on her work, the British army instituted reforms that led to dramatically lower rates of typhus among its soldiers.

At the time of these early epidemiologic studies, it was still unclear that microorganisms could cause disease. Even though bacteria had been discovered, the germ theory of disease (see Chapter 6, pp. 126–127) was still more than 20 years away. Nowadays we understand the role played by microorganisms in disease, and this understanding greatly facilitates the work of epidemiologists as they attempt to track down possible causes for disease outbreaks. In addition to investigating the cause and source of unknown illnesses, modern epidemiologists also try to predict epidemics and develop organized responses to them.

An epidemic is a sudden increase in the number of cases of a specific disease

If a particular disease abruptly increases in numbers above what is considered normal, an **epidemic** of that disease may be in progress. "Normal," though, varies from disease to disease. A dozen cases of viral encephalitis in a large city during the summer would be cause for alarm and would probably result in health warnings and cancelled outdoor activities. During the flu season, if a dozen people in the same city contracted influenza, health authorities would breathe a sigh of relief about the surprisingly low number of cases.

Not all epidemics are caused by infectious microorganisms. Environmental toxins, for example, might result in cancer epidemics. Exposure to heavy pollution can cause epidemics of respiratory disease. Here, we focus on infectious disease epidemics—those clearly caused by microorganisms.

A single contaminated site can give rise to a common source epidemic

CASE: FOOD-BORNE MISERY

Since the 1980s *Escherichia coli* O157:H7 has been making headlines as a serious cause of food-borne illness, resulting in bloody diarrhea and kidney damage. Since then, other dangerous, new strains of this species have arisen as well. In 2015, for instance, *E. coli* O26 came to our attention when it caused two outbreaks, both traced to Chipotle Mexican Grill, a popular restaurant chain. In October of that year, the first outbreak began in Washington State and Oregon, and eventually spread to nine other states (**Figure 14.3**). Fifty-five people were affected, 11 of which required hospitalization. The smaller, second outbreak, also linked to Chipotle, affected five individuals in three states.

An investigation demonstrated that those affected had previously eaten at one of the 17 Chipotle restaurants in the week or so prior to the outbreak. The epidemiologic evidence suggested that a common meal item or ingredient served at the restaurants was a likely source of both outbreaks. However, Chipotle serves foods with several ingredients that are mixed and used

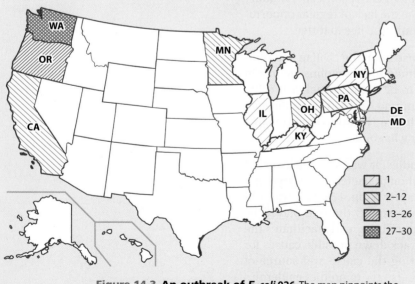

Figure 14.3 An outbreak of *E. coli* O26. The map pinpoints the states affected in the larger of the two outbreaks, which began in mid-October, 2015. The second, smaller outbreak occurred in December. Both outbreaks were traced to a popular Mexican restaurant chain.

together in multiple menu items; this complicated the identification of a specific ingredient, and the outbreaks could not be linked to any single food item. It was assumed, however, that the initial contaminated product was shipped to the restaurants from a single distribution source. Once the source of the outbreaks was determined, the restaurants were closed and all food items discarded. The outbreak subsided, with the last cases reported about 10 days after the restaurants were closed.

1. **What type of epidemic was this?**
2. **_E. coli_ is usually described as a nonpathogenic member of the normal microbiota. Why is this strain pathogenic?**
3. **Why did the number of cases rapidly peak and then quickly decline? What explains the observation that the final cases were reported about 10 days after the restaurants were closed?**

As the name implies, a **common source epidemic** occurs when numerous individuals become infected from the same original source. The cholera epidemic described in the previous section is an example; the common source was the Broad Street pump. A 2013 hepatitis A outbreak, traced to contaminated pomegranate seeds imported from Turkey and used in a health drink is another example, as is the _E. coli_ 026 outbreak in our case. Common source epidemics often occur because of a breakdown in sanitation or due to contamination at a central distribution point for food or water. _E. coli_, including the strain responsible for the Chipotle outbreak, is usually also contracted via food- or water-borne transmission. In this instance, the common source is assumed to have been a food distribution center that shipped a contaminated food item to different Chipotle restaurants.

Various animals, including cattle, can serve as reservoirs for _E. coli_, including the 026 strain. Infection is consequently often associated with contaminated meat. Alternatively, because it is released in the feces of infected animals, soil can become contaminated. If such soil comes in contact with produce, that produce can serve as a source of infection. Unpasteurized apple juice, onions, and bagged spinach have all been identified as the source of pathogenic _E. coli_ outbreaks.

There is no mystery as to why _E. coli_ 026 is pathogenic. Many strains of _E. coli_ live peacefully in our intestines, causing no problem. _E. coli_ 026, as well as the notorious _E. coli_ O157:H7, produce a powerful toxin that is lacking in nonpathogenic strains. The gene for this toxin, called Shiga toxin, is very similar to a toxin-encoding gene found in the bacterium _Shigella enteritis_. It is believed that pathogenic _E. coli_ acquired this gene from _S. enteritis_ via the process of transduction (see Chapter 7, pp. 156–157).

When a common source epidemic begins, there is usually a very rapid rise in the number of individuals affected, with all of them becoming ill within a relatively brief period of time (**Figure 14.4**). Once the source is eliminated, cases will continue to be reported for a time period approximately equal to the duration of one incubation period (the time between exposure to the organism and the first onset of symptoms—see Chapter 11, p. 265). The incubation period for Shiga toxin-producing strains of _E. coli,_ including both the 026 and the O157:H7 strains, is usually about 3 or 4 days, although it can sometimes be as short as 1 day or as long as 10 days. In the Chipotle Mexican Grill outbreak, the number of cases peaked just as the source of infection was discovered. The restaurants were then closed, and the number of cases declined. The last few cases, infected just before restaurant closure, cropped up about 10 days later—a length of time corresponding to the maximum incubation period. When the common source is not identified, cases may be reported indefinitely.

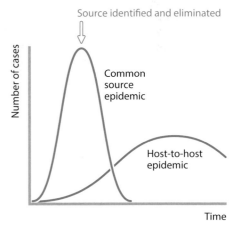

Figure 14.4 Common source and host-to-host epidemics. In common source epidemics, there is a steep rise in the number of cases. Once the source is eliminated, the number of cases declines, reaching zero in a time corresponding to one incubation period of the pathogen in question. In host-to-host epidemics, there is a slow increase in cases, as the disease agent begins to spread through the population, and a more gradual decline. If an epidemic begins, epidemiologists frequently determine from the shape of the curve what type of epidemic is in progress.

Host-to-host epidemics spread from infected to uninfected individuals

Some infectious agents are propagated through a population when they are transmitted from infected individuals to susceptible ones. In such a situation there is no single source. As the agent spreads through the population, there are many potential sources of infection, since every infected individual can serve as a source. This type of epidemic, in which the disease is transmitted from infected to uninfected individuals, is called a **host-to-host epidemic**. Epidemics of this type can start with one infected person, but this individual may infect many other people, each of whom, in turn, can infect many others. A flu epidemic is a common example.

Host-to-host epidemics start out slowly (see Figure 14.4). As the infectious agent spreads and the number of potential sources of infection grow, the epidemic picks up steam until the number of cases reaches a peak. The epidemic then slowly wanes, continuing as long as some susceptible individuals become exposed and infected. When confronted with an unexplained epidemic, epidemiologists frequently look at the shape of the curve shown in Figure 14.4 to determine which type of epidemic they are dealing with.

The reasons that a host-to-host epidemic begins are often complex. In the next section we discuss some of the most important ones.

Epidemics can occur for biological, environmental, and/or social reasons

---

CASE: TRACKING DOWN A FEATHERED FELON

Up through the 1990s there was a curious outbreak of diarrhea in England each May and June. Carol Phillips, a microbiologist from Northampton, decided to investigate. She knew that the diarrhea was being caused by *Campylobacter jejuni*, Gram-negative bacteria commonly found in many animal reservoirs. Humans become infected when they consume contaminated food or water. Intriguingly, Phillips found that most of the late spring/early summer victims had home milk delivery, with milk and other dairy products delivered to their door early each morning.

1. **What was the link between home milk delivery and infection with *C. jejuni*?**
2. **Why was this a particular problem in May and June?**

As previously described, common source epidemics usually occur because of a breakdown of some sort in sanitation that suddenly allows transmission to occur. Host-to-host epidemics, on the other hand, can be due to multiple factors that often interact.

Sometimes, weather is the key. For instance, the periodic warming of the eastern Pacific Ocean called El Niño has been linked to epidemics of malaria in South America and hantavirus in the Southwestern United States. El Niño causes major changes in prevailing weather patterns; some areas experience drought, while others are deluged with rain. In South America, increased precipitation results in lots of standing water, which is ideal for mosquito breeding. With a boom in mosquito populations, the incidence of malaria, which is transmitted by mosquitoes, also rises. For hantavirus, most data suggest that the probability of an epidemic is greatest when populations of deer mice, the reservoir for the virus, are at their peak. Deer mouse populations rise and fall with food supply. When winter and spring are relatively wet, as they are during an El Niño event, there is a bonanza of seeds and other plant foods for the mice, and their population skyrockets. With more deer mice, the frequency of human–mouse interaction rises, resulting in increased disease transmission.

Other, surprising factors sometimes facilitate disease transmission, thereby setting the stage for an epidemic. A hepatitis C epidemic in Japan during the 1960s and 1970s provides a useful example. Hepatitis C is caused by a blood-borne virus that replicates in liver cells. Once the liver becomes infected, there is a high probability that a chronic condition will develop that can result in significant liver damage. Often, however, symptoms of disease do not appear for decades. Consequently, a person showing symptoms of hepatitis C infection may have been infected years earlier.

During World War II, Japanese soldiers were sometimes injected with methamphetamines to "improve their fighting spirit." Unfortunately, the syringes used to administer the injections were often reused without being properly sterilized. And unsterile syringes are ideal vehicles for hepatitis C transmission. At the time, though, no one gave a thought to hepatitis C infection. The result was an unwelcome war souvenir for many Japanese veterans.

Now let's return to the odd incidence of diarrhea in England that began late each spring. Dr Phillips found that most people who became ill had home milk delivery. Furthermore, the delivered milk bottles often had damaged foil caps. She suspected that an animal of some sort, acting as a reservoir, was contaminating the milk early in the morning before the human residents took the milk inside. To find out, she set up video cameras at four homes to record what happened on the doorsteps in the early morning hours. On 13 occasions, she caught magpies and crows in the act, pecking through the foil-capped bottles to get at the milk, something that's also been observed in other birds such as blue tits (**Figure 14.5**). When she screened the 13 pecked bottles for *C. jejuni*, nine were positive for the bacteria. Why May and June? That's when eggs are hatching and Phillips thought the birds drank the milk to regurgitate it back to their babies at their nests. The problem was easily solved by replacing the foil with bird-proof caps. But until their milk was rendered bird-proof, the English, renowned as animal lovers, presumably bore up with their characteristic stiff upper lip, even if it did mean the occasional loose lower bowel.

Just because some factors, such as weather conditions or even thieving animals, make an outbreak more likely, there is no guarantee that an epidemic will occur. One of the most important variables is the immune status of the human population.

Figure 14.5 Caught in the act. Birds in England with a taste for dairy products have learned to obtain milk by peeling the foil caps off milk bottles. The pictured bird is a blue tit. Other avian species, including crows and magpies, were behind the *C. jejuni* outbreaks in England each spring.

Epidemics become more likely when fewer people in a population are resistant

In any population, we typically expect that some people will be susceptible to a particular microorganism, while others will be immune. As we saw in Chapter 12, immunity follows an initial infection, due to the development of immunologic memory. This immunity may last months, years, or even a lifetime. Individuals who have never been exposed to the microorganism in question remain susceptible.

The proportion of individuals in any population that is immune to a specific pathogen is always changing. When people recover from an infection, they are likely to be immune to subsequent infections with the same pathogen

(**Figure 14.6**). Vaccination achieves the same result, without the need for actual disease. Whether individual immunity is generated by infection or vaccination, the population's overall immunity increases. Immunity, however, can wane over time. When it does, the proportion of immune individuals in the population declines. This proportion also decreases as babies are born, or when susceptible newcomers enter the population.

When immunity in a population is sufficiently high, an epidemic becomes unlikely or even impossible, because there are too few susceptible people in the population to sustain a large number of cases. Even susceptible individuals are less likely to get sick when many individuals are immune since immune individuals do not transmit the disease to susceptible people. When a high proportion of the population is immune, there are not many routes of transmission available for the pathogen to reach those few susceptible people (**Figure 14.7**). When few individuals are immune, most people are vulnerable and if they become infected they can transmit the pathogen to other susceptible individuals. Transmission is consequently relatively easy for the pathogen.

The ever-changing immune status of populations explains why so many host-to-host epidemics occur in cycles (**Figure 14.8**). If an epidemic occurs, many people become ill, and as they recover, they become immune. At some point, so much of the population is immune that the number of new cases starts to decline and the epidemic wanes. The proportion of immune individuals will continue to rise as the epidemic subsides, but at a much slower rate, because fewer and fewer people are getting ill and subsequently developing immunity. Once the epidemic has finally ended, immunity in the population is at its peak. Another epidemic is not possible at this point. As time passes, however, and as immunity gradually wanes in previously immune people, or as new, susceptible individuals enter the population, another epidemic is eventually possible. This may take months, years, or decades, depending on the rate at which this decline in the population's immunity occurs. Sporadic cases might occur even when immunity in the

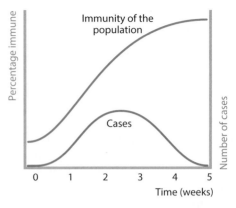

Figure 14.6 Immunity in a population and epidemics. If the proportion of immune individuals in a population to a particular pathogen is low, when that pathogen enters the population, an epidemic is possible because there are so many susceptible potential hosts. If an epidemic begins, as it does in this figure at time zero, immunity in the population starts to rise, as infected people recover and are subsequently immune to a second infection. At some point the proportion of susceptible individuals is too low to permit effective pathogen transmission (approximately week 3 in this example). At that point the number of cases begins to fall and the epidemic wanes. The proportion of immune individuals will continue to slowly rise, reaching its highest level as the last few people become ill at the end of the epidemic (week 5).

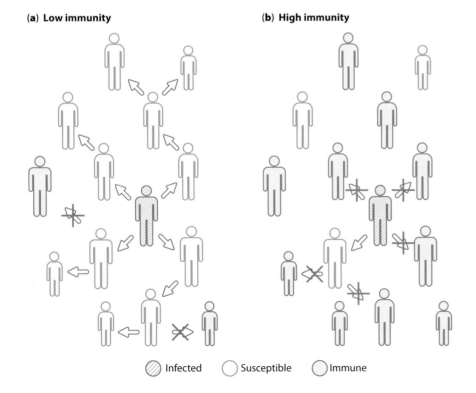

Figure 14.7 Immunity in a population and rate of transmission. (a) If few individuals are immune and an infected individual (in red) enters the population, there are many possible transmission routes to susceptible individuals (in green). Resistant individuals (in blue) are too scarce to block transmission and prevent an epidemic. (b) When most individuals are immune, the few susceptible individuals are protected from transmission by the large number of resistant individuals. Because transmission between infected and susceptible people is unlikely, the number of new cases will be low.

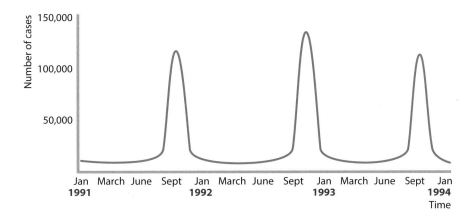

Figure 14.8 Cyclic epidemics of chickenpox. Chickenpox epidemics traditionally occurred at the beginning of the school year in the fall. Before the advent of the chickenpox vaccine, almost all children starting school at age 5 were susceptible. Consequently, each fall a new group of susceptible hosts came together in crowded classrooms, where the chickenpox virus, which is spread via respiratory transmission, could thrive. A single infected child could start a large epidemic. Once an epidemic began, increasing numbers of children developed immunity, resulting in lower rates of transmission. Finally, viral transmission was so low that the epidemic subsided. Because children who were infected became immune to future infections, another epidemic was not possible until another new group of young students started school again the following fall. Since the introduction of the vaccine in 1995, the number of annual chickenpox cases has dramatically declined.

population is high, but not the large number of cases associated with an epidemic.

How high must a population's level of immunity be to prevent an epidemic? The approximate level, known as the **epidemic threshold**, varies for different pathogens. With influenza, for example, it is estimated that an epidemic cannot occur when 90–95% of the population is immune. For polio the value is about 70%. In other words, as long as at least 70% of the population is immune, transmission is too poor to permit a polio epidemic. Different threshold values usually reflect, at least in part, how easily a given pathogen is transmitted to new hosts. Pathogens such as the influenza virus, which spread easily, generally have higher threshold values. **Figure 14.9** traces the timing of epidemic cycles for a hypothetical pathogen, with epidemics possible only when the population's immunity is below the epidemic threshold.

For many diseases, public health officials rely on vaccination programs to maintain immunity above threshold and prevent epidemics. As long as the percentage of vaccinated individuals is maintained above threshold, an epidemic cannot occur. This is an easier task to accomplish for diseases such as polio, with relatively low thresholds. For diseases with high thresholds, it is considerably harder to maintain levels of immunity high enough to prevent outbreaks. The flu, for example, with a threshold of 90–95%, is an expected, if unpleasant, part of life in many parts of the world, and regular epidemics are to be expected. Let's now take a look at the epidemiology of this relatively common viral disease, which in some respects is anything but common.

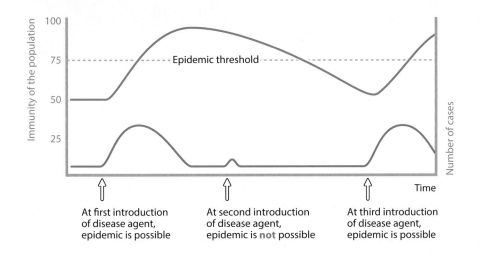

Figure 14.9 Epidemic cycles and immunity in populations. In this hypothetical example the disease in question has an epidemic threshold of 75%. Initially, an epidemic is possible because the proportion of immune individuals is below the threshold. When the disease organism enters the population for the first time, the number of cases and immunity in the population both rise. As the population's immunity rises above the threshold, the number of cases begins to fall. At this point, even if the disease agent is reintroduced, a second epidemic is not possible. Over time, however, a population's immunity will gradually decline because of waning individual immunity or an influx of susceptible individuals. When it eventually falls below the threshold, a second epidemic (third introduction on the graph) is once again possible.

Influenza epidemics are a regular occurrence

CASE: FLU SEASON

One day in February, Sheila, a healthy 60-year-old woman, returns home feeling achy and feverish. After a fitful night, she goes to the doctor and her suspicions are confirmed—she has the flu. She cannot remember the last time she had the flu, but she clearly recalls the winter of 1969 when it seemed like everyone had the flu. An uncle of hers even died.

Sheila has heard that this newest flu strain arose in China, that it is particularly nasty, and that the chances for a flu epidemic this year are high. She is worried about her elderly parents, who live in her neighborhood and whom she sees frequently. Her doctor recommends that they come in right away for a flu shot, even though they had one last year. As for Sheila, the doctor says that she will probably feel better in a week or so but that other than getting plenty of rest and taking pain relievers like aspirin, there is not much that can be done for her.

1. **What determines the likelihood of a flu epidemic in any given year?**
2. **Why are some strains of the flu more dangerous than others? Why was the 1968/1969 strain so severe?**
3. **Why do Sheila's parents need a flu vaccination when they had one the previous year?**

Influenza certainly makes anyone's top 10 list for major epidemic diseases in Europe, North America, and other developed parts of the world. The disease is no newcomer. An epidemic described by Hippocrates from the fifth century BC is thought to have been influenza, and in the last 900 years or so, there are more than 300 records of flu-like epidemics—an average of one epidemic somewhere in the world approximately every 2.4 years. Furthermore, since the early 1700s, there have been 22 recorded influenza **pandemics**. A pandemic literally means a "worldwide epidemic." The term is used more commonly to imply that an epidemic is spread over a very wide geographic area, not necessarily the entire world. There were three influenza pandemics in the twentieth century. Most recently, the World Health Organization declared in June 2009 that swine flu had attained pandemic status.

The most deadly influenza pandemic in recent history was the 1918 Spanish flu, which affected much of the globe and killed an estimated 50 million people worldwide (**Figure 14.10**). In 1957, 70,000 people in the United States alone were killed by the Asian flu. The 1968 pandemic, caused by the Hong Kong flu, killed a million people worldwide, including Sheila's uncle. The World Health Organization estimates that 284,500 people died worldwide as a result of the 2009 pandemic. In years when there is neither an epidemic nor a pandemic, influenza is said to be **endemic**. An endemic disease is one that is always present in a particular geographic area, usually occurring at a low level.

Why is influenza sometimes endemic and sometimes epidemic, and why does it occasionally explode into a bona fide pandemic? To answer this and the other questions presented in Sheila's case, we first need to review a few details of the structure and genetic makeup of the influenza virus.

Influenza is an RNA virus with a segmented genome

Three types of influenza virus affect humans: influenza A, B, and C. They are related viruses, but they differ in a number of details. Influenza A causes the most severe illness, and it is the only one of the three viruses for which animals are essential reservoirs. The use of animal reservoirs makes this

Figure 14.10 The Spanish flu pandemic. The Spanish flu was the most serious pandemic of the twentieth century. (a) Soldiers of the 505th Service Battalion en route to Europe aboard the USS *President Lincoln*. It is believed that the massive movements of troops and refugees during World War I helped to spread the virus. (b) A streetcar conductor in Seattle checking boarding passengers for masks. During the pandemic, although the cause of the disease was unclear, it was understood that transmission occurred via the respiratory route, prompting the widespread use of masks to limit infection.

virus particularly dangerous and it is one of the key reasons that influenza A is the only type of influenza that causes pandemics. Due to its overriding importance in human health, we will confine our present discussion to influenza A.

Influenza virus is an RNA virus; it encodes its genome in RNA rather than DNA (for a complete review of RNA viruses, see Chapter 5, pp. 99–100). The viral RNA genome is actually composed of eight segments, each of which carries different genes (**Figure 14.11**). Each newly produced viral particle must have all eight of these segments if it is to replicate.

Two viral genes, located on different segments of the genome, are of particular importance in understanding both the virulence and epidemiology of influenza. These are the genes that code for hemagglutinin (H) and neuraminidase (N) (see Figure 14.11). Hemagglutinin is the viral envelope glycoprotein, which allows an influenza viral particle to latch onto and then enter host cells in the respiratory tract. Neuraminidase is an enzyme found in the viral envelope, which allows newly formed viral particles to leave a host cell. Consequently, both H and N are essential to the completion of the viral replicative cycle (see Chapter 5, pp. 95–97). They are also the principal antigens recognized by the host immune system; when you are infected with influenza, you produce antibodies against the H and N antigens. Memory for these antigens is protective against a second infection by the same strain. Different strains of influenza, however, differ in their H and N antigens, and immunologic memory against one type of H or N does not guarantee protection against other types. There are 16 known variations of hemagglutinin, all of which cause different immune responses. Likewise, there are nine neuraminidase immunological types. The different forms of these antigens are called H1 through H16 and N1 through N9, respectively. Different influenza strains are designated by their specific type of hemagglutinin and neuraminidase. A strain with H4 and N3, for instance, is designated influenza H4N3. If you are infected with H4N3, you will be resistant to further infections with this strain. You would remain susceptible, however, to different strains such as H5N6. You would have partial immunity to strains H4N6 or H5N3.

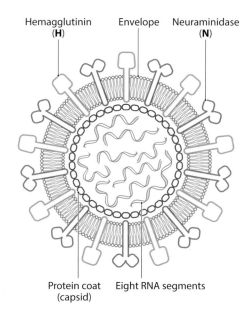

Figure 14.11 Structure of the influenza virus. The eight RNA segments carry the genes used by the virus for synthesizing various proteins. Among those proteins are the envelope glycoproteins hemagglutinin (H) and neuraminidase (N), which are the primary antigens recognized by the immune system.

Influenza virus undergoes regular genetic change

Remember that influenza has a segmented RNA genome and that the genes for H and N are on separate segments. This segmented RNA genome has two important consequences for influenza epidemiology. First, like all RNA viruses, influenza virus mutates rapidly—much faster than DNA viruses, bacteria, or eukaryotes. Organisms that rely on DNA have repair mechanisms that ensure that many mutations are corrected. These repair mechanisms are lacking in RNA viruses. In Chapter 7, for example, we learned that DNA polymerase has proofreading ability—if it makes a mistake, the enzyme can cut out a mismatched nucleotide and replace it. The RNA polymerase used by RNA viruses cannot do this. Consequently, if the enzyme makes mistakes, those mistakes remain in the newly synthesized RNA. With large numbers of mutations constantly being introduced, RNA viruses are constantly changing. Some of these changes might affect proteins that are recognized by a host immune system as antigens. Therefore, if memory is developed against a particular antigen, and if that antigen mutates sufficiently, immunological memory against the original antigen may not be protective against mutated strains of the virus. This phenomenon is known as **antigenic drift**. For influenza virus, changes in both the H and N antigens due to antigenic drift can mean that exposure to one viral strain does not provide protective immunological memory against different strains.

Second, sometimes a cell can be infected by two influenza viruses at the same time. Each of these viruses will replicate all eight of its RNA segments, but when new progeny viruses assemble (see Chapter 5, pp. 97–98), they may get some of the eight segments from one parent virus and some from the other. As long as they obtain all eight segments, the source does not matter and the virus can replicate successfully. This reshuffling of the genetic material can cause dramatic new viral strains to emerge to which few if any humans are immune. This phenomenon is called **antigenic shift**.

For influenza A, antigenic shift is most likely to occur in a duck, pig, or other animal reservoir. When multiple viral strains infect the same animal, the mixing and matching of the RNA segments can begin, occasionally producing new combinations (**Figure 14.12**). Not all of these newly created strains can easily infect humans. Some can be transmitted from the animal reservoir to the human but cannot be transmitted from human to human. Occasionally, however, the new strain can move not only from animal to human but between humans as well. Furthermore, shift can result in new combinations of the H and N antigens, for which a human population has little or no immunity. For example, if a cell in an animal reservoir is simultaneously infected with influenza H2N2 and H8N8, new viral progeny could be produced that are either H2N8 or H8N2. If a population's immunity to this newly created strain is low or nonexistent, the stage is set for an epidemic or even a pandemic.

Because of antigenic drift and shift, different strains of influenza virus circulate from year to year. This explains why Sheila's doctor recommended a new vaccination for her parents, even though they were vaccinated last year. The previous year's vaccination probably offers little or no protection against this year's strain. Because influenza is a moving target, epidemiologists at the World Health Organization continually monitor the virus in an attempt to predict what strain will be in circulation during the coming year. Such predictions are then used to develop that year's vaccine.

Genetic drift and shift set the stage for influenza epidemics and pandemics

The constantly changing nature of the virus also explains in large part why the virus can vary so much from year to year in severity. When a person is infected with influenza, the severity of symptoms is determined by how well the virus is replicating. In individuals with no established immunity, the virus initially replicates quickly, causing more pronounced symptoms. If there is partial immunity to either H or N, there will be less viral replication and consequently milder symptoms. When Sheila read that this year's strain would be particularly severe, it meant that the viral strain would have hemagglutinin and neuraminidase antigens to which there was little or no immunity. Replication in infected hosts would be rapid, resulting in a more pronounced illness.

The 1918 Spanish flu was caused by an H1N1 strain of the virus. The 1957 Asian flu pandemic was caused by an H2N2 strain. In 1969, the Hong Kong flu pandemic that killed Sheila's uncle was caused by an H3N2 strain. The 1968 pandemic killed fewer people than the 1918 or the 1957 pandemic, partly because in 1969 many people still had partial immunity to the N2 antigen, which was present on the 1957 strain, just 11 years earlier.

To illustrate just how unpredictable influenza can be, and how the prediction of epidemics remains in part a guessing game, we need look no further than very recent history. Up through 2008, epidemiologists were concerned

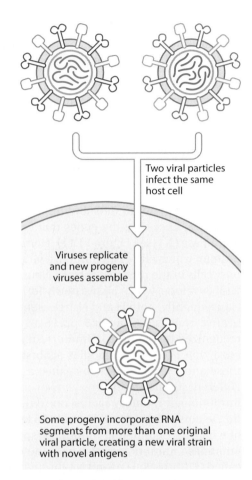

Two viral particles infect the same host cell

Viruses replicate and new progeny viruses assemble

Some progeny incorporate RNA segments from more than one original viral particle, creating a new viral strain with novel antigens

Figure 14.12 Antigenic shift. When influenza viruses infect a cell, they make copies of all eight of their RNA segments. Each new progeny virus must contain all eight segments. Occasionally, when a single cell is co-infected by more than one virus, a progeny virus may obtain RNA segments from more than one original virus. The new hybrid virus has undergone antigenic shift. Immunity in the population to this new viral strain is probably very low.

that the next influenza pandemic might involve the H5N1 strain, known as avian influenza, or bird flu. The first cases of H5N1 infection in humans occurred in 1997, when 18 people in Hong Kong contracted this strain of influenza, previously known to infect only birds. Six of the patients died. A second rash of outbreaks in Southeast Asia began in 2003. Since then, the virus has spread across Asia, Europe, and North Africa, killing millions of birds. The number of human cases has remained relatively low because most transmission to humans has been through contact with infected birds (**Figure 14.13**). Human-to-human transmission, while possible, has been infrequent. Nevertheless, the H5N1 virus remains extremely pathogenic, killing up to 60% of those humans who become infected.

With attention focused on avian influenza, public health officials were caught off guard when a new strain of influenza reared its head in March 2009. The virus, commonly called swine flu, was first detected in Mexico and identified as H1N1 in April 2009. Although H1N1 virus was known from the past, most notably from the Spanish flu pandemic, in some respects this was a completely novel form of influenza, created by genetic shift. It is thought to be a re-assortment of four known strains: one that normally infects humans, one endemic in birds, and two endemic in pigs. The virus is highly transmissible between humans, although it appears no more pathogenic than other familiar influenza strains. In June 2009, as the virus continued to spread around the world, the World Health Organization declared a global pandemic, the first such declaration in half a century. When the pandemic was declared over in August 2010, up to 200 million people worldwide were estimated to have been infected.

When will the next new influenza strain make its debut? That is difficult to predict, but the appearance of novel strains is almost a certainty. A new avian influenza, H7N9, for instance, was reported in China in March 2013. The H7N9 strain is highly virulent, but does not spread well between humans. Whenever a new strain does appear, however, or should a strain like H7N9 suddenly begin to spread well from human-to-human, it will be up to epidemiologists to detect such events. In fact, any time there is a suspicious disease outbreak, epidemiologists swing into action. Next we explain exactly what they do when a medical mystery crops up.

Figure 14.13 Avian influenza reservoirs. A marketplace in Vietnam. Many domestic birds, including chickens and ducks, can serve as reservoirs for H5N1, commonly called bird flu.

Epidemiological investigations may reveal the causes of new disease outbreaks

We opened this chapter with a description of the Disneyland measles outbreak. Although an amusement park might sound to some like an odd place for an epidemic to begin, the pathogen was well known. Once measles virus was identified, epidemiologists knew exactly what they were dealing with, and they could zero in on the source. All patients were carefully questioned, as epidemiologists looked for something that they all had in common. Once it was realized that they had all visited Disneyland, authorities were able to pinpoint the source of the problem and control measures could be implemented. Other epidemics described in this chapter would be similarly investigated.

Yet sometimes, outbreaks are more mysterious and health authorities cannot quickly identify a specific cause. Recall, for instance, the enigma of seasonal canine illness from Chapter 13. In these instances, the problem becomes trickier and the task of epidemiologists is more complex. Nevertheless, when they are confronted with such a situation, epidemiologists rely on a defined protocol, in which clues and evidence are gathered in a step-by-step manner, to help them understand what is happening. We will next see how this is done with an example from the not-too-distant past.

CASE: A NEW BUG ON THE BLOCK

In June 1999, residents of Queens, New York, began reporting that crows with no obvious injuries were dying. On August 23, the city's health department was notified that three elderly patients had symptoms of a neurological condition, suggesting some sort of disease outbreak. By August 31, with six cases now reported, the Centers for Disease Control and Prevention sent epidemiologists to New York to investigate an "encephalitis of unknown origin." Their first guess was that this was a type of insect-transmitted encephalitis virus, most likely St Louis encephalitis (SLE), in part because SLE uses birds as its reservoir and can cause neurological symptoms in humans. New York City mounted a major mosquito abatement program to control the outbreak, using malathion sprayed from helicopters.

Meanwhile, a veterinarian at the Bronx Zoo had doubts that this was SLE. Many exotic birds at the zoo were dying, but their symptoms did not match those of SLE. Furthermore, SLE does not usually cause such massive mortality and most birds remain asymptomatic. On the basis of the symptoms and the high mortality, other diseases, such as Newcastle disease, avian influenza, and eastern equine encephalitis (EEE), seemed more likely candidates. Yet if Newcastle

disease or avian influenza were the cause, the chickens at the children's zoo should have been the first to die. The chickens, however, were all healthy. Furthermore, the zoo's emus were fine, even though emus are highly vulnerable to EEE (**Figure 14.14a**).

On September 9, the zoo's veterinarian sent samples from dead birds to the National Veterinary Services Laboratory in Ames, Iowa. The laboratory found virus in the central nervous system of these birds, and when it took electron microscope images, the viral particles were found to be only 40 nm across—too small for SLE (**Figure 14.14b**). On September 24, the virus was confirmed as West Nile virus (WNV) (**Figure 14.14c**). This was surprising because this virus had never been reported in North America before. On September 30, WNV was isolated from mosquitoes in Queens. By late October, New York City reported 62 cases and seven deaths, and infected mosquitoes, birds, and horses had been found as far away as Connecticut and Maryland.

1. **What exactly did epidemiologists do to track down the cause of this unknown outbreak?**
2. **Where do seemingly new diseases like West Nile virus come from? What factors cause new diseases to appear?**

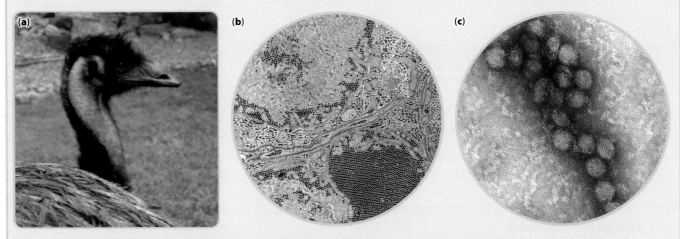

Figure 14.14 Crucial clues in a medical mystery. (a) Healthy emus at the Bronx Zoo helped to rule out Eastern equine encephalitis as a possible cause of an epidemic of unknown origin. (b) St Louis encephalitis virus, first implicated as the epidemic's cause, was the wrong size to be the virus isolated from dead birds from the zoo. The small black spots in the cells' cytoplasm are newly replicated particles. Each of these viral particles is about 50 nm across. (c) Electron micrograph of West Nile virus. Each of these viral particles (seen here as spherical structures in the cytoplasm of an infected cell) is approximately 40 nm across. This is significantly smaller than SLE and this fact helped rule out SLE as the cause of the 1999 epidemic in New York.

A case definition helps health authorities determine if unusual cases are related

When the first cases of encephalitis of unknown origin appeared, health authorities wanted to be certain that the cases were all related. In a city as large as New York, there will always be patients with unusual neurological symptoms in any given week, and they might all be suffering for different reasons. If so, then there would not be any specific epidemic in progress. However, if all patients had developed encephalitis for the same reason, a red flag would be raised.

To find out whether an epidemic was under way, health authorities first developed a **case definition**. A case definition is essentially a list of common symptoms, obtained from doctors who have treated the patients and by questioning the patients themselves. After careful questioning and examination, a case definition for the encephalitis in New York emerged. It included changes in mental status, headache, stiff neck and back, fever, and malaise. With this case definition now in hand, epidemiologists instructed local doctors to be on the lookout for patients with similar symptoms.

Time, place, and personal characteristics of a new disease provide clues to the disease's identity

Now, aware of the problem and armed with case definitions, local doctors in New York began to identify new patients who seemingly suffered from the same condition. As the list of patients grew, it became possible to start defining some of the important epidemiological parameters of the outbreak, each of which might provide clues as to what was happening.

One such parameter is the **time characteristic**—a description of when people are getting sick. In the present example, the first cases were appearing in the late summer and early fall (**Figure 14.15**). What are people doing at this time of year? Are they all going to the same places, eating the same foods, or engaging in the same activities? Or, are they all being bitten by insects? This later possibility, combined with the case definition suggestive of encephalitis, pointed at candidate pathogens such as SLE or EEE, both of which are mosquito-transmitted viruses.

A second important parameter is the **place characteristic**, which describes where affected individuals live, or where they were when they became ill. Do patients live in rural or urban settings? Do they all work in a similar sort of environment? In this example, all early patients lived in Queens, New York, and surrounding areas. As more cases were reported, it became obvious that this particular disease could strike in rural areas as well.

Personal characteristics are also assembled. The personal characteristics portray who is getting sick. Are the sick people mainly of one sex, age, or cultural background? If so, epidemiologists would investigate what it is about a particular group of people that places them at elevated risk, in the

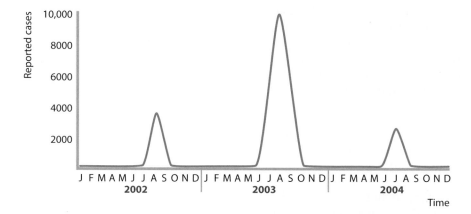

Figure 14.15 Time characteristic for West Nile virus. Numbers of US human West Nile cases for the 2002–2004 seasons. The incidence of West Nile virus mirrors the seasonal abundance of its mosquito vectors, rising sharply in the summer and falling to zero every winter. Similar time characteristics would be seen for other mosquito-vectored diseases such as Eastern equine and St Louis encephalitis.

hope that this will help them uncover the underlying cause of the disease. The first WNV patients were all elderly men and women. We now know that younger people are equally likely to become infected with WNV, but in younger individuals, encephalitis is much less likely to develop. Most people infected with WNV remain asymptomatic or develop a much milder flu-like illness. Because so many of these mild or asymptomatic cases went unreported, it is likely that during those few summer months in 1999, many more people than the 62 reported cases were actually infected.

As the number of encephalitis patients grew, time, place, and personal characteristics of the epidemic became increasingly clarified, and epidemiologists could begin to concentrate on possible causes. Certainly the case definition, along with the fact that cases were concentrated in the late summer and early fall, pointed to a viral form of encephalitis transmitted by mosquitoes. Other arthropods such as ticks are also more common in the summer, but since most of the early cases were urban rather than rural, mosquito transmission seemed far more likely. SLE was proposed as a likely cause, but other questions had to be answered before a definite diagnosis could be nailed down. First of all, epidemiologists wanted to know if this disease could be transmitted from infected to susceptible humans. If so, care givers, family members, and others who spent time with affected individuals would be more likely to become sick themselves, compared to people who had no contact with patients. Careful investigation determined that those who spent time with patients were not at elevated risk. This indicated that the disease agent could not pass from human to human and further pointed to SLE, which can infect humans only through the bite of an infected mosquito.

Case–control studies can pinpoint a common risk factor among affected individuals

Once time, place, and personal characteristics are defined, the next step is to develop a **case–control study**. A case–control study is used to determine what factor or factors link the affected individuals and distinguish them from unaffected individuals. In such a study, the affected people are the cases. Each case is matched with a control—an individual who has not become ill. The matched case and control should be as similar as possible in age, sex, place of residence, and other factors that investigators consider important. The only significant difference should be whether or not they are affected by the disease. Epidemiologists then look for any other difference between the cases and controls that could explain why the cases got sick and the controls did not. It turned out that those who developed encephalitis were far more likely to have spent time outside in the evening than those who remained healthy. Because the mosquitoes that transmit encephalitis viruses tend to feed at night, epidemiologists were further assured that they might be on the right track.

Case-control studies often provide the valuable missing clue that helps epidemiologists crack a difficult mystery. Before anyone understood what Lyme disease was, for instance, the case–control study showed that those who had become ill were far more likely to own household pets. This led investigators to believe that they were dealing with something brought into the home by dogs or cats, perhaps a pathogen transmitted by fleas or ticks. Eventually a newly discovered tick-borne bacterium called *Borrelia burgdorferi* was identified as the causative agent. In the early investigation of AIDS, the case-control study suggested a sexual mode of transmission by revealing that the main factor separating the cases from the control group was having multiple sexual partners. When hantavirus first struck the American Southwest, the case–control study also helped elucidate the disease source. Most of those affected had recently cleaned out buildings such as garages and storage sheds. This knowledge helped investigators zero in on rodent droppings, which are common in such structures.

Although the case–control study in the early West Nile virus epidemic strongly indicated an encephalitis virus such as SLE, as we saw in our description of the early epidemic, the final identification of the virus did not go smoothly. As it turns out, it was the astute veterinarian at the Bronx Zoo who finally set the wheels in proper motion, when she suspected that the cause of the epidemic was perhaps something new. Only after she sent samples to the veterinary diagnostic laboratory in Iowa was it clear that this was not SLE after all. Fortunately, because WNV is also mosquito-borne, control measures instituted for what was thought to be SLE were appropriate for WNV as well. This episode nicely illustrates the importance of communication and the involvement of all interested parties when we are confronted by a new health threat. Veterinarians are not routinely consulted when there is an outbreak of human illness. In this case, their participation was essential. And it is worth remembering that most emerging diseases are zoonoses. The recognition that animal and human health are often part of the same larger picture has led to the **one health** movement—the collaborative effort of multiple disciplines to attain optimal health for people, animals, and the environment. The concept that both human and animal health are mutually dependent, and that both are impacted by environmental quality is increasingly integrated into public policy by agencies such as the European Union, the CDC, and the World Bank.

West Nile virus is just one of many new diseases that have sent epidemiologists and other health authorities scrambling in recent decades. Since 1970, diseases such as Lyme disease, Ebola hemorrhagic fever, acquired immune deficiency syndrome (AIDS), and severe acute respiratory syndrome (SARS) have taken us by surprise. In each instance, epidemiologists proceeded through the same series of steps until the nature of the new epidemic was revealed. Next, we discuss where new diseases come from in the first place.

Emerging and re-emerging diseases are new or changing diseases that are increasing in importance

New diseases often seemingly come out of nowhere on an almost annual basis (**Figure 14.16**), and there are no doubt newer ones lurking in our future. On the other hand, sometimes an old microbial foe, thought to be vanquished, comes back to haunt us. We next consider where these diseases, called **emerging** or **re-emerging** diseases, come from and why they crop up with such alarming frequency.

Repeatedly throughout this text we have emphasized that diseases can change as conditions change. Pathogens may become more or less deadly as transmission becomes easier or more difficult. As the environment is altered, reservoirs might become more or less common. Even changes in human behavior can inadvertently alter the disease equation, making some outbreaks more or less likely.

These changes sometimes cause new diseases to appear, or they may cause the resurgence of diseases that were thought to be controlled. Such emerging diseases receive considerable media attention, but they are not strictly a modern problem. Syphilis was an emerging disease of sixteenth-century Europe, thought to have been transported from the Americas by returning explorers. As we discussed in Chapter 6, yellow fever emerged as a major problem during the first attempt to construct the Panama Canal in the late

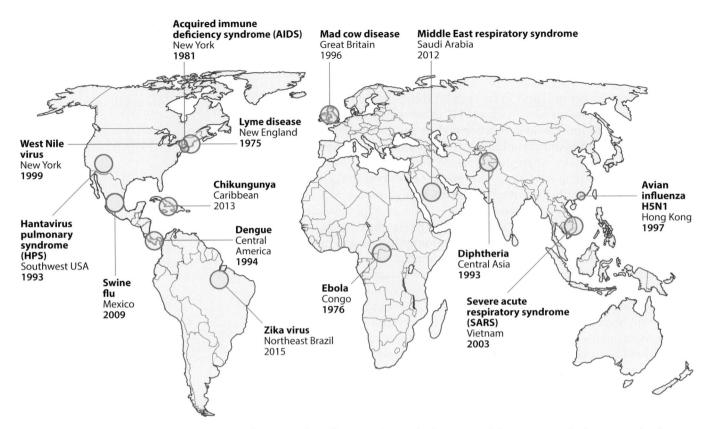

Figure 14.16 Representative emerging diseases. Geographic locations and dates represent the first report of each disease as a new and emerging or re-emerging threat.

1800s. During the mid-twentieth century, polio, which had been present as an endemic disease for thousands of years, roared to prominence as a serious emerging disease of epidemic proportions (**Figure 14.17**).

At first, the reason for polio's emergence sounds counterintuitive. Polio is transmitted through contaminated food and water. Before improved sanitation became common, most people were exposed to polio as very young children. Although infected children may develop a mild intestinal illness, they rarely develop paralysis, by far the most serious consequence of polio. Newborns are protected by maternal antibodies and there is some indication that the virus has a more difficult time invading the nervous system of infants. In many parts of the world, however, sanitation vastly improved during the twentieth century. Consequently, the number of people exposed to polio as infants declined dramatically. If these individuals contracted the virus later in life, paralysis was a more likely consequence. Thus, while polio was historically endemic throughout the world, epidemic paralytic polio was largely the result of improved sanitation.

Increasing human population and urbanization can also set the stage for the emergence of disease. Even changes in travel and trade influence the distribution of diseases around the world. And we don't have to look very hard to provide excellent examples.

Zika virus illustrates several important points about disease emergence

Zika virus was first isolated from monkeys in Uganda in 1947. Over the next 60 years there were only 14 cases diagnosed in humans. In 2007, an outbreak

(a)

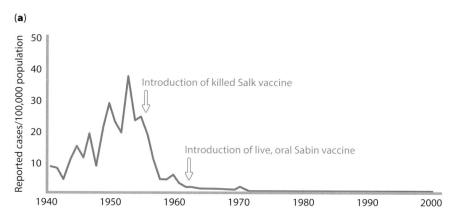

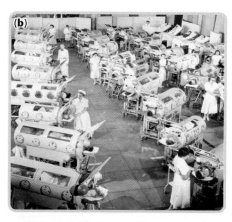

Figure 14.17 Paralytic polio in the twentieth century. (a) Incidence of paralytic polio in the United States from 1940–2000. Following a peak in the early 1950s, the number of cases plummeted with the introduction of the Salk vaccine in 1955. Today, thanks to vaccination, polio has been eliminated from much of the world. (b) Children with paralytic polio breathing with iron lungs during the height of the polio epidemic.

of Zika virus occurred on the Pacific Ocean island of Yap. Those affected suffered only mild illness, associated with fever and a pinkish rash. In 2013 another even larger outbreak occurred in French Polynesia, where there were over 8000 suspected cases.

It turns out that those outbreaks were a harbinger of worse to come. In 2014, doctors in northeast Brazil began to see illness in which patients presented with a pinkish rash, bloodshot eyes, fever, joint pain, and headaches. These cases were initially diagnosed as dengue, for which symptoms are similar. Diagnostic testing, however, indicated that most of these patients were negative for dengue. By March 2015, the illness had been reported from three Brazilian states, and in May of that year, the diagnosis of Zika virus was confirmed. This is the first report in the Americas for what was originally an African virus. By mid-2016 the virus had spread throughout large portions of Latin America and the Caribbean (**Figure 14.18**) with an estimated 1.5 million infected individuals in Brazil alone. Most of those infected remain asymptomatic or, as in previous Zika outbreaks, suffer only minor symptoms. Yet in this most recent epidemic, Zika has taken a sinister new turn; if a woman becomes infected while pregnant, the virus can be vertically transmitted to her fetus. Here it can cause serious impairment of neurological development, resulting in the tragic condition called microcephaly (**Figure 14.19**). Since the start of the epidemic, approximately 3500 babies have been born in Brazil with microcephaly as of mid-2016.

Although Zika virus can be transmitted vertically or sexually, most transmission occurs via infected mosquitoes. The main mosquito vector, *Aedes aegypti*, is most common in the tropics, where it prefers to feed on humans, inside buildings. Consequently, Zika virus might be best described as an urban disease of the tropics.

What accounts for Zika's upsurge? Part of the answer is that in the last 150 years, the human population in the tropics has gone from 2% to 50% urban. As more and more people have moved into cities, previously undeveloped land has become urbanized. More people living in crowded cities is good news for *Aedes* mosquitoes. Not only is there more of their preferred food, but along with people comes their garbage. Much of this garbage, such as old tires and cans, makes ideal mosquito breeding sites because it holds rainwater. With more people and more mosquitoes, an increase in a mosquito-borne disease is not surprising. And because the appropriate *Aedes*

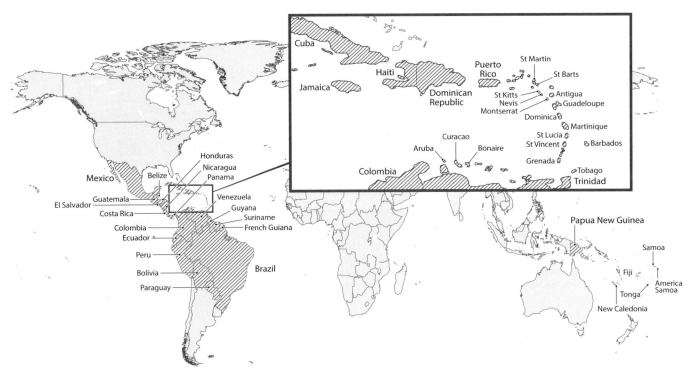

Figure 14.18 Distribution of Zika virus. Since its arrival in Brazil, probably in late 2013, Zika virus has spread to 55 countries and territories in the Americas. Because the appropriate mosquito vectors are also found in other parts of North America, including the southern United States, it is expected that Zika virus will continue to spread. Countries with reported active transmission are highlighted in pink.

mosquitoes are found elsewhere, including North America, there is no reason to believe that Zika has finished spreading. In fact, in 2016 the World Health Organization stated that the virus is likely to spread throughout the Americas, wherever the *Aedes* vectors are found.

Increased travel and trade no doubt are part of the Zika story as well. International jet travel, common only since the 1960s, has provided a handy

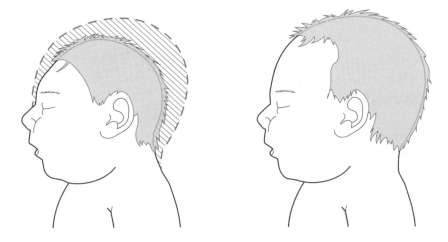

Figure 14.19 Microcephaly. A newborn baby with microcephaly is depicted on the left, compared with a normal newborn on the right. In 2016 it was confirmed that if a woman becomes infected with Zika virus during pregnancy, the virus can be transmitted vertically to the developing fetus, in which it can cause microcephaly. The brain does not develop properly in affected fetuses, resulting in a smaller-than-normal head, along with other neurological problems, such as poor motor function and intellectual disability. As of mid-2016, over 3500 cases of this previously very rare condition were reported from Brazil.

way for Zika virus and other diseases to reach new destinations. A genetic analysis of Zika virus indicates that the strain causing problems in Latin America is very similar to the strain that caused the 2013 outbreak in French Polynesia (**Figure 14.20**). Both of these strains are significantly different from strains that still circulate in Africa. By using such genetic comparison, researchers have put together a plausible route of Zika's migration from Africa to Asia, to Polynesia to Brazil. And by using molecular clock techniques (see Chapter 4, pp. 70–72), they have homed in on the date of Zika's arrival in Brazil—it was probably around late 2013. Because symptoms are so mild in most people, the virus was able to spread surreptitiously and it took over a year before the medical community had a clear idea of what was taking place.

This is probably the way other mosquito-borne viruses such as dengue virus, West Nile Virus, or more recently, chikungunya virus, arrived in the Americas. All of these viruses are endemic to the tropics and are transmitted by mosquitoes. West Nile virus, discussed in our case, made its New York debut in 1999, most likely when an infected mosquito crossed the ocean in a jetliner. The same is probably true of chikungunya virus, which first appeared in the Caribbean in late 2013. With plenty of the reservoirs, they need to persist (birds in the case of West Nile Virus, mainly mammals for dengue and chikungunya virus), and with abundant mosquitoes required for transmission, these new arrivals have made themselves at home in the Americas and elsewhere.

The importance of factors such as increasing urbanization and international travel points out an important fact about emerging diseases. Human society is changing and the environment is being altered at an unprecedented rate. Consequently, although new diseases have periodically arisen throughout human history, such diseases are emerging today as never before.

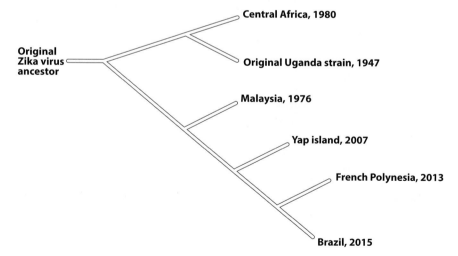

Figure 14.20 Phylogeny of Zika virus. The RNA genome of Zika virus from various geographic locations has been sequenced, and a comparison of those sequences has allowed researchers to track the virus's spread. The genome sequence from the original strain isolated in Uganda is most similar to other strains in Africa. It appears that between the mid-1940s and the mid-1970s, Zika virus spread to Asia, as evidenced by the 1976 Malaysia strain. At some later point the virus made its way to the South Pacific. The close similarity between the 2013 French Polynesia strain and the Brazilian strain, isolated in 2015, indicates that the South American epidemic began with the introduction of the virus from Polynesia, some time in late 2013 or early 2014.

Emerging diseases can be categorized as one of four basic types

Emerging diseases are placed in one of four categories. These can be briefly described as follows:

1. invasion of a new host population by a known pathogen

2. appearance of a completely new, previously unknown disease

3. association of a well-known disease with a new pathogen

4. increased virulence or a renewed problem with a well-known but previously less virulent or well-controlled pathogen.

When a known pathogen reaches a new host population and therefore changes its overall distribution, it is considered to be in category 1. Most emerging diseases, including dengue, West Nile virus, and Zika virus, are in this category. As described, international jet travel, with no place now more than a few hours away from any other place, plays an important role in emerging diseases of this type.

Some pathogens exist in the environment undetected for many years. They may occasionally cause illness or death, but if cases are rare or sporadic, such pathogens may remain unrecognized. At some point a convergence of events finally brings a "new" disease to our notice. Such an apparently new disease is placed in category 2. Examples include AIDS (first detected in 1981), hantavirus (unknown before 1991), and SARS (2003).

Middle East respiratory syndrome (MERS), first described in Saudi Arabia in 2012, provides a recent example. MERS is caused by a coronavirus, similar to the virus that causes SARS. The primary symptoms are problems with breathing. As of 2016 over 1000 people have being diagnosed with MERS and around 400 have died; most cases have occurred in the Arabian Peninsula, although occasional cases elsewhere have been reported. A large outbreak in Korea in 2015 reminds us that this virus is likely to continue its global spread. The Korean outbreak (close to 200 cases and 20 deaths) might best be considered a category 1 emerging disease, because it has occurred outside of what was thought to be the normal distribution of MERS.

No discussion of previously unknown diseases is complete without mentioning Ebola hemorrhagic fever, perhaps the most spectacularly gruesome emerging disease of all. This disease, caused by Ebola virus, burst on the scene in 1976 in Congo and Sudan. Since then it has occasionally come out of hiding, causing outbreaks across Africa (**Figure 14.21a**). Ebola can cause up to 90% mortality, although some strains of the virus are somewhat less lethal. A strain of this African virus, found in Reston, Virginia, in 1989 in imported African primates, turned out to be nonpathogenic in humans, although highly virulent in monkeys.

Ebola virus causes massive hemorrhaging in its victims. Bleeding may be internal or external, which results in a dangerous loss of blood pressure. Human-to-human transmission occurs via contact with infected material such as blood or contaminated needles. The preparation of deceased victims for burial by relatives has been a major factor in transmission between people.

The reservoir of Ebola virus remains unknown, although bats are thought to be the most likely candidate (**Figure 14.21b**). There is still no accepted drug treatment that is both safe and effective. Current treatment basically focuses

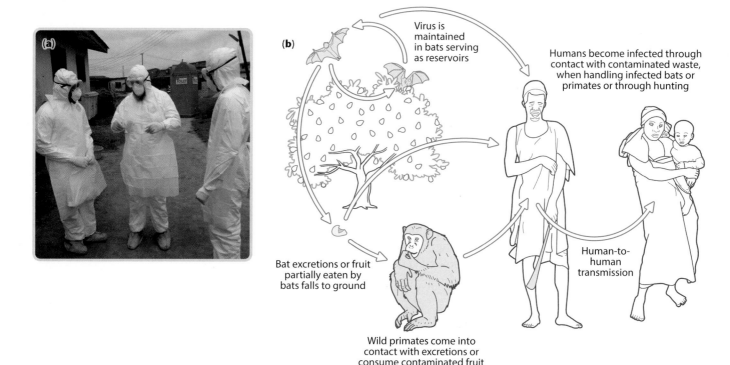

Figure 14.21 Ebola in Africa. The recent West African epidemic was the largest in history. The epidemic began in Guinea in December 2013. As of mid-2016, although the epidemic has subsided, there are still occasional new cases. There were over 28,000 reported cases and more than 11,000 deaths, although the World Health Organization (WHO) considers this to be a gross underestimate. (a) Nigerian and WHO physicians don protective gear prior to entering an Ebola ward. (b) Although not yet confirmed, bats are thought to be the principal reservoir for the virus. Nonhuman primates, along with certain other mammals, may become infected after interacting with contaminated wastes or fruits. Some primates, including chimpanzees and gorillas, are vulnerable to serious disease. Humans interacting with either bats or infected animals, often due to hunting, may become infected. Transmission between humans can subsequently occur.

on maintaining blood pressure and replacing lost blood and fluids, and containment of epidemics is attempted by isolating affected individuals and otherwise trying to limit transmission. Several promising, experimental vaccines are under investigation.

Many of the diseases in both categories 1 and 2 are zoonoses—namely, diseases of animals that can be transmitted to humans as well. West Nile virus, which is a disease of birds, and Zika, a disease of nonhuman primates, certainly fall into this category. The natural transmission cycle of MERS is thought to involve primarily bats as reservoirs, as is that of Ebola virus (see Figure 14.21b). As human development continues to encroach on natural habitats, and as interactions with natural animal reservoirs increases, new zoonoses are certainly something to expect in our future.

Category 3 consists of those emerging diseases in which a well-known disease is newly associated with a particular pathogen. Gastric ulcers are an example. Such ulcers were originally believed to be caused by excess stomach acid. Earlier in this text we discussed the discovery of *Helicobacter pylori* and its role in gastric ulcers. Kaposi's sarcoma is another example. This once very rare form of skin cancer is now recognized as an opportunistic infection in AIDS patients, caused by a type of human herpesvirus.

Re-emerging diseases are placed in category 4. Re-emerging diseases are those that were considered to be under control but have more recently been causing increasingly serious problems. Many re-emerging diseases are caused by pathogens that have become drug resistant. In Chapter 13, we discussed drug resistance and how once easily treated infections have in some cases become untreatable. Multi-drug-resistant *Staphylococcus aureus* and *Mycobacterium tuberculosis* are examples. *M. tuberculosis* also reminds us of how the problem of immunosuppression can lead to disease re-emergence. It is no coincidence that the worldwide resurgence of tuberculosis began soon after the problem of HIV became a global phenomenon. Tuberculosis is now a leading cause of death in untreated AIDS cases.

The above examples demonstrate just how complex the reasons for disease emergence can be. As we have seen, increased travel, changes in weather, and increased contact with animal reservoirs can all tip the scales in favor of emergence. Sometimes even politics is the primary culprit. An epidemic of diphtheria in the former Soviet Union that started in 1993 is widely attributed to the collapse of the government in 1990; the newly independent nations that emerged from the breakup of the Soviet Union failed to maintain anti-diphtheria vaccination programs. With human society and the environment continuing to change at breakneck speeds, epidemiologists will have plenty to keep them occupied in the foreseeable future.

Bioterrorism is an ongoing threat

CASE: THE ANTHRAX SCARE

In late 2001, soon after the terrorist attacks of September 11, 2001, the United States was in the grip of a new terror threat. Between October 4th and November 20th, 22 cases of anthrax were reported. Five of the patients died. The source was traced to letters contaminated with *Bacillus anthracis* endospores. The letters were deliberately sent to several prominent individuals, including members of the media and the US Senate (**Figure 14.22**). Most of those affected were mail handlers or those who entered workplaces where contaminated mail was processed or received.

1. **Why is *Bacillus anthracis* such a potent bioterrorism agent?**

Figure 14.22 The 2001 anthrax mail attacks. This letter, which contained anthrax spores, was sent to Senator Tom Daschle of South Dakota. Other letters were sent to several other prominent individuals. The letters were believed to be sent by a disgruntled scientist at a government biodefense laboratory in Maryland.

It is an unfortunate sign of the times that the most fearsome emerging diseases may not always arise in nature. With the ongoing focus on terrorists and radical fringe groups, **bioterrorism** has become part of our vocabulary. Bioterrorism refers to the use of biological agents to kill or otherwise cause fear and mayhem. As with other, naturally occurring emerging diseases, epidemiologists must be able to recognize a bioterrorism event, elucidate its cause, and develop a response strategy.

Compared with nuclear weapons or other weapons of mass destruction, biological agents are cheap. Unlike a chemical poison, some microorganisms used as weapons can spread from person to person, complicating control efforts. Because only a small quantity of a biological agent is necessary to wreak havoc, transport is less of a problem. For example, in 1979, 66 people suddenly died in the Russian city of Sverdlovsk when a nearby biological weapons research facility accidently released anthrax spores into the air. This large loss of life was precipitated by the release of less than 1 gram of spores.

But not any microorganism will do. Spore-forming organisms such as *Bacillus anthracis,* the causative agent of anthrax, are among the most opportune candidates (see Chapter 3, pp. 57–59 for a description of endospores). The resistant endospores can be aerosolized and spread through the environment, where they persist until they happen to infect someone. This bacterium can also be extremely lethal. Untreated pulmonary anthrax, caused when spores are respired into the lungs, has a mortality rate of close to 100%.

Viruses, especially those that spread well from person to person, also pose risks as terrorist weapons. Smallpox is a particular concern, because of its high pathogenicity and its ability to spread through a population rapidly. Although an effective smallpox vaccine was developed many years ago, it is no longer widely used. Many people are consequently susceptible. Smallpox was eradicated in the wild in the late 1970s, but the virus is still maintained in laboratories in the United States and Russia. The fear is that terrorists might somehow gain access to these stocks. Yet biological weapons are not as easy to use as they may sound. Creating an effective biological weapon usually requires that the potential agent be produced in a weaponized form. This may involve making the agent more lethal, more able to enter the host, more easily transmissible, or better able to persist in the environment. Fortunately, this requires a level of knowledge and technology that is unavailable to every would-be terrorist.

In light of ongoing terrorist activity, many governments now have response plans in place in order to take quick action to contain any outbreak. In a world where political agendas are often pursued through less than peaceful means, however, bioterrorism remains a threat. Understanding the biology of the involved microorganisms is crucial to our protection.

Looking back and looking forward

In this chapter we have investigated how diseases behave in populations and the factors that are responsible for disease outbreaks. Such outbreaks can occur for a variety of reasons, not all of which are strictly biological. We have also investigated why so many epidemics tend to occur cyclically and we have become familiar with how epidemiologists investigate disease outbreaks. We should not be surprised when new diseases seemingly arise out of nowhere.

In the last few chapters, including Chapter 14, we have focused on disease, casting microorganisms in a less-than-favorable light. In the interest of fair play, our final three chapters emphasize some of the more benign aspects of the human–microorganism interaction. In Chapter 15, we concentrate on biotechnology and the vital role that microorganisms play. When diabetics need insulin, or when toxic waste spills must be cleaned up, genetically modified microorganisms more and more frequently lend a hand. In Chapter 15, we explain how they do it.

Garland Science Learning System

- http://garlandscience.rocketmix.com/students/

- Discover why flu is a problem year after year after year

- Test your knowledge of this chapter by taking the quiz

- Familiarize yourself with the terminology used in this chapter by using the vocabulary review

- Get help with the answers to the Concept questions

Concept questions

These questions are designed to help you start thinking like a microbiologist. The answers are not always simply found in the text. Instead, you will need to take the concepts about which you have learned and apply them to new situations. Some of the questions may not even have just a single correct answer. Help is provided as part of the GSLS resources, which can be accessed through http://garlandscience.rocketmix.com/students/.

1. Consider the following statistics for the number of cases of a particular infectious disease in a given geographic area.

MONTH	NUMBER OF CASES
January	0
February	4
March	101
April	611
May	2404
June	900
July	313
August	48
September	9
October	0
November	0
December	0

In which month is the proportion of immune individuals the highest? In which month would it be the lowest? Is this proportion increasing or decreasing between February and May? Between May and September? Between October and December? Would you guess that this is a common source epidemic or a host-to-host epidemic? Explain your reasoning.

2. Discuss how the proportion of immune individuals in a population affects the likelihood of epidemics. How are vaccines used to increase immunity in a population and thereby reduce the likelihood of an epidemic?

3. Imagine that you want to write a screenplay for a movie in which a new, exceedingly virulent virus emerges from the Brazilian rainforest and within a few years has wiped out a sizable proportion of the human population. You want your movie to be as biologically realistic as possible in terms of factors such as your virus's mode of transmission, its reservoir, etc. You also want as much realism as possible in terms of why and how the virus emerged and how your movie ends. Provide a plot summary for your movie.

4. We have learned that weather is frequently an important contributing factor to when an epidemic will occur. Global warming certainly constitutes an enormous change in our weather. Are there any diseases you have learned about in this text that you think will become more or less likely to occur in epidemics in your country? What about emergent or re-emergent diseases? Can you think of diseases that are more likely to emerge where you live due to a warming atmosphere?

5. You are an epidemiologist working for the World Health Organization. If a series of severe cyclones were to hit India, what types of epidemics might you become worried about and why?

6. A number of measles cases are suddenly reported in a large city. At elementary school X, with a student population of 500, 485 students have been vaccinated against measles. At elementary school Y, also with 500 students, only 250 have been vaccinated. Explain why, at elementary school X, the 15 unvaccinated students are less likely to get measles than the 250 unvaccinated students at school Y.

7. Here in Chapter 14, we learned about the pioneering work of Florence Nightingale in determining the risk factors for typhus. Knowing what you know now about how a case–control study is designed and used, imagine that you could go back in time and assist Nightingale in developing a case–control study to help elucidate how typhus was transmitted. What would your advice be?

THE FUTURE IS HERE: MICROORGANISMS AND BIOTECHNOLOGY

15

Man's best friend has become man's brawnier friend. In late 2015, Chinese scientists announced that they had successfully used a new gene-editing technique called CRISPR/Cas to create customized beagles that were especially muscular (**Figure 15.1**). The researchers say the dogs will not be bred to sell commercially, but that owing to their robust physique, such dogs may be especially valuable for their police and military applications.

The scientists used CRISPR/Cas to disable the gene in the dogs that codes for the protein myostatin. This protein inhibits the growth of muscle cells. Accordingly, with their myostatin gene essentially disabled, muscles grew to double their normal size.

These dogs illustrate just some of what is possible in the era of **genetic engineering**, the direct and purposeful manipulation of DNA to create essentially new living things with certain qualities. Yet you may be asking yourself, as jaw dropping as this announcement is, what does it have to do with microbiology? As we will see in this chapter, microbes have been front and center in all aspects of the genetic revolution. The CRISPR/Cas technology, for example, has been adopted from a natural process that occurs in many bacteria. Consequently, if we wish to gain an understanding of modern biotechnology, we need to understand the tools and tricks that have been provided to us by microorganisms.

Some of the topics we will discuss are controversial. Using CRISPR/Cas to create hulking canines can evoke strong opinions and raises concerns that this same technology might be contemplated to create designer human babies with certain characteristics. Other issues, such as the use of genetically modified food, have direct relevance for our lives and those of our families. This underscores perhaps the most important reason for this chapter; all of us, scientists and nonscientists alike, are living in the age of biotechnology. The thoughtful person might wonder if genetically modified organisms pose a risk to the environment. Or we may be asked to consider if insurance companies or employers should have access to genetic information indicating a person's likelihood of developing cancer or certain other diseases. Each of us must come to our own conclusions on such issues, but in order to make rational, well-informed decisions, we must understand the science behind the issues.

Figure 15.1 The world's first gene-edited dogs. Hercules (left) and Tiangou (Mandarin for "Heaven Dog") have twice the normal amount of muscle compared with typical beagles, thanks to gene editing using CRISPR/Cas technology. This technology was adapted from a naturally occurring, microbial process.

The genetic revolution began through an understanding of microbial genetic exchange mechanisms

Recall from Chapter 7 (see pp. 154-157) that bacteria are adept at exchanging DNA with each other via various mechanisms, collectively called

All case studies have a few questions at the end, the answers to which will become apparent as you read the sections following the case.

horizontal gene exchange. In the 1940s, scientists began to understand the processes of transformation, conjugation, and transduction, recognizing for the first time that these processes involved the transfer of genetic material. Indeed, up until that time, it was not even clear what DNA did. The first unambiguous evidence that DNA was in fact the genetic material came in 1944, when it was recognized that DNA transfer occurred during bacterial transformation.

Once scientists understood how gene exchange worked in bacteria, they began to wonder whether it might be possible to deliberately use these microbial processes to place certain genes exactly where they wanted them. And in the early 1970s, researchers at Stanford University and the University of California accomplished a feat that has shaken the world ever since; they spliced a gene from a virus into the DNA of *E. coli* bacteria. The field of **recombinant DNA** was born, and the term genetic engineering entered our vocabulary. Recombinant DNA refers to the process in which DNA from one source is introduced into the DNA of a different organism or cell. The result is hybrid DNA with DNA from two distinct sources.

Other accomplishments followed rapidly. Before long DNA from two different types of bacteria was successfully combined. In 1973, researchers managed to insert a gene from an animal, the African clawed toad (*Xenopus laevis*), into *E. coli* plasmid DNA.

Since then, molecular biologists have hardly looked back. In 1982, the first recombinant DNA product became available commercially, when Humulin® was approved for medical use (**Figure 15.2**). Humulin® is human insulin, used to treat diabetics. It is produced by bacteria into which the human insulin gene has been introduced. Prior to its production, insulin was obtained from the pancreases of slaughtered farm animals. Although this insulin allowed many diabetics to control their condition, some people developed allergic reactions to the animal protein. Other human proteins supplied by recombinant bacteria soon followed, and by the end of the 1990s, over 125 such proteins, including clotting factors, used to treat people with hemophilia, and interleukins, used to modulate immune responses, were available.

With the advent of new technologies such as the gene-editing process noted in the introduction, the pace of change in the field of biotechnology seems destined to accelerate. We will now investigate some of the principal techniques employed by molecular biologists as they manipulate the genetic material.

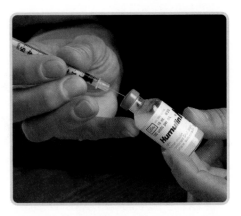

Figure 15.2 The first recombinant DNA medical product. Since 1982, diabetics have had access to Humulin®: pure human insulin produced by bacteria into which the human gene for insulin production has been spliced.

Sequencing techniques can be used to reveal the sequence of nucleotides in a DNA sample

We have had many opportunities throughout this text to refer to specific genes found in microorganisms or their hosts. In Chapter 7 we learned that a gene is a specific sequence of DNA nucleotides that codes for the production of a particular protein. In many cases that specific DNA sequence is known.

To the nonscientist, the ability to determine the order of nucleotides in a molecule of DNA may seem like magic, but the techniques are actually straightforward. To ascertain a gene sequence, we first need to obtain enough of the DNA in question. Once sufficient DNA is on hand, we can determine the actual nucleotide sequence. Thus, the entire business of

CASE: TB—BC (BEFORE COLUMBUS, THAT IS)

"In fourteen hundred and ninety two, Columbus sailed the ocean blue." Did he bring tuberculosis with him? In Chapter 6 we learned about the various infectious diseases that European explorers introduced into the New World. Tuberculosis was always thought to be on that list. Some experts were doubtful, however, because mummies that predated the arrival of the Spanish in Peru sometimes had lung lesions that looked similar to tuberculosis. To find out if tuberculosis in the Americas predated the arrival of Europeans, researchers obtained lung tissue from a 1000-year-old Peruvian mummy (**Figure 15.3**). DNA was extracted from lung lesions and subjected to the polymerase chain reaction (PCR). This process is used to make many copies of, or amplify, target DNA (in this case, *M. tuberculosis* DNA). A short segment (approximately 100 base pairs) of bacterial DNA was amplified. When this DNA was sequenced, the researchers found that the DNA collected from the lung lesions contained base sequences identical to DNA from modern *M. tuberculosis*, clear evidence that tuberculosis was present in Peru at least 500 years before the first Spaniard set foot in the New World. Columbus had nothing to do with it.

1. **What is PCR and how is it used to amplify DNA?**
2. **Once the DNA from the mummy was amplified, how was it sequenced and compared with DNA from modern *M. tuberculosis*?**

Figure 15.3 Exonerating evidence? Peruvian mummies, over 1000 years old. DNA removed from lung lesions in such mummies indicates that tuberculosis was in the New World for hundreds of years before the arrival of European explorers.

sequencing a gene is often a two-step process. Here we discuss one way those two steps might be carried out.

Large amounts of specific DNA sequences can be obtained with the polymerase chain reaction

Early attempts to sequence DNA were often hampered by a limited supply of the DNA in question. This problem was brilliantly solved in the 1980s with a technique known as the **polymerase chain reaction (PCR)**. Starting with a small sample of DNA, PCR allows scientists to quickly copy any genetic sequence in the DNA until they have a sample large enough to sequence.

PCR is carried out in a number of repeated cycles. Each cycle doubles the amount of the DNA of interest (the target gene). If we begin with 100 copies of a particular gene that we wish to amplify, we will have 200 copies after one PCR cycle, 400 copies after two cycles, and so on. Because usually 30 or so cycles are run in a PCR reaction, we end up with a very large number of copies.

The PCR reaction is prepared in a small reaction tube. In this tube, the gene is copied essentially in a similar manner to the DNA replication discussed in Chapter 7. As with DNA replication, a supply of A, G, T, and C nucleotides and the enzyme DNA polymerase are required to copy the target DNA. But not just any DNA polymerase will do. As we will see, during each PCR cycle, the DNA is heated to the point at which the two DNA strands separate into single strands. If DNA polymerase from most sources were used, the enzyme would denature when heated and would be unable to copy the target. Consequently, DNA polymerase from a thermophilic (heat-loving) bacterium is used. The bacterium, *Thermus aquaticus*, grows only in extremely hot water, so it has an unusually heat-stable DNA polymerase. The enzyme, called Taq polymerase (from *Thermus aquaticus*), is a key component in PCR reactions.

Appropriate **primers**, short DNA sequences of approximately 20 nucleotides that are complementary to the ends of the target DNA (**Figure 15.4**), are also needed. Once the target DNA is separated into single strands, a primer attaches to the end of both strands. The primers essentially act as start signals, indicating to the DNA polymerase where the ends of the target gene are located. The DNA polymerase then elongates the primer, synthesizing a new daughter strand of the target DNA. The primers used to amplify a 100-base-pair length of *M. tuberculosis* DNA would be short sequences complementary to the ends of this 100-base-pair region.

Once the reaction tube contains all of the necessary components, it is placed into a *thermocycler*, a device that can be programmed to alternate among various temperatures. Each cycle of the PCR reaction passes through three temperature stages (see Figure 15.4). The first stage heats the DNA up to the point where it separates into single strands. In stage 2, the temperature is lowered, allowing the primers to attach to the ends of the target. Finally, in stage 3, the DNA polymerase (Taq) copies each of the single strands, using the supply of nucleotides added to the reaction mix. All subsequent cycles pass through the same three stages. Because the amount of the target DNA doubles after each cycle, the DNA of interest will have been amplified to literally billions of copies by the end of a PCR procedure.

Newer forms of PCR have been developed for specific purposes

Like other aspects of biotechnology, PCR technology has not stood still. Rather, there are now many variations of PCR available for specific purposes. Reverse transcription PCR (RT-PCR), for example, is used to first amplify DNA, starting with RNA. Using the enzyme reverse transcriptase, the RNA is reverse-transcribed to DNA. This DNA can then be amplified by conventional PCR, as described above. RT-PCR is commonly used to reveal what genes are being expressed (being transcribed into mRNA; see Chapter 7, pp. 142–144) in any given situation. It might also be utilized to determine what gene actually gave rise to a specific RNA transcript. Our ability to perform RT-PCR relies on yet another microbial enzyme—namely, reverse transcriptase, which is characteristic of retroviruses (see Chapter 5, pp. 99–100). This fact provides us with yet another reminder of how much various microorganisms have contributed to our ability to analyse and manipulate the genetic material.

Gel electrophoresis can be used to determine if a PCR was successful

Once a PCR is complete, the researcher needs to know if the DNA in question has been successfully amplified. To find out, a technique called **gel electrophoresis** is often utilized.

As the name suggests, a sample from the reaction tube is placed on a gelatinous material, or gel. The gel contains microscopic pores through which DNA can pass. An electrical current is run through the gel, and since DNA itself carries negative charges, the DNA moves through the gel, away from the

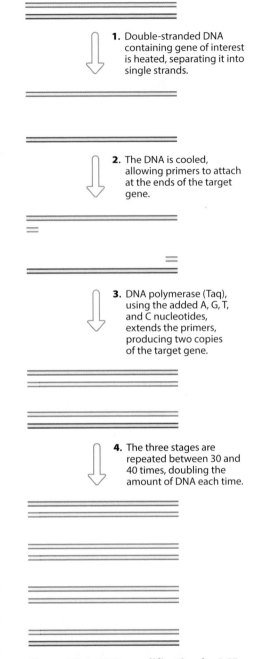

1. Double-stranded DNA containing gene of interest is heated, separating it into single strands.

2. The DNA is cooled, allowing primers to attach at the ends of the target gene.

3. DNA polymerase (Taq), using the added A, G, T, and C nucleotides, extends the primers, producing two copies of the target gene.

4. The three stages are repeated between 30 and 40 times, doubling the amount of DNA each time.

Figure 15.4 DNA amplification by PCR. Using the polymerase chain reaction (PCR), a particular gene can be copied billions of times. Each reaction is carried out in a reaction tube, into which the DNA to be copied is added. DNA polymerase (Taq), nucleotides, and primers, which are short nucleotide sequences complementary to the ends of the target gene, are also added. Each cycle of a PCR reaction goes through three stages. In the first stage, the reaction mix is heated, causing the DNA to separate into single strands. In the second stage, the reaction is cooled to allow the primers to attach to the ends of the gene. In the third stage, the DNA polymerase uses the added nucleotides to extend the primer and produce new, double-stranded DNA molecules. After each cycle of these three stages, the amount of DNA is doubled.

(a)   Amplified DNA

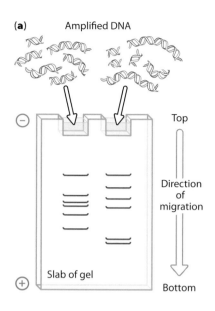

Top

Direction
of
migration

Slab of gel

Bottom

Figure 15.5 **Gel electrophoresis.** (a)
Following PCR, a sample from the reaction tube is
loaded onto a gel. The sample is placed in a well
at the top of the gel, and DNA migrates within
the gel from the negative electrode toward
the positive electrode. The rate of migration
is determined by the size of the fragment. (b)
Following migration, the DNA is stained with a
dye that is visible under ultraviolet light. Each
band is composed of amplified DNA of the same
length. The length can be determined by the
distance the DNA has migrated, and if the length
is correct, the researcher knows the amplification
was successful.

negatively charged electrode and toward the positively charged electrode
(**Figure 15.5a**). The rate at which the DNA moves through the gel depends on
the length of the DNA that was amplified; smaller fragments move readily,
whereas larger fragments, impeded by the fibers of the gel matrix, move
more slowly. Therefore, if allowed to run for a given period of time, short
DNA molecules migrate the greatest distance, while the largest ones migrate
the shortest distance from the starting point. If the researcher knows how
long the gene under investigation is, in terms of the number of nucleotides,
he or she can accurately predict how far it will migrate in the gel. After a specified amount of time, any DNA on the gel is stained with a specific dye. An
easily visible, stained band of DNA that has migrated the correct distance
indicates to the researcher that the PCR was successful (**Figure 15.5b**).

Amplified DNA can be sequenced

Following PCR amplification, it is sometimes possible to proceed directly to
DNA sequencing, in which the exact nucleotide sequence of the target DNA is
determined. Today, researchers may utilize any one of various techniques,
according to their specific research goals. These techniques continue to
become easier, cheaper, and faster. Although the details of these techniques
are beyond the scope of this text, DNA sequencing is now routine and is carried out daily in laboratories around the world. Once an actual base sequence
for a particular stretch of DNA is obtained, the information is analyzed by computer, yielding a sequential list of the nucleotides in the DNA (**Figure 15.6**).

Rapid and efficient gene sequencing techniques have opened the door to
tremendous research opportunities. As described in Chapter 4, for example,
gene sequences are now used in modern taxonomy to deduce evolutionary
relationships. In the case of tuberculosis in a Peruvian mummy, we see how
the technology was used to answer an interesting question about an important infectious disease. When researchers sequence a new gene, they submit
the data to an ever-growing online database called GenBank that now contains billions of sequences. Any newly sequenced gene can be compared
with known sequences in the database, and within seconds, a researcher
can find out how similar or different it is from the same gene in other organisms. If you sequence a gene from a lung lesion obtained anywhere in the

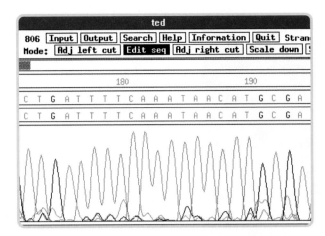

Figure 15.6 Reading a DNA sequence. Following DNA sequencing, the DNA sequence is analyzed by computer. The nucleotide sequence is translated into a chromatograph, which corresponds to the DNA sequence of interest. Each color peak (blue = C, red = T, green = A, and black = G) represents one of the four nucleotides in a specific order.

world and want to know whether it is from *M. tuberculosis*, you merely go to GenBank and enter your sequence. In almost no time you find out whether your gene matches *M. tuberculosis* genes in the database. Sequenced genes from other microorganisms can be similarly identified.

For many biologists, the ability to sequence genes has merely whetted their appetites. Rather than settling for the sequence of individual genes, why not sequence the entire genome of an organism? With the entire DNA blueprint in hand, biologists can start to answer questions they had previously only dreamed about. Whole genomes could be compared to provide an even clearer picture of evolutionary history. One could determine how the genes of a bacterium and a eukaryotic organism differ or how they are the same. With the complete genome of a pathogenic microorganism in hand, the time-consuming and difficult task of determining which genes are responsible for virulence might be greatly simplified. In fact, questions such as these are now being investigated routinely, thanks to the modern field of **genomics**—the sequencing and study of entire genomes.

The entire genomes of many organisms have been sequenced

CASE: *E. COLI'S* DARK SIDE

On the surface, *E. coli* might seem to have multiple personalities. We have frequently referred to this species throughout the text, usually as a harmless member of our normal microbiota. But in Chapter 14 we were introduced to *E. coli* 026, which caused an outbreak of food-borne illness, traced to a popular restaurant chain. Even more malevolent is *E. coli* O157:H7, a highly pathogenic strain of the bacteria, which first reared its head in the 1990s, when it was detected in contaminated hamburgers. Since then, outbreaks have been traced to contaminated produce, meat, and unpasteurized juices. *E. coli* O157:H7 is unusual because it carries genes that code for potent toxins, the most virulent of which is Shiga toxin. An infection with *E. coli* O157:H7 can cause severe diarrhea, kidney damage, and occasionally even death.

1. What is Shiga toxin and how did normally benign *E. coli* acquire the ability to produce it?
2. How has the study of microbial genomes permitted this question to be answered?
3. Do answers to the above questions suggest any practical applications for the treatment of dangerous *E. coli* infections?

The first genomes to be sequenced were understandably some of the smallest. In 1976 the first complete genome was obtained—it was a small bacteriophage with a total genome of only 5386 bases. It was not until 1995 that a complete bacterial genome was sequenced. It was from a small prokaryote—the 1,830,137 bases that make up the entire genome of *Haemophilus influenzae*.

Since the mid-1990s the pace of sequencing has been mind-boggling. A year after *H. influenzae*, the genome of the first eukaryote, *Saccharomyces cerevisiae*, or common brewer's yeast, was sequenced. Many animal, plant, and microorganism genomes followed. In 2001, a rough draft of the human genome was published, and things have only accelerated since then (**Figure 15.7**). Thousands of prokaryotic and eukaryotic genomes have now been sequenced, and genome projects for many other living things are in the works. Even our extinct human relative, the Neanderthal, has had its genome sequenced. The project was completed in 2013, using DNA extracted from a 130,000 year-old bone found in Siberia.

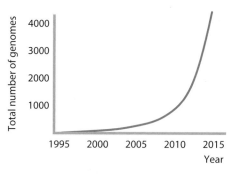

Figure 15.7 The genomics explosion.
The red line traces the exponential growth in the number of bacterial and archeal genomes that have been completed since 1995.

Traditionally, sequencing has been accomplished by first enzymatically cleaving the genome of interest into a large number of fragments, and then amplifying each of those fragments into many copies. The fragments are then sequenced and sophisticated computer programs identify regions on each fragment that overlap with those of other fragments. The various fragments are then ordered into a complete genetic map. Fragments that are too large to easily sequence can be digested into even smaller fragments and subsequently sequenced. Once all of these smaller fragments are sequenced, they are reassembled by the computer, and yet another genome has been completed.

Today, the genomics researcher may choose from a wide variety of next-generation sequencing techniques that allow genomes to be sequenced more efficiently than ever before. All have various advantages and disadvantages, which influence the researcher's choice. One very new technique being considered, to provide just one example, is nanopore sequencing. A nanopore is simply a small pore approximately 1 nanometer in diameter. The pores, which are actually made of certain bacterial proteins, are attached to a sugar polymer. The pore is placed in a fluid that conducts electricity and a current is applied. This current can be measured as ions flow through the pore. Next, single-stranded DNA to be sequenced can be fed through the pore, much as a thread might be passed through the eye of a needle. As the DNA passes through, it disrupts the flow of ions, and this change in the electrical current can be measured. And because each of the four DNA nucleotides—A, G, C, and T—disrupts the electrical flow in a characteristic way, the exact nucleotide passing through the pore at a particular moment can be determined.

The potential for the field of genomics is enormous, and applications for medicine, agriculture, and industry are now being realized. Our case provides an excellent example. To deduce the origins of the *E. coli* Shiga toxin gene, researchers began by comparing the genomes of virulent and nonvirulent strains. They found several genes present in the virulent bacteria that were absent in the nonvirulent *E. coli*. When these genes were entered into the GenBank database and compared with other genes already in the database, it was found that the novel *E. coli* O157:H7 genes were very similar to genes found in *Shigella*, a genus of bacteria long associated with severe dysentery. One of these genes was the one coding for Shiga toxin. The genetic similarity strongly suggested that *E. coli* had somehow acquired the Shiga toxin gene from *Shigella*.

How did the toxin-producing virulence gene get from *Shigella* to *E. coli*? It is currently believed that a bacteriophage shuttled the genes from one species to the other. In other words, the genes were transferred by transduction (see Chapter 7, pp. 156–157). Aside from demonstrating yet again the versatility and diversity of microorganisms, this information has at least one potential practical application. If an antibiotic could be developed that inactivated or interfered with *Shigella* toxin, it should also work against pathogenic *E. coli*.

Genomic analysis has led to a better understanding of other diseases as well. For example, examination of the genome of *Borrelia burgdorferi*, the bacterium that causes Lyme disease, reveals that *B. burgdorferi* has an unusually large number of genes that code for surface proteins. It is believed that by regularly changing the proteins on its cell surface, this species can evade host response and thereby establish a chronic infection. By better understanding how these genes are switched on and off, it may be possible to interfere with the process, rendering the bacteria more susceptible to immune destruction.

Not all genomic research necessarily deals with pathogens. For example, if bacteria are found living in habitats contaminated with toxic wastes, it is possible that these organisms might digest the wastes and convert them to less harmful substances such as carbon dioxide. If these bacteria digest benzene, dioxin, or other nasty chemicals, they must have genes that code for the enzymes that permit them to do so. Genomic analysis can help identify such genes.

Genetic material taken directly from environmental samples can be sequenced

Microbial genomics typically relies on cultured microorganisms as a source of DNA. In recent years, however, techniques have been developed to sample genetic material directly from the environment. Early work suggested that in any given environment such as sea water or soil, only a small proportion of the microbes found there could be cultured and subsequently sequenced. As techniques have improved, this suggestion has been borne out; there is a vast world of microorganisms in the environment that we knew little or nothing about because we could not grow them in a laboratory. This realization has given rise to the new field of **metagenomics**—the study of genetic material taken directly from environmental samples. Metagenomics has the potential to dramatically change our view of the living world as it uncovers previously unknown microbial diversity in the environment. It also enhances our understanding of environmental processes. To provide just one example, a recent metagenomic study of the bacteria found in sea lion feces suggests that these feces may be an important nutrient source in coastal ecosystems. The bacteria discovered in the sea lion waste were especially good at breaking down nutrients in the feces, making them available to plants and other organisms at the base of marine food chains.

Even our own bodies are fair game for those interested in metagenomics. In 2008 the *Human Microbiome Project* was initiated with a goal of determining if there is a core human microbiome that can be correlated with health. The data from this long-term project are still being analyzed, but there have already been a number of interesting findings. The researchers estimated that there are more than 10,000 microbial species inhabiting the human body. Surprisingly, it was also concluded that microbes contribute more

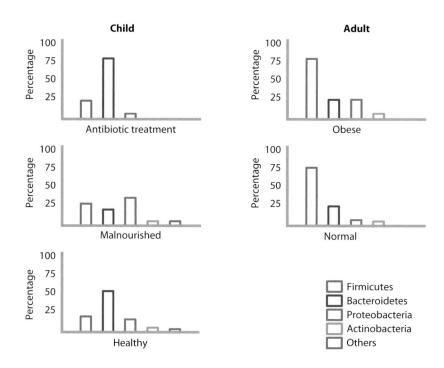

Figure 15.8 Changes in the human microbiota. Thanks to advancing techniques in metagenomics, it is now possible to trace changes in our microbiota at different life stages and under different conditions. Metagenomics is used to sequence the microbiota DNA and consequently identify the organisms present under various circumstances. Each group of microbiota measured here is a large taxon within the domain Bacteria. It is increasingly thought that characteristics, such as malnutrition in children or obesity in adults, is at least in part determined by the makeup of the microbiota.

genes necessary for human survival than we do. They estimate that bacterial genes are 360 times more abundant than human genes. Or maybe this should not be so surprising in light of our discussion in Chapter 11 (pp. 255–257) about the many ways the microbiota affects our health and wellbeing (**Figure 15.8**).

When unique, new genes with potentially valuable properties are discovered, it is now possible to transfer these genes to more easily cultured microbes or even directly into animals or plants. This creates, in effect genetically modified or "engineered" organisms with valuable new attributes. We examine some applications of genetic engineering later in this chapter. Meanwhile, let's now look at how genetic engineering is accomplished.

Foreign genes can be introduced into the genomes of different organisms

CASE: BIOLOGICAL BLACKMAIL

In the Clive Cussler thriller *Dragon*, the hero, Dirk Pitt, does battle with a group of evil industrialists bent on world domination. As part of their nefarious scheme, they have genetically engineered special bacteria capable of rapidly digesting petroleum. Their intention is to introduce the bacteria into the United States' Strategic Oil Reserves. Brought to its knees by a lack of oil, America will have no choice but to give in to the demands of the villains. Fortunately, Dirk saves the day, but the creation of such oil-eating bacteria is not so far-fetched.

1. **How can a gene that codes for the ability to digest petroleum be introduced into bacteria?**

As discussed earlier, recombinant DNA technology involves the cloning of genetic information from one source (the donor) into an entirely different organism (the recipient). As we saw, initial successes involved placing viral genes into bacteria, bacterial genes into different bacteria, or a frog gene into a bacterium (**Figure 15.9**). Now, such techniques are commonplace and are

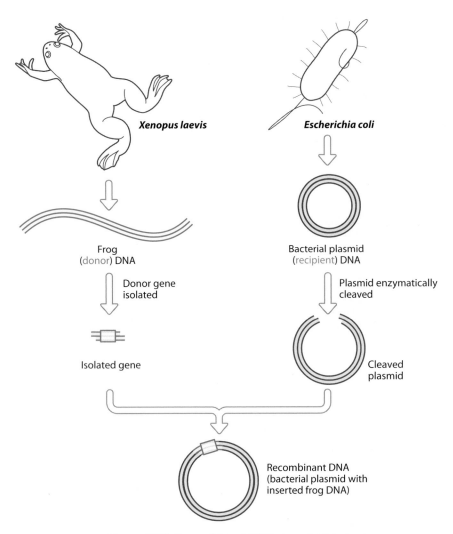

Figure 15.9 Recombinant DNA. An early biotechnology success involved the splicing of animal DNA into a bacterial plasmid.

utilized in many practical applications. To provide just one example, consider the human protein called AT3, which inhibits blood clot formation and is often used to treat people in whom dangerous blood clotting occurs. To create a relatively inexpensive and renewable source of AT3, the human gene for AT3 was inserted into goat DNA, producing goats that secreted AT3 human protein in their milk. Animals with genes from at least one other species, like these goats, are called **transgenic animals**.

We next explore some of the basic techniques of genetic engineering that make marvels like transgenic animals possible, and we also discuss other remarkable uses of this technology.

Bacterial restriction enzymes are useful for cutting DNA at specific sites

We have already learned that even bacteria have their own pathogens. And just as we are protected from potentially dangerous microorganisms by our immune system, bacteria are also armed against bacteria-specific viruses (bacteriophages). One interesting defensive mechanism is a group of proteins called **restriction enzymes**, which literally chop up the DNA of invading bacteriophages. After their discovery in the 1970s, biologists began to

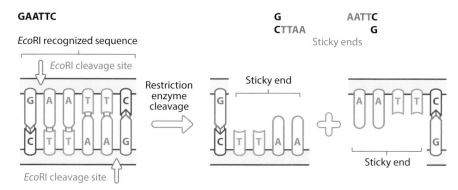

Figure 15.10 A representative restriction enzyme. Many restriction enzymes have very specific cleavage sites. *Eco*RI cuts DNA on both strands only at the sequence GAATTC, always cleaving between the terminal G and A. Because the same cut is made on both DNA strands, overhanging sticky ends that readily bind to complementary bases are produced. The name *Eco*RI indicates that this restriction enzyme comes from *Escherichia coli*. Other restriction enzymes also cut at specific cleavage sites.

study the properties of restriction enzymes and found that many of them did not cut DNA randomly. Instead, they cut only at very specific places. For example, one restriction enzyme, called *Eco*RI, cuts DNA only at the nucleotide sequence GAATTC (**Figure 15.10**). Remember that DNA is a double-stranded molecule and that the two strands are complementary. Therefore, the strand opposite the target base sequence has the complementary sequence CTTAAG. When *Eco*RI finds this sequence of paired bases, it makes identical cuts on both DNA strands, between the terminal G and the adjacent A. This means that the cuts on the two opposing strands are staggered. Because the cut is staggered, the two remaining DNA fragments have short, unpaired, overhanging ends (see Figure 15.10). These are called **sticky ends**, because they readily bond with (stick to) complementary bases. To date, almost 1000 different restriction enzymes have been identified, all of which cleave DNA at different sites.

The discovery of restriction enzymes gave biologists a powerful new tool that has figured prominently in recombinant DNA technology. By using the same restriction enzyme to cut both the donor and recipient DNA, complementary sticky ends on the donor and recipient DNA molecules are produced (**Figure 15.11**). If donor and recipient DNA are then mixed, the sticky ends allow the donor DNA to insert itself into the recipient DNA molecule. This capacity to cut and paste DNA from one source into the DNA from another source allows for the production of recombinant DNA from two different sources.

Genes of interest can be cloned by inserting them into bacteria

Cloning refers to the creation of copies of a cell, a gene, or an entire organism. When a genetic engineer wants to insert a foreign gene into the DNA of a different organism, it is frequently necessary to first clone the gene to generate many copies. Genes are often cloned by introducing them into bacteria. The bacteria then do all the work, replicating the foreign gene of interest as they replicate their own DNA.

First, the gene of interest is incorporated into the bacterial DNA. Usually the plasmid, rather than the main bacterial chromosome is used, because the much smaller plasmids are easier to work with. Recall from Chapter 7

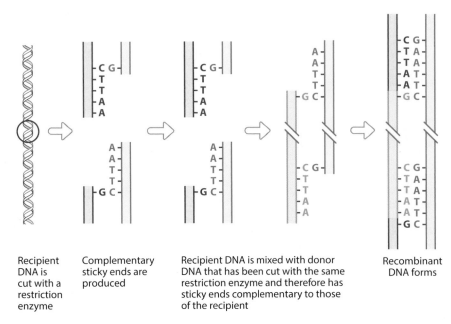

Recipient DNA is cut with a restriction enzyme

Complementary sticky ends are produced

Recipient DNA is mixed with donor DNA that has been cut with the same restriction enzyme and therefore has sticky ends complementary to those of the recipient

Recombinant DNA forms

Figure 15.11 Use of restriction enzymes to recombine DNA from different sources. If both recipient and donor DNA are cut with the same restriction enzyme, they will form complementary sticky ends. When DNA from two sources (for example, from two different bacterial species or from a prokaryote and a eukaryote) that has been cut in this way is then mixed, donor DNA may insert into recipient DNA, forming recombinant DNA. Following insertion, the breaks in the phosphate–sugar backbone are sealed with a different enzyme called ligase that forms new covalent bonds between adjacent nucleotides. The gray breaks in the donor DNA indicate that the donor DNA may be of any specified length.

(p. 157) that plasmids are small loops of DNA with their own origin of replication. They can therefore replicate independently of the bacterial chromosome and they are often exchanged by bacteria during conjugation. In genetic engineering, the plasmid serves as a **cloning vector**—that is, a self-replicating DNA molecule into which foreign DNA can be inserted.

The first step is to obtain the necessary plasmids and a suitable restriction enzyme. Both the plasmid and the restriction enzyme must be selected with care. An appropriate plasmid must have two key genes: an antibiotic resistance gene and a lethal gene, for example **Figure 15.12**. The lethal gene is so-called because it kills any bacterial cell in which it is functioning. The selected restriction enzyme must be able to cut the plasmid within the lethal gene. If any of these requirements are not met, it will be difficult to determine later in the process which bacteria ultimately carry the recombinant plasmid and which do not.

To insert the foreign donor gene into the plasmid, the same restriction enzyme is used to cut both the foreign DNA and the plasmid (see Figure 15.12). This treatment produces complementary sticky ends in both the donor DNA and the plasmid. The two types of DNA are then mixed together, and some of the donor DNA pairs with the cut ends of the plasmid. Observe in Figure 15.12 that recombinant plasmids—those that have incorporated the foreign DNA—have had their lethal gene disrupted. In other plasmids, the sticky ends simply rejoin. These plasmids have failed to incorporate the foreign gene and their lethal gene remains intact. At this stage we do not know which plasmids are recombined and which are not.

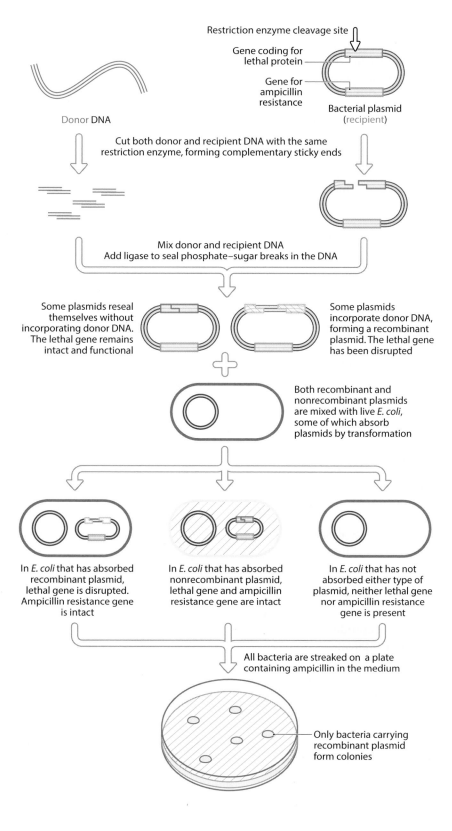

Restriction enzyme cleavage site

Gene coding for lethal protein

Gene for ampicillin resistance

Bacterial plasmid (recipient)

Donor DNA

Cut both donor and recipient DNA with the same restriction enzyme, forming complementary sticky ends

Mix donor and recipient DNA
Add ligase to seal phosphate–sugar breaks in the DNA

Some plasmids reseal themselves without incorporating donor DNA. The lethal gene remains intact and functional

Some plasmids incorporate donor DNA, forming a recombinant plasmid. The lethal gene has been disrupted

Both recombinant and nonrecombinant plasmids are mixed with live *E. coli*, some of which absorb plasmids by transformation

In *E. coli* that has absorbed recombinant plasmid, lethal gene is disrupted. Ampicillin resistance gene is intact

In *E. coli* that has absorbed nonrecombinant plasmid, lethal gene and ampicillin resistance gene are intact

In *E. coli* that has not absorbed either type of plasmid, neither lethal gene nor ampicillin resistance gene is present

All bacteria are streaked on a plate containing ampicillin in the medium

Only bacteria carrying recombinant plasmid form colonies

Figure 15.12 Producing and identifying recombinant bacteria. A foreign gene that is to be cloned (the donor DNA) is randomly incorporated into copies of a replicating bacterial plasmid. In this example, the bacterial plasmid has a resistance gene for ampicillin. There is also a cleavage site for a restriction enzyme in the lethal gene. After donor DNA is incorporated into some plasmids, the plasmids are absorbed by *E. coli* cells. The bacteria that contain the recombinant plasmid survive. Bacteria with nonrecombinant plasmids are killed by the lethal gene. Those with no plasmid are killed by ampicillin.

Next, an enzyme called **DNA ligase** is added. This enzyme binds, or ligates, the breaks in the sugar–phosphate backbone. The result is a sealed, continuous plasmid, which may or may not contain an inserted foreign gene.

Now that the foreign gene has been inserted into the cloning vector (the plasmid), we need to get the cloning vector into a cell where it can replicate. This cell, called the **cloning host**, provides the environment in which many copies of the foreign gene are made. A cloning host must be able to accept the cloning vector. It should also replicate quickly and must be able to maintain the foreign gene for many generations. Frequently, *E. coli* is the bacterial species of choice. The cells and the plasmids are mixed together, and the bacteria are chemically induced to absorb the plasmid by transformation (see Chapter 7, p. 156). Some of the bacterial cells will absorb a recombinant plasmid, while others will absorb a plasmid without the foreign DNA. Some will fail to absorb plasmids of any type. The problem now is to identify the bacterial cells with a recombinant plasmid. The secret to accomplishing this task lies in the lethal and antibiotic resistance genes.

Returning again to Figure 15.12, note that there are now three types of bacterial cells, defined by the plasmids they may have absorbed. First are those cells that absorbed a plasmid lacking donor DNA. They are resistant to the antibiotic but they carry a lethal gene. Second are the cells that did not absorb a plasmid. They do not carry the lethal gene, but they remain sensitive to the antibiotic. Third are the cells that absorbed a recombinant plasmid containing donor DNA. They are resistant to the antibiotic, and the donor DNA has been inserted in the middle of the lethal gene. The lethal gene has therefore been rendered nonfunctional and does not kill the cell.

The cloning hosts are next identified and isolated. This is accomplished by spreading the bacteria over a culture medium containing the antibiotic. Only bacteria containing the recombinant plasmid survive and grow. Bacteria that absorbed a nonrecombinant plasmid die because the functional lethal gene codes for a lethal protein that kills them. Bacteria that did not absorb a plasmid are killed by the antibiotic. Those bacteria that absorbed the recombinant plasmid are protected from these threats and grow into observable colonies that can be maintained indefinitely.

As the colonies of the cloning hosts grow, their recombined plasmids replicate with them. Each of these bacteria can now produce a foreign protein and with literally billions of bacteria now on hand, this protein is made in large amounts. If you remember the fictional example in our case, this protein would have been an enzyme that allowed the genetically engineered bacteria to digest oil. Presumably these bacteria would have been slipped into petroleum storage facilities where they would have reproduced in astronomical numbers, destroying our oil supplies as they replicated. Although this example was just fiction, genetically engineered bacteria are now used routinely to produce commercial or medical products. For example, at one time, the only way to get human growth hormone was to remove it from the pituitary glands of human cadavers. Now, transgenic bacteria with the gene for this human hormone churn it out in substantial quantities.

Other types of cloning vectors may be used in certain situations

Bacterial plasmids cannot carry very much foreign DNA, and so their usefulness for certain purposes is limited. Eukaryotic genes, for instance, which often contain many introns, are often too big for a typical bacterial plasmid. Consequently, other types of cloning vectors are required.

Bacteriophages (bacteria-specific viruses; see Chapter 5, pp. 102–104), for example, make ideal cloning vectors in some cases. Certain bacteriophages, called phage lambda, are specific for *E. coli* and can be engineered to carry more foreign DNA than a plasmid. Large sections of the phage lambda

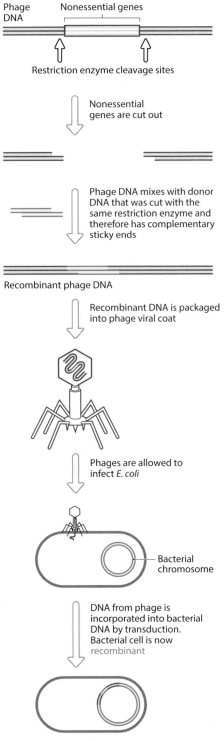

Figure 15.13 Phages as cloning vectors. Phage lambda is a useful tool for inserting donor DNA into *E. coli* DNA.

genome are removed and replaced with the desired donor genes. These recombinant phages are then allowed to infect *E. coli*, inserting their recombined DNA by viral transduction (see Chapter 7, pp. 156–157). In effect, the phage acts as a microscopic syringe to inject foreign DNA into the bacteria. Because so much of the phage DNA has been deleted to make room for the donor DNA, the viral infection does not kill the bacterial cell. Rather, recombined viral DNA is incorporated into the bacterial DNA as in normal transduction (**Figure 15.13**). Once incorporated, donor DNA is replicated along with the bacterial DNA.

Powerful new tools for altering genes are becoming available

We began this chapter with a reference to the **CRISPR/Cas system**, used to create especially muscular dogs. CRISPR/Cas is actually a type of prokaryotic immune response, but it has earth-shaking potential for the field of genetic engineering. But just what is CRISPR/Cas and how might it change our approach to altering genomes for specific purposes?

In 1987 an odd thing was noticed in bacterial genomes; there were often repeated DNA sequences separated by unique sequences in the chromosome. This feature of the genome was named *clustered regularly interspersed short palindromic repeats* or CRISPR. Their function remained a mystery until it was realized that the unique sequences matched sequences in phage DNA. It was subsequently learned that these sequences are fragments of DNA from viruses that had previously tried to attack the cell. In 2007, it was demonstrated that certain of these sequences introduced into bacteria could protect the bacteria from phage attack. In other words, amazingly, prokaryotes with CRISPR/Cas have a type of immunological memory that enhances their ability to defend themselves against viruses they have encountered in the past.

Vital to this immune protection is an enzyme called Cas (Crispr associated protein). When activated, Cas digests the DNA from an invading virus. The mechanism by which CRISPR and Cas destroy invading viruses is depicted in **Figure 15.14**.

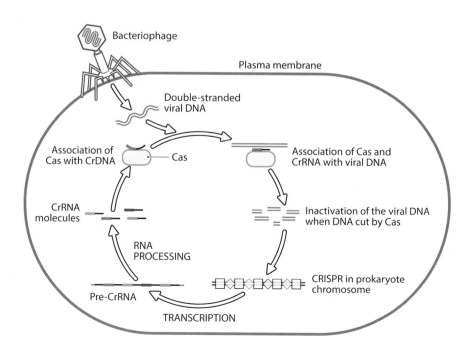

Figure 15.14 The CRISPR/Cas system.
This immune response to phage has been found in the genomes of approximately 40% of bacteria and 90% of archaea. CRISPR consists of repeated DNA sequences (black squares) and unique viral sequences (colored diamonds) aligned in the bacterial chromosome. The CRISPR DNA is transcribed into a long molecule of RNA (Pre-CrRNA). This long RNA transcript is cleaved into individual RNA molecules, each consisting of one repeat and one sequence of unique viral RNA (CrRNA). Each CrRNA associates with a Cas protein. If a phage with a DNA sequence complementary to the CrRNA injects its DNA into the cell (in this example, the red phage DNA has a complementary sequence with the red CrRNA), the Cas–crRNA complex binds the phage DNA and cleaves it. The phage can no longer replicate, allowing the cell to survive the attempted phage attack.

So, why is CRISPR/Cas viewed as such a revolution in genetic engineering? We now realize that this system can be used to precisely edit the genome of any organism. Potentially it may hold the key to curing any genetic disease. Suppose a researcher wishes to introduce a foreign gene into the DNA of a different organism at a precise point in the organism's genome (**Figure 15.15**). All that is needed is the foreign gene, Cas, and a short RNA sequence (termed the guide RNA), which are simultaneously injected into cells of the host organism. Cas then uses the guide RNA to pinpoint its site of action, where it will cut the targeted host sequence in a precise spot. Then, the host cell's DNA repair mechanisms (see Chapter 7, pp. 162–163) may fill in the DNA break with the introduced, foreign DNA. This technique might be used, for instance, to correct a genetic defect. The defective gene could be targeted for disruption and inactivation by Cas and the introduced gene might be a normally functioning version of the defective gene. This technique was first shown to work in 2012 and has since been used on a wide variety of organisms, including fungi, plants, and animals. New breakthroughs and innovations seem to appear almost daily. Researchers have treated muscular dystrophy in mice using CRISPR/Cas technology, and they have created pigs with conditions like diabetes and high cholesterol so that new treatments can be tested. In 2015, a media sensation erupted when Chinese scientists announced that they had successfully used the technique on nonviable human embryos.

Whatever you might think about such announcements, one thing is abundantly clear: We have truly entered the biotechnology age. Next, we look at just a few of the new discoveries and applications that are a direct result of our ability to analyze and manipulate DNA.

DNA technology has provided a better understanding of genes and how they function

The availability of genomic information has already changed our thinking about many aspects of biology. For example, prior to the publication of the human genome, it was generally believed that humans had about 100,000 genes. We now know that the number of human genes is much smaller, probably between 20,000 and 25,000. Also surprising, most of our DNA apparently does not code for protein; much of it is involved in regulating expression of other genes or has no known function. Another revelation is how similar we humans are. Two unrelated humans differ by only about one nucleotide in a thousand. In other words, your DNA is approximately 99.9% the same as that of any other human.

Our similarity extends well beyond even our own species. Human DNA is more than 98% identical to that of our closest nonhuman relative, the chimpanzee, and 80% to that of the mouse. We even share more than half of our DNA, about 60%, with rice plants.

Examination of microbial genomes has yielded new insights as well. For example, we now know that the smallest bacterial genomes are exactly where you might expect them to be—in intracellular pathogens such as *Mycoplasma pneumoniae* (causative agent of atypical or "walking" pneumonia). These are highly fastidious bacteria (see Chapter 8, p. 191) that let the host cell do much of their work. Other bacterial genomes are usually about 10 times bigger. Nonfastidious microbes have the largest genomes, reflecting their ability to carry out many metabolic reactions that other, more fastidious bacteria cannot. Comparison of bacterial genomes also confirms a point we made in Chapter 7; microorganisms are incredibly diverse—certainly as

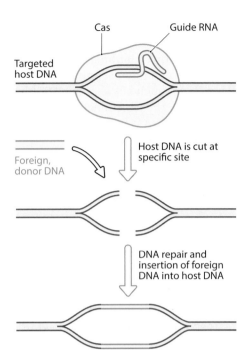

Figure 15.15 Gene editing with CRISPR/Cas. Gene editing is a type of genetic engineering in which DNA is inserted, replaced, or removed from a genome. By introducing Cas and an appropriate guide RNA into a cell, an organism's genome can be easily and cheaply modified at any desired location. Here, Cas and an RNA molecule that specifically base pairs with a DNA sequence in a specific host gene have been introduced. CAS binds the double-stranded host DNA at the precise site, as dictated by the base pairing of the guide RNA and the host DNA. The DNA is separated into single strands and both strands are cut. The host cell's DNA repair mechanisms then repair the cut. If a foreign DNA sequence has also been introduced, as it has been here (in blue), the foreign DNA may be incorporated into the host DNA. If the host DNA is part of a defective gene, this defective gene is now rendered inactive and may be replaced with a normally functioning version of the gene, potentially curing a genetic defect in the host.

genetically diverse as eukaryotes, in spite of the fact that they do not reproduce sexually. On average, between 10% and 20% of the genes in any bacterial species appear to be unique to that species.

Information obtained as a consequence of genetic technology continues to provide us with a better understanding of biological processes. For example, genome comparison between microorganisms can provide insight into a particular organism's evolution. Many of the basic principles that we have already reviewed in this text have been either confirmed or uncovered through such comparisons. In Chapter 9, for instance, we discussed the theory of endosymbiosis. The mitochondrion and the chloroplast, in particular, are believed to have originated when free-living aerobic and photosynthetic bacteria entered into a symbiosis with anaerobic, non-photosynthetic primitive eukaryotes. Genome comparisons of the DNA of mitochondria and chloroplasts with bacterial DNA have allowed us to determine which bacteria specifically gave rise to these organelles. For mitochondria, sequence similarity is greatest with the ancestor of *Rickettsia prowazekii*, the intercellular pathogen that causes typhus (see Chapter 6, pp. 117–118). Recall that mitochondria have their own DNA, and this mitochondrial DNA shares many similarities with the DNA of *R. prowazekii*.

Ecological adaptations are likewise often better understood following genome analysis. In Chapter 4 (see p. 75), for example, we discussed the bacterium *Deinococcus radiodurans*, which is at the top of the heap when it comes to the ability to withstand massive doses of radiation. This species is unusually adept at repairing radiation-induced damage to its DNA and it was assumed that *D. radiodurans* must have some remarkably unique DNA repair mechanisms. Analysis of its genome demonstrates that this is not the case. What it does have is simply more of the mechanisms that are already found in other bacteria. For example, most bacteria have one gene called *mutT* that can repair damaged nucleotides. *D. radiodurans* comes equipped with about 20 of these genes.

Might scientists actually create living cells in the laboratory one day? Such an idea is not as far-fetched as it sounds, and genome analysis is helping biologists determine exactly what genes would be crucial for any such artificial cells. By analyzing the smallest bacterial genomes—those of *Mycoplasma*, for instance—we are starting to understand which genes constitute the bare minimum necessary for life. It has already been estimated that, to sustain a cellular mode of life, somewhere around 260 proteins would constitute the rock bottom for existence.

The practical applications of genetic engineering run the gamut from the ridiculous to the sublime (**Figure 15.16**). At one extreme are innovations such as squirt guns that are loaded with a solution containing jellyfish genes that glows in the dark when squirted onto skin. Or consider transgenic cats, in which the gene responsible for cat allergies has been deleted. At the other end of the spectrum are procedures that prevent disease and save lives. We next look at a few such innovations that are already available or that may soon be.

DNA technology has numerous medical applications

As discussed earlier in this chapter, diabetics may now control their condition with pure human insulin, produced by bacteria (see Figure 15.2). In the manner previously described, the human gene for insulin production was inserted into a plasmid and then absorbed by bacteria. The result was living microbial insulin factories that churned out pure human insulin in an essentially limitless supply. We also mentioned clotting factors used to treat

Figure 15.16 Fluorescent fish. Zebra fish, marketed under the name GloFish®, have been genetically engineered to express a jellyfish gene, which allows them to fluoresce. Recently, genetically modified fluorescent tetras and tiger barbs have been added to the roster of available fish.

hemophiliacs and interleukins, available to regulate immune responses, both available courtesy of recombinant bacteria.

Other medically valuable proteins are produced not by bacteria but by transgenic animals. The most common way to insert foreign, donor genes into animals has been to deliver the genes by way of a virus into fertilized eggs or very early embryos (**Figure 15.17a**). In other words, the virus acts as a sort of microbial shuttle to carry foreign genetic material into the recipient animal. Alternatively, the donor gene can be injected into eggs or embryos, where it may be incorporated into the host DNA. The goats we mentioned earlier that produce anticlotting agents are one example (**Figure 15.17b**). Other proteins made by transgenic animals include human albumin, produced by transgenic cows, which is used to increase blood volume, and a protein made by transgenic sheep that is used to treat hereditary emphysema. Very recently, transgenic cows have been created that produce human antibodies against hantavirus. These antibodies can be used to treat patients suffering from infection with this virus. This new therapy is still experimental, but offers hope against a serious disease for which no reliable drugs are currently available. The notion of using transgenic animals in this way to produce medically important compounds is often somewhat amusingly called "pharming."

What is the advantage of using transgenic animals over simply letting transgenic bacteria make the protein? One straightforward reason is that, in transgenic animals, the protein is often secreted in milk or semen, where it is easy to collect. Furthermore, in eukaryotes, many proteins are modified in certain important ways following their production. These modifications take place in the endoplasmic reticulum and Golgi apparatus. Prokaryotes, lacking these organelles, do not necessarily modify proteins in the same way.

Vaccines may now be created using recombinant DNA technology

To illustrate the impact that biotechnology has had on vaccine development, consider the vaccine that is used to protect chickens and turkeys from two diseases at once: fowlpox and Newcastle disease. To make the vaccine, the

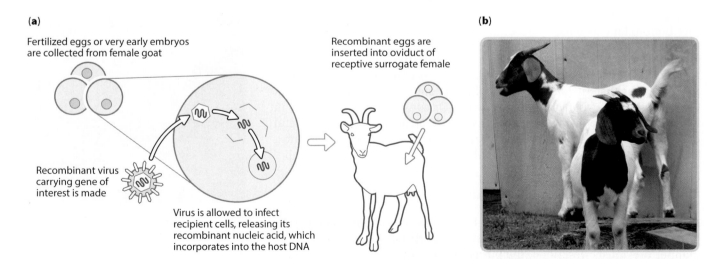

(a)

Fertilized eggs or very early embryos are collected from female goat

Recombinant eggs are inserted into oviduct of receptive surrogate female

Recombinant virus carrying gene of interest is made

Virus is allowed to infect recipient cells, releasing its recombinant nucleic acid, which incorporates into the host DNA

(b)

Figure 15.17 Transgenic goats. (a) The virus used in this technique must be a virus that incorporates its DNA into that of the host. Often a retrovirus is used for this purpose. (b) Down on the "pharm." These baby transgenic goats will secrete a human anticlotting protein in their milk or semen.

fowlpox virus is modified by removing the genes that cause disease symptoms in birds. The genes that code for proteins eliciting immunity (antigenic proteins; see Chapter 12, pp. 285–286), however, are left intact. Meanwhile, the genes that code for immunity-inducing proteins from Newcastle virus are added to the fowlpox nucleic acid. The result is a recombinant virus that cannot cause either disease. It can, however, be used as a **recombinant vaccine** to induce immunity against both diseases at once (**Figure 15.18**).

In Chapter 12 we described the **subunit vaccine**, currently used to protect against hepatitis B. Subunit vaccines, unlike traditional vaccines, consist of neither a killed nor an attenuated microorganism. Rather, those microbial proteins that best stimulate an immune response are identified. Once the genes that code for these proteins are pinpointed, they are cloned, resulting in large amounts of the immune-stimulating proteins. These proteins are then utilized as a vaccine to boost immunologic memory. In addition to hepatitis B, subunit vaccines are now in development for various other pathogens. Not only are these vaccines safer than traditional vaccines, but they can usually be produced much faster as well. Recall, for instance, our discussion of influenza epidemics in Chapter 14. A subunit vaccine against a new strain of influenza might be available with a few months' warning. This is in contrast with the 6–9 months needed to mass-produce traditional influenza vaccines—an important consideration if a new epidemic is looming.

Figure 15.18 A nontraditional vaccine produced through biotechnology. Recombinant vaccines are used to protect poultry against more than one disease at once.

Genetic conditions in humans may one day be corrected with gene-editing techniques

With the human genome now available, more and more of the mutant genes responsible for various genetic diseases are being identified. And since the 1990s, scientists have wondered if it might be possible to treat some of these diseases by introducing normal copies of the gene into affected individuals. The normal gene then produces the necessary protein that the mutant gene was unable to produce. This form of treatment, known as **gene therapy**, was first used successfully to treat a child with a serious immunological deficiency caused by a single mutant gene. The disease is called severe combined immunodeficiency or SCID. Although gene therapy has been successful in the treatment of SCID and some other conditions, in most cases the technique is still experimental. Getting the host cells to take up foreign DNA and then to express this DNA in significant amounts still presents a formidable challenge.

The arrival of CRISPR/Cas technology, however, has breathed new life into the idea of correcting genetic abnormalities through the use of gene editing (see Figure 15.15). We have already mentioned how animals have been created with diseases mimicking human diseases, which can then be studied in order to better understand the diseases themselves and to test new treatments. Because the use of gene editing on humans raises important ethical issues, the technique is not currently used on people. But that day may be fast approaching. To highlight one recent breakthrough, melanoma, a dangerous form of skin cancer, is often treatable with certain anti-cancer drugs. In some people, however, the drugs are ineffective because these individuals carry a mutant gene that makes tumor cells resistant to the drug. In 2015, it was shown that these mutant genes can be inactivated using CRISPR/Cas. With the mutant gene silenced, the tumor cells once again become vulnerable to the drugs. We also mentioned earlier that Chinese scientists created a stir in 2015 when they announced that they had edited the genome of nonviable fetuses. The scientists were attempting to correct a mutation that causes thalassemia, an often-lethal genetic disorder that reduces the ability of red blood cells to carry oxygen. Although they were only partially successful, we are well on the way to a point where such interventions are not only possible, but even routine.

Genetically modified crops are now a reality

CASE: THE MONARCH BUTTERFLY: THREATS REAL AND IMAGINED

In 1999, monarch butterflies (**Figure 15.19**) found themselves at the center of a biotechnology controversy. These insects lay their eggs on milkweed and the caterpillars feed exclusively on this plant. But concerns about the monarch's survival arose when it was discovered that milkweed growing in the vicinity of genetically modified Bt corn was dusted with corn pollen. Monarch caterpillars that were fed pollen from Bt corn in the laboratory died in large numbers. A public outcry about the potential danger of genetically modified crops resulted and several investigations to further assess the risk to butterfly populations were launched.

1. **What is Bt corn? How might it pose a risk to butterflies?**
2. **What did subsequent investigations show?**

Figure 15.19 Butterflies at risk? The controversy surrounding genetically modified food crops was highlighted by reports that Bt corn posed serious hazards for monarch butterflies.

Over 90% of the soybeans grown in the United States have been genetically altered. The same is true for corn and certain other crops. Many other countries have similarly increased their use of genetically modified (GM) crops (**Figure 15.20**). These transgenic plants have been created in a manner similar to the transgenic animals discussed earlier. One common technique has involved the modification of plasmids taken from the bacterium

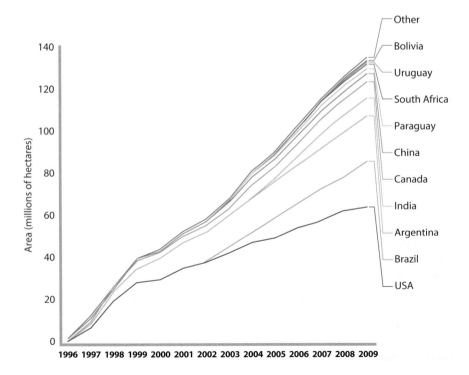

Figure 15.20 Increased used of genetically modified (GM) Crops. Data are presented for various countries between 1996 and 2009 (x axis). The y axis measures millions of hectares used for such crops. In 2011, 160 million hectares (1.6 million square kilometers) were devoted to GM crops, representing over 10% of the world's croplands.

Agrobacterium tumefaciens. This unusual species infects many plant species, causing the plants to grow tumors commonly called galls (**Figure 15.21a**). When *A. tumefaciens* infects a plant, it can insert some of its plasmid DNA directly into the plant cells. This bacterial DNA is subsequently incorporated into the plant DNA, where it codes for proteins initiating gall formation. But this bacterium can be made to play a very different role if plasmids are obtained from the bacteria and modified by previously described techniques. Gall-inducing genes are disrupted, and a foreign gene coding for a desirable characteristic is inserted (**Figure 15.21b**). The plasmids retain the ability to enter plant cells, but now, instead of gall-forming genes, they introduce a donor gene, creating a transgenic plant embryo. Some plants will not absorb *A. tumefaciens* DNA, and for them, a gene gun

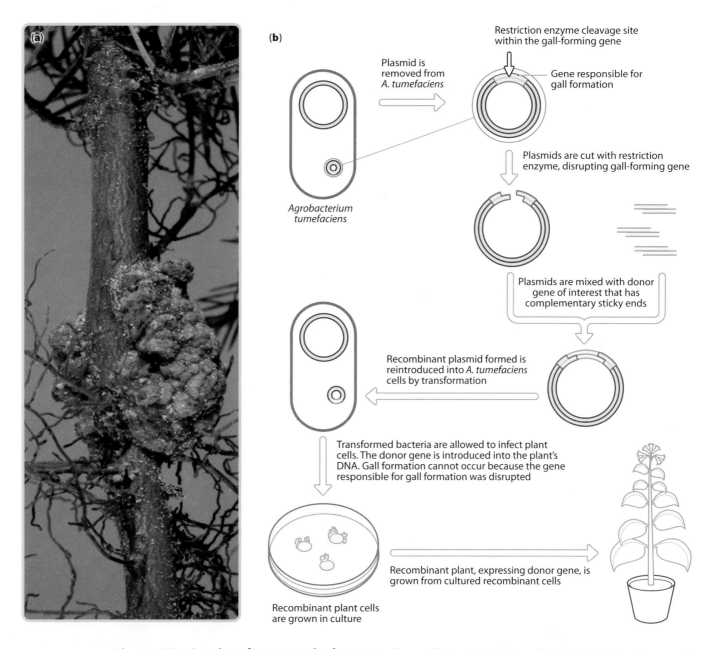

Figure 15.21 Creation of a transgenic plant. (a) A gall, caused by *Agrobacterium tumefaciens*. This bacterium is commonly used to introduce foreign genes into plants, because when it infects plant cells it inserts plasmid DNA into the host plant DNA. (b) A protocol for creating transgenic plants by means of modified *A. tumefaciens* plasmids.

can be used to introduce foreign genes of interest. The genes are made to adhere to tiny metal bullets, which are then fired into the cells with a blast of compressed gas.

Foreign genes that are introduced in this way may come from other plants that naturally have a desirable characteristic, such as high oil content or an increased tolerance for drought. Other times the genes do not come from plants at all. For example, the Bt corn in our case carries a bacterial gene from *Bacillus thuringiensis*. The gene codes for the production of a powerful toxin that can kill insects. Corn modified in this manner essentially makes its own insecticide, greatly reducing damage from insect pests. Likewise, a gene from *Salmonella* is currently used in transgenic cotton to make the cotton plants resistant to weed-killing chemicals.

Is all this a good thing? It depends on whom you ask. The genetically modified soybeans are lower in saturated oils, and the altered corn has increased resistance to insect pests. Yet many people are strongly opposed to genetically modified (GM) foods. Some worry about the consumption of GM foods and about currently unknown effects they might have on health. Many European countries and other nations have either banned GM foods entirely or require the placement of labels on all GM foods, identifying them as such.

One major environmental concern is how GM foods might affect wild plants. Many food plants can hybridize with native plants, and if such hybridization occurs between GM plants and native plants, new genes might be introduced into wild populations. The consequences are difficult to predict. If weeds, for example, obtain a gene for resistance to herbicides from a resistant GM crop, what will be the result? Our case of potentially endangered butterflies shows that it is not even necessary for GM plants to hybridize with wild plants to cause mayhem. Simply releasing genetically modified plant parts into the environment may be enough to create problems. Certainly the laboratory evidence suggested that Bt corn pollen could harm the butterflies. Such pollen might be carried by the wind to the milkweed plants, where monarch caterpillars might consume it. Subsequent investigations, however, showed that, in the field, monarch caterpillars rarely if ever ingested enough Bt pollen to cause death. Monarch butterfly populations have seriously declined in recent years, but it is believed that other problems, such as use of pesticides and loss of habitat are the principal culprits. Bt corn is highly unlikely to be a factor in this instance, but controversy over GM crops will no doubt continue. Although capable of great benefit, genetic engineering always includes an element of risk. The risks require careful analysis before any genetically modified product is introduced.

Biotechnology has become routine in many industrial processes

The next time you slip into a pair of stone-washed jeans, you might use the occasion to ponder the marvels of biotechnology and its use of microorganisms (**Figure 15.22**). The cotton used to make jeans is essentially composed of cellulose, a complex carbohydrate found in plant material. Cellulose can be made softer by treating it with cellulose-digesting enzymes called cellulases. Some of these enzymes originally came from fungi, while others were first isolated from bacteria that colonize clams and oysters. The genes that code for the commercially valuable enzymes have been introduced into other microorganisms that are easier to rear in the laboratory. These microbes then produce the cellulase enzymes in large quantities. Not only do these enzymes produce fashionable pants, but because they are

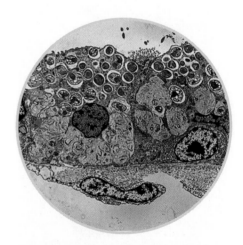

Figure 15.22 No stones about it. Certain clams and other bivalves harbor bacteria that produce the enzyme cellulase (seen here inside the cells of a mussel). The genes for this cellulose-digesting enzyme have been introduced into other microorganisms, which are then utilized in the production of stone-washed jeans.

biodegradable, they are also environmentally friendly. Thus, stone-washing has nothing to do with stones. Apparently, jean manufacturers did not think "microbially washed" had marketing potential.

Recombinant microorganisms now also play an important role in the new field of **bioremediation**, or the use of biological processes to reduce pollution. Like the bacteria in our earlier case, transgenic microorganisms that degrade petroleum and other environmental pollutants have been produced. A species of *Flavobacterium*, for instance, actually produces enzymes that degrade organophosphates, a group of chemicals that includes highly toxic nerve gases. The gene that enables these bacteria to digest organophosphates has been introduced into *E. coli*, which can then degrade the toxin in the event of a toxic chemical spill. Should such a spill occur, recombinant *E. coli* stand ready to swing into action as important first responders. We discuss bioremediation and other industrial uses of microorganisms further in Chapter 17.

Biotechnology raises important ethical and safety concerns

Just because we *can* do something, does this mean we should?

New technology that threatens to change lives has always generated controversy. Protests arose during the Industrial Revolution among those who believed that new mechanized technology posed a serious danger to workers and their way of life. We still debate whether or not the development of nuclear energy was wise. Certainly DNA technology is no exception. From its very inception, genetic engineering has raised safety and ethical issues (**Figure 15.23**). In the 1970s, scientists worried whether recombinant bacteria might escape into the environment, and if so, what problems they might cause. Today we ask ourselves about the safety of GM foods. Are they safe to consume? What ecological havoc might they create? Is the ability to grow more food, perhaps using fewer pesticides, worth the risk?

Figure 15.23 Genetic engineering: the continuing controversy. A protest against genetically modified food highlights the strong opinions stirred by this issue.

We now have the ability to screen people for genetic disorders. If such screening is carried out, who should have access to the results? And the lightning-fast pace of development of gene-editing technology means that it may not be long before the debate begins about whether or not to apply these techniques to humans. Is tinkering with human genetic information something we wish to do? Certainly a strong case can be made when it comes to correcting genetic disorders. But how will we proceed when it becomes possible to create designer babies in which genes for desired characteristics—increased intelligence or athletic skill, for example—might be spliced into developing embryos. This ethical dilemma does not begin and end with humans. Is it proper to create new forms of life such as transgenic animals? Is it morally defensible to develop animals with human diseases like diabetes, just so we can study them?

Answers to these questions will come not only from scientists but also from politicians and the general public. Unfortunately, it is a sign of the times that frequently politicized arguments often generate more heat than light. We live in the "sound bite" era, in which arguments, whether or not they are valid, are used to advance a political or social agenda. The emphasis is too often on winning rather than being right.

What is the solution? The first step is making sure that the public has the necessary information and genuinely understands the issues under consideration. Only then can we cast votes and make decisions that are based on rational analysis rather than emotion or political agenda. It is certainly true that everyone is entitled to his or her opinion, but opinions mean little when unsupported by knowledge and understanding.

Thus, we conclude Chapter 15 very much where we started it. Biotechnology has enormous potential for good, but it also raises important safety and ethical issues. Each of us, armed with an understanding of what is involved, can help encourage wise policy decisions, allowing this powerful new technology to be used in the most thoughtful, beneficial, and ethical manner.

Looking back and looking forward

The capacity to create organisms with recombinant DNA or to alter the genomes that organisms inherit is now a reality. DNA technology has substantially increased our knowledge of basic biological processes and it has given us the ability to confer new and beneficial characteristics to living things. It is true that biotechnology offers enormous benefits, but it also raises important safety and ethical issues. With an understanding about the science behind this life-altering technology, we are now better positioned to make decisions about its wise use. In the future, as the pace of advancement continues to accelerate, it will become even more important that the general public know something about what goes on in laboratories around the world.

Chapter 16 introduces another way that microbes have impacted our daily lives—in terms of our diet. Ever since the first bread rose and the first wine flowed, microorganisms have contributed to the food we eat. In fact, it would be difficult to get through a meal without consuming something that comes to you courtesy of bacteria or fungi. Many of these food items have been around for thousands of years. Others have found a place on our dinner table relatively recently. In Chapter 16 we look closely at the hand that microorganisms have in our diet and how much less interesting that diet would be without microbial assistance.

Garland Science Learning System

- http://garlandscience.rocketmix.com/students/

- Discover how you know bacteria are there if you can't see them and you can't grow them

- Test your knowledge of this chapter by taking the quiz

- Familiarize yourself with the terminology used in this chapter by using the vocabulary review

- Get help with the answers to the Concept questions

Concept questions

These questions are designed to help you start thinking like a microbiologist. The answers are not always simply found in the text. Instead, you will need to take the concepts about which you have learned and apply them to new situations. Some of the questions may not even have just a single correct answer. Help is provided as part of the GSLS resources, which can be accessed through http://garlandscience.rocketmix.com/students/.

1. How would you describe the term "biotechnology" to someone with no scientific background?

2. Explain why the production of sticky ends in DNA, when the DNA is cut with a restriction enzyme, is an important reason why such enzymes are valuable when creating recombinant DNA.

3. Name three ways that microorganisms have contributed to our ability to manipulate or alter DNA.

4. Here in Chapter 15 we found that parasitic bacteria, which live intracellularly in host cells, typically have smaller genomes than other bacteria. Why does this observation make sense?

5. Look through a newspaper or magazine, or look online to find an article or advertisement for a genetically engineered product. Then provide a brief outline that describes how you think this product may have been produced.

6. We concluded Chapter 14 with a discussion of bioterrorism. Should we be concerned about the use of genetic engineering by terrorists to create more lethal bioterrorism weapons? How might genetic technology be used for such a purpose?

GUESS WHO'S COMING TO DINNER: MICRO-ORGANISMS AND FOOD

16

For the true beer connoisseur, there is really no place quite like Belgium, home to some of the world's best brews, and certainly to some of the most unusual. And no trip to Belgium would be complete without sampling the Trappist ales. The Trappists, an order of Catholic monks, have been brewing beer in their monasteries for centuries. But be alert for imitations. While other beers might use the name "Trappist," only 11 authentic Trappist monasteries in the world still produce the genuine article (**Figure 16.1**).

Perhaps the most unusual of the Trappist ales is lambic beer, sometimes called Belgian champagne. This ale is definitely not for the casual beer drinker or anyone who prefers the crisp, clean taste of a traditional German-style lager. Lambic beer is decidedly stronger both in alcohol content and in flavor. It strikes many as overly sour. The secret to its unusual flavor is the manner in which it is made.

Most beer is produced when commercial yeast strains, almost invariably members of the genus *Saccharomyces*, anaerobically metabolize the sugars in grain, releasing ethyl alcohol as a fermentation waste product (see Chapter 8, pp. 178–179, for a discussion of fermentation). The production of lambic ales, like that of most beers, begins by adding a conventional *Saccharomyces* strain to a boiled mix of grain, water, and hops. Once fermentation is under way, the

Figure 16.1 Trappist ale. A monk oversees brewing at a Trappist monastery in Belgium. Genuine Trappist ale is made only in Trappist monasteries.

All case studies have a few questions at the end, the answers to which will become apparent as you read the sections following the case.

monks open the windows of the brewery and let nature take its course. Wild yeast cells are wafted in on the Belgian breeze, and as they settle in the fermentation vats, they also participate in fermentation. The result is somewhat variable, because each type of wild yeast releases different metabolic waste products, imparting its own special contribution to the final flavor. The final brew consequently ends up quite different from what we normally think of as beer. Regular or straight lambic beer consists of this simple, wild, fermented beverage. If that sounds a bit intimidating, you may wish to sample kriek beer instead. This is lambic ale that is flavored with cherries. There are also pêche (peach-flavored); framboise, into which raspberries have been added; and even varieties flavored with apricots and blackcurrants.

Although lambic ale may seem a bit unusual, the use of microorganisms in the production of foods and beverages is commonplace. Consider, for instance, a fairly typical meal. Perhaps you would like to start with a salad. If your favorite dressing is oil and vinegar or blue cheese (or any other dressing containing vinegar or cheese), you will adorning your salad with a food item produced by microbes. Likewise, if you select cheese enchiladas, a hearty stew with summer sausage, or an Asian dish flavored with soy sauce as your entrée, microorganisms contributed to the recipe. The same is true for your dinner rolls, or almost any other bread product that might complement your meal. And to drink? Beer, wine, most soft drinks, and even coffee are produced with microbial assistance.

Not all microbial activity is so benign when it comes to our food. Throughout this text we have had ample opportunity to discuss food- and water-borne diseases and the manner in which microorganisms cause food to spoil. We have similarly spent plenty of time investigating how food-borne diseases are controlled and treated and how microbially induced food spoilage may be reduced. So, enough with the bad news. Here in Chapter 16 the spotlight is on microorganisms as partners rather than adversaries. After a bit of history we look at the ways in which fungi and bacteria are either consumed directly as food or used in the production of everything from pickles to pinot noir to provolone cheese. The production of some nutritional supplements, including vitamins and amino acids, also relies on microbial assistance, and we investigate the part microorganisms play in stocking the shelves of the local health food store. Finally, we list a few recipes involving microorganisms. Try one, raise a glass to your microscopic culinary colleagues, and say, "Bon appétit"!

The beginnings of a beautiful friendship

CASE: FLIGHT FROM EGYPT

Microorganisms have contributed to at least some of humanity's dietary needs since ancient times. Even the Bible refers to foods in which microbial processes are crucial, such as in the story of the Israelites and their hasty departure from Egypt (Exodus 12:5): "Seven days shall ye eat unleavened bread...."

1. **What is the connection between this biblical quotation and food microbiology?**

Microbial activity can help to preserve the quality of some foods

Imagine what life would be like without refrigeration. Suddenly, eating would become literally a hand-to-mouth affair. Most foods spoil rapidly,

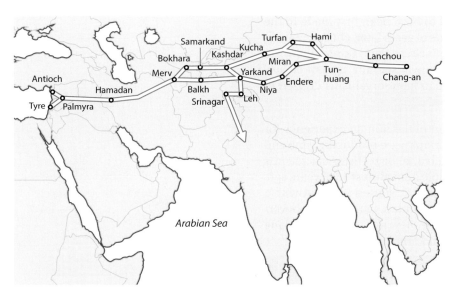

Figure 16.2 The spice trade. This map highlights the main routes of the Silk Road between Europe and Asia, along which many goods, including spices, were transported. The quest for spices was a response in part to their ability to inhibit microbial growth and consequently food spoilage.

especially in warm weather, so in general, unless you were willing to simply hold your nose and force down rotten food, you would have to buy, hunt, harvest, or gather your meals almost daily. Or, if you were rich enough, you could buy spices. Not only do spices make food tastier, but many contain compounds that inhibit microbial growth, reducing food spoilage. Garlic, oregano, thyme, cumin, and capsaicin (found in chili peppers) are just a few commonly used spices that have been found to exert strong antimicrobial activity. Prior to the development of more modern food preservation techniques, spices were one of the few reliable ways to slow down food spoilage. In fact, through much of history, spices were big business, in part for just this reason. Many early explorers and traders were motivated by the high price that spices commanded (**Figure 16.2**). Marco Polo, for instance, set off for China with his father and uncle in search of marketable spices found only in East Asia.

Or, as an alternative, you might actually encourage the spoilage of certain foods. As we learned in Chapter 8, many microorganisms are fermenters, able to break down biological molecules to a variety of waste products in the absence of oxygen. In some cases, these wastes serve a valuable purpose; they add flavor to food as well as prevent additional spoilage caused by less savory microbes. Fermented foods have formed an important part of the human diet for thousands of years. Those producing the foods had no idea that organisms too small to see were involved—they simply knew that the process worked.

Fermented dairy products and grains have been used for thousands of years

The earliest production and consumption of fermented foods is lost in antiquity. Fermented dairy products have probably been around ever since a horseman in Central Asia filled up a goatskin with milk, tied it to his pack, and hit the trail. Taking a drink at the end of the day, the rider found that the milk, which had sloshed around all day in the hot sun, had turned into a semisolid curd. The first yogurt had been made, all because of the bacteria that had fermented sugars in the milk (see Chapter 8, pp. 193–195). Cheese, another fermented milk product, is believed to have been first made in the Middle East at least 8000 years ago. Thousands of years later, cheese-making became both an industry and an art in ancient Greece and Rome. There are hundreds of types of cheese today, and as we will see, the type of cheese produced is in large measure due to the type of microorganisms utilized in the cheese-making process.

Alcoholic beverages have likewise been a part of the human experience for thousands of years. In 4000 BC, ancient Sumerians were producing wine and beer from fermented grapes or barley (**Figure 16.3a**). Historical records from ancient Egypt show that beer was a taxable item almost 5500 years ago, and Romans, living in pre-caffeine times, frequently started their day with a flask of wine. In fact, because water was often polluted, wine was the beverage of choice for many early Europeans. By the eleventh century, drinking establishments were a fixture across Britain and other parts of Europe.

Bread, which like alcohol relies on yeast fermentation (see Chapter 8, pp. 178–179), also got its start in the Middle East. Ancient Egyptian hieroglyphics clearly depict bread-making in about 2500 BC (**Figure 16.3b**), and large excavated Egyptian bakeries are believed to have supplied thousands of people each day with the staff of life. The preference of ancient people for bread that had risen (leavened bread) as opposed to bread dough that was baked without first rising (unleavened bread) is well illustrated by the biblical quote in our case. The Israelite slaves, abruptly freed by the Egyptian Pharaoh, were in such haste to abandon Egypt that they left before allowing their bread to rise. The hard, flat, unleavened matzos eaten by observant Jews at Passover commemorate this event. The New Testament (in I Corinthians 5:6) also makes reference to the use of yeast in bread-making: "Know ye not that a little leaven leaveneth the whole lump?"

While fermented foods have been produced and consumed throughout human history, it was not until the late 1800s that the role of microorganisms was at last understood. As we learned in Chapter 6, at this time, Pasteur and Koch and other scientists first described the role of microorganisms in processes such as disease and food spoilage. Once early microbiologists understood the importance of microorganisms, they began to work out the relationships between particular food products and specific microorganisms. With the advent of pure culture techniques, the right microorganism for the job could be cultivated and introduced into food items, reducing the chance of contamination with unwanted microbial species. The early twentieth century witnessed the development of large-scale industrial food production, and microbial fermentation became big business.

Figure 16.3 Fermented foods in ancient history. (a) The Sumerians were among the first to brew beer. The photograph shows a brewing recipe written in cuneiform script, the earliest known writing system. (b) Egyptian illustration depicting bread-making.

Some fungi and bacteria are consumed directly as food

CASE: WHAT'S FOR TUCKER, MATE?

On his first trip to Australia, Bill arrives in Sydney for a professional conference. On his first morning in town, he goes down to his hotel's buffet for breakfast. Bill picks up a couple slices of toast, and after finding a table, he notices that along with butter, there are several jams and jellies available as condiments. There is also an unfamiliar dark brown spread that Bill suspects might be something like apple butter. Eager to try something new, he spreads a thick layer on his toast. Yet when he takes a bite, he quickly realizes that the unknown spread is certainly not apple butter. The strong, salty flavor is almost more than he can take, and after forcing it down, he quickly gulps a glass of orange juice to get rid of the taste. Later, he notices that many people are eating this spread, and they seem to like it. Asking someone at a neighboring table, he finds out that the brown spread is called "Vegemite" and that it is practically an Australian national dish.

1. **What exactly is Vegemite, and how is it made?**

The impact of most microorganisms used in food preparation is indirect. Microbial activity in products like milk, cheese, and vinegar results in the release of compounds that alter the taste, smell, or consistency of the food. However, a few microorganisms have a more immediate connection to human gastronomy. These are the ones we simply eat as is. Next, we describe some of the ways that microorganisms themselves are on the menu.

To many people, the thought of dining on fungi or bacteria sounds revolting. Yet throughout this text we have been reminded that most single-celled organisms pose no threat and are often beneficial. Furthermore, every bite of food or swallow of liquid that we take is laden with microorganisms. So why should it be unappetizing to consume them intentionally? Nevertheless, for some people a food product that is obviously microbial raises a red flag. Perhaps it is a case of "out of sight, out of mind." Bacteria in our lunch that we cannot see and usually do not think about are one thing. Intentionally seeking out microorganisms for lunch is something else altogether.

Yet we readily eat mushrooms, the larger cousins of microscopic fungi. Indeed, some mushrooms are considered to be a true delicacy and often demand a high price. When you include mushrooms in your pizza or omelet, you are actually eating the fruiting bodies: the large reproductive structures of even larger filamentous fungi living under ground (see Chapter 4, pp. 83–84). Each mushroom is formed by numerous hyphae that together make up the reproductive body, called the mycelium. Of the more than 300,000 tons of mushrooms consumed in the United States annually, most are grown commercially in temperature- and humidity-controlled buildings, in an organically rich bed of compost (**Figure 16.4a**).

Figure 16.4 The mushroom business. (a) White button mushroom production at a California mushroom farm. (b) Chanterelles are especially prized by gourmets and mushroom hunters.

Mushrooms are appreciated for their taste and texture. Although they are not a rich source of vitamins or protein, mushrooms can easily form part of a healthy diet. They are essentially fat- and cholesterol-free, and they provide a good source of dietary fiber. For many people, they are also fun to collect. During the summer months, taking to the woods to harvest morels, chanterelles, and other highly prized varieties is an age-old tradition (**Figure 16.4b**). Unfortunately, it can also cause friction between collectors, landowners, and conservationists who decry the destruction of mushroom populations at the hands of overzealous mushroom hunters. Part of the problem, however, is based on a general misunderstanding regarding basic mushroom biology. The fungus is not killed when a mushroom is picked; it has merely lost a fruiting body. This means that picking mushrooms is more like picking berries from a bush than it is like overfishing a lake. The subterranean fungus that produces the mushroom continues to survive. Indeed, mushroom harvesting likely encourages the production of a new crop of fruiting bodies. On commercial mushroom farms, systematic picking actually increases the production of additional mushrooms.

So, if we are willing to eat mushrooms, why shouldn't we eat smaller, microscopic fungi? In certain places like Australia, we already do. Vegemite is a commercial product made from an extract of *Saccharomyces cerevisiae* (brewer's yeast) to which vegetable flavorings and spices have been added (**Figure 16.5**). The yeast used to make Vegemite is often left over from beer brewing. Vegemite has a dark reddish-brown color and it is one of the richest sources of essential B vitamins. Although Vegemite has the consistency of peanut butter, it has a flavor all its own. As Bill found out, Vegemite is definitely an acquired taste. It is very salty, and to the uninitiated, the flavor can be overpowering. Yet for Australians, Vegemite is an essential part of the culture. It is found in more than 90% of homes, and children are brought up on it from the time they are infants. A Vegemite sandwich is the quintessential Aussie snack.

Vegemite and other consumable forms of yeast are examples of **single-cell protein**—bulk microorganisms that are used as a food source for humans or animals. Because of its high vitamin content, yeast can be added to wheat or other grains to enhance nutritional value. When yeasts are grown on high-nitrogen media, they are also fairly high in protein content. Simple powdered yeast can also be added to foods as a nutritional supplement. Too much yeast in the diet, however, may be a problem for some people because of the high nucleic acid content, which can lead to gout.

Bacteria have been less successful at working their way into the human diet. A few decades ago, cyanobacteria in the genus *Spirulina* were hailed as the answer to food shortages (**Figure 16.6**). It is estimated that 10 tons of protein

Figure 16.5 Lunchtime in Australia. A Vegemite advertisement from the 1960s. Older Australians probably recall a commercial jingle for this product from their childhood: "We are happy little Vegemites as bright as bright can be. We all enjoy our Vegemite for breakfast, lunch, and tea. Our mummy says we're growing stronger every single week. Because we love our Vegemite, we all adore our Vegemite, it puts a rose in every cheek!"

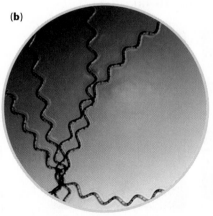

Figure 16.6 Bacteria as food. (a) *Spirulina* tablets. Although it is a rich source of protein, *Spirulina* has yet to catch on as a mainstream food. It is, however, readily available in health food stores as a dietary supplement. (b) *Spirulina platensis* cells.

per acre could be harvested from *Spirulina*, compared with 0.16 tons per acre for wheat and 0.016 tons per acre for beef. *Spirulina*, however, never really gained broad acceptance as a food item, although it is available as a nutritional supplement in health food stores. If you really want to sample *Spirulina*, a good place to go is the North African country of Chad. Here, *Spirulina platensis* is collected from the shallow waters of Lake Chad. The bacterial mats are sun-dried and cut into protein-rich cakes called dihé (**Figure 16.7**).

As previously mentioned, however, it is microbial metabolic products rather than the microorganisms themselves that are far and away the principal route by which bacteria and fungi affect our diet. Most of these by-products are the result of fermentation. Next we take a look at how this basic aspect of microbial metabolism is so crucial to many of the foods we eat.

In the absence of oxygen, some microorganisms undergo fermentation, releasing specific waste products

Fermentation was described in detail in Chapter 8 (pp. 178–179). As we learned, in the absence of oxygen, many cell types continue to produce ATP via glycolysis alone. The final product of glycolysis is pyruvate, but without oxygen, pyruvate cannot be broken down further. Instead, it is reduced to a final waste product in order to oxidize NADH to NAD^+. The NAD^+ is required for glycolysis to continue.

As previously mentioned, humans have taken advantage of microbial fermentation for thousands of years, using it to preserve food or enhance its flavor. If fermenting microorganisms are either added or naturally gain access to a food product, the fermentation waste products change both the chemical and textural qualities of the food, extending its shelf-life and/or creating new desirable flavors and aromas. We next take a closer look at how some of these foods are made, highlighting the indispensable services provided by fermenting bacteria and fungi.

Flour from grains provide a sugar source in bread-making

Considering its central role in the diet of so many people, bread is a good place to start our survey of microbially produced foods. To make bread, the flour, obtained from wheat, rye, or other grains serves as a sugar source for yeast. As the yeast ferments the sugars in the flour, it causes the bread to rise and also contributes to the taste and aroma of the final baked product.

To make bread, a strain of the single-celled fungus *S. cerevisiae* is added to the bread dough (**Figures 16.8a** and **16.8b**). Surrounded by dough, some of the yeast cells find themselves in an anaerobic environment. Consequently, in order to grow, they must ferment the sugars in the dough.

The pyruvate produced at the end of glycolysis is reduced to ethanol (ethyl alcohol) and carbon dioxide. The CO_2 gas causes the dough to rise, giving the bread its light, fluffy texture (**Figure 16.8c**). Small amounts of other fermentation wastes contribute to the flavor of the bread. All traces of ethanol evaporate from the bread as it is baked. If yeast is not added to bread dough, the final baked product will remain flat and hard.

Occasionally, other microorganisms are employed to produce more unusual breads. Sometimes a complex assemblage of microorganisms is called for

Figure 16.7 Dihé preparation. A woman pours Spirulina that she has collected from Lake Chad into a round depression for drying. Once dried, it will be sold in local markets as dihé.

that contains both yeast and bacteria. In sourdough bread, *Lactobacillus sanfrancisco* contributes to the flavor by releasing lactic acid, giving the bread its characteristic acidic or sour flavor. The flavor of rye bread is due in part to the use of lactic-acid-producing bacteria such as *Lactobacillus bulgaricus*, *Lactobacillus plantarum*, or *Streptococcus thermophilus*.

Wine is produced from fermented fruit juice

CASE: HOMEMADE WINE

Joe has recently taken up the hobby of winemaking, and he has already made a variety of wines that he enjoys sharing with family and friends. The final products vary considerably in character and taste, depending on the type of wine he is making. For one thing, the sweet wines tend to be lower in alcohol content, while the dry wines have a higher final alcohol content. Although Joe has always liked wine, he never noticed this correlation between sweetness and alcohol content before he started making wine himself.

1. **Why is there a relationship between the sweetness and alcohol content of homemade wine?**
2. **Why is this trend less valid for commercially made wine?**

Just about any type of plant juice can used to make wine. If you look hard enough, you can find wine made with fermented plums, oranges, or pomegranates. Typically, however, most wine is produced by grape fermentation.

To start, grapes are first crushed to obtain the grape juice, called the **must**, which contains most of the sugars (**Figure 16.9a**). The must is usually treated with sulfur dioxide to rid it of any microorganisms that are naturally present on the grapes. If these bacteria and fungi are not eliminated, their own fermentation waste products will end up in the wine, resulting in an unpredictable and usually undesirable final product.

Once contaminating microorganisms have been removed, an appropriate strain of *S. cerevisiae* (brewer's yeast) is added. As the yeast ferments the sugars in the must, ethyl alcohol is produced (**Figure 16.9b**). During fermentation, which usually lasts about 4 days for red wine and 10 days for white wine, the temperature is maintained at approximately 25°C. Eventually, the yeast cells are killed off by the alcohol waste they produce, resulting in a final alcohol content between 7% and 18%. The now-fermented must is placed in large vats to allow sediments to settle. It is also filtered and flash-pasteurized to kill any remaining microorganisms. The wine can then be stored in casks (**Figure 16.9c**), and it is during this stage that aging or the development of the wine's final bouquet, or distinctive and characteristic fragrance, occurs.

Wines may be red or white, sweet or dry, wonderful or undrinkable. This variety is the result of many factors. Filtered must is always clear (or white) in color, regardless of the type of grape. The grape skin, on the other hand may be red or white, based on the presence or absence of reddish pigments. To make a white wine, either white grapes are used or the skin of red grapes is removed. If a red wine is being made the grape skin is simply permitted to stay in the must until fermentation starts, when the chemical components of the skin disolve, imparting the familiar red color.

As Joe observed, the sweetness and alcohol content of wine are actually opposite sides of the same coin. To make a sweet wine, fermentation is

Figure 16.8 How bread is made. (a) A commercial example of compressed brewer's yeast. (b) Bread dough is prepared, mixed with yeast, and kneaded thoroughly. (c) Risen bread is the result of carbon dioxide being released as a waste product.

stopped relatively early, before the yeast can ferment many of the sugars in the must. The remaining sugars give the wine its sweet taste and the alcohol content is relatively low, because of the short fermentation period. To make a dry wine, on the other hand, fermentation is allowed to go to completion. All sugars have been broken down during the extended fermentation period and alcohol content is relatively high. The length of the fermentation period can be regulated by various means. One way is to select a strain of yeast that is either relatively tolerant or intolerant of alcohol. Alternatively, the wine-maker can simply intervene, stopping fermentation at any point. When wine is commercially produced, the alcoholic content can be adjusted after the fact, simply by adding or removing alcohol.

Champagne and other sparkling wines are produced like still wines, except that fermentation is allowed to continue after the wine is bottled. Sealed inside the bottle, the carbon dioxide produced as a fermentation waste cannot escape, resulting in the bubbly final beverage.

If all of this sounds fairly simple and straightforward, then why does wine vary so much in quality? Winemaking is both a science and an art. Because so many factors can affect the final product, the making of a truly fine wine is a complex affair. The climate and soil in the grape-growing region can affect the chemical composition of the grapes, which in turn influences the flavor. The precise time at which the grapes are harvested determines their sugar content. The preparation of the must, the strain of yeast used for fermentation, the duration of fermentation, and the length of aging all contribute to the wine's ultimate flavor. One can even get a degree in **oenology**, the science of winemaking, at certain universities.

Beer making relies on the fermentation of grains

Beer brewing also consists of a number of steps. Grains, most commonly barley, serve as the source of sugar for yeast fermentation. However, unlike grapes, grains cannot be directly used for fermentation, because grains store their sugar in the form of starch. Yeast cells lack the enzymes necessary to metabolize starch, so an additional trick or two is needed to permit yeast fermentation.

First, the barley is **malted** (Figure 16.10a). This simply means that the barley seeds are allowed to sprout, because sprouted barley naturally contains starch-digesting enzymes. The sprouted barley, or malt, is then dried and roasted and mixed with water in a step called mashing (Figure 16.10b). During mashing, the naturally occurring enzymes in the malt break down the starch, converting it into simple sugars that the yeast cells can ferment. Once the starch has been degraded, the remaining solids are removed, and the liquid portion containing the sugars, known as the **wort**, is mixed with hops. Hops are the flower portion of the hop plant, which account for much of the variation of a beer's flavor. Hops have a bitter taste and the brewer adjusts the amount of hops depending on the type of beer desired. The mixture is then boiled. During boiling, the mixture is concentrated, flavor-imparting chemicals are leached out of the hops, and potentially contaminating microorganisms are killed. The liquid wort is then separated from the solid hops, in preparation for fermentation in a fermentation vessel (Figure 16.10c).

Once again, strains of *Saccharomyces* are usually the yeast of choice. The particular species selected depends on what type of beer the brewer is making. To brew a lager-style beer, a bottom-fermenter such as *Saccharomyces carlsbergensis* is used. The cells of this species settle to the

Figure 16.9 Wine production. (a) A small wine press is used to extract the grape juice known as must. (b) Fermenting must releases ethyl alcohol. (c) Casks, used to store wine, can contribute to the wine's final aroma.

Figure 16.10 Beer brewing. (a) Malted barley, white sprouts visible. (b) An empty mash tun used in the mashing process. The integrated mash rake churns the mash to make sure thorough mixing occurs and all mash is wet, thus ensuring that no sugars are wasted. (c) Modern fermentation vessels where wort is fermented by added yeast.

bottom of the fermentation tank. Fermentation by *S. carlsbergensis* takes about 10 or 12 days at a temperature of about 8 or 10°C. Or, if an ale is desired, the brewer selects *S. cerevisiae*, a top-fermenting yeast. The less dense cells of this species are carried to the surface of the wort by carbon dioxide bubbles. Fermentation takes only about 6 days, and it is carried out at a higher temperature of about 20°C. Pilsner or lager beer is a bottom-fermented beverage (**Figure 16.11**), whereas stouts, India pale ale, and the like are top-fermented beers.

Either bottom or top fermentation yields an alcohol content of around 3.5–6%. Because carbon dioxide was allowed to escape during fermentation, CO_2 must be added to produce the beer's familiar carbonation. Remaining yeast cells or other sediments are allowed to settle, and any contaminating microorganisms are removed by either filtration or pasteurization. The beer is now ready for bottling or canning.

Various other plant materials can be used to produce liquors

Making a liquor like bourbon or vodka is not much different from making beer. The main difference is the source of starch. Corn is the grain of choice when making bourbon, while scotch is primarily fermented from barley. A single malt scotch uses only barley. A blended scotch contains whisky from fermented barley blended with another whisky in which wheat, corn, or other grains were used. Rye whiskey, as the name suggests, contains a minimum of 51% rye. Vodka traditionally relied on potatoes as the initial starch

Figure 16.11 Bottom and top fermentation. Bottom fermentation results in the lighter colored beers seen in the photo. Darker beers are brewed with top fermentation.

source, although it can be made from various grains. Rum is made from fermented sugar cane or molasses, and tequila comes from fermented agave. Unlike beer and wine, liquors are **distilled** after fermentation is complete. This simply means that the fermented liquid is boiled, which concentrates the liquid, yielding a final beverage with a higher alcohol content and a higher concentration of the chemicals responsible for specific flavors.

A long list of other plant products are fermented to produce popular food items

What do pickles, sauerkraut, and Greek olives have in common? Along with many other, sometimes less-familiar foods, they are all examples of fermented fruits or vegetables (**Figure 16.12**). Unlike most of the fermentation we have discussed thus far, the fermentation of vegetables (or pickling) does not usually depend on adding a specific starter culture for fermentation. Rather, most pickling relies on bacteria that are naturally found on the surface of these plant products. Under the correct conditions, these bacteria produce a lactic acid fermentation waste that both prevents spoilage and gives the food its unique flavor.

Sauerkraut, which literally means "sour cabbage" in German, is made from shredded cabbage. The cabbage is placed in an anaerobic fermentation chamber, along with bacteria found naturally on the cabbage itself. Salt is added to eliminate Gram-negative bacteria, thus allowing the lactic-acid-producing Gram-positive species to thrive. Some of the important fermenting species are *Leuconostoc mesenteroides*, *Lactobacillus plantarum*, and *Lactobacillus brevis*. As fermentation continues, the increased lactic acid production causes the pH to drop, eventually killing off the fermenting organisms. By the time the pH drops to about 3.5, all microorganisms have been killed, and the sauerkraut is ready to eat. The low pH ensures that this food product is slow to spoil, because few contaminating microorganisms can tolerate such an acidic environment.

Pickles result from cucumber fermentation. Cucumbers are placed in a brine (that is, a salt solution) and allowed to ferment. The bacteria found naturally on cucumbers are sometimes used, but the brine can also be inoculated with specific fermenting species. Different styles of pickle can be produced by adding dill, garlic, or other flavorings to the fermentation vat. Olives, as well as other fruits and vegetables, are made in a similar way.

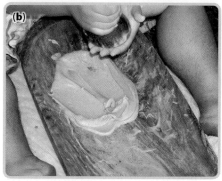

Figure 16.12 Fermented vegetables found in different cultures. (a) Kimchi, popular in Korea, is composed of cabbage and other vegetables, fermented by lactic-acid-producing bacteria. (b) Traditional Hawaiian method of pounding taro root into poi. The mashed material is then allowed to ferment to yield the final product.

Microbes produce various products used to season or add flavor to food

CASE: VINEGAR'S ILLUSTRIOUS PAST

For such a commonplace product, vinegar certainly has an exotic history. Legend has it that the Babylonians discovered vinegar before 5000 BC and used it as a cleaning agent. Only later did they learn that it could be used to preserve food and that it tasted good. Helen of Troy is said to have bathed in vinegar as a way to relax, while Ancient Egypt's Cleopatra allegedly used vinegar to win a bet with Marc Anthony. She had already learned that certain objects, including pearls, would dissolve in vinegar. After she wagered that she could prepare a meal that cost 10 million Sesterces (approximately $30 million today), Cleopatra dropped one of her gigantic

and extremely rare pearl earrings into vinegar, waited for the pearl to dissolve, and drank the concoction in a single gulp. Ancient Greeks and Romans used vinegar extensively in their meal preparation and made it simply by introducing air into wine casks.

1. Why did vinegar work as a cleaning agent or digest pearls?

2. How is vinegar different from most of the other microbially produced foods we have discussed?

3. How did the introduction of air allow wine to turn into vinegar?

Several microbial products have earned a spot in our kitchen as flavor enhancers or as vital parts of many recipes. Vinegar and soy sauce are two common examples. Additionally, the flavor of many commercially produced foods is at least in part due to citric acid, also produced by microorganisms.

Vinegar is an acetic acid solution. And because it is acidic, it can be used as an effective solvent, useful for cleaning away lime deposits on a faucet, or even, if desired, to dissolve pearls. It is also used on everything from salad to spaghetti. The acetic acid is produced by bacteria of the genus *Acetobacter*. These strictly aerobic, Gram-negative bacilli convert the ethyl alcohol produced by yeast fermentation into acetic acid by oxidizing the alcohol. In other words, the bacteria used to make vinegar are not themselves fermenters. Rather they are aerobes that convert a fermentation product, ethyl alcohol, into a different molecule with a distinctly different taste. So, when ancient peoples like the Romans introduced air into wine casks, they were simply providing these aerobic bacteria with the oxygen they required. The name "vinegar" comes from the French "vin" (wine) and "aigre" (sour).

Today, one vinegar-making technique involves allowing *Acetobacter* to grow on wood shavings as a biofilm (**Figure 16.13**). A fermented liquid containing alcohol is sprayed onto the bacterially coated wood, and as the alcohol seeps through the shavings, the bacteria oxidize it into acetic acid. Different types of vinegar are made by selecting different sources of alcohol. Any fermented liquid that contains ethyl alcohol is suitable; some common sources are fermented apple juice (to make cider vinegar), beer (malt vinegar), or wine (wine vinegar).

Soy sauce is an important part of many Asian dishes. To make it, a mixture of cooked soybeans and roasted wheat is inoculated with a mold in the genus

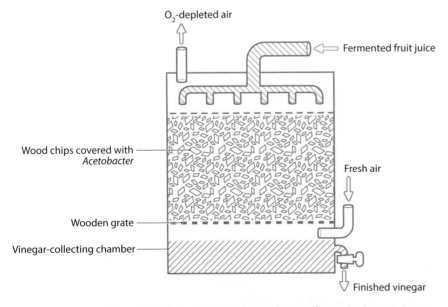

Figure 16.13 A vinegar generation tank. Aerobic bacteria (genus *Acetobacter*) convert ethyl alcohol into acetic acid (vinegar) in the presence of oxygen. Although vinegar production starts with a fermentation waste (ethyl alcohol), the bacteria involved in vinegar generation are not themselves fermenters. The bacteria are allowed to grow on wood chips as biofilms inside the generation tank. The chips are sprayed with a fermented liquid containing ethyl alcohol, and as the liquid percolates through the wood chips, the bacteria growing on the chips convert the alcohol to vinegar. The finished product is collected at the bottom of the tank. Different fermented liquids are used to produce different types of vinegar.

Aspergillus. The fungus digests the proteins and carbohydrates in the soybeans and wheat, resulting in a solution full of simple sugars and amino acids. This solution is then mixed with a salty brine, and salt-tolerant bacteria and yeasts are introduced. After up to a year of fermentation, the liquid soy sauce is removed and is ready for use.

If you think the price of gasoline is high, be thankful that a soda still costs only a dollar or so. It would cost a lot more without microbially produced citric acid. In addition to soft drinks, citric acid is used to flavor candy, jams, jellies, ice cream, and other foods. Citric acid was originally obtained from citrus fruit, but in 1923, a much cheaper microbial fermentation technique was developed. Most citric acid is produced by allowing the fungus *Aspergillus niger* to ferment the sugars in molasses. The sucrose in the molasses is first digested to glucose and fructose, which are then metabolized to citric acid. More than 130,000 tons of citric acid are produced annually around the world; most of it is used to enhance the flavor of commercially produced foods.

Fermented milk is the basis of making cheese and yogurt

One of the enjoyable things about international travel is trying out new and exotic foods, and some of the more exotic are made from fermented milk. If in Russia, for example, you may want to sample koumiss, or fermented mare's milk. The same beverage, called airag in Mongolia, was popular at the court of Genghis Khan (**Figure 16.14**). If your travels take you to Kazakhstan, on the other hand, don't miss out on the chance to try fermented camel's milk (shubat). Not all fermented dairy products are so out of the ordinary, and here we discuss a couple of the more familiar ones that can be enjoyed without using your passport.

The production of yogurt was described in Chapter 8 (see pp. 193–195). Briefly, yogurt is made by allowing milk to ferment. The milk is made thicker by either removing water or adding powdered milk. Two types of fermenting bacteria are then added. *Streptococcus thermophilus* releases

Figure 16.14 Mare's milk for airag.
In central Asia, horse milk, used to produce fermented milk products, is still obtained in the traditional way.

lactic acid, giving the yogurt its tart taste. The acids also denature proteins in the milk, converting it to a semisolid. *Lactobacillus bulgaricus* releases lactic acid as well, but it also produces aromatic compounds that give yogurt its distinct smell.

As mentioned earlier, cheese has been a part of the human diet for thousands of years. Although there are hundreds of varieties, cheeses are easily classified into one of a few groups based on moisture content (**Table 16.1**). Any type of milk can serve as a starting material, and today there are cheeses made from the milk of cows, goats, sheep, horses, and camels.

The first step in cheese manufacture is the conversion of milk into a semisolid **curd**. Traditionally, this is done by inoculating the milk with lactic-acid-producing bacteria. As described above for yogurt, the released acid causes the pH to drop, thereby denaturing milk proteins and causing the curd to develop. Today, when cheese is made commercially, the curd is formed by adding a clotting enzyme called rennin to the milk. Rennin was originally obtained from the stomachs of calves. Today rennin is synthesized by bioengineered microorganisms.

Once the milk has clotted and the curd has formed, it is separated from the remaining liquid, called **whey**. The pH, as well as salt and moisture content of the curd, is adjusted and the curd is physically molded into a desired shape. At this point, some cheeses are essentially ready to eat. Cream cheese and cottage cheese, for instance, are soft, high-moisture cheeses that have a short shelf-life. No additional aging is required. Many other cheeses must undergo an additional aging process, in which various bacteria and molds are essential. Appropriate microorganisms are generally allowed to colonize the surface of the curd. Enzymes produced by the organisms seep into the curd, digesting carbohydrates, lipids, and proteins. Various by-products remaining behind give the cheese its unique flavor, aroma, and texture. To make Camembert, a soft cheese, for instance, the mold *Penicillium camemberti* is allowed to coat the surface of the curd. Protease enzymes produced by the mold give the cheese its creamy consistency. Typically, a soft cheese requires only a few weeks of aging. Blue cheese, which is somewhat less soft, takes several months to age properly. The blue veins that form are due to the growth of *Penicillium roqueforti*, used to inoculate the curd (see Figure 1.9b).

Hard cheeses may be aged for a year or more. They also owe their particular flavor to aging with specific microorganisms. Swiss cheese, for example, owes its easily identifiable sharp, nutty taste to *Propionibacterium*. Not only do these bacteria alter the cheese's flavor, but trapped carbon dioxide gas released during fermentation forms the characteristic holes.

Table 16.1 Cheese variety. A convenient way to classify cheeses is by their moisture content. Softer cheeses, which have a higher moisture content, are aged a relatively short time. Hard cheeses, relatively low in moisture content, may be aged up to a year or more.

TYPE OF CHEESE	EXAMPLES
Soft	Brie, Camembert, mozzarella
Semisoft	Blue, Muenster, Roquefort
Hard	Cheddar, Colby, Gouda, Swiss, Edam, feta
Very hard	Parmesan

Certain meat products require fermentation

Compared with plant and dairy products, the list of fermented meats is short. Summer sausage, as well as salami and country-cured hams, are examples of meat products involving fermentation. When sausage or salami is made, ground meat is inoculated with commercially available fermenting bacteria. Cured hams rely on fungi that naturally colonize the surface of the ham. For the more adventurous, you might wish to try fermented seal flipper, the next time you find yourself north of the Arctic Circle. While you are at it, sample a fermented fish head (locally known as "stinky head"). But be careful: Alaska has the highest rates of botulism in the United States for a reason. The anaerobic environments created to encourage fermentation of such local delicacies can also permit the growth of *Clostridium botulinum* (see Chapter 11, p. 269).

Bacteria help to prepare coffee beans for roasting

Microbial involvement in other aspects of our diet is more subtle. Coffee, for instance, owes little or nothing of its taste or aroma to microorganisms, but the beans are processed with microbial assistance. Before coffee beans can be roasted, the outer pulplike coating of the bean must be removed. Much of this pulp is composed of the carbohydrate pectin, and although it can be removed in other ways, the bacterium *Erwinia dissolvens* is especially well suited to the task. This species produces a pectin-digesting enzyme, permitting easy removal of the pulp.

Microbes produce many vitamin supplements

CASE: MICROBIAL METABOLITES FOR A FEATHERED FRIEND

When Judy's scarlet macaw Vladimir suddenly becomes listless and develops diarrhea, she becomes very concerned. After all, Vladimir has been a member of the family for years and he has always been very active and healthy. Judy assumes that Vladimir must have some sort of infection, so she takes him to the veterinarian. The veterinarian begins by giving Vladimir a complete examination. He allows the bird to grip his finger with his claws and notes that Vladimir is very weak. He also notices that one of Vladimir's toes is twisted or curled out in an unnatural manner. When he asks about the bird's diet, Judy responds that he mainly eats a commercial parrot food available at the supermarket. To Judy's surprise, the veterinarian says that he does not think Vladimir has an infection. His diagnosis is a condition called curly toe, which is treated by increasing the bird's dietary vitamin intake. Riboflavin is especially necessary because very few commercial bird foods contain enough of this vitamin. Judy thanks the veterinarian profusely, and within a week, Vladimir seems to be doing much better. After a few more weeks he appears to be completely recovered.

1. What exactly is riboflavin, and why is it so crucial for birds?
2. How do microorganisms contribute to the commercial production of riboflavin?

Pernicious anemia, characterized by low numbers of red blood cells, is caused by a lack of vitamin B12. Human cells cannot synthesize this essential vitamin, and consequently, we obtain most of our vitamin B12 from food. Some bacterial strains, on the other hand, are able to make vitamin B12 in abundance, and some of these organisms are part of our normal intestinal microbiota. We therefore go through life with our vitamin B12 needs met in part by our symbiotic microbial partners. Vitamin B12 is also produced commercially for both human and animal consumption. Much of this supply is synthesized by bacteria of the genera *Pseudomonas* and *Propionibacterium*. To ensure high levels of vitamin B12 production, these microorganisms are grown on media supplemented with cobalt, which is an essential structural part of the vitamin. Vitamin B12 is regularly added as a supplement to hog and poultry feed, or it can be bought over the counter in any health food store or pharmacy. It is also commonly added to foods such as bread and breakfast cereal.

Vladimir's problem was principally due to a deficiency of riboflavin (vitamin B2), which is also produced commercially by microorganisms. Animals require riboflavin to form enzymes essential for carbohydrate metabolism. In humans, a riboflavin deficiency can cause eye problems or lesion formation on the skin or in the mouth. Horses and other domesticated animals may show similar signs if they are fed a low-quality diet.

Birds are especially vulnerable to riboflavin deficiency. Only a few commercially available bird foods contain enough riboflavin to meet nutritional requirements, especially for growing chicks (**Figure 16.15**). In some long-lived birds like macaws, riboflavin deficiency can take years to develop. Symptoms can then appear quite rapidly. Vladimir's symptoms, including listlessness, diarrhea, and the curled toe, are typical. In serious cases, there is significant neurological damage. Fortunately, as Judy found out, the problem can often be remedied by adding sufficient riboflavin to the diet, although complete recovery may take several months. Like vitamin B12, riboflavin is often added to bread, cereal, and other foods. Much of the commercially available riboflavin is synthesized by yeasts such as *Ashbya gossypii*.

Figure 16.15 Curly toe in a chicken. Like parrots, chickens diets deficient in riboflavin (vitamin B2) are susceptible to this condition. Other symptoms include listlessness and flaky skin. The condition can be corrected by supplementing the bird's diet with the vitamin, produced commercially by yeast.

Amino acids are commercially produced with microbial assistance

Amino acids are used as additives to enhance nutritional quality and flavor. The low-calorie sweetener NutraSweet® is also made synthetically from amino acids. Lysine is produced commercially by bacteria in the genus *Brevibacterium*. Lysine is an example of an essential amino acid. Humans cannot synthesize lysine and it must therefore be included in our diet. This can be a particular problem for vegetarians, since many of the best lysine sources are animal products. Consequently, lysine is often added to plant-derived foods to improve their nutritional quality. It can also be bought as a supplement in health food stores. Glutamic acid is also produced by bacteria. This amino acid is used to make the flavor enhancer monosodium glutamate (MSG), common in Asian cuisine.

Certain food items may contribute to a healthy microbiota

In Chapter 11 we reviewed the use of Preempt®, a solution containing normal microbiota that is sprayed on baby chickens. As the chicks preen, they ingest these bacteria, establishing a healthy intestinal microbiota that can prevent colonization by pathogens such as *Salmonella*. Preempt® is a good example of a **probiotic**, a product consisting of living bacteria that promotes overall health.

Probiotics, though, are no longer just for chickens. Indeed, the development of probiotic products for humans is a rapidly growing part of the health food industry. Many of the claims made by probiotic advocates have yet to be substantiated, and those relatively few scientific trials that have been conducted often provide contradictory results. If probiotics are in fact some way beneficial, it remains unclear in many cases exactly how they exert their beneficial effects. As discussed in Chapter 11, many potential beneficial roles have been ascribed to a healthy microbiota. The microbes found in probiotics may consequently contribute to our health in a variety of ways.

If you choose to try probiotics, you may wish to do what you can to spur on the growth of the beneficial organisms they include. To do so, you may avail yourself of **prebiotics**—foods that are believed to contain compounds that stimulate growth of a healthy microbiota. Many such foods contain certain carbohydrates that we often collectively refer to as dietary fiber. Although we do not have enzymes to digest these carbohydrates, many members of the microbiota can use these molecules as a source of energy. We might then reap the benefits of a more robust and quickly growing microbiota. Some of the foods thought to be especially rich in prebiotics include gum Arabic, chicory root, dandelion greens, and Jerusalem artichoke.

Time to eat: a few microbial recipes

Would you care to try your hand at making some of the foods we've been discussing? Here is your chance. Some of the following recipes require little or no special equipment. For others, you may need to pick up a few supplies. All of them involve microorganisms in some capacity.

Sauerkraut

Sauerkraut is fermented cabbage. The fermentation relies on Gram-positive bacteria already found on the surface of the cabbage leaves (**Figure 16.16a**). Salt is added both for taste and to inhibit the growth of undesirable Gram-negative species.

Ingredients
cabbage
salt

Preparation
1. Wash the cabbage and cut it into quarters. Remove the core, and shred the cabbage finely with a sharp knife.

2. Place a layer of cabbage in a wide-mouthed jar or crock, sprinkle with salt, and press down firmly. Continue to add additional layers of cabbage and salt until the jar is full.

3. Cover the top with a clean cloth, and place a plate on top. Add a weight to the plate to hold it down. Place the jar in a warm spot to ferment. After a few days, remove the froth on the top of the cabbage, replace the cloth, plate, and weight, and allow to stand for another 3 days. Then repeat the process.

4. Move the jar to a cool spot for about 2 weeks. Then, enjoy fresh, old-fashioned homemade kraut. To continue the microbial theme, you may wish to serve with summer sausage and a German lager.

Pickles

There are many different varieties of pickles, and the following is just one example. If you like this recipe, you might try different variations of the brine solution next time. To increase the quantity, you need only adjust the

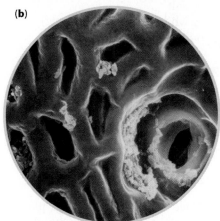

Figure 16.16 Home fermentation. (a) Homemade sauerkraut is tasty and easy to make. (b) Like sauerkraut, pickles can be made by use of fermenting organisms found naturally on the surface of the plant. The electron micrograph shows *Lactobacillus plantarum* growing naturally on the surface of a cucumber.

amounts of all ingredients. In this basic recipe, we simply rely on microorganisms normally inhabiting the surface of cucumbers to do the fermenting (**Figure 16.16b**).

Ingredients

1 tablespoon (tbsp) salt per quart of water
1 clove garlic
1 stalk sliced celery
1 sliced carrot
1 chili pepper
1 bay leaf
1 bud dill
1 lb (0.45 kg) washed pickling cucumbers

Preparation

1. Combine all the ingredients except the cucumbers in a pot and boil. Following a brief boiling, allow the pickling solution to cool.

2. Puncture the cucumbers with a fork and add to the pickling solution. Add to wide-mouthed jars and seal tightly.

3. Let stand for 3 weeks, then enjoy.

Bagels

As an example of a homemade bread, you might try bagels, the only type of bread in which the dough is boiled before it is baked (**Figure 16.17**). Bagel making is a bit more complicated than the recipes presented so far, but you still will need no or few special utensils, and all ingredients are readily available. As with most breads, *S. cerevisiae* (baker's yeast), once in the anaerobic environment of the bread dough, will perform the necessary fermentation.

Ingredients

7.6 fluid ounces (225 ml) scalded milk (boiled and hot)
2 ounces (50 g) butter
1 ounce (25 g) sugar
2 teaspoons (tsp) dried baker's yeast
½ tsp salt
1 egg white (medium-sized egg)
14 ounces (400 g) plain flour

Preparation

1. Put the milk, butter, and sugar into a mixing bowl and mix thoroughly. The milk should still be hot enough to melt the butter completely.

2. When the mixture has cooled and is lukewarm, sprinkle on the yeast and allow to sit for 45 minutes. Then stir in the egg white and salt and add the flour, kneading it into a soft dough. Knead the dough until it is smooth and elastic.

3. Place the dough in a large plastic bowl with a snap-on lid, or seal the dough inside the bowl with plastic wrap. Allow the dough to rise until it has doubled in size. This should take about 1 hour. Then knead the dough briefly to remove large gas bubbles.

4. Divide the dough into roughly 15 lumps, each about the size of a tennis ball. Moisten your finger and cover it with flour. Then drive your finger through each lump to make the bagel hole. Twirl the bagel around your

Figure 16.17 The bread with the hole.
Bagels were supposedly invented in the late 1600s by an Austrian baker. He wanted to honor Austria's king and because the king was an enthusiastic equestrian, the baker made his gift in the shape of stirrups (German, bügel).

finger on the work surface until the hole is at least one-third the diameter of the entire bagel.

5. Place the bagels on a tray and set in a damp place. To prevent drying, you may place a rack over the bagels, and lay a damp towel over the rack. Allow them to rise until they look puffy. This takes only an additional 10 or 15 minutes. Do not let them double in size. Preheat your oven to 390°F (200°C). While the oven is warming up, prepare a large shallow pan of gently boiling water.

6. Place the bagels in the water a few at a time so that they don't touch each other. They will float about halfway submerged in the water. Leave them for 15 to 20 seconds, then remove with a slotted spoon. Arrange them on a greased baking tray.

7. Bake for about 20 minutes, until they are golden brown and have a hollow sound when you tap them. Allow them to cool, and serve with your favorite topping.

8. Once you've mastered the ABCs of bagel baking, try adding other ingredients such as chopped onions or raisins to the dough after kneading.

Yogurt

Homemade yogurt makes a delicious and nutritious breakfast or snack. Here are the basics. What you add in terms of fruit flavoring is up to you.

To ferment the lactose sugar in the milk, we need two types of bacteria, both of which will release lactic acid as a fermentation waste. This acid gives yogurt its tart flavor, and it also converts the liquid milk to the semisolid yogurt as it denatures the milk proteins. To obtain the required *S. thermophilus* and *L. bulgaricus*, we will simply use a small amount of plain yogurt with active cultures as a starter culture. Once fermentation begins, *S. thermophilus* initially digests the lactose. As levels of lactic acid rise and the pH drops, bacterial growth is inhibited. The more acid-tolerant *L. bulgaricus* then takes over, fermenting the remaining milk sugar. As the pH continues to drop, the growth of other contaminating microorganisms that might otherwise spoil the yogurt is inhibited. *L. bulgaricus* also releases certain aromatic chemicals, such as acetaldehyde, that give yogurt its pleasing bouquet.

Ingredients
1 quart (about 1 L) whole milk
2 tbsp natural yogurt with live cultures
1/3 cup nonfat dried milk

Preparation
1. Be sure that all containers and utensils are thoroughly cleaned before use to prevent contamination. The milk must first be sterilized by boiling (remember that even pasteurized milk contains some microorganisms).

2. Add about 1 ounce (20 g) of the milk powder to 3.5 fluid ounces (100 ml) of water. Add this to the whole milk in a cooking pan. Heat the milk to boiling, allowing it to boil for 1 minute (no more), and remove from the heat.

3. Place a kitchen thermometer in the milk and allow the milk to cool until the temperature falls to between 100 and 89°F (38 and 32°C). Pour the cooled milk into ½ quart (½ L) jars. Canning jars with screw-on lids make good yogurt containers.

4. Mix about 0.5 ounce (14 g) of the plain yogurt starter culture into the milk. Cap the jar with its lid. Place the jar in a warm oven (between 100 and 89°F or 38 and 32°C). Do not disturb the jar for 3 hours, then check every hour for yogurt to form. This may take half a day. If you move or shake the container during the incubation process, the yogurt may not coagulate properly.

5. When the milk has thickened, remove the jar of yogurt from the oven. Taste it, and if a tarter flavor is desired, allow the yogurt to incubate for another hour or so.

6. Add fruit or flavouring like vanilla extract, as desired. Once the yogurt is completed, it can be refrigerated. It should be eaten within the next 4–5 days.

Greek feta cheese

Feta is a popular semihard cheese that features prominently in Greek cuisine (**Figure 16.18**). It is also a reasonably easy way to introduce yourself to the ancient art of cheese-making. Although feta cheese can be made with whole cow's milk, we will stick with tradition, relying on goat's milk.

Cheese-making requires more ingredients and effort than the recipes provided so far. All ingredients, however, are available from gourmet shops, natural foods outlets, or online cooking supply companies. To ferment the milk sugars, we need *Lactococcus lactis*, which is most easily obtained as the commercial product Mesophilic-A. The clotting enzyme rennin, available commercially as liquid rennet, and calcium chloride are also necessary.

Ingredients
2 gallons (7.6 L) pasteurized whole goat milk
¼ tsp Mesophilic-A
½ tsp calcium chloride
½ tsp liquid rennet
2 tbsp cheese salt

Figure 16.18 Feta cheese. Traditional feta cheese, made with goat's milk, is used widely in Greek cooking.

Preparation
1. Warm the milk to 85°F (30°C) in a large pot. In a separate dish, mix the calcium chloride (½ teaspoon) with 2 tablespoons of cool distilled water. Add this mixture to the milk and stir gently for 30 seconds. Maintain the milk's temperature at 85°F (30°C).

2. Add ¼ teaspoon of Mesophilic-A and stir gently. Then, allow the milk to stand at 85°F (30°C) for 1 hour.

3. Add ¼ teaspoon of liquid rennet to 4 ounces (120 ml) of cool distilled water. Stir in gently. Cover and allow the milk to sit undisturbed for 30 minutes or until the milk forms a solid curd that shows a clean break. If, after 30 minutes, the consistency is more like that of soft yogurt, continue to wait until the solid curd forms.

4. While leaving the curd in its pot, cut the curd into ½ inch (1.25 cm) slices all the way across the pot. Rotate the pot 90° and repeat, as if forming a checkerboard of thin slices.

5. Stir the curd slowly every 10 minutes for 1 hour, gradually heating the pot to 95°F (35°C). After 1 hour of slow cooking, drain the liquid portion (the whey) by pouring the contents of the pot into a colander. Allow to drain for 1 hour. The curd will form back into a solid mass.

6. Cut the curd into small blocks and turn it over in the colander. Allow to drain an additional 30 minutes. Add salt to taste.

7. Pack the curd into a sterilized quart jar. Prepare a brine of 2 cups distilled water and 2 tablespoons cheese salt. Pour the brine over the curd and seal tightly. Refrigerate and allow the jar to sit for at least 2 weeks. The longer it ages, the better it tastes.

Once you are ready to try your cheese, make yourself a Greek salad, complete with other fermented foods such as Greek olives, and a dressing made with balsamic vinegar.

Vegemite sandwich

Last but not least, here is your introduction to this Australian staple (**Figure 16.19**). Although real Aussies love Vegemite, the spread is definitely an acquired taste.

Preparation

1. Spread two pieces of your favorite bread with butter or margarine. Either toasted or untoasted bread is appropriate.

2. Cover the buttered bread with a thin layer of Vegemite. Only a small amount is necessary. Some connoisseurs mix the Vegemite into the butter before spreading. Put the two slices together and give it a try. If you actually like it, consider yourself fair dinkum.

Figure 16.19 Down Under cuisine. An essential part of Australian culture.

Looking back and looking forward

Microorganisms have contributed to the human diet for millennia. While a few microbes are consumed directly as food, by far the largest contribution of microorganisms is based on their release of metabolic wastes. These waste compounds alter the flavor, consistency, and shelf-life of various plant, dairy, and meat products. Many food additives and health food products also owe their production to microorganisms. And partly for fun, but also to highlight important microbiological principles, you might try one of the recipes we have provided, and really sink your teeth into food microbiology.

Food production is just one area in which humans have enlisted microbial help. Various other industrial processes also rely on bacteria and fungi. In Chapter 17, we take a look at how microbial metabolism is harnessed in a range of industries and aids in the production of everything from detergents to pesticides to pharmaceuticals. And when it comes to dealing with some of the problems we face in the twenty-first century, including issues as complex as pollution control and renewable energy, microorganisms are increasingly coming to our assistance. In Chapter 17, we explain how.

Garland Science Learning System

- http://garlandscience.rocketmix.com/students/

- Discover why a lack of oxygen can be a good thing, especially if you like beer, wine, yoghurt, cheese, or bread

- Test your knowledge of this chapter by taking the quiz

- Familiarize yourself with the terminology used in this chapter by using the vocabulary review

- Get help with the answers to the Concept questions

Concept questions

These questions are designed to help you start thinking like a microbiologist. The answers are not always simply found in the text. Instead, you will need to take the concepts about which you have learned and apply them to new situations. Some of the questions may not even have just a single correct answer. Help is provided as part of the GSLS resources, which can be accessed through http://garlandscience.rocketmix.com/students/.

1. We learned that the cyanobacterium *Spirulina*, if raised for food as single-cell protein, could produce up to 10 tons of protein per acre. For wheat and beef, the corresponding figures are 0.16 tons per acre and 0.016 tons per acre, respectively. Recalling what you learned in Chapter 10 about the way that energy moves through an ecosystem, explain the above observation.

2. After drinking part of a bottle of wine, you recork the bottle, saving the rest for another time. A few months later, you pour a glass from this bottle and find it to be undrinkable. What do you think happened microbiologically during the interim?

3. Go back through Chapter 16 and find three particular species of bacteria or fungi used in a food-making process. Are the involved microbes respiring anaerobically or aerobically during the food-making process?

4. In Chapter 8 we discussed the concept of the growth curve as it applies to microorganisms. Imagine that you introduced a population of yeast cells into grape must, for the purpose of making wine. Then imagine how those yeast cells would grow as they fermented the sugars in the must. What factors would cause the population to leave the log phase of growth and enter the stationary phase? The decline phase?

5. Here in Chapter 16 we discussed the use of probiotics, or foods containing live microorganisms to provide certain benefits. After having also read Chapter 15 on biotechnology, can you think of any specific way that microorganisms in probiotics could be genetically modified to provide enhanced benefits?

6. In our recipe for sauerkraut, salt is added to eliminate any Gram-negative bacteria. Why do you think it is necessary to eliminate these bacteria?

7. Imagine you are planning a dinner party. Prepare a sample menu highlighting dishes in which microbes played a crucial role. Next to each menu item, specify the microorganism involved and what that microorganism contributed.

8. Figure 8.27 illustrates how different microorganisms respond to oxygen. Go back and review that figure and then consider the microorganisms discussed here in Chapter 16. Most of the microbes used in food production, because of the metabolic wastes they produce, fall into one of the categories illustrated in Figure 8.27. Which category is it?

BETTER LIVING WITH MICROORGANISMS: INDUSTRIAL AND APPLIED MICROBIOLOGY

Air and water pollution are harmful for living things, but less obvious is the adverse impact pollution has on some of our most beloved historical monuments. In cities around the world, acid rain and pollution compound the effects of natural erosion, turning architectural masterpieces into dust. A principal problem is the way in which contaminants in the air and rain affect the calcium carbonate ($CaCO_3$) found in marble, limestone, and other traditional building materials. As pollution dissolves the $CaCO_3$, the structures slowly degrade, until previously intricate details erode away and become featureless lumps (**Figure 17.1a**).

Here to help is *Mycococcus xanthus*. This common soil bacterium secretes $CaCO_3$ under appropriate conditions in a process called **biomineralization**. And at the burial place of Spanish kings and queens, the Royal Cathedral in Granada (**Figure 17.1b**), these microbial stonemasons are already proving their worth.

University of Granada researchers discovered that when pieces of limestone from the cathedral were placed in a liquid broth containing *M. xanthus*, the material secreted by the bacteria seeped into the stone, significantly strengthening it within 2 weeks. Biomineralization replaces the $CaCO_3$ that had been lost and helps to bind together other mineral components in the stone. Furthermore, unlike resins used to protect stonework, the bacteria do not clog pores in the stone, which would later trap water, accelerating decomposition. They also cause less fracturing and color change. The next step was to either spray the bacteria on the cathedral or to wrap the building's walls in a sterile nutrient broth to encourage the growth of *M. xanthus* already growing on the stone façade. The restoration project is still ongoing, but recent results in 2015 indicate that both of these techniques provide significant improvement in stone strength and durability that have lasted since the project began in 2009. Similar projects are now in the works for other endangered monuments around the world.

This is just one way that microorganisms offer practical solutions to real-world problems. Microbes are involved in the production of everything from pharmaceuticals to plastics to pesticides. They offer possible solutions to our energy needs, and they can be used to degrade toxic waste or to clean up polluted water. Here in Chapter 17 we explain how microorganisms are used in industry and survey some of the ways that the industrial use of microorganisms affects our daily lives. We first look at how microorganisms are utilized in industrial and commercial processes to produce valuable products, including antibiotics, insecticides, or enzymes with industrial applications. We then turn our attention to some of the ways that microorganisms may be harnessed to clean up toxic sludge, put fuel in our gas tanks, or solve other sticky problems.

All case studies have a few questions at the end, the answers to which will become apparent as you read the sections following the case.

Microorganisms have numerous commercial applications

Industrial microbiology began with the use of microorganisms for food, as described in Chapter 16. Once the role of microorganisms in processes such as fermentation was fully understood, it was discovered that many microbial metabolic products had useful, nondietary applications. In Chapter 6, for instance, we discussed how bacterially produced acetone played an important part in World War I. In Chapter 13, we learned that many antimicrobial drugs originally came from microorganisms themselves. With the commercialization of these and other products, the modern discipline of industrial microbiology was born. In the past few decades, the biotechnology revolution, as described in Chapter 15, has opened the door to a host of new potential industrial applications for microorganisms. Yet whether the microorganism in question is a common soil fungus or a genetically engineered bacterium expressing foreign genes, many of the basic considerations for finding, cultivating, and utilizing the organism are the same. We begin by reviewing a few basic concepts and examining some of the problems that must be dealt with before a microorganism is ready to go to market.

Figure 17.1 Bacteria to the rescue.
(a) Air pollution and acid rain can wreak havoc on the stonework of historical buildings. (b) Granada's Royal Chapel (La Capilla Real), where a new strategy to control pollution-induced erosion with a common soil bacterium is aiding restoration efforts.

CASE: UNEARTHING A NEW ANTIBIOTIC

Sam, a student in a microbiology laboratory course, is instructed to find a microorganism in the environment with potential commercial or industrial applications. He decides to screen soil samples taken from his garden for bacteria or fungi that produce new antibiotics. He first prepares bacterial lawns (see Figure 1.10a) of *Escherichia coli* (a Gram-negative species) and *Staphylococcus epidermidis* (a Gram-positive species). He then places small amounts of soil on each of his media plates. The plates are incubated for 24 hours at 35°C. When Sam inspects his plates, he finds that there are small, clear halos without bacterial growth surrounding some of the soil particles on the plates inoculated with *S. epidermidis*. There are no such bacteria-free zones surrounding the soil particles on the plates inoculated with *E. coli* (**Figure 17.2**). On the basis of these results, Sam concludes that his soil

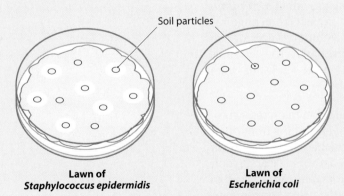

Halo that is free of bacterial growth indicates that this soil particle contains microorganisms that secrete an antibiotic effective against *S. epidermidis*

Bacteria are able to grow up to the soil particles, indicating no antimicrobial activity against *E. coli*

Soil particles

Lawn of Staphylococcus epidermidis

Lawn of Escherichia coli

Figure 17.2 Discovering a new antibiotic. Some of the soil particles in the plate on the left are surrounded by zones of inhibition or halos where the bacteria are unable to grow. Microorganisms in these soil particles are secreting an antibiotic that inhibits growth. No such activity is seen in the plate on the right.

sample contains a microorganism of some type that secretes a compound with antibiotic properties, which is apparently more effective against Gram-positive organisms.

1. **Why would soil microorganisms produce antibiotics?**
2. **Might such an antibiotic be of commercial or medical value? If so, what other factors would have to be considered, and what other steps would have to be followed, before this or any newly discovered, potentially valuable microbial product could be deemed useful?**

Many microorganisms produce metabolites with commercial potential

As we saw in Chapter 8, all living things, including microorganisms, carry on their life processes through various metabolic pathways. Each pathway consists of separate steps, in which a starting material is transformed into a series of intermediate compounds and is ultimately converted into one or more final products. These intermediate compounds or final products, produced as a consequence of metabolic activity, are called **metabolites**. Many metabolites have useful applications. Antibiotics are just one important example. In the experiment described in our case, the halos that Sam observed suggest that a soil organism of some sort is releasing a metabolite that inhibits the growth of the Gram-positive bacteria. Microorganisms often produce such metabolites because, as we learned in Chapter 10, competition between soil organisms can be especially fierce. Many microbes consequently produce metabolites that inhibit the growth of potential rivals. Other metabolites are produced by microorganisms as a consequence of their normal metabolic activity. The initial goal of the industrial microbiologist is to identify metabolites that may have practical applications.

To find microbes that produce metabolites with commercial potential, microbiologists have traditionally scrutinized the natural environment, isolating organisms from soil, water, spoiled food, plants, or animals. These organisms are then screened for potentially valuable metabolites, in a manner similar to what Sam did in his hunt for new antibiotics. Although such searching (or, as it is sometimes called, "bioprospecting") has been in progress for years, we have literally only scratched the surface; most environmental microorganisms remain unclassified and unstudied. Undoubtedly, efforts to find useful microorganisms in the environment will continue because the potential payoff is enormous.

Microbial production of citric acid provides an excellent example of how careful (and sometimes simply lucky) bioprospecting can yield colossal profits. Citric acid, common in many food products (see Chapter 16, p. 401), is also used as a preservative in stored blood and in medicinal ointments. Citric acid is also used in detergents to replace phosphates, which are a known water pollutant (see Chapter 10, pp. 233–234). Originally, all citric acid was obtained from citrus fruit, principally from Italy. However, political instability in Italy and the advent of World War I prompted the search for alternative sources of the lucrative compound. In 1917 scientists at Pfizer discovered that *Aspergillus niger*, a common fungal food contaminant, produced citric acid under appropriate environmental conditions. By 1929, techniques had been developed for the large-scale industrial production of citric acid. Suddenly, this valuable industrial product could be produced at a small fraction of the original cost, propelling Pfizer to the forefront of industrial microbiology. The company was well positioned to use a similar technology in the 1940s to mass-produce the antibiotic penicillin from the fungus

Penicillium notatum (also known as *Penicillium chrysogenum*). Pfizer's place as one of the world's largest pharmaceutical manufacturers was now assured.

Microorganisms producing promising metabolites usually must undergo strain improvement

But the odds of bioprospecting success are long. As Sam's experience demonstrates, finding an organism that produces a useful substance is not necessarily difficult. Finding one with real industrial potential is not so easy. First of all, most potentially useful metabolites are made in only minute amounts. Microorganisms generally tightly regulate their metabolism through processes such as induction and repression, discussed in Chapter 7 (see pp. 149–151). Strains that overproduce an enzyme or metabolite tend to be selected against because they waste energy by producing substances that are unnecessary. Random mutations ensure that overproducing strains are occasionally found, but as a rule, any strain that has industrial potential requires **strain improvement**—a process of increasing the production of the metabolite in question far beyond what a naturally occurring strain would synthesize. If Sam wishes to pursue his project further, his next task would be to isolate his soil microorganism in pure culture and subsequently subject it to strain improvement.

How are strains improved? There are several strategies that might be employed. One common technique is to subject a promising strain to mutagens such as ultraviolet (UV) light to induce random mutations. Many of these mutations will be of no significance. Others will be harmful. Occasionally, however, a mutation will occur that results in overproduction of a particular metabolite. Such mutations often disrupt regulatory mechanisms that ordinarily limit metabolite production. The microbiologist then uses a process of artificial selection (see Chapter 9, p. 212) to isolate and select for the mutated organisms of interest. The isolated organisms are then grown on fresh medium. Additional random mutations make it likely that among these cells, some will produce even higher levels of the metabolite, and those organisms are once again isolated. Repeated rounds of such selection can result in strains that produce remarkably high levels of a particular substance. For example, the original strains of the *Penicillium* mold, from which penicillin was first discovered, produced about 5 mg/L of the antibiotic. By artificially selecting those strains that produced the most penicillin, scientists raised the yield to 60,000 mg/L—a 12,000-fold increase. Likewise, in Chapter 16, we learned about the mold *Ashbya gossypii*, utilized to produce riboflavin (vitamin B2). Selection for overproduction has resulted in strains that synthesize approximately 20,000 times the amount of the vitamin that the microorganisms need to meet their own metabolic needs.

Sometimes, metabolite production can be enhanced by adjusting temperature, pH, oxygen levels, or other environmental parameters. As we learned in Chapter 8, for example, all microorganisms have an optimum temperature at which they grow the fastest. For some metabolites, called **primary metabolites**, synthesis closely mirrors population growth (**Figure 17.3**). In such cases, simply providing this optimum temperature may result in substantially increased metabolite yields. For other metabolites, called **secondary metabolites**, synthesis occurs only after the microbial population has ended its period of exponential growth and entered the stationary phase (see Chapter 8, p. 192). In this case, the environment may be altered in such a way that entry into stationary phase is encouraged. Sometimes careful regulation of nutrients can alter metabolite synthesis. Penicillin-producing molds, for instance, produce more of the antibiotic when provided with lactose instead of glucose.

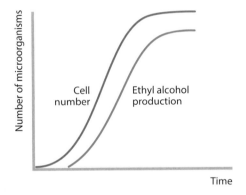

(**a**) Primary metabolite: ethyl alcohol

Number of microorganisms

Cell number

Ethyl alcohol production

Time

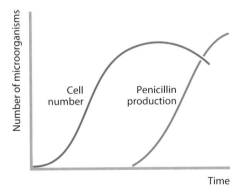

(**b**) Secondary metabolite: penicillin

Number of microorganisms

Cell number

Penicillin production

Time

Figure 17.3 Primary and secondary metabolites. (a) Ethyl alcohol (ethanol) is an example of a primary metabolite. Its production by fermenting yeast cells closely mirrors yeast cell population and increases fastest during the yeast's exponential growth phase. As the yeast enter the stationary phase, alcohol production slows. (b) Penicillin is produced by the mold *Penicillium chrysogenum* as a secondary metabolite. Penicillin synthesis does not begin until the mold enters the stationary phase of population growth.

Increasingly, the techniques of genetic engineering discussed in Chapter 15 are being utilized to create microorganisms that produce a particular metabolite in abundance. Sometimes, for example, an organism is discovered that, although it produces a potentially valuable metabolite, is difficult to culture and grow in the laboratory. If the gene responsible for the metabolite synthesis is identified, it may be possible to isolate and clone the gene, introducing it into a more manageable organism such as *E. coli*.

And the best is no doubt yet to come. With new techniques such as the gene-editing tools discussed in Chapter 15 becoming available, it seems that the sky is now the limit when it comes to manipulating the genetic material of potentially useful microorganisms.

Potentially valuable microbes must grow well in an industrial setting and must not pose undue risks to humans or the environment

High levels of metabolite synthesis are certainly necessary, but this alone does not guarantee a smooth transition to industrial application. The organism in question also must be able to grow in large-scale cultures. This is not necessarily a simple criterion to satisfy. As we have learned, microorganisms can be quite specific in their environmental requirements, and required conditions are not always easy to duplicate in the laboratory. It is not uncommon, for instance, for bacteria and fungi to form mutualistic relationships with other microorganisms in their environment (see Chapter 10, pp. 245–246). Sometimes these relationships are essential if a given organism is to produce a particular metabolite. One species may produce a desired compound only when it has access to a different compound produced by the symbiotic partner (**Figure 17.4**). Complex microbial assemblages of this sort are difficult to recreate in a laboratory, and they can greatly complicate the process of developing a strain of commercially viable microorganisms.

Rate of population growth is also an important consideration. Ideally, a microorganism of interest will grow and rapidly produce the metabolite in question. Preferably, the organism will also thrive in a relatively inexpensive liquid medium. In some cases, waste products from one industrial process, such as the liquid whey formed during cheese-making (see Chapter 16, p. 402), are used to culture microorganisms for an entirely different industrial application.

Finally, an industrial microorganism must be safe, both for humans and for the environment. It is difficult, if not impossible, to prevent the release of industrial organisms into the environment completely. Consequently, human, animal, or plant pathogens are ruled out, even if they produce valuable metabolites. Fortunately, safety standards are easier to satisfy, because by the time most industrial strains of microorganisms have been developed, they are so highly specialized and "domesticated" that they cannot survive in the natural environment (**Figure 17.5**). Selected for rapid and high metabolite production, as well as specific environmental conditions, they are a far cry from the wild strain that was isolated initially. Yet this does not mean that we can assume such strains are absolutely safe, and stringent testing and examination of any microorganism are necessary before its use in industrial or other processes can be condoned.

A defined series of steps are followed to move production from the laboratory to the factory

Suppose we have developed a strain of microorganisms that produces a valuable product in commercially viable quantities. We have determined

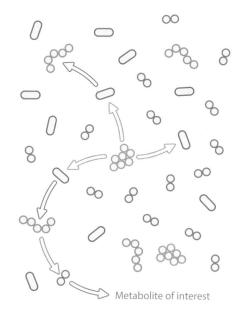

Metabolite of interest

Figure 17.4 Metabolite production in the environment. Microorganisms often produce certain metabolites only in close proximity to other microbial species, because one species produces a metabolic waste used by other species as a substrate for metabolite synthesis. In this example, there are four microbial species indicated by the four colors, living together in their natural environment. Green cells produce a metabolic waste (green arrows) that can be used by red cells. Only those red cells growing close to green cells can ultimately produce their own metabolite (red arrows). Orange cells growing nearby absorb the metabolite from the red cells, and produce yet a third metabolite (orange arrows), which in turn is absorbed by blue cells. Only those blue cells absorbing the necessary substrate from orange cells can produce the final, commercially valuable metabolite. Such complex relationships are not easy to elucidate and are often extremely difficult to reproduce in a laboratory setting.

the environmental parameters that permit optimum growth for our purposes, and we have unambiguously demonstrated that the microorganism in question poses no risks to humans or the environment. Certain problems must still be addressed before commercial production can begin. Metabolic processes that proceed satisfactorily in a few fluid ounces of nutrient broth in a laboratory flask do not necessarily occur efficiently in a 100,000-gallon industrial vessel. To cite just one problem, many metabolic processes of commercial value occur best under aerobic conditions, but proper aeration of a very large volume of liquid nutrient broth is not necessarily a simple task.

Microbial production of industrial metabolites generally takes place inside a device called an **industrial fermenter**. Basically, an industrial fermenter is a large cylinder made of stainless steel. The fermenter is outfitted with equipment to maintain constant temperature and pH, to aerate and adequately mix the liquid medium, to add nutrients as required, and to harvest the metabolite of interest (**Figure 17.6**). Ideally, those conditions that provided optimum metabolite production in a laboratory flask are recreated in the industrial fermenter.

Industrial fermenters can be quite large, holding thousands of gallons. Running and maintaining such a device requires a considerable input of time and money. Consequently, the transfer from the laboratory flask to a large industrial fermenter usually involves certain relatively low-cost

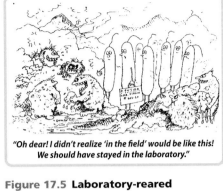

"Oh dear! I didn't realize 'in the field' would be like this! We should have stayed in the laboratory."

Figure 17.5 Laboratory-reared microorganisms returned to their original environment. Bacteria that have been subjected to strain improvement frequently no longer survive well in the environment. Should such bacteria be inadvertently released, they are usually at a competitive disadvantage with naturally occurring wild-type strains.

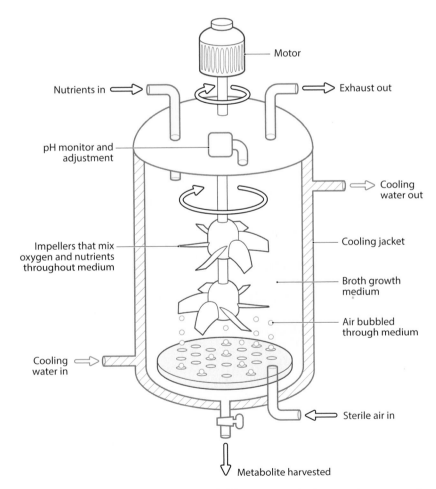

Figure 17.6 An industrial fermenter. The fermenter is constructed so that temperature, aeration, pH, and nutrient levels can be carefully controlled, in order to provide optimum conditions for metabolite harvest.

Figure 17.7 The scale-up process. (a) Initial research for commercially valuable products is conducted with standard laboratory equipment, and reactions are carried out in flasks or test tubes. Microbial processes that show promise are next carried out in a sequence of progressively larger fermenters, in which production problems can be worked out. (b) A small-scale laboratory fermenter. (c) A typical pilot plant fermenter. (d) A large-scale industrial fermenter.

intermediate steps, in which problems can be addressed and conditions can be adjusted as necessary.

First, a process that appears to have commercial or industrial potential in a small laboratory flask or test tube is carried out in a **laboratory fermenter**, which may hold 5 gallons or so of nutrient broth. Here, various combinations of nutrients, temperature, oxygen, and pH can be tested inexpensively, in order to determine which combination is optimal. If results in the laboratory fermenter are satisfactory, the next step is often to move up to a larger scale of perhaps 500–1000 gallons in a **pilot plant fermenter**. Should things proceed smoothly at this stage, the process is finally transferred to the large-scale industrial fermenter. This sequence of stages is referred to as **scale-up** (**Figure 17.7**).

Once the industrial fermentation process is successfully under way, the metabolite of interest is recovered by filtration, settling, or other methods. Depending on the metabolite, it may now be ready to use or it may require further treatment such as drying or purification. Next we take a look at some of these valuable substances that come to us courtesy of microbial activity.

Antibiotics are microbial metabolites that are produced commercially

As we learned in Chapter 13, antibiotics were originally isolated from microorganisms themselves, and many are still commercially produced by bacteria and fungi. Soil bacteria of the genus *Streptomyces* have proven to be especially adept at producing valuable antibiotics. Examples include erythromycin, streptomycin, tetracycline, vancomycin, and rifampin. Other antibiotics such as penicillin and cephalosporin are synthesized by fungi.

Commercial antibiotic production takes place in large fermenters that may hold up to 100,000 gallons. As in other fermenters, oxygen levels, pH, and temperature must be carefully monitored and controlled. Provided nutrients usually include a sugary molasses-like substance or a combination of sugars, proteins, and vitamins.

Although thousands of antibiotics from various microorganisms have been described, relatively few actually make it to commercial production. In some cases it has not been possible to produce the antibiotic in sufficient quantities. Other potential antibiotics may be shelved when it is determined that while they work well in a laboratory setting (*in vitro*), they are ineffective in a living animal body (*in vivo*). Others are discarded because they are found to have undesirable side effects on animal hosts. An antibiotic is produced and marketed only after research has demonstrated that it is effective against certain microorganisms *in vitro* and that it can be produced in sufficient quantities, and after it has passed stringent testing for effectiveness and safety, first in animals and finally in clinical human trials.

So clearly Sam, the student in our case, has a long way to go before he makes his fortune in pharmaceuticals. There is a high probability that the microorganism in his soil sample belongs to a group of fungi or *Streptomyces* that has already been investigated. Even if his organism is previously unknown, its value may be limited, unless the compound it secretes works differently from other antibiotics. Furthermore, the production problems discussed in this section must be overcome.

Medically valuable steroid hormones can be produced by microorganisms

CASE: MENOPAUSE, MARES, AND MICROBES

Women who have undergone it know that menopause can be unpleasant. Hot flashes and trouble sleeping are just two of the many symptoms that some women experience. For some, symptoms are so severe that the steroid hormone estrogen may be prescribed. This is a form of hormone replacement therapy in which hormones no longer produced by an individual in sufficient amounts are administered to replace those that are lacking. For over 50 years an important source of this estrogen has been the urine of pregnant horses. Marketed under the name Premarin® (from <u>pre</u>gnant <u>mare</u> ur<u>ine</u>), the hormone helps to relieve the symptoms of menopause.

Yet this is a relatively expensive process that is not particularly pleasant for the horses. In fact, many would argue that using horses in this way is a serious form of animal abuse. Factors cited include the manner in which urine is collected, the fact that both water and exercise are restricted, and the premature death of these horses.

1. **What options does a patient have if she would like to use hormonal replacement to relieve symptoms of menopause, but is also concerned with animal welfare?**

Various steroid hormones, including cortisone, used to treat pain and inflammation, and progesterone, used to prevent miscarriages, can be synthesized commercially by microorganisms. The process involves providing an appropriate microorganism with a commonly available sterol compound that it can convert into the medically valuable hormone (**Figure 17.8**). Such minor chemical alterations by bacteria and fungi, transforming a common substrate into a valuable product, are called **bioconversions**.

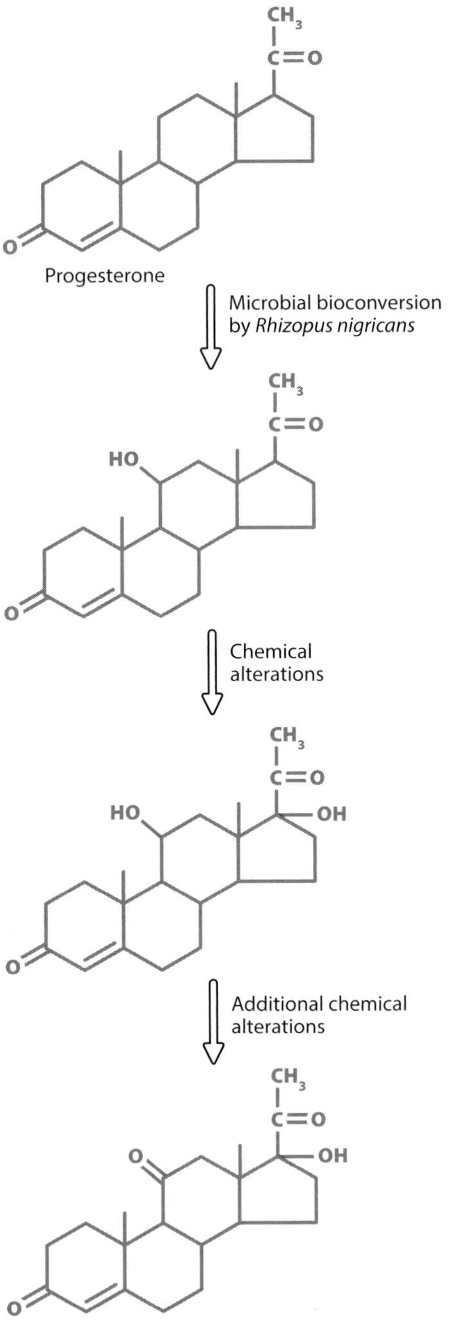

Microbial bioconversion of steroids into medically useful hormones provides an inexpensive and quick alternative to either chemically synthesizing the hormone in question or recovering it from an animal source. While Premarin® may help menopausal patients feel better, those interested in animal welfare might be glad to learn that more benign alternatives exist. Some of these alternative sources of estrogen are derived from plants, while others are produced by microbial bioconversion. Like other steroid hormones, estrogen production starts with cholesterol. Several steps are required to convert cholesterol to a form of estrogen called estradiol: the form of estrogen produced in the ovaries. Bacteria, specifically *Arthrobacter simplex* and

Figure 17.8 Steroid production by bioconversion. Minor chemical alterations by specific bacterial or fungal species can result in the production of medically valuable steroid hormones at a small fraction of the cost and labor of obtaining the hormone from other sources. In this example, the female hormone progesterone is being converted into cortisone, useful for the suppression of inflammation. The first step in the process involves the bioconversion of progesterone to an intermediate compound by the fungal microorganism *Rhizopus nigricans*.

Table 17.1 Representative medical products other than antibiotics produced by microorganisms.

PRODUCT	USE	MICROBIAL SOURCE
Pravastatin	Lowering cholesterol	*Penicillium citrinum* (a fungus)
Cyclosporin	Prevents rejection in organ transplant patients	*Tolypocladium inflatum* (a fungus)
Ergot alkaloids	Induction of labor	*Claviceps purpurea* (a fungus)
Mitomycin	Anti-cancer drug	*Streptomyces caespitosus* (a bacterium)

species of *Mycobacterium*, are crucial to the bioconversion of cholesterol into estradiol.

The list of medically useful products synthesized industrially by microorganisms does not end with antibiotics and steroid hormones. Microbial products are used to lower cholesterol, induce labor, fight cancer, and prevent organ rejection in transplant patients. Table 17.1 lists a few such products.

Microbial metabolites have many other, nonmedical uses

Microorganisms produce a large variety of novel enzymes, many of which have industrial applications. In Chapter 15 we learned that stone-washed jeans have more to do with bacterial enzymes and less to do with stones than the name implies. Likewise, the next time you tenderize a steak before grilling, keep in mind that the tenderizer you sprinkle on the meat may be a bacterial protease enzyme that initiates the process of protein digestion. Such proteases in drain cleaners help to digest clogs caused by hair, which is largely protein. They also lend a hand in laundry detergents, where they digest protein-containing stains, like blood stains. A grass stain, on the other hand, is often dissolved by bacterial cellulases in the detergent, which digest the cellulose in the plant material into simple sugars that are rinsed away.

These are just a few of the microbial enzymes that find their way into common products. As with those microbe-based products already discussed, the first step is to find or develop a mutant strain that produces abnormally high levels of the particular enzyme. Increasingly, the genetic techniques discussed in Chapter 15 allow scientists to engineer organisms to suit specific industrial needs. Subsequently, the production is scaled up as previously described, ultimately resulting in quantities sufficient for commercial purposes.

A newer and exciting group of enzymes is the **extremozymes**, produced by microorganisms normally found in extreme environments. Thermophiles, for example, thrive in very hot water (see Chapter 8, pp. 188–189), and their enzymes are active at temperatures that would inactivate typical enzymes. In Chapter 15, for example, we discussed how the polymerase chain reaction (PCR), so vital in modern genetic analysis and biotechnology, involves a series of high-temperature steps. Most DNA polymerase enzymes would quickly be denatured and therefore of no use at these temperatures. Accordingly, PCR relies on a type of DNA polymerase known as Taq polymerase. The enzyme is named after the thermophilic bacterium *Thermus aquaticus*, from which it is obtained. Because thermophilic bacteria can be difficult to maintain in the laboratory, genes for thermophilic enzymes are now routinely spliced into more easily cultured mesophilic bacteria. The desired enzyme is then produced by these now-recombinant bacteria.

Many industrial processes that work best at high temperature might benefit from such heat-tolerant enzymes. Likewise, other procedures might require a very cold, acidic, or halophilic environment. Enzymes produced by psychrophilic (cold-loving), acidophilic (acid-loving), or halophilic (salt-loving) bacteria may then be most appropriate.

Biopesticides are now readily available in any garden supply store to assist those troubled by insect pests. Biopesticides are simply proteins produced by specific bacteria that are toxic to certain insects. *Bacillus thuringiensis*, for instance, produces a biopesticide that is highly toxic to many types of insects, particularly the moth larvae that can wreak havoc on the leaves of flowers, tomatoes, and other commonly cultivated plants. A subspecies, *Bacillus thuringiensis israelensis*, is especially effective against mosquito larvae and is often used in mosquito control programs. *B. thuringiensis* is a spore-former. Spores containing the protein toxin are ingested by the insect, and after the cells lining the gut are destroyed, the toxin enters the insect's blood, causing paralysis and death. The protein is most lethal against insects with a basic pH in their digestive system. Fortunately, dogs, cats, and other animals, as well as humans, have either a neutral or acidic pH in their digestive system, making the toxin safe to use.

After the bacteria are grown in large numbers in a fermenter, the conditions are altered to induce spore formation. The spores are then collected and mixed with inert compounds, for marketing as a pest control agent that can be dusted on plants. When sold commercially it is often referred to simply as "Bt" (**Figure 17.9a**).

In many cases, farmers may now dispense with spreading bacterial spores, and simply grow crops into which the Bt toxin gene has been spliced (**Figure 17.9b**). Such crops essentially produce their own pesticides, and consequently suffer less from insect damage (see Figure 17.9b). In the United States, for instance, genetically modified corn that expressed the Bt toxin was first approved in the mid-1990s. This drastically reduced damage from the European corn borer and related pests. In India, by 2015, more than seven million cotton farmers were growing Bt cotton. Such genetically modified crops are still banned by many countries, and the use of these crops raises all of the safety concerns discussed in Chapter 15.

Some microorganisms themselves are used as materials

Increasingly, microorganisms are being considered not just for what they produce but as materials themselves in the growing field of **nanotechnology**—technology or the use of materials at an extremely small scale. For example, researchers have converted bacteria into humidity detectors by coating the bacteria with tiny gold particles (**Figure 17.10**). As the environment dries out and humidity falls, the bacteria lose water, moving the gold particles on their surface closer together. Electrodes apply a voltage across the bacteria, and as the gold beads get closer together, an increased current is detected. The scientists found that if the humidity dropped from 20% to 0%, there was a 40× increase in current. These microscopic barometers, which function even if the bacteria die, work best in dry environments. They might even be useful on future space missions to detect humidity on other planets such as Mars. Similar metal-plated microorganisms have been investigated for many other nanotechnology applications, including as semiconductors and optical devices. Magnetic bacteria, to provide one very recent, final example, have been targeted to stomach tumors. Once they

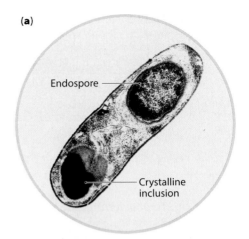

(a)

Endospore

Crystalline inclusion

(b)

Figure 17.9 Biopesticides. (a) *Bacillus thuringiensis* is the source of the biological pesticide Bt. The bacillus has already formed its spore. Such spores contain a protein, which is toxic to many insect pests. *B. thuringiensis* is grown in fermenters and when the bacteria have reached high numbers, conditions within the fermenter are altered to induce spore formation. The spores can then be sold as Bt, which is used on plants to reduce insect damage. (b) The bacterial gene for Bt has been spliced into the genome of various crop plants. Here, Bt in recombinant peanut plants (bottom) protects the plants from damage caused by the lesser cornstalk borer (top).

have arrived at the tumors, the patient lies in a magnetic resonance imaging (MRI) device. The MRI scan causes the iron in the microbial magnets to spin, and the heat generated by the spinning kills the malignant cells.

In addition to the commercial applications we have been discussing, microorganisms can help us solve some of today's thorniest problems. These include water pollution and toxic waste, solid waste reduction, and the always worrisome energy shortage. Here we next examine some of the ways that microbial partners can help us get out of some of the jams we have created for ourselves.

New strategies are required to combat environmental pollution

CASE: THE SWEET SMELL OF SUCCESS

Sewage stinks. The rotten-egg smell comes from hydrogen sulfide, released by bacteria as they digest organic material in the sewage. Sewage plants often combat this problem by using devices called chemical scrubbers that filter the hydrogen sulfide through lye and bleach. Although this technique works, it is at best a partial solution, because new environmental problems are created in the process. Fortunately, a cheaper and more environmentally friendly sustainable strategy is now available thanks to hydrogen-sulfide-digesting bacteria (**Figure 17.11**). The sewage is trickled through biofilms of these bacteria, growing on polyurethane foam in a process called bioremediation. The microorganisms convert the offending hydrogen sulfide into odorless sulfates, which are carried away by water seeping over the foam. Although it is expensive to convert a traditional sewage plant using scrubbers to one using biofilms, low operating costs mean a plant can make up the difference in about 2 years, once the switch is made.

1. What environmental problems are caused by the use of chemical scrubbers?
2. What exactly is bioremediation, and how does this process help alleviate the environmental problems caused by the scrubbers?
3. Why are the bacteria grown in biofilms on polyurethane foam?

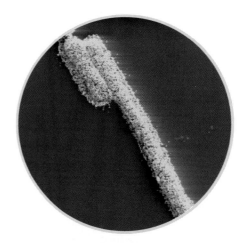

Figure 17.10 Microbial humidity sensors. The bacteria in the photograph, *Bacillus cereus*, have been coated with microscopic gold particles. Electrodes then apply a current, transferred by the gold particles. As the humidity drops, the bacteria lose water, reducing their volume and bringing the gold particles closer together. This is detected as an increase in current.

Figure 17.11 Microbial biofiltration. At wastewater treatment plants, bacteria can eliminate smelly hydrogen sulfide by percolating the sewage through a tower filled with bacteria-covered foam blocks. Arrows indicate the direction in which the sewage is pumped. As sewage trickles through the foam, the bacteria convert the hydrogen sulfide into odorless sulfates.

Modern industry and chemical production have improved our standard of living in many ways. But this improvement comes with a cost: increased environmental pollution. While some pollutants are degraded relatively quickly, others can last for years or decades, causing significant environmental degradation. Not only are natural ecosystems damaged, but humans may also be at risk if pollutants, including pesticides or heavy metals, are concentrated in the food chain (see Chapter 10, pp. 222–223). Toxic material that enters ground or surface water may find its way into our drinking water, creating further health hazards.

Since the environmental dark days of the mid-twentieth century, there has been considerable progress, at least in more developed countries, as the problems of water and toxic waste pollution have been recognized and addressed. Important legislation, like the Clean Water Act, enacted in 1977, ensures that active measures are taken to prevent such pollution and violators are punished. Nevertheless, these problems have not vanished, and as human populations continue to increase, pressure on the environment persists. In the United States, for example, there are thousands of designated hazardous waste sites. Approximately 1300 of these heavily polluted areas have been designated as "Superfund sites" by the US Environmental Protection Agency. These are areas so hazardous that their clean up, which will cost billions of dollars, is an especially high priority.

Historically, the basic approach to the release of pollutants into the environment was out of sight, out of mind. People believed that the environment was so vast that whatever we released into it would eventually be absorbed, diluted, and broken down. Hard lessons taught us that this is not the case, and various chemical and physical procedures to remove dangerous materials from water and soil have been subsequently developed. These procedures are often expensive and require constant monitoring. And frequently, they simply transfer the environmental burden elsewhere. Some chemicals, for instance, can be detoxified by incineration, but this increases air pollution. Certain toxins can be concentrated and prevented from entering water, but they must then be stored indefinitely in a secure, leak-proof container. For example, the chemical scrubbers used in the treatment of sewage, as described in our case, are highly effective at removing hydrogen sulfide, but they rely on toxic chemicals such as lye or bleach. Were these chemicals to leak into groundwater, there would be serious environmental repercussions. Sewage plants, consequently, must go to great effort and expense to prevent such contamination.

Current technology will continue to play a vital role in environmental protection and the control of pollution, but increasingly, scientists are turning to microorganisms as a way to break down pollutants in an environmentally friendly manner.

Microorganisms are used to digest harmful chemicals through the process of bioremediation

The use of living organisms to degrade environmental contaminants is called **bioremediation**. This technique is already being used in many ways, including oil-spill clean up, drinking water treatment, and toxic chemical degradation. Bioremediation is often performed by allowing certain microorganisms to grow in a thick biofilm over a solid substrate. A biofilm, as described in Chapter 10 (pp. 226–227), is a growth of microorganisms held together by a secreted carbohydrate matrix. A fluid solution containing contaminants is allowed to seep through the carbohydrate matrix, placing the individual bacteria within the biofilm into direct contact with the fluid. The bacteria then degrade the toxic waste as the fluid percolates through the biofilm. In other words, biofilms are uniquely constructed to serve as biological filters, placing the degrading microorganisms into slow and steady contact with toxic materials. Ideally, the fluid leaving the biofilm is contaminant-free.

Not just any bacterial species can be used in this way. To break down a toxic chemical, the microorganism in question must produce enzymes capable of such digestion. As we have seen throughout this text, different microorganisms have different enzymes, and therefore different digestive capabilities. It should not be too surprising that microorganisms, which have evolved to live almost everywhere and utilize almost every possible resource, can be found that digest even the most toxic materials.

Often the most promising species for such toxin digestion are nonfastidious organisms. Recall from Chapter 8 (p. 191) that nonfastidious organisms are those that can digest a very large range of organic molecules, because they are especially well endowed with many different digestive enzymes. *Pseudomonas* is a good example. These bacteria are already commonly used to help clean up oil spills, because some *Pseudomonas* species have enzymes that allow them to break down the complex hydrocarbons found in petroleum. A fastidious species, on the other hand, would likely be a poor choice for bioremediation. Fastidious organisms have very specific nutrient requirements, because they produce relatively few digestive enzymes.

Both the environmental context and the microbe being used determine how bioremediation is conducted

In certain cases, bioremediation is a simple affair because the necessary organisms are already present in the environment and all that is necessary is to encourage their growth. Certain *Pseudomonas* species, for instance, as previously mentioned, are able to digest petroleum. Following an oil spill, contaminated water can be enriched with nutrients such as nitrogen- and phosphorus-containing compounds to spur on the growth of *Pseudomonas* already found in marine environments (**Figure 17.12a**). When microbial growth is accelerated in this way, the clean up of the oil proceeds much faster, with less environmental impact, and at a reduced cost.

In other situations, bioremediation requires more effort. Sometimes, although a strain of organisms that digests a particular pollutant can be isolated and identified (**Figure 17.12b**), it either cannot digest enough of the pollutant or it does not survive well in the contaminated site. In such cases, laboratory selection, as described previously, must be used to develop strains with both the capacity to degrade large amounts of a particular pollutant and to survive and grow quickly once released into the environment. As an alternative to isolating and selecting for a useful strain, the genes responsible for degradation in one organism can be introduced into the genetic material of a more manageable species, by use of the techniques described in Chapter 15. Such genetically modified organisms may have

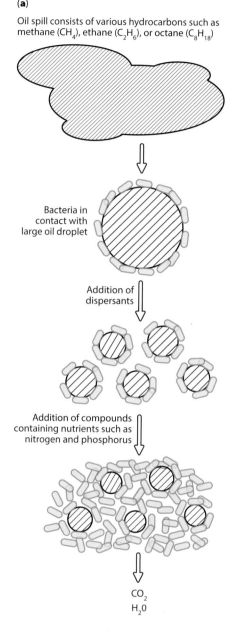

(a)

Oil spill consists of various hydrocarbons such as methane (CH_4), ethane (C_2H_6), or octane (C_8H_{18})

Bacteria in contact with large oil droplet

Addition of dispersants

Addition of compounds containing nutrients such as nitrogen and phosphorus

CO_2
H_2O

Figure 17.12 Bioremediation of oil spills. (a) Bioremediation often begins by spreading dispersants onto the oil, which break large oil droplets into smaller ones. This provides greater surface area for oil-digesting microbes. The addition of nutrient-containing compounds spurs microbial growth, which then digest the oil, releasing CO_2 and H_2O as wastes. (b) Cultures of different oil-degrading bacterial species under evaluation for their bioremediation potential.

limited use, however. They are often unlikely to survive for long if released at a contaminated site, and because they are genetically altered, their release into the environment carries substantial risk. With newly created genetically modified organisms, it is not possible to anticipate all of the environmental problems this organism might actually cause rather than solve. In most cases, the release of such organisms is illegal. This is not to suggest that such microbes are of no value. Genetically modified organisms can be of use in carefully controlled facilities, such as water treatment plants, provided suitable precautions are taken to prevent their spread to the environment.

Sometimes, the problem is not so much finding the right bacteria as it is finding the correct way to expose these bacteria to the material to be digested. The idea of using bacteria to remove hydrogen sulfide from sewage, as described in our case, has been around for some time. But it has always been considered too inefficient compared with chemical scrubbers. To be degraded, hydrogen sulfide needs to be in contact with bacteria for at least 10 seconds. Until recently, there was no practical and economical way to accomplish this. With the use of polyurethane foam as described in our case, researchers vastly increased the surface area of their biological filter by filling the silo-shaped filtering tower with 4-cm^3 porous-foam blocks. The bacteria grow as a biofilm throughout the nooks and crannies of the blocks, and as the sewage trickles through the tower, the microorganisms have ample opportunity to convert the malodorous hydrogen sulfide to less offensive compounds. This strategy may be appealing for other, similar problems, such as the disposal of animal wastes in large-scale agriculture.

Bioremediation can prove valuable in many different settings

Polychlorinated biphenyl compounds (PCBs) are double-ringed compounds containing chlorine atoms that were once used extensively in heavy-duty electrical equipment and as industrial solvents. At the time, it was not understood that these compounds posed a significant environmental hazard. We now know that PCBs accumulate in the fatty tissue of animals that consume them, where they can cause a number of problems. Many are known carcinogens.

PCBs are now banned in most industrialized countries, but many of these compounds can persist for years in the environment, where they continue to work their way into food chains. The Hudson River in New York State was particularly contaminated because of the large electrical equipment plants located along its shores. In the 1980s, however, came the surprising and welcome news that PCB concentrations in the river sediments were much lower than expected. Research subsequently showed that anaerobic bacteria in the sediments were able to break the compounds down to smaller, simpler compounds, which could then be degraded completely by other bacteria. This finding highlighted the potential value of bioremediation for the elimination of these stubborn toxic chemicals. It also demonstrated that specific environmental conditions, in this case an anaerobic environment, were required for successful bioremediation to take place.

Uranium- and plutonium-containing radioactive waste is another daunting problem at many toxic waste sites. One of the difficulties with such wastes is that they are often mixed with other chemicals and are present at low concentration. Bioremediation may one day offer a solution. Recently, a process for concentrating uranium has been developed. The technique relies on bacteria that reduce uranium, making it insoluble in water. The now-insoluble uranium can then be separated out from other, soluble compounds.

Additional research focuses on *Deinococcus radiodurans*, a bacterial species with a remarkable tolerance for radiation that we discussed in Chapter 4. Mercury compounds are common in radioactive waste sites, and the goal is to genetically engineer strains of *D. radiodurans* that reduce mercury to a less toxic form. Because of its inherent resistance to radiation, *D. radiodurans* is uniquely qualified to work in such hot zones.

Military bases, in addition to their obvious purpose, often provide buffers to overdevelopment. In this sense many military bases also serve as wildlife refuges, especially important in light of increasing urbanization and the encroachment upon open space. But often, such sites are heavily contaminated with many toxic compounds or explosives. Until about 1970, one of the most commonly used explosives was trinitrotoluene (TNT) and tons of TNT still pose a danger at military instillations and weapons factories around the world. *Clostridium bifermentans*, however, has recently been found to readily digest TNT. The bacteria can adhere to carbon compounds, which are then inoculated into the contaminated soil. Starch is then added to the soil to serve as an energy source for the growing microorganisms. The explosive is then degraded to harmless materials. A number of contaminated sites have already been rendered TNT-free by this technique.

Bacteria may help solve the solution of plastic waste

The world is awash in plastic. Countless tons of the stuff end up littering the landscape and polluting the oceans. Indeed, a recent study by the World Economic Forum (WEF) estimates that almost one-third of all plastic ends up in the environment and only 14% is recycled. Indeed, the WEF report predicts that by 2050 there will be more plastic in the ocean than fish! Until recently plastic was considered to be nonbiodegradable; no known microbes could digest it.

But talk about bioprospecting success! Japanese researchers announced in 2016 that by sampling soil bacteria in the area around a bottle recycling plant they isolated a new species called *Ideonella sakaiensis*, which efficiently digests the common plastic used in beverage bottles and household cleaning products. The bacterium uses two enzymes: the first to break down the plastic and the second to digest an intermediate compound. The result is small molecules that the bacteria can then use to meet their own nutritional needs. This major announcement offers at least some hope that *I. sakaiensis* might help insure that the WEF's grim predictions do not become reality.

Fungi can be effective bioremediation agents

One of the main points that we have made about fungi in this text is that when it comes to decomposition, fungi take top honors. Indeed, their ability to degrade organic material is their primary contribution to a healthy ecosystem. Fungi release extracellular enzymes and acids that break down complex molecules like cellulose. The degraded components of these molecules are then absorbed by the fungi to meet their own nutritional requirements.

So, if fungi are so good at decomposition, why not let them join bacteria on the list of microbes that can be used in the bioremediation of pollutants? In fact, fungi are already contributing in just this way, thanks to the expanding field of **mycoremediation**—bioremediation utilizing fungi.

As with bacteria, the key to success in mycoremediation is to select the right fungi for the job. Not only must the fungus be able to produce the enzymes necessary to degrade a specific pollutant, but it must be able to survive and grow under the specific environmental conditions found where their use is contemplated. Fungi that naturally digest wood have been found to be particularly useful. Such fungi have been found, for instance, that can digest toxic compounds found in petroleum and in pesticides.

Microorganisms can help reduce solid waste and improve soil quality through composting

CASE: WASTE REDUCTION BEGINS AT HOME

After buying their first home, Hal and Lylette are excited to start a vegetable garden, something that was never possible in their rented apartments. Although they have little experience, they know that kitchen waste can be turned into a valuable soil amendment by composting. After reading up on the subject, they learn that composting is a bit more complicated than merely dumping vegetable waste into a pit and waiting for it to decompose. Rather, they note that proper compost must regularly be turned, either by hand or by a special mechanical device (**Figure 17.13**). Once they begin to compost, they are surprised to observe that after they turn it, the compost heats up.

1. **What exactly is composting?**
2. **Why does compost have to be periodically turned?**
3. **Why does the composting material generate heat?**

Figure 17.13 Backyard composting. Kitchen and yard waste provide the starting material for compost. Such waste is added to a compost container, such as the homemade one shown here, where regular turning of the compost introduces oxygen. Following decomposition, compost is removed. Such decomposed material can then be used as mulch in the yard or garden.

Microorganisms can easily be enlisted to quickly reduce the amount of solid organic waste through **composting**. In this technique, familiar to many homeowners, kitchen and yard waste is decomposed by encouraging the growth of certain microorganisms. Not only does this result in less solid waste, but the degraded products can be used to improve soil quality in gardens or yards. Composting works best when the soil is well oxygenated, because this promotes the growth of aerobic microorganisms. As we learned in Chapter 8, aerobes grow much faster than anaerobes and consequently degrade organic material more quickly. Although anaerobes can eventually degrade organic wastes, their much slower metabolism and reduced growth rate greatly slow the process (**Figure 17.14**).

Consequently, one of the keys to successful composting is to ensure that the waste remains well oxygenated, so that rapid aerobic microbial degradation can occur. As Hal and Lylette learned, this means that the waste must periodically be turned in order to introduce a fresh supply of oxygen (see Figure 17.13). Compost oxygenation also explains why the decomposing material tends to heat up. Within a few days, the temperature of the compost pile rises as aerobic soil microorganisms initiate the decomposition process. As oxygen levels within the composted material fall, the temperature also declines as microbial growth slows. The compost is then turned, oxygen is replenished, and the process repeats until the digestion of the organic waste is complete, typically in several weeks. Not only is the bulk of the waste decreased by about two-thirds, but a rich, mulch-like material is produced. As an added bonus, the regular reintroduction of oxygen reduces growth of

Figure 17.14 Decomposition under anaerobic conditions. Because their rate of growth is slow, the ability of anaerobic bacteria to decompose waste can be slow. This newspaper, a copy of *The Daily Mirror*, was buried and therefore persisted under anaerobic conditions for decades. Note the June 16, 1967, publication date.

Figure 17.15 Municipal composting. Composting is increasingly used by cities, both to reduce their solid waste and to provide soil amendments for city parks or for homeowners. In this photo, organic wastes are being oxygenated mechanically.

anaerobic methanogens (see Chapter 8, p. 176), meaning that less polluting and hazardous methane gas is released.

Composting is no longer merely an avocation for the home gardener. In an ever-increasing number of municipalities, residents can place leaves, grass clippings, and other yard waste into special receptacles for collection. These wastes may be deposited in long, low piles at the landfill site where they can be regularly oxygenated mechanically (Figure 17.15). The benefits of such city-sponsored composting are many. Landfills, now reserved for nonbiodegradable materials, fill up more slowly, reducing the need for yet newer landfills. The final organic product can be used on city-owned parks and golf courses, or it can be made available for home or agricultural use. With less methane production, there is a greatly reduced risk of explosions caused by inadvertent methane combustion in a landfill. Finally, as discussed in Chapter 10, methane is an important greenhouse gas. With their tree trimmings being composted rather than being added to landfills, residents can rest easily, knowing that their yard waste is not contributing to global warming.

Plastics may be replaced by biodegradable, microbially produced alternatives

We recently mentioned the problem of plastic waste and how a newly discovered bacterial species might help us to cope with this environmental headache. Bacteria might assist in other ways as well; they might be able to produce easily degradable plastic alternatives.

Research into such alternatives is beginning to bear fruit. Starch-based plastics, for example, use starch molecules to link together other biodegradable molecules. These plastic-like compounds can be digested first by starch-digesting soil bacteria. Other microorganisms next degrade the starch-linked molecules. Bacteria themselves have also been investigated as sources of substitutes for conventional plastics. Some bacteria store carbon in the form of a molecule called poly-β-hydroxyalkane, or PHA (Figure 17.16a). When environmental sources of carbon are low, these bacteria, using an enzyme called depolymerase, digest their supply of PHA to meet their metabolic needs. Some bacteria secrete an extracellular form of this enzyme that can degrade any PHA in the immediate environment.

PHA has many of the same properties as plastic and its use as a biodegradable plastic alternative is under investigation (Figure 17.16b). Providing

Figure 17.16 PHA: a microbial source of biodegradable plastic. (a) The bacterium *Rhodobacter sphaeroides* stores carbon as an energy reserve in the form of poly-b-hydroxyalkane (PHA) in storage granules. The granules are visible in the cytoplasm of the cell in the center of the photo. (b) Because they are biodegradable, PHAs could be an attractive alternative to the petroleum-derived plastics that pollute our environment. The sheet of PHA on the left has not yet been degraded. By 6 weeks, degradation is almost complete.

bacteria with different carbon sources induces them to synthesize different forms of PHA. This means that specific bioplastics could be produced for plastic bags, squeeze bottles, rigid plastic materials like pipes, or other purposes. When discarded, such products can be degraded by bacteria that secrete the depolymerase enzyme.

A few products are already made from bacterially produced plastic. The use of PHA, however, has been slow to catch on because, compared with conventional plastics made from petroleum, PHA-based plastics are still quite expensive. Yet if oil prices rise significantly, the time may come when biodegradable plastics become the rule rather than the exception.

Microorganisms may help meet the demand for alternative energy

Technically speaking, petroleum and other hydrocarbon fuels are renewable. Yet the geological processes necessary to form new hydrocarbon deposits take millions of years. For all practical purposes, once we deplete our supply of hydrocarbon fuels, they are gone for good. Unless, that is, we increase our use of renewable, biologically produced hydrocarbons.

For example, ethanol is a biological fuel already in widespread use. It is produced by fermenting agricultural products, including corn or sugarcane, and then frequently blended with gasoline for use in motor vehicles. Ethanol, however, is not without problems. When made from corn, it diverts an important food item from livestock feed or the dinner table to ethanol production. With so much corn being used for fuel, the price we pay for food has risen accordingly. Consequently, there has been increased interest in cellulosic ethanol, for which plant wastes such as stalks and leaves are the source instead of edible plant parts like corn kernels. Recall from Chapter 2 (p. 31) that cellulose is a long-chain polysaccharide composed of many glucose molecules, and it is the principal component of plant cell walls. Although cellulose is energy-rich, very few non-microbes can tap this energy because most organisms lack the enzymes necessary to digest cellulose. Many microorganisms, on the other hand, can digest cellulose, and if used in ethanol fermentation, it might be possible to use plant wastes as a sugar source. In fact, in a few places, the fungus *Trichoderma reesi* is doing just that, and it is proving its value in converting agricultural wastes into ethanol. In a sense, perhaps this fungus is making up for a notorious reputation; in the damp

Figure 17.17 Methane: an alternative form of energy. Methane gas released at this Austin, Texas, landfill is collected and converted into electricity. The structure in the photo is the electrical generator. The gas starts to accumulate within months after a landfill is sealed and gas may be produced for 5 years or more.

tropics it can cause jungle rot—the destruction of clothing and other products made from cotton, which is mostly cellulose.

Methane, produced by methanogenic microorganisms, is another potential source of biofuel. Previously, we have cast methane in a bad light; it acts as a greenhouse gas, and it can cause fires in landfills. But it can also be burned as a hydrocarbon fuel. In many places, methane released from landfills is already being used as a source of inexpensive energy (**Figure 17.17**). Another potential source of methane is the large amount of manure, filled with methanogens, that is an abundant waste product of animal feedlots.

Or, consider hydrogen gas (H_2). When oxidized, H_2 releases a relatively large amount of energy as described by the following chemical equation: $2H_2 + O_2 \rightarrow 2H_2O +$ energy.

Note that the only waste product is water, meaning that this source of energy is entirely pollution-free. And the hydrogen used to generate this energy comes from water, meaning that there is an essentially limitless supply. But there's a catch; hydrogen can be extracted from water by chemical or physical means, but this requires a great deal of energy. The large amount of energy needed to get the hydrogen in the first place makes the overall process inefficient. So why not let photosynthetic microbes produce the hydrogen, using only sunlight as an energy source?

Indeed, although still experimental, it has been found that under anaerobic conditions, certain photosynthetic bacteria or algae release hydrogen as a fermentation waste. While the process has been successful under laboratory conditions, it has not yet proven feasible to produce hydrogen in this way on a large scale. Once a proper methodology is worked out, however, biological hydrogen production might prove invaluable in helping us meet future energy needs in an environmentally responsible way.

Algae in particular have attracted a great deal of attention as possible biofuel sources (**Figure 17.18**). Algal fuels have several advantages over other biofuels. Algae grow quickly compared to other crops considered as fuel sources, and they produce hundreds of times more energy per unit of area than conventional crops. Furthermore, whereas crops like corn need plenty of fresh water, algae can thrive on waste or brackish water, and they can be grown on land that is otherwise unsuitable for agriculture. Depending on the species of algae used, the part of the algal cell that is harvested, or the technique employed, various types of fuels can be produced. For example, extracted

Figure 17.18 Algae as a source of biofuel. The structure photographed is a photobioreactor, in which algae produce biofuels. The algal cells grow in nutrient-rich water pumped through the tubes, while the transparency of the tubes allows sunlight to enter.

lipid can be converted into ethanol, while the carbohydrate portion of the cell may be utilized for biodiesel production.

Algal biofuel technology is still relatively new and hurdles must be overcome before it is cost effective and commercially viable. While the technology to harvest the algae and convert it into fuel already exists, the process is still quite expensive and contamination remains a problem.

Other, more exotic uses of microorganisms for energy production may await us in the future. For example, bacteria of the genus *Shewanella* are facultative anaerobes. Recall from Chapter 8 (pp. 187–188) that facultative anaerobes use oxygen as a final electron acceptor when it is available, releasing the reduced oxygen as water. In the absence of oxygen, a different final electron acceptor is used and a different end product is released. *Shewanella*, however, is unusual because when it is growing under anaerobic conditions in the soil, electrons released at the end of respiration do not reduce a final electron acceptor inside the bacterial cell itself. Rather, the bacteria export the electrons outside their cells, where the electrons attach to metallic ions. In the laboratory, researchers have found that they can substitute an electrode for these naturally occurring ions, effectively turning the bacteria into microscopic electric generators. Such bacteria could be used to create microbial fuel cells capable of generating energy. In a conventional fuel cell, hydrogen is combined with oxygen to create electricity, with water as a by-product. Microbial fuel cells would involve placing bacteria like *Shewanella* in an anaerobic environment, along with glucose as an energy source. The efficiency of such fuel cells ranges from 60–80%. In other words, between 60% and 80% of the energy available in a glucose molecule's electrons is ultimately transferred to the electrodes by the bacteria. By comparison, the efficiency of a typical internal combustion engine in a modern car is rarely better than 30%. The study of microbial fuel cells is in its infancy, but this emerging technology may one day help us to meet our ever-growing demand for energy in an environmentally sustainable way.

Looking back and looking forward

The capacity of microorganisms to help us solve complex problems or to produce new and useful products is seemingly limited only by our imagination. Yet numerous hurdles must be overcome before any promising microbe can be used industrially or to produce a commercial product. When successful, however, various microorganisms are or may one day be of assistance as we confront issues as diverse as waste disposal, pollution control, and the need for cleaner, renewable energy.

As for looking forward, who knows? We have reached the end of this book, but the tale of microorganisms and how they affect us continues. What will be the next new emerging disease to strike? What new breakthrough in biotechnology is just over the horizon? When will new vaccines or drugs lift the burden of diseases that threaten us today? How will microbes help us solve complex, twenty-first century environmental problems? Answers to these questions are not a matter of if, but when. The next microbial story is right around the corner, and it will no doubt provide more evidence, if more evidence were needed, that these smallest inhabitants of our planet are among the most remarkable.

Garland Science Learning System

- http://garlandscience.rocketmix.com/students/

- Discover how to grow crops with added pesticides built in

- Test your knowledge of this chapter by taking the quiz

- Familiarize yourself with the terminology used in this chapter by using the vocabulary review

- Get help with the answers to the Concept questions

Concept questions

These questions are designed to help you start thinking like a microbiologist. The answers are not always simply found in the text. Instead, you will need to take the concepts about which you have learned and apply them to new situations. Some of the questions may not even have just a single correct answer. Help is provided as part of the GSLS resources, which can be accessed through http://garlandscience.rocketmix.com/students/.

1. Many people suffer from gas after they eat certain fruits and vegetables like beans, broccoli, and cabbage. For these individuals, a product called Beano® is available. The major ingredient in this product is an enzyme called galactosidase, produced by the fungus *Aspergillus niger*. This enzyme digests the carbohydrates in the plant material that are responsible for much of the gas. Provide a simple description of how you think scale-up for the production of Beano® may have progressed.

2. Suppose the production of a potentially valuable metabolite is regulated by an inducible operon. After reviewing how such operons function (Chapter 7, pp. 149–151) and how CRISPR/Cas technology can be used to silence certain genes (Chapter 15, pp. 377–378), discuss what regions of the operon in question might be silenced using CRISPR/Cas to cause overproduction of the desired metabolite.

3. Look through the food items, cleaning products, medicine chest, etc., of your home. Which products can you find that are commercial microbial products?

4. Here in Chapter 17 we have reviewed various environmental problems, which may, in part, have microbial solutions. Can you think of an environmental or ecological problem that was not specifically addressed? Furthermore, can you foresee any way that microorganisms might be involved in addressing the problem that you have thought of?

5. Alcohol, when produced by yeast cells, is an example of a primary metabolite. Penicillin, when produced by mold, is an example of a secondary metabolite. How would the production of alcohol and penicillin differ in terms of the environmental conditions provided for the microorganisms?

6. If a microorganism is removed from nature and subjected to strain improvement in the lab, it will usually not be able to survive if returned to its original environment. Why not?

7. Look closely at the garbage in the wastebaskets in your home. Then make a ballpark estimate about the percentage of the garbage that could be degraded by microorganisms through composting.

8. A microbiologist who is also a big sports fan makes the statement that "without microorganisms there would be no steroid problem in sports." Do you think there is any truth at all to this statement? If the use of microorganisms in bioconversion, as discussed here in Chapter 17, were somehow impossible, would the steroid problem in sports go away? Would it be reduced? Would it become more serious?

9. We have learned that that certain bacterial species are unusually resistant to radiation because of their highly efficient DNA repair mechanisms. Can you think of any potential commercial or industrial applications for such bacteria?

FIGURE ACKNOWLEDGMENTS

Figure Number	Credit Line	Figure Number	Credit Line
1.1a	Courtesy of Tony Brain & David Parker/Science Photo Library	2.11a	Courtesy of Manuel Amieva, Shumin Tan & Lydia Marie Joubert
1.1b	Courtesy of Christine L. Case, Skyline College	2.17	© Courtesy of Flagstaffotos
1.1c	Courtesy of the National Park Service	2.18	© Courtesy of James G. Howes
1.2a	Courtesy of Scimat/Science Photo Library	2.20	Courtesy of CDC/Larry Stauffer, Oregon State Public Health Laboratory
1.2b	Courtesy of University of Newcastle upon Tyne	3.4	From www.microbiological-garden.net. With permission from Heribert Cypionka
1.2c	Courtesy of Marc Brodkin	3.6a	Courtesy of CDC/W.A. Clark
1.4a	Courtesy of CDC/Stuart Brown	3.6b	Courtesy of CDC/Joe Miller
1.4b	Courtesy of CDC/Francis W. Chandler	3.6c	Courtesy of CDC
1.4c	Courtesy of CDC	3.6d	Courtesy of CDC/Richard Facklam
1.5	Courtesy of NIAID	3.6e	Courtesy of CDC/V.R. Dowell, Jr.
1.6a	From Prusiner S (1998) *PNAS* 95:13363–13383. With permission from © National Academy of Sciences	3.6f	Courtesy of CDC
		3.6h	Courtesy of Gary E. Kaiser, The Community College of Baltimore County
1.6b	Courtesy of USDA APHIS/Michelle Crocheck	3.7	Courtesy of John Ruby, University of Alabama
1.7a	Courtesy of CDC/Maryam I. Daneshvar	3.8	Courtesy of Gary E. Kaiser, The Community College of Baltimore County
1.7b	Courtesy of USDA ARS/Peggy Greb		
1.8a	From Ranalli G, Alfano G, Belli C et al. (2004) *J Appl Microbiol* 98:73–83. With permission from John Wiley and Sons	3.9	From Lane MC & Mobley HL (2007) *Kidney Int* 72:19–25. With permission from Macmillan Publishers Ltd.
1.8b	Courtesy of Christine L. Case/Skyline College	3.10	Courtesy of Dr Aizawa
2.1	Courtesy of Carl S. Kirby	3.15a	Courtesy of CDC/Mike Miller
2.5	Courtesy of William Herring, www.LearningRadiology.com	3.15b	Courtesy of CDC/George P. Kubica
		3.18	CNRI/Science Photo Library

Figure Number	Credit Line	Figure Number	Credit Line
3.19b	Courtesy of CDC/Larry Stauffer, Oregon State Public Health Laboratory	4.22	Courtesy of Henry Mühlpfordt
3.20a	Courtesy of Caroline C. Philpott, NIH	4.23	Courtesy of Jay W. Pscheidt, Oregon State University
3.20b	Courtesy of CDC/Mae Melvin	4.24	Courtesy of USDA Forest Service/Robert L. Anderson
3.20c	Courtesy of Gene Gushansky	4.26	Courtesy of Craig W. Stihler, WVDNR
3.23	From Buey RM, Calvo E, Barasoain I et al. (2007) *Nat Chem Biol* 3:117–125. With permission from Macmillan Publishers Ltd.	5.2b	Courtesy of Pavel Hrdlička published under CC BY-SA 3.0
		5.7b	Courtesy of Perelman School of Medicine, University of Pennsylvania
3.24a	Courtesy of Leilo Orci, University of Geneva, Switzerland	5.11a	From Lam E, Kato N & Lawton M (2001) *Nature* 411:848–853. With permission from Macmillan Publishers Ltd.
3.24b	Courtesy of D.L. Schmucker, Department of Anatomy, UCSF	5.11b	Courtesy of Ben Schumin/The Schumin Web
3.27	Courtesy of CDC/Edwin P. Ewing, Jr.	5.12b	Courtesy of Larry Goodridge
4.1	Courtesy of NASA	5.15	Courtesy of Jean-Loup Charmet/Science Photo Library
4.4a	Courtesy of CDC/Shirley Maddison		
4.4b	Courtesy of WHO/TDR/Ane Haaland	5.16	Courtesy of WHO
4.4c	From Lewis FA, Liang Y-s, Raghavan N & Knight M (2008) *PLoS Negl Trop Dis* 2:e267. With permission from Public Library of Science.	5.19	Courtesy of Peter Hooper, CSIRO published under CC BY-SA 3.0
		6.3	Painting by Michiel Sweerts, Plague in an Ancient City. LACMA
4.6	Courtesy of CDC/Stan Erlandsen	6.7	Courtesy of the Oriental Institute of the University of Chicago
4.9	From Daly MJ (2009) *Nat Rev Microbiol* 7:237–245. With permission from Macmillan Publishers Ltd.	6.8	Courtesy of The US National Archives and Records Administration
4.10a	Courtesy of CDC/Ewin P. Ewing	6.9	Courtesy of CDC/Frank Collins
4.10b	Courtesy of CDC/Susan Lindsley	6.10	Courtesy of USDA/Scott Bauer
4.11	Courtesy of NASA	6.12a,b	Library of Congress
4.12a	Courtesy of Mike Dyall-Smith	6.14	Library of Congress
4.12b	From Loker ES & Hofkin BV (2015) Parasitology: A Conceptual Approach. Garland Science.	6.18	Courtesy of the National Library of Medicine
		6.19	Courtesy of the National Library of Medicine
4.15a,b,d	Courtesy of Steve Upton	6.20	Library of Congress
4.15c	Courtesy of Yvonne R. Vaillancourt, Biology Department, University of Massachusetts at Boston	6.21	Courtesy of the National Library of Medicine
		6.22	Courtesy of the National Library of Medicine
4.18b	Courtesy of CDC/Robert Simmons & Janice Haney Carr	6.25a	Courtesy of the National Library of Medicine
		6.25b	Courtesy of the National Library of Medicine
4.21	Courtesy of US Fishery and Wildlife Service/Charles H. Smith, vergrößert von Aglarech	7.4a	Courtesy of G. Murti/Science Source

Figure Number	Credit Line	Figure Number	Credit Line
7.5a	Courtesy of Brian Wells	9.14	Adapted from Embley TM & Williams TA (2015) *Nature* 521:169–170. With permission from Macmillan Publishers Ltd.
7.5b	Courtesy of Smith College and the Center for Microscopy and Imaging (CMI)	9.16	Courtesy of Raymond E. Goldstein, University of Cambridge
7.17b	From Dragon F, Gallagher JEG, Compagnone-Post PA et al. (2002) *Nature* 417:967–970. With permission from Macmillan Publishers Ltd.	9.17	Courtesy of CDC/Ed Ewing
		9.20a	Courtesy of CDC/George Kubica
7.23	From Miller Jr. OL, Hamkalo BA & Thomas Jr. CA (1970) *Science* 169:392–395. With permission from American Association for the Advancement of Science	9.20b	Courtesy of Tim Berger, UCSF School of Medicine
		9.23	Courtesy of M.W. Mules
		9.24b	Courtesy of N.C. Hinkle
7.26a	Courtesy of Chris Frazee & Margaret McFall-Ngai published under CC BY 4.0	9.25b	Courtesy of LifeStraw® Family, Vestergaard Frandsen, 2008
7.32b	Courtesy of Charles Brinton	9.27	From O'Neill SL, Hoffmann AA & Werren JH (1997) Cytoplasmic incompatibility in insects. Influential Passengers: Inherited Microorganisms and Arthropod Reproduction, 2nd ed. Published by and permission from Oxford University Press
8.2	Courtesy of Toksave published under CC BY-SA 3.0		
8.3	Courtesy of Nick Hobgood published under CC BY-SA 3.0		
8.16	Courtesy of Sandie Smith		
8.19a	Courtesy of NASA	10.1	Courtesy of Peggy Greb, USDA Agricultural Research Service
8.19b	Courtesy of Marshall D. Sundberg, in collection of Botanical Society of America, 1998	10.2	Courtesy of Nick Evans, Webmaster-Bishops Offley, UK village website, www.bishopsoffley.co.uk
8.26	Courtesy of Lourdes Norman-McKay	10.6a	Courtesy of US Geological Survey
8.29	From Cantu D, Vicente AR, Labavitch JM et al. (2008) *Trends Plant Sci* 13:610–617. With permission from Elsevier	10.6b	Courtesy of B. Pitts, Montana State University Centre for Biofilm Engineering
		10.6c	© Reprinted with permission from Water Pik, Inc., Fort Collins, CO.
8.30	Courtesy of Archives and Special Collections, Queen Elizabeth II Library, Memorial University of Newfoundland	10.9	From Dazzo FB, Truchet GL, Sherwood JE et al. (1984) *Appl Environ Microbiol* 48:1140–1150. With permission from the American Society for Microbiology
8.32	Courtesy of SCIMAT/Science Source		
9.2	Courtesy of National Park Service/ J. Schmidt	10.10a	Courtesy of Peter West, National Science Foundation
9.4a	Courtesy of OAR/National Undersea Research Program (NURP), NOAA	10.10b	Photo by Russ Kinne, courtesy of National Science Foundation
9.4b	Courtesy of J. Kuyken	10.13	Courtesy of NOAA
9.8a	Courtesy of NASA	10.14b	Courtesy of NASA/GSFC, MODIS Rapid Response
9.8b	Courtesy of J.W. Schopf, UCLA		
9.9b	Courtesy of Wayne Lanier	10.16a	Courtesy of Miriam Godfrey
9.10	Courtesy of Stuart Edwards published under CC BY-SA 3.0		

Figure Number	Credit Line	Figure Number	Credit Line
10.16b	Courtesy of Woods Hole Oceanographic Institution	11.20b	Courtesy of Preston Maxim, Department of Emergency Medicine, San Francisco General Hospital
10.17	Courtesy of NASA/JPL/DLR	11.21	Courtesy of F.A. Murphy, University of Texas Medical Branch
10.18	Courtesy of NASA/JSC/Stanford University		
10.20	Courtesy of Mark Willis, www.marksveg-plot.blogspot.co.uk	11.24	Courtesy WHO/TDR/Wellcome
10.22	Courtesy of USDA-ARS/Markus Dubach	12.5c	From Champion JA & Mitragotri S (2006) *PNAS* 103:4930–4934. © With permission from National Academy of Sciences, USA.
10.24	Courtesy of Richard Ling published under CC BY-SA 3.0		
10.26	Courtesy of Acropora published under CC BY-SA 3.0	12.8	Courtesy of David Scharf/Science Source
		12.12b	Courtesy of W. van Ewijk
10.28a	Courtesy of Gary E. Kaiser, The Community College of Baltimore County	12.22	Courtesy of WHO
		13.1	Courtesy of Kai Troester
10.28b	Courtesy of CDC	13.3	Courtesy of St. Vrain Sanitation District
10.31	SEM by Daniel Kadouri & George O'Toole, and colorized by Russell Monds, Dartmouth Medical School	13.4	Courtesy of US Fish and Wildlife Service
		13.6	Courtesy of Christine L. Case, Skyline College
10.32a	Courtesy of Bruce Jaffee	13.10	Courtesy of Animal Health Trust
10.32b	Courtesy of Joey Spatafora	13.15	Courtesy of CDC
11.1	From Taylor H (1999) *Nat Med* 5:492–493. With permission from Macmillan Publishers Ltd.	13.19	Courtesy of Stephanie Dellis, College of Charleston
		13.22	Courtesy of Farm Sanctuary
11.2	Courtesy of Gregory Dimijian/Science Photo Library	13.23b	Library of Congress
		14.1	Courtesy of Tuxyso published under CC-BY-SA 3.0
11.4	Courtesy of CDC/John Molinari & Sol Silverman, Jr.	14.3	Courtesy of CDC
11.8a	Courtesy of WHO	14.5	Courtesy of Colin F. Sargent
11.8b	Courtesy of CDC/WHO	14.13	Food and Agriculture Organization of the United States/Photo by P. Johnson
11.9a	Courtesy of CDC/Brian Judd		
11.11a	Courtesy of USDA-ARS/Photo by Scott Bauer	14.14a	Courtesy of Leslie Y. Kranz
11.11b	Courtesy of © Dorling Kindersley Images/Photo by Frank Greenaway	14.14b	Courtesy of CDC/Fred Murphy & Sylvia Whitfield
		14.14c	Courtesy of CDC/P.E. Rollin
11.15	Courtesy of CDC	14.17b	Property of Rancho Los Amigos National Rehabilitation Center. Printed with permission.
11.16	Courtesy of Robert Colebunders		
11.18	Courtesy of Allergan Pharmaceuticals. (The BOTOX® Cosmetic is a prescription medication that should be used only under the supervision of a physician)	14.18	Courtesy of CDC
		14.21a	Courtesy of CDC
11.20a	Courtesy of CDC/Richard Facklam	14.22	Courtesy of FBI

Figure Number	Credit Line	Figure Number	Credit Line
15.1	Courtesy of Guangzhou Institutes of Biomedicine and Health, Chinese Academy of Sciences	16.8a	Courtesy of Hellahulla published under CC BY-SA 3.0
15.2	Will & Deni McIntyre/Photo Researchers, Inc.	16.9a	Courtesy of Nicubunu published under CC BY-SA 3.0
15.3	Courtesy of Colegota	16.9b	Courtesy of Agne27 published under CC BY-SA 3.0
15.5b	Courtesy of Pacific Northwest National Laboratory	16.9c	Courtesy of Sanjay Acharya published under CC BY-SA 3.0
15.7	Adapted from work courtesy of Estevezj published under CC BY-SA 3.0	16.10a	Courtesy of SJB published under CC BY-SA 3.0
15.8	Adapted from Ottman N, Smidt H, de Vos WM & Belzer C (2012) *Front Cell Infect Microbiol* 2:104. doi: 10.3389/fcimb. 2012.001040, published under CC BY 3.0	16.10b	Courtesy of Henna published under CC BY-SA 1.0
		16.10c	Courtesy of Kafziel published under CC-BY-SA 3.0
15.14	Adapted from work courtesy of James Atmos published under CC BY-SA 3.0	16.11	Andrew Brookes, National Physical Laboratory/Photo Researchers, Inc.
15.16	Courtesy of GloFish® Fluorescent Fish, www.glofish.com	16.12	Courtesy of Scott Crawford, Kipahulu Ohana
15.18	Courtesy of USDA-ARS/Photo by Bruce Fritz	16.14	Courtesy of Alan K. Outram
15.20	Courtesy of Fafner published under CC BY-SA 3.0	16.16a	Courtesy of Alina Zienowicz
		16.19	Courtesy of Kraft
15.21a	Division of Plant Industry Archive, Florida Department of Agriculture and Consumer Services, www.bugwood.org	17.1a	Courtesy of © University Corporation for Atmospheric Research
15.22	Courtesy of Charles Fisher	17.5	From Heitzer A & Sayler GS (1993) *Trends Biotechnol* 11: 334–352. With permission from Elsevier
15.23	Courtesy of FoEI, www.foei.org		
16.1	Courtesy of St Sixtus Abbey, Westvleteren Brewery, Belgium	17.7a	Courtesy of David B. Fankhauser
16.4a	© 2006 by the California Integrated Waste Management Board. All rights reserved.	17.7b,c	Courtesy of Bioengineering AG, Switzerland
		17.7d	Courtesy of Joe Mabel
16.4b	Courtesy of Chris Schnepf, University of Idaho, www.bugwood.com	17.9a	From Federici BA, Park HW, Bideshi DK et al. (2003) *J Exp Biol* 206:3877–3885. With permission from Company of Biologists Ltd.
16.5	Courtesy of Kraft		
16.6b	Courtesy of © UTEX, The Culture of Algae at The University of Texas at Austin	17.10	Berry V & Saraf RF (2005) *Angew Chem Int Ed* 44:6668–6673. With permission from Wiley
16.7	From Batello C, Marzot M & Touré AH (2004) The Future Is An Ancient Lake. www.fao.org/docrep/010/y5118e/ y5118e00.htm. With permission from Food and Agriculture Organization of the United Nations	17.11	Courtesy of Marc Deshusses, Duke University
		17.12b	Jerry Mason/Photo Researchers, Inc.

Figure Number	Credit Line	Figure Number	Credit Line
17.13	Courtesy of Red58bill published under CC BY 3.0	**17.16a,b**	Courtesy of K. Sudesh, Ecobiomaterial Research Laboratory, USM
17.14	Courtesy of Epping Forest District Council	**17.17**	Courtesy of the City of Austin Solid Waste Services
17.15	Courtesy of the City of Modesto Compost Facility	**17.18**	Courtesy of IGV Biotech published under CC BY-SA 3.0

GLOSSARY

-cidal A suffix indicating that a compound or process kills. For instance, "bactericidal" compounds kill bacteria and "fungicidal" compounds kill fungi.

-static A suffix indicating that a process or compound inhibits microbial population growth. For instance, "bacteriostatic" processes prevent bacterial replication.

A subunit The portion of a bacterial exotoxin that interferes with normal host-cell function.

abiotic factor A physical environmental parameter such as temperature or oxygen concentration that affects a living organism.

abiotic synthesis The proposition that life began through the formation of biological molecules from simple precursor compounds.

acid A compound that, when placed in water, causes the concentration of protons (H^+) to increase.

acid-fast bacteria Bacteria with a characteristic cell wall type, consisting of long-chain fatty acids interlaced with peptidoglycan, from which stains cannot be removed by washing in acid.

acid-fast stain A staining technique allowing acid-fast and non-acid-fast bacteria to be distinguished.

acidophile An organism that grows best in conditions of low pH.

activation energy The energy needed to initiate a chemical reaction.

active site The part of an enzyme that temporarily binds the substrate in a biochemical reaction. This binding alters the shape of the substrate, facilitating its conversion to product.

acute disease A disease in which symptoms develop rapidly but last for only a short time.

adaptive immune response Type of response made by the immune system that is directed specifically against the infecting pathogen and will not protect against any other pathogen.

adenosine triphosphate (ATP) A molecule containing chemical bonds whose hydrolysis releases large amounts of energy; used by cells as an energy source for activities in which energy is required.

adsorption The attachment of a bacteriophage to the surface of a bacterial cell.

aerobic respiration Cell respiration in which oxygen is the final acceptor of electrons produced by the oxidation of "fuel" molecules in energy generation.

aerosol transmission *see* **airborne transmission**.

aerotaxis Movement toward (positive aerotaxis) or away from (negative aerotaxis) oxygen.

aerotolerant anaerobe An organism that does not use oxygen but is not harmed by its presence.

agglutination The clumping of large aggregations of antibody and antigen.

airborne transmission Also called aerosol transmission. Transmission of pathogens in small droplets through the air from an infected host to an uninfected, susceptible host.

algal bloom A sudden explosion of growth in a population of aquatic algae.

allele A specific form of a particular gene.

amino acid A type of biological molecule that constitutes the basic unit from which proteins are made.

ammonification The process by which decomposers release ammonium ions into the environment, through the degradation of nitrogen-containing organic molecules.

anaerobic In the absence of oxygen. When used to describe respiration, it refers to cell respiration in which a final electron acceptor other than oxygen is being used.

anatomical barriers Barriers to entry by physical or chemical means, that prevent colonization of a host by a microorganism.

antibiotic An antibacterial agent, either synthetic or produced naturally by fungi or bacteria.

antibody An immune system protein produced by plasma cells in an adaptive immune response, which is able to neutralize the pathogen or otherwise help to control the infection.

anticodon A three-base sequence in a tRNA molecule, which permits the tRNA to interact temporarily with the correct mRNA codon during translation.

antigen A piece of a molecule that induces an adaptive immune response.

antigenic drift A minor genetic change in the antigenic makeup of certain viruses, due to random mutations.

antigen-presenting cell (APC) A type of leukocyte that engulfs foreign microbes and presents fragments of their proteins as antigens to T cells.

antigenic shift A major change in the antigenic makeup of viruses with a segmented genome, caused when gene segments from different strains of the virus combine to create a new strain.

antiparallel   Describes the opposite orientation of associated DNA strands in a double helix, in which one strand is in the 5′ to 3′ direction while the opposing strand is inverted in the 3′ to 5′ direction.

antisense molecule   Very short segments of DNA or RNA that have a base sequence complementary to a crucial viral gene. The antisense molecule binds this complementary nucleic acid and blocks its expression or replication.

antiseptic   A chemical that can be used to eliminate microorganisms from living tissue without harming it.

apicomplexans   A group of mostly non-motile protozoa, characterized by an intracellular structure known as the apical complex.

applied science   The use of basic science to solve a specific practical problem.

Aquifex/Hydrogenobacter group   A primary lineage of bacteria, noted by their tolerance of very high temperatures.

Archaea   One of the two domains of prokaryotic microorganisms; the other is the Bacteria. Members of this domain are called archaea (singular: archaeon).

artificial selection   The purposeful direction of animal or plant breeding, in order to develop breeds with certain beneficial characteristics.

aseptic technique   Laboratory or medical processes carried out in a prescribed manner intended to prevent contamination by unwanted microorganisms.

asexual reproduction   Reproduction from a single parent by budding or cell division.

assembly   Of viruses, the production of new virus particles by the association of newly produced viral proteins and nucleic acid.

atom   The most basic chemically recognizable unit of an element. It can be divided into smaller subatomic particles, but it is then no longer distinguishable as the specific element.

atomic mass   The total number of protons and neutrons present in the nucleus of a given atom.

atomic number   The number of protons present in the nucleus of a given atom.

ATP synthase   An enzyme complex through which protons in higher concentration on one side of a membrane can return to the other side of the membrane. Energy released as the protons move down their concentration gradient is used by the enzyme to synthesize ATP from ADP and phosphate.

attachment   The first step in the viral replicative cycle in which a virion makes contact with a host cell.

attenuation   The process of weakening a pathogen to the point at which, although it can still infect a host, it can no longer cause disease.

autoclave   A device in which objects can be sterilized through a combination of moist heat and high pressure.

autoinducer   The signaling molecule released by bacteria that induces quorum sensing when its concentration passes a certain threshold level.

autotroph   An organism that can synthesize organic molecules by using inorganic precursors.

B cell   A type of lymphocyte that, upon activation, matures into an antibody-secreting plasma cell.

B subunit   The portion of a bacterial exotoxin that allows the exotoxin molecule to bind to and enter specific host cells.

bacillus   General name for any bacterium with a rodlike shape (plural: bacilli). *Bacillus* is the name of a genus of rod-shaped bacteria.

Bacteria   One of the two domains of prokaryotic microorganisms; the other is the Archaea. Members of this domain are called bacteria (singular: bacterium).

bactericidal   Describes a procedure, chemical, drug, or other process that kills bacteria.

bacteriophage   A virus that infects bacteria.

bacteriostatic   Describes a procedure, chemical, drug, or other process that inhibits bacterial replication.

barotolerance   The ability to withstand high pressure.

base   A compound that, when placed in water, causes the concentration of protons (H$^+$) to decrease.

basic science   Scientific inquiry focused on revealing the manner in which fundamental processes in nature proceed.

basidiocarp   The spore-producing reproductive structure produced by fungi of the Basidiomycota.

benthic zone   The deepest zone in an aquatic environment, including the bottom sediments.

bilayer   A double layer of phospholipids found in cell membranes, in which phosphate groups form the two surfaces of the membrane, while fatty acids are sandwiched between the two phosphate layers.

binary fission   Mode of reproduction of prokaryotic cells by division into two daughter cells.

binomial nomenclature   A system of naming living things in which each organism receives a Latin name consisting of two terms, the first of which identifies the genus of the organism, and the second of which identifies the species within that genus.

bioconversion   The change of one compound into another by microbial action.

biofilm   A microbial community in which the cells are encased in a layer of polysaccharide.

biological vector transmission   Transmission of a pathogen to the host in which it causes disease via an invertebrate vector, in which the pathogen replicates and/or develops.

biomass   The collective mass of all organisms in an ecosystem.

biomineralization   The process by which some bacteria secrete calcium carbonate ($CaCO_3$) under appropriate conditions.

biopesticides   Toxic proteins produced by specific microorganisms, which can be used to kill certain insect pests.

bioremediation   The use of microorganisms to degrade harmful chemicals.

bioterrorism   The use of disease-causing biological agents to kill or cause fear in the general population.

biotic interaction   An ecological interaction between living organisms, such as competition or predation.

broad-spectrum   The general name given to antimicrobial drugs that are effective against a wide range of microorganisms.

budding   The release of enveloped viruses from a host cell by a process in which the newly assembled viral particle protrudes through the plasma membrane and acquires a coating of plasma membrane—the viral envelope.

Calvin cycle A series of reactions used by photosynthetic eukaryotes, as well as some photosynthetic prokaryotes, including the cyanobacteria, to fix carbon dioxide into organic molecules.

capsid The protein coat of a virus particle.

capsomere One of the individual protein subunits making up the complete capsid of a virus particle.

capsule A polysaccharide-rich outer layer secreted by some bacteria, forming a thick, regular shell. The capsule may aid in adherence or immune evasion.

carbohydrate A class of biological molecule that always contains carbon, hydrogen, and oxygen. Carbohydrates are important in living things as sources of energy and as structural components. They are hydrophilic in nature.

carbon fixation The autotrophic process of incorporating carbon dioxide into organic carbon compounds.

carnivore A meat eater. A secondary or higher consumer in a food web that feeds upon lower level consumers.

carrier proteins Transmembrane proteins that transport molecules across cell membranes.

case definition A list of common symptoms for a particular medical condition.

case–control study An epidemiological study used to determine the cause of a specific illness, in which a group of individuals suffering from the illness are compared with a similar population of healthy individuals.

cell The functional unit of life. A highly organized, typically microscopic structure, always surrounded by a cell membrane, and containing genetic material in the form of DNA.

cell lysis The rupturing of the cell membrane, resulting in cell death.

cell-mediated response The component of an adaptive immune response in which cytotoxic T cells detect and kill infected host cells.

cell membrane Also called the plasma membrane. A membrane forming the outer boundary of all cells, composed of a bilayer of phospholipids in which proteins are embedded. The cell membrane regulates the transport of molecules and ions into and out of the cell.

cell nucleus The subcellular structure in eukaryotic cells that contains the genetic material. The nucleus is bounded by a double membrane.

cell respiration The gradual breakdown of "fuel" molecules, such as sugars, resulting in the production of ATP.

cell theory The scientific theory that states that all living things are composed of one or more cells. The theory also states that cells are the basic units of structure and reproduction in all organisms.

cell wall A protective structure exterior to the cell membrane that surrounds the cell. Found in all plant, algal, and fungal cells as well as almost all prokaryotic cells. The composition depends on the type of cell.

chemical bond The general term for any type of linkage between atoms.

chemically defined medium Growth medium for microorganisms of a defined composition, containing specific amounts of particular nutrients required for growth.

chemoautotroph An autotrophic organism in which the energy source for the synthesis of organic molecules is chemical energy obtained from inorganic chemicals.

chemotaxis Movement toward a favorable chemical stimulus (positive chemotaxis) or away from a harmful chemical stimulus (negative chemotaxis).

chemotherapeutic agent A compound that can be used to treat a medical condition.

chlamydiae A lineage of bacteria, all of which are parasitic and live within the cytoplasm of host cells.

chloroplast The organelle in photosynthetic eukaryotic cells in which photosynthesis takes place.

chromosome A structure consisting of DNA and protein, which carries genetic information. Prokaryotic cells usually contain a single chromosome, whereas eukaryotic cells have multiple chromosomes.

chronic disease A disease in which symptoms are present for a prolonged period.

cilium Short motile cellular projection composed of protein, present on some eukaryotic cells (plural: cilia). Cilia can be involved in the locomotion of the cell or in the generation of a current of fluid over the cell surface.

class A taxonomic level between order and phylum.

cloning The creation of copies of a cell, a gene, or an entire organism.

cloning host A cell into which a cloning vector is introduced and in which it can replicate.

cloning vector A self-replicating DNA molecule into which foreign DNA can be inserted for the purpose of making many copies of that DNA.

coccus A bacterium with a spherical shape (plural: cocci).

co-infection Simultaneous infection of a host with more than one pathogen.

coding strand The strand of DNA in a gene that is transcribed into RNA during transcription.

codon A three-base sequence in mRNA that specifies either a particular amino acid or a stop command in the translation of mRNA into protein.

coevolution hypothesis A hypothesis proposing that viruses evolved along with cellular life as obligate intracellular parasites.

commensalism A relationship between two organisms of different species living in close association, in which one of the organisms is benefitted and the other is neither benefitted nor harmed.

common source epidemic An epidemic caused when a large number of individuals become infected from the same original source.

competition An interaction between two species in the same habitat that are trying to acquire the same resources.

competitive exclusion The phenomenon observed in some competitive interactions, in which the better-adapted organism drives the less-well-adapted organism to extinction.

complex capsid   A type of viral capsid, neither completely icosahedral nor helical in shape, and often consisting of multiple protein layers.

complex medium   A growth medium for microorganisms, to which a variety of nutrients has been added.

composting   A technique in which organic solid waste is degraded though the activity of microorganisms.

condensation reaction   A reaction in which the individual building blocks of biological molecules (amino acids for proteins or monosaccharides for carbohydrates, for example) are linked together. An H atom is removed from one molecule and a hydroxyl (OH) group is removed from the other, allowing the two molecules to make a chemical bond, with the release of a molecule of water.

conjugation   The transfer of DNA from one bacterial cell to another involving cell-to-cell contact.

constitutive gene   A nonregulated gene that is continuously expressed, independently of any environmental conditions.

contamination   The unwanted presence of microorganisms.

control group   A group of treatments in an experiment in which the experimental variable is not manipulated. The control group is compared with the experimental group, in which the experimental variable is manipulated. This permits the scientist to assess the effect, if any, of the experimental variable on the treatments.

controlled variable   A factor that is maintained the same in both experimental and control groups. Controlled variables are those variables not being tested in the testing of a specific hypothesis.

coral bleaching   The loss of mutualistic zooxanthellae by coral polyps. Following this loss, the remaining polyp appears whitish in color.

coupled reactions   Two chemical reactions that occur simultaneously, such that an energy-releasing reaction provides the energy needed for an energy-requiring reaction to proceed.

covalent bond   A type of chemical bond in which two atoms are linked through the sharing of electrons.

CRISPR/Cas   A prokaryotic immune system that confers resistance to phages and provides a type of acquired immunity. Used as a biotechnology tool to add, disrupt, or change the sequence of specific genes in other organisms.

cross-reactivity   The ability of some antibodies generated in response to a specific antigen to also bind an entirely different antigen.

curd   The solid portion of milk that has been separated from the liquid portion during the production of certain dairy products such as cheese.

cyanobacteria   A lineage of photosynthetic bacteria that release oxygen as a photosynthetic waste product.

cyst   A metabolically inactive stage in many protozoan life cycles.

cytokine   Biologically active proteins released by cells that act on the same or other cells to regulate an immune response.

cytopathic effect   Any adverse effect that a virus has on its host cell.

cytoplasm   The material forming the bulk of a cell inside the cell membrane. The cytoplasm is composed of a semifluid mix of proteins and other biological molecules (the cytosol) in which are located various subcellular organelles (in eukaryotic cells).

cytoskeleton   A network of protein fibers in eukaryotic cells that provide structure to the cell and enable movement of and within cells.

cytotoxic T cell (Tc cell)   A type of T cell that detects and kills infected host cells that are displaying foreign antigen on their surfaces.

Darwinian medicine   The idea that principles of natural selection can be used to artificially select for reduced pathogen virulence.

daughter cell   Either one of the two newly formed cells arising from the division of a "parent" cell.

daughter DNA   A newly synthesized DNA molecule produced by replication of a "parent" DNA molecule.

decline phase   The phase in microbial growth during which the overall number of cells is declining.

decomposer   An organism that absorbs organic matter from dead organisms and degrades it into inorganic molecules, which are returned to the environment.

definitive host   The host in a parasite's life cycle in which sexual reproduction of the parasite occurs.

denatured protein   A protein that has unfolded, losing its specific three-dimensional shape. Because they are not folded into their proper conformation, denatured proteins show decreased or no activity.

dendritic cell   The main immune-system cell that presents antigen to T cells and activates them.

deoxyribonucleic acid (DNA)   The principal genetic material of all living things, with the exception of some viruses. It is composed of chains of covalently linked nucleotides, and carries genetic information encoded in the order, or sequence, of the nucleotides. This information specifies the proteins that a given organism can make. In cells, DNA is always found as a double-stranded double helix.

designer drug   A drug specifically constructed to bind with or otherwise interfere with a particular target, for which the structure is already known.

direct contact transmission   Transmission via physical contact such as kissing, shaking hands, or sexual contact.

disaccharide   Two monosaccharides linked together.

disease   The disruption in some way of normal host function.

disinfectant   A chemical agent appropriate for the elimination of microorganisms from nonliving material.

distilled   Describes beverages or other liquids in which certain substances in the beverage are concentrated by evaporating off a portion of the liquid.

DNA ligase   An enzyme that forms bonds between adjacent DNA nucleotides, when there is a break in the phosphate-sugar backbone between the nucleotides.

DNA mismatch repair enzymes   Enzymes involved in the recognition and removal of improper bases that were incorporated into DNA by mistake during DNA replication.

DNA polymerase   The enzyme that adds nucleotides to the newly synthesized growing DNA strand in DNA replication, using a strand of the parental DNA molecule as a sequence template.

DNA sequencing   Any technique by which the order of nucleotides in a DNA molecule can be determined.

DNA virus   Virus in which the genetic information is encoded in DNA.

domain The most inclusive taxonomic category. All living things belong to one of three domains—the Bacteria, the Archaea, and the Eukarya.

double helix Double-stranded DNA, in which two helical DNA strands are wound around each other, forming a double-helix structure.

eclipse phase The time between the uncoating of a virus and the assembly of new viral particles, during which the virus exists as nucleic acid only.

ecological community The various populations of different species within a defined geographic area.

ecology The study of the relationship of organisms to each other and to their environment.

ecosystem An ecological community and the physical environment with which the community interacts.

effector T cell A T cell that has been activated to perform its immunological function.

egg The gamete produced by female animals.

electron A fundamental subatomic particle that carries a negative charge. Electrons are distributed around the nucleus of the atom in a series of concentric "electron shells."

electron affinity A measure of the strength of attraction that an atom has for electrons.

electron shell A specified distance from the nucleus of an atom, where that atom's electrons are found. Each shell holds a specific number of electrons. Larger atoms have more electron shells, increasingly distant from the nucleus, to accommodate the larger number of electrons.

electron transport A process in which electrons are serially transferred from one compound to another, generating the conditions in which ATP can be synthesized.

element A substance that cannot be broken down to simpler components by chemical processes.

emerging disease A new or changing disease that is increasing in importance.

endemic disease A disease that is always present in a particular geographic area, usually occurring at low level.

endergonic reaction A chemical reaction that requires an input of energy.

endocytosis A mechanism used by eukaryotic cells to import molecules into the cell by engulfing them to form internal membrane-enclosed endosomal vesicles.

endomembrane system A group of membrane-enclosed organelles in eukaryotic cells that are involved in both endocytosis and exocytosis.

endoplasmic reticulum A network of membrane-enclosed tubules in eukaryotic cells that is involved in a variety of cellular processes, including intracellular transport and exocytosis. It is part of the endomembrane system.

endospore A structure formed by some bacteria that allows the cell to persist in a metabolically inactive state when the environment is not conducive to growth.

endosymbiosis The theory that some eukaryotic cell organelles, specifically the chloroplast and the mitochondrion, were originally prokaryotic cells that entered into a symbiotic relationship with larger cells.

endotoxin The lipid portion of the lipopolysaccharide of the outer membrane of Gram-negative bacteria, which has toxic properties when released upon the death of the bacterial cell.

envelope A phospholipid bilayer surrounding the capsid of some viruses.

enzyme A biological catalyst; usually a protein, but some enzymes are RNAs. Enzymes speed up the rate of biological reactions so that they can occur at a useful rate in the conditions inside living cells.

epidemic A sudden increase in the number of cases of a specific disease, beyond what is considered to be a normal number of cases.

epidemic threshold The minimum proportion of a population that must be immune to a specific pathogen to prevent an epidemic.

epidemiology The study of diseases in populations.

ergot A toxin produced by the fungus *Claviceps purpurea*.

escaped-gene hypothesis A hypothesis proposing that viruses were originally fragments of prokaryotic or eukaryotic nucleic acids that evolved the capacity to replicate independently.

Eukarya The taxonomic domain consisting of all organisms that have eukaryotic cells.

eukaryotic cell A cell in which the DNA is enclosed in a membrane-enclosed organelle known as the nucleus. Animals, plants, fungi, protozoa, and algae have eukaryotic cells and are often collectively referred to as eukaryotes.

eutroph A microorganism requiring relatively high nutrient levels.

eutrophication The depletion of oxygen in an aquatic environment, often caused by microorganisms in response to an input of excess nutrients.

evolution The genetically based development of new properties in organisms over successive generations.

exergonic reaction A chemical reaction that releases energy.

exocytosis A mechanism in eukaryotic cells by which material can be released from the cell. It involves the fusion of a transport vesicle with the plasma membrane and the release of the vesicle's content to the extracellular environment.

exon A block of nucleotide sequence within a eukaryotic gene that codes for an amino acid sequence. Genes generally have more than one exon. Exons are separated by introns, nucleotide sequences that do not code for amino acids.

exotoxin A general term for the protein toxins produced by many pathogenic bacteria.

experimental group A group of treatments in an experiment in which the experimental variable is manipulated. The experimental group is compared with the control group, in which the experimental variable is not manipulated. This permits the scientist to assess the effect, if any, of the experimental variable on the treatments.

experimental variable The factor being tested in the testing of a specific hypothesis. The experimental variable will be manipulated in the experimental group, and any effect of this variable will be detected by comparison with a control group, in which the experimental variable was not manipulated.

extracellular environment The environment outside a cell.

extremozyme   An enzyme that retains its ability to function under certain extreme physical conditions.

F⁻ cell   A bacterial cell able to act as a recipient during conjugation.

F⁺ cell   A bacterial cell able to act as a donor during conjugation.

facultative anaerobe   An organism that can grow with or without oxygen.

family   A taxonomic level between genus and order.

fastidious   Describes organisms that make relatively few of the biological compounds that they require for growth.

fermentation   The anaerobic breakdown of organic molecules such as carbohydrates for the purpose of ATP synthesis, in which the final electron acceptor is an inorganic molecule.

fermentation waste product   The reduced form of pyruvate, formed to regenerate NAD⁺ in the absence of oxygen, allowing cells to continue to undergo glycolysis. The precise nature of this waste varies according to cell type.

filamentous fungus   A fungus made up of an extensively branched system of tube-like filaments.

fimbria   Also called a pilus. A short protein fiber extending beyond the cell wall of many bacteria (plural: fimbriae). Fimbriae aid in adherence to specific surfaces.

final electron acceptor   A molecule that accepts electrons at the end of the electron transport chain.

flagellum   A long motile protein fiber, extending beyond the cell wall of many bacteria (plural: flagella). Flagella-bearing bacteria move by rotating their flagella. Some eukaryotic cells use flagella for movement, but the structure and mechanism of action of eukaryotic flagella are quite different from those of bacterial flagella.

flavobacteria   A lineage of bacteria, all of which are strict anaerobes.

fleshy fungi   A group of multicellular fungi characterized by the production of reproductive structures composed of tightly packed hyphae.

fluid mosaic model of membranes   The accepted model of membrane structure, in which membrane proteins are embedded in a fluid phospholipid bilayer.

fomite   An inanimate object serving as the vehicle of transmission during indirect contact transmission of a pathogen.

food- or water-borne transmission   Transmission of a pathogen via contaminated food or water.

food chain   The path of nutrient flow between different species in an ecological community, such as the route linking primary producers and predators.

food web   A network diagram that aims to show all the routes of nutrient flow between the species, or a group of species, in an ecosystem. A network results, rather than a food chain, because some species can feed at different trophic levels.

frameshift mutation   A mutation caused by either the deletion of a nucleotide or the insertion of an extra nucleotide into a DNA sequence, which upsets the correct amino acid coding.

free radical   A highly reactive, electrically charged compound that can bind to and damage other molecules.

fusion   A mechanism used by some enveloped viruses to enter host cells, in which the viral envelope fuses with the host cell's plasma membrane, releasing the capsid into the host cell's cytoplasm.

gamete   A specialized sex cell produced by eukaryotic organisms for sexual reproduction.

gel electrophoresis   A technique used to separate a mixture of molecules on the basis of their movement in a gel-like material. How far it moves depends on the size and electrical charge of the molecule.

gene   A stretch of DNA carrying the encoded genetic information necessary to specify the structure of one protein or portion of a protein.

gene expression   The overall process of converting the encoded information in a gene's nucleotide sequence into a specific protein.

generation time   The time required for a cell or individual organism to divide or replicate.

gene therapy   The treatment of genetic disease by introducing normal copies of a gene that is defective in the individual affected.

genetic barriers   A barrier to infection based on genetic incompatibility between a potential host and a pathogen.

genetic engineering   The purposeful direct manipulation of DNA to create living things with novel genetic characteristics.

genetic recombination   A process in which an exchange of genetic information occurs between two chromosomes or DNA molecules, resulting in the production of new genotypes.

genome   The entire sum of an organism's genetic material.

genomics   The field in biology that attempts to sequence and study the complete genomes of organisms.

genotype   The specific genetic makeup of an individual for any particular characteristic.

genus   A group of closely related species. The level of classification between species and family.

germ theory of disease   The concept developed in the nineteenth century that microorganisms could cause disease.

glycocalyx   The gel-like layer composed of polysaccharides, secreted by many bacteria to form a layer exterior to the cell wall.

glycolysis   The first stage of cell respiration in which carbohydrate substrates are oxidized to pyruvate without a requirement for oxygen.

glycoprotein   A protein that is covalently linked to carbohydrates.

Golgi apparatus   An organelle found in many eukaryotic cells. Part of the endomembrane system that is involved in protein modification and transport.

Gram-negative bacteria   Bacteria that have both a thin peptidoglycan cell wall and an outer membrane, and which do not retain Gram stain.

Gram-positive bacteria   Bacteria with a cell wall composed of multiple peptidoglycan layers but lacking the outer membrane of Gram-negative bacteria, and which retain Gram stain.

Gram staining   A staining technique allowing differentiation between Gram-positive and Gram-negative bacteria.

green nonsulfur bacteria   A lineage of anaerobic, photosynthetic bacteria.

green sulfur bacteria   A lineage of anaerobic, photosynthetic bacteria that use sulfur compounds during photosynthesis.

halophile An organism that grows best at relatively high concentrations of salt.

helical capsid A common type of viral capsid in which the capsomeres form a spiral around the nucleic acid.

helminth A general term for parasitic worms.

helper T cell (Th cell) A type of T cell responsible for coordinating an adaptive immune response.

herbivore A member of an ecological community that eats producers (for example, plants or photosynthetic algae).

heterotroph An organism that obtains the materials for biosynthetic reactions by consuming organic molecules rather than constructing them from inorganic precursors.

homeostasis Process by which cells or organisms maintain an internal environment that is compatible with their survival.

horizontal gene transfer The transfer of genes between organisms in some way other than traditional reproduction.

host The organism in or on which a parasite lives.

host range The spectrum of hosts that a pathogen can infect.

host-to-host epidemic An epidemic in which the disease is transmitted from infected to non-infected individuals.

humoral response That component of an adaptive immune response in which antigen-specific antibodies are produced by plasma cells.

humus The top layer of soil, rich in organic material.

hydrocarbon A compound composed of carbon and hydrogen only.

hydrogen bond A type of noncovalent bond common in or between proteins. It can be formed between a hydrogen with a partial positive charge and certain atoms with negative charge in another chemical group.

hydrolysis reaction A reaction in which a covalent bond is broken by the involvement of a molecule of water, adding H to one reaction product and OH to the other. For example, hydrolysis breaks down polymeric biological molecules such as proteins and polysaccharides into their component building blocks.

hydrophilic Property of a molecule indicating that it will interact with water. Hydrophilic molecules have partial or complete electrical charges, which interact with the partial charges on water molecules.

hydrophobic Property of a molecule indicating that it will not interact with water, because it has no charges to interact with the partial charges on water molecules.

hydroxyl group A hydrophilic chemical group consisting of OH.

hypha A tube-like filament that is the typical structural unit of most fungi (plural: hyphae).

hypothesis A tentative explanation for a particular question about the natural world. A proper hypothesis must be used to generate predictions about future events, which can then be tested by experiment or observation, allowing the hypothesis to be either accepted or rejected.

icosahedral capsid A common type of viral capsid in which each of its 20 faces is an equilateral triangle.

immunological memory The ability of an adaptive immune response to grow stronger with repeated exposure to the same pathogen.

immunology The scientific study of host defenses against pathogens.

inactivated virus A virus that is no longer able to infect host cells and replicate.

incidence of disease The number of new disease cases over a specified length of time.

inclusion body A darkly staining body seen in the cytoplasm of cells infected with certain viruses. It is composed of viral proteins that have failed to assemble correctly.

incubation period The time between initial exposure to a pathogen and the onset of signs and symptoms.

indirect contact transmission Transmission of a pathogen via an inanimate object.

inducer A molecule that can stimulate expression of an operon by binding a repressor protein and removing it from the operator.

Inducible Describes an operon in bacteria in which genes are expressed following the removal of the repressor from the operator, by a molecule called the inducer.

inductive period The time between infection and the onset of an adaptive immune response, during which lymphocytes are dividing and developing into effector cells.

industrial fermenter A large vat in which the microbial production of a desired metabolite, such as an antibiotic, takes place on a commercial scale.

infection A situation in which microorganisms are reproducing in or on a host.

infectious dose-50 (ID_{50}) The number of microorganisms that must enter a host to establish an infection in 50% of exposed hosts.

inflammation A host response to infection and tissue damage that is characterized by vasodilation and increased permeability of small blood vessels at the site of damage.

innate immunity Inborn host defenses that are not specific for a particular type of pathogen.

intercalating agent A chemical that produces mutations by inserting itself into a replication fork during DNA replication.

interferon An antiviral cytokine released by cells in response to viral infection.

intermediate host The host in a parasite's life cycle in which sexual reproduction of the parasite does not occur.

intertidal zone An ecological zone in a marine environment located between the low and high tide marks.

intron A block of nucleotide sequence within a eukaryotic gene that does not code for an amino acid sequence. An intron separates two exons, which are the regions within a gene that encode amino acids.

in vitro A term denoting that a process was carried out under laboratory conditions.

in vivo A term denoting that a process was carried out in a living organism.

ion A charged atom, produced by the gain or loss of one or more electrons.

ionic bond A chemical bond formed by the mutual attraction of negatively charged and positively charged ions.

isomers   Molecules with the same chemical formula but different structures.

kingdom   A taxonomic level between phylum and domain.

Koch's postulates   Criteria that must be demonstrated to show that a particular microorganism causes a particular disease.

Krebs cycle   A pathway that can occur in cell respiration, in which two-carbon molecules are oxidized using NAD^+ and FAD to produce NADH, $FADH_2$, and CO_2. It is also called the citric acid cycle.

laboratory fermenter   A relatively small vessel in which combinations of physical and nutritional factors can be manipulated in the laboratory to determine the optimum conditions for microbial metabolite production.

lag phase   The phase in microbial growth, before the log phase, during which there is no apparent population growth.

latent infection   A type of viral infection, during which the virus does not replicate.

lethal dose-50 (LD_{50})   The number of pathogens that must be used to infect a host in order to kill 50% of the infected hosts.

leukocyte   General name for a white blood cell. There are several different types of leukocytes, which are involved in defense against infection and immune responses.

light reaction   In photosynthesis, the processes that capture the energy of sunlight and convert it into chemical energy carriers, and which can only occur in the light.

lipase   Any enzyme that breaks down lipids by hydrolyzing ester bonds.

lipid   A class of biological molecules that are primarily used in energy storage and as structural components of cells. Fats, oils, and sterols are examples of lipids. They are largely hydrophobic in nature.

lipid A   The lipid portion of the lipopolysaccharide of Gram-negative bacteria. It anchors the lipopolysaccharide to the outer membrane.

lipopolysaccharide   A complex molecule composed of lipid and polysaccharides, found in the outer membrane of Gram-negative bacteria.

liposome   A small, hollow spherical structure, formed by phospholipids when placed in water. Liposome-like structures may have been the precursors of the first cells.

littoral zone   The shallowest part of the photic zone, nearest to the shore, through which light penetrates right to the bottom.

log phase   The phase in microbial growth during which population growth is exponential.

lumen   The interior space or cavity within a structure.

lymph   The fluid present within the lymphatic system.

lymphatic system   A network of vessels that returns extracellular fluid to the blood and is involved in carrying pathogens from the tissues to lymph nodes.

lymph node   A small organ rich in lymphocytes, in which adaptive immune responses are generated. Lymph nodes are connected to each other by the lymphatic system.

lymphocyte   A type of white blood cell that generates adaptive immune responses.

lysis   The bursting of a cell membrane, resulting in cell death.

lysogeny   Infection of a bacterium by a bacteriophage in which the bacteriophage does not replicate.

lysosome   An organelle of eukaryotic cells that is involved in the digestion of worn-out cellular components and of foreign material imported into the cell.

lytic cycle   The replicative cycle of a bacteriophage inside an infected bacterium, which results in lysis and death of the infected bacterium and release of newly formed bacteriophage.

macrophage   The main type of phagocytic cell in vertebrates. It scavenges dead cells; it recognizes, ingests, and kills microorganisms; and it can also act as an antigen-presenting cell.

major histocompatibility complex (MHC)   A cluster of genes encoding proteins known as MHC molecules, which are required for the presentation of antigen to T cells.

malt   Germinated barley seeds in which the starch is digested to simpler sugars.

matrix   In a mitochondrion, the compartment enclosed by the outer membranes and in which the reactions of the Krebs cycle take place.

maximum growth temperature   The highest temperature at which an organism can grow.

mechanical vector transmission   A mode of pathogen transmission by an invertebrate vector, in which the pathogen adheres to the invertebrate's body without replicating or developing in any way.

memory B cell   A long-lived B cell that responds to a second infection of the host by a given pathogen by generating a secondary humoral immune response.

memory T cell   A long-lived T cell that responds to a second infection of the host by a pathogen by generating a secondary immune response.

mesophile   An organism that grows best at intermediate temperatures.

messenger RNA (mRNA)   An RNA molecule transcribed from a protein-coding gene.

metabolism   The biochemical reactions carried out by a cell or organism.

metabolite   An intermediate compound or final product produced in a metabolic pathway.

metagenomics   The study of genetic material taken directly from environmental samples.

methanogen   A member of a lineage of anaerobic archaea that produce methane gas as a metabolic waste product.

microaerophile   An organism that grows best at low oxygen concentrations.

microbe   A living thing too small to be seen with the unaided eye. Microbes include bacteria, archaea, fungi, protozoa, algae, and viruses.

microbial barrier   A type of barrier to entry of a potential host by a pathogen in which the host's normal microbiota block infection by the pathogen.

microenvironment   The environment immediately surrounding a microorganism.

microorganism A living thing too small to be seen with the unaided eye. Microorganisms include bacteria, archaea, fungi, protozoa, algae, and viruses.

microRNA Small RNA molecules, used by eukaryotes to regulate gene expression.

minimal medium A growth medium for nonfastidious microorganisms, containing a carbon source and inorganic compounds.

minimum growth temperature The lowest temperature at which an organism can grow.

mismatch repair The process in which DNA repair enzymes recognize, remove, and replace mispaired or damaged DNA nucleotides in a molecule of DNA.

missense mutation A mutation in which a change in a single nucleotide results in a change in the amino acid encoded by the affected codon.

mitochondrion An energy-generating organelle found in most eukaryotic cells in which most of the cell's ATP is produced.

model organism A species that has been intensively studied in the laboratory. Biological principles revealed through the study of model organisms may then be extrapolated to other living things.

mode of transmission The manner in which a pathogen travels from an infected host to new, uninfected, and susceptible hosts.

mold A type of fungus characterized by a loosely organized mycelium.

monoculture An agricultural area dominated by a single crop.

monosaccharide Any of the sugars that serve as the basic carbohydrate building blocks. Monosaccharides can be linked together to form larger molecules, such as disaccharides (two monosaccharides), trisaccharides (three monosaccharides), or polysaccharides (many saccharides).

motility The capacity for active movement.

multicellular Consisting of many cells.

must The juice extracted from crushed grapes at the first step in winemaking.

mutagen A chemical or a physical agent (such as radiation) that increases the likelihood of mutations in DNA.

mutation A change in the nucleotide sequence of DNA.

mutualism A biological interaction between two organisms that provides benefits for both.

mycelium The entire network of hyphae in a filamentous fungus.

mycoremediation Bioremediation utilizing fungi.

mycorrhizal association A mutually beneficial relationship between a fungus and the roots of a plant.

mycosis A fungal infection in animals.

NADPH The reduced form of **nicotimamide adenine dinucleotide phosphate** (NADP⁺), which is formed at the end of electron transport in the light reactions of photosynthesis. NADPH functions as a source of electrons, with which to reduce carbon in the Calvin cycle.

nanotechnology Technology or the use of materials at an extremely small scale.

narrow-spectrum The general name given to antimicrobial drugs that are effective only against very specific microorganisms.

native protein conformation The active form of a protein when it is folded up into its specific three-dimensional shape.

natural selection The mechanism through which biological evolution is thought to occur.

negative taxis Movement away from an unfavorable stimulus.

neutralization The inactivation of pathogens or toxins by the binding of a specific antibody.

neutral mutation A mutation in which the change in a particular DNA sequence does not change the sequence of amino acids specified by the DNA.

neutron A fundamental subatomic particle present with protons in an atom's nucleus. Neutrons do not carry an electrical charge.

neutrophil A type of phagocytic cell that engulfs and kills microorganisms in an immune response.

nicotinamide adenine dinucleotide (NAD⁺) A molecule that acts as a hydrogen or electron acceptor in many biochemical reactions.

nicotinamide dinucleotide phosphate (NADP⁺) A molecule similar to NAD⁺ that is used as a final electron acceptor in the electron transport chain of the photosynthetic light reaction.

nitrification The microbial conversion of ammonium to nitrite and nitrate.

nitrogen fixation The conversion of atmospheric nitrogen into a form that plants can absorb. In nature, only a relatively small number of bacterial species can fix nitrogen.

nonenveloped virus A virus lacking the outer phospholipid bilayer present in enveloped viruses.

nonfastidious Describes organisms that synthesize many or most of the biological compounds that they require for growth.

nonpolar covalent bond A type of covalent bond in which the two bonded atoms have an equal affinity for electrons. The electrons are consequently shared equally, resulting in a lack of any partial charges on the bonded atoms.

nonsense mutation A change in a single nucleotide that results in the conversion of a readable codon into a stop codon.

normal microbiota Those microorganisms that colonize a host without causing disease.

nosocomial infection An infection acquired in a hospital.

nuclear envelope The double-layered membrane surrounding the nucleus in eukaryotic cells.

nucleic acid DNA or RNA, which are together involved in the storage and readout of genetic information. Nucleic acids are formed of chains of covalently linked nucleotides. In all cellular forms of life and in many viruses, DNA serves as the store of genetic information, while RNA assists in converting the information encoded in DNA into proteins.

nucleocapsid The capsid and associated nucleic acid of a virus particle.

nucleoid area The area within the cytoplasm of a prokaryotic cell, in which the DNA is found.

nucleotide The building block of a nucleic acid, consisting of a sugar molecule, covalently bound to a phosphate group and to a nitrogen-containing base (either a purine or a pyrimidine).

nucleus   Of a eukaryotic cell, *see* **cell nucleus**. Of an atom, the central core of an atom, composed of protons and neutrons. The nucleus makes up almost all of an atom's mass.

obligate aerobe   An organism that requires oxygen for survival and growth.

O polysaccharide   A molecule forming the polysaccharide portion of the lipopolysaccharide found in the outer membrane of Gram-negative bacteria.

obligate anaerobe   An organism for which oxygen is toxic.

obligate intracellular parasites   Parasites that must enter the cell of a host to complete their life cycles.

oenology   The science of winemaking.

oligotroph   A microorganism adapted to low nutrient levels.

omnivore   An animal that eats both plants and other animals.

oncogene   A cancer-causing gene, of either viral or host origin.

one health   The collaborative effort of multiple disciplines working at the local, national, or global level, to attain optimal health for people, animals, and the environment.

oocyst   The infective, cyst-like structure in the life cycle of some protozoa.

operator   A regulatory nucleotide sequence present in an operon. When the operator is bound by a repressor protein, the operon is not expressed.

operon   A type of gene organization found mainly in prokaryotes in which a group of adjacent genes is induced or suppressed as a unit. When induced, the genes are transcribed together as a single mRNA from which the individual proteins are translated.

opportunistic infection   Disease caused by a microbe that does not usually cause disease. It occurs in unusual circumstances, such as when the host's immune system is weakened.

opsonin   A molecule that forms a physical link between a phagocytic cell and the material to be engulfed, facilitating phagocytosis.

optimum growth temperature   The temperature at which an organism's growth rate is greatest.

order   A taxonomic level between family and class.

organ   In multicellular eukaryotes, a discrete structure composed of more than one tissue working together to carry out a particular function.

organelle   General term for a membrane-enclosed structure within a eukaryotic cell that carries out a distinct function.

organic molecule   Any molecule built on a skeleton of two or more linked carbon atoms.

origin of replication   The point on a DNA molecule where the two parental strands are separated in preparation for DNA replication.

outer membrane   A second membrane present in Gram-negative bacteria, located beyond the plasma membrane and the thin cell wall.

oxidation   The addition of oxygen to a molecule, or the loss of electrons or hydrogen.

oxidizing agent   Any substance that readily oxidizes other compounds by removing electrons from those compounds.

pandemic   Technically, a worldwide epidemic. Commonly used to describe any widely distributed, multi-continent epidemic.

parasite   An organism that lives in or on another organism, from which it obtains a place to live and reproduce, and a source of nutrients.

parental DNA   Refers to a double-helical DNA molecule before DNA replication, each strand of which gives rise to a new molecule of double-helical daughter DNA.

parent cell   The original cell, before cell replication, which gives rise to two daughter cells.

pasteurization   A process that eliminates most pathogens or organisms causing spoilage from liquids through brief exposure to high temperature.

pathogen   A disease-causing microorganism.

pathogenesis   The capacity of a microorganism to cause disease.

pattern recognition   The ability of cells of the innate immune system to recognize certain general types of molecules that are unique to microorganisms.

pelagic zone   The **photic zone** in the open ocean or a large lake, where light penetrates. Below the pelagic zone is the **profundal zone**, into which light does not reach.

penetration   The stage in the viral replicative cycle, during which a viral particle enters a host cell, either by endocytosis or fusion.

pentaglycine bridge   A short peptide consisting of five glycine residues, which links one layer of peptidoglycan to another in bacterial cell walls.

peptide   A short chain of linked amino acids, either occurring independently or forming a portion of a protein.

peptide bond   The covalent bond linking two adjacent amino acids in a peptide or protein.

peptidoglycan   A complex molecule composed of carbohydrate and peptides, found in the cell walls of many bacteria.

peripheral protein   A membrane protein that is attached to the membrane surface or is embedded within one layer of the membrane without passing through to the opposite side.

permeability   The capacity of a barrier to permit the movement of substances across the barrier. A freely permeable barrier allows all substances to pass. A selectively permeable barrier only permits certain substances to pass.

personal characteristics   A description of the personal attributes of those individuals who become ill with a particular medical condition.

phage therapy   The use of bacteriophage to destroy bacteria causing disease in an animal host.

phagocytosis   A process by which relatively large particles (such as bacterial cells) are engulfed by a eukaryotic cell, with the plasma membrane extending around the particle to form a transport vesicle that is brought into the cell.

phenotype   The outward appearance or other detectable property of an organism in regard to a specific trait. It is the result of the interaction of the organism's genotype for that trait and environmental influences.

phosphate group   A functional group with the chemical formula PO_4^{3-}.

phospholipid   A type of lipid consisting of glycerol covalently linked to two fatty acids and one phosphate group.

phospholipid bilayer A structure consisting of two adjacent layers of phospholipids, both with the hydrophilic phosphate groups of the phospholipids projecting outward into an aqueous environment, and the hydrophobic fatty acid tails projecting inward, away from water. Biological membranes are composed of phospholipid bilayers.

phosphorylation The process of adding a phosphate group to a molecule.

photic zone The vertical zone in an aquatic environment through which light can penetrate, permitting photosynthesis.

photoautotroph An autotrophic organism in which light is used as the original energy source for the synthesis of organic molecules.

photosynthesis The process by which certain prokaryotes and eukaryotes capture energy from sunlight and use it for the synthesis of organic molecules from inorganic precursors.

pH scale A numerical scale used to represent the concentration of protons (H^+) in solution. The pH scale thus provides a relative measure of the acidity of a solution.

phylogenetics A technique used to classify living things based on their similarity in DNA sequence, which reflects evolutionary relationships.

phylum A taxonomic level between class and kingdom.

phytoplankton Photosynthetic, planktonic microorganisms.

pilot plant fermenter A fermenter in which microbial metabolite production is tested, before the use of a large-scale industrial fermenter.

pilus Also called a fimbria. A short fiber made of protein that extends beyond the cell wall of many bacteria (plural: pili). Pili aid in bacterial adherence to specific surfaces.

place characteristics A description of where individuals who become ill with a particular medical condition are found.

plasma cell An antibody-producing cell that develops from a B cell in the adaptive immune response.

plasma membrane Also called the cell membrane. A phospholipid bilayer membrane forming the outer boundary of cells. It also contains proteins and regulates the transport of materials into and out of the cell.

plasmid A small self-replicating circle of DNA that is present in many bacterial cells in addition to the main chromosome.

polar covalent bond A type of covalent bond in which the two atoms involved have different affinities for the shared electrons. The electrons are consequently shared unequally, imparting a partial negative charge to the atom with higher affinity and a partial positive charge to the atom with lower affinity.

polymerase chain reaction (PCR) A technique used to make a very large number of copies of a particular DNA "target" sequence.

polysaccharide A large carbohydrate molecule composed of many monosaccharides covalently linked together into a branched or unbranched chain.

population All the members of a particular species that live in a defined geographic area.

porin A channel-like protein found in the outer membrane of Gram-negative bacteria, through which small, hydrophilic molecules can cross the outer membrane.

portal of entry The route by which a pathogen enters a host.

portal of exit The route by which a pathogen leaves a host.

positive taxis Movement toward a favorable stimulus.

prebiotic Food that is believed to contain compounds that stimulate growth of a healthy microbiota.

primary consumer Also called a herbivore. An organism that eats producers.

primary immune response The adaptive immune response that occurs upon first exposure to a pathogen.

primary metabolite A metabolite for which the rate of synthesis closely mirrors population growth.

primer A short, single-stranded DNA sequence used in PCR that is complementary to one end of the target sequence and is used to start DNA replication at the beginning of the required region of the DNA.

prion An infectious protein, thought to lack all nucleic acid. Prions are thought to reproduce by forcing normal proteins to take on the shape of the aberrant prion protein.

probiotic A food product containing live microorganisms that are thought to promote good health.

producer An organism that converts inorganic carbon into biological molecules through the process of carbon fixation.

product The substance formed in a chemical reaction. In living organisms, the conversion of a substrate into a product is catalyzed by an enzyme.

profundal zone The vertical zone of water below the photic zone in an aquatic environment, in which photosynthesis cannot occur because light does not penetrate.

prokaryotic cell A cell in which the DNA is not enclosed within a membrane-enclosed nucleus. Bacteria and archaea have prokaryotic cells and are often collectively referred to as prokaryotes.

promoter A nucleotide sequence found at the beginning of a gene that indicates where RNA polymerase binds and transcription of the gene begins.

proofreading The ability of DNA polymerase to recognize and remove a mismatched nucleotide in DNA, and replace it with the correct nucleotide.

prophage Bacteriophage DNA when it is part of the bacterial chromosome during lysogeny.

protease Enzymes that digest proteins by hydrolyzing the peptide bonds that link the amino acids together.

protease inhibitor A type of drug that inhibits the protease of retroviruses, thus inhibiting replication and transmission of the virus.

protein A class of large biological molecules, each composed of a folded chain of linked amino acids. There are many different kinds of proteins, each of which performs distinct functions in the cell and organism, for example as components of cell structure, enzymes, transporters, and gene regulators. Many bacterial toxins are proteins.

proteobacteria A lineage of bacteria characterized by a Gram-negative outer membrane.

proton A fundamental subatomic particle found in the nucleus of an atom. Protons carry a positive charge, and the number of protons determines what type of element an atom is.

proton gradient   The buildup of protons (H⁺ ions) on one side of a membrane, relative to the other side. Proton gradients across membranes can be used as an energy source to transport materials across the membrane, and also as a source of energy for ATP synthesis.

psychrophile   An organism that grows best at relatively low temperatures.

psychrotolerant   Describes organisms that can continue to grow at relatively low temperatures.

pure culture technique   The process of growing microorganisms in the laboratory in a prescribed manner such that cultures contain only a single, desired type of organism.

purine   A class of double-ringed, nitrogen-containing organic bases that form part of some nucleotides found in DNA and RNA. The two purines present in both DNA and RNA are adenine and guanine.

pyrimidine   A class of single-ringed, nitrogen-containing bases that form part of some nucleotides found in DNA and RNA. Thymine and cytosine are the two pyrimidines present in DNA; uracil and cytosine are the two pyrimidines present in RNA.

pyruvate   The three-carbon molecule that is the end product of glycolysis.

quorum sensing   A mechanism of regulating gene expression that depends on the density of cells in a given area. The bacteria release signaling molecules and when the concentration of these molecules passes a certain threshold, genes dependent on quorum sensing are expressed.

R group   The functional group bound to the central carbon of amino acids. The R group is different in all amino acids and it confers upon the amino acids its specific chemical properties.

radioresistant cocci   A lineage of bacteria, notable for their ability to withstand high levels of radiation.

reactivation   Of viruses, the return of a latent virus to active replication.

reactive   In describing atoms, an atom that is likely to interact, and therefore form chemical bonds with other atoms.

recombinant DNA   A DNA molecule formed by combining DNA from two different sources.

recombinant vaccine   A vaccine produced through the use of recombinant DNA technology.

redox reaction   A reaction in which one molecule is oxidized and another molecule is reduced.

Red Queen hypothesis   The hypothesis that in host–parasite interactions, an evolutionary adjustment on the part of one species requires a subsequent evolutionary adjustment by the other species. Consequently, continuing evolution is required by both species to ensure survival.

reduction   The addition of hydrogen or electrons to a molecule, or the loss of oxygen.

re-emerging disease   A disease that was considered to be under control but is now causing increasingly serious problems.

regressive hypothesis   A hypothesis proposing that viruses evolved from intracellular prokaryotic organisms.

regulated gene   A gene whose expression can be controlled, as environmental conditions warrant.

regulatory T cell (T_R cell)   A type of T cell that may be involved in inhibiting adaptive immune responses.

release   In the replicative cycle of a virus, the stage at which newly assembled virus particles are released from the host cell.

replication fork   The exact site on a replicating DNA molecule where nucleotides are being added to the growing, newly synthesized DNA strands.

repression   The inhibition of operon expression in which a molecule called the co-repressor binds the repressor, allowing the repressor to bind the operator and therefore block transcription.

repressor   Any protein that inhibits expression of an operon in prokaryotes and of individual genes in eukaryotic cells.

reservoir   A place where a pathogen can survive indefinitely without causing disease and from which it can infect new hosts when they become available.

restriction enzyme   An enzyme that cuts DNA at specific nucleotide sequences.

retrovirus   A type of RNA virus that makes a DNA intermediate that is integrated into a host-cell chromosome. The virus is replicated by assembly of viral proteins and genomic RNA expressed by the viral DNA.

reverse transcriptase   A retroviral enzyme that produces DNA by transcription of an RNA template, a process called reverse transcription.

reverse transcriptase inhibitor   A type of drug that inhibits the activity of retroviral reverse transcriptase, thus preventing viral replication.

reverse transcription   The process of converting RNA into DNA, utilizing the enzyme reverse transcriptase. An important step in the replicative cycle of retroviruses.

rhizosphere   The soil environment immediately surrounding plant roots.

ribonucleic acid (RNA)   Along with DNA, one of the two types of nucleic acid found in cells. RNA functions mainly to convert the genetic information stored in DNA into proteins and RNAs that carry out the functions of the cell.

ribosomal RNA (rRNA)   A type of RNA that, along with specific proteins (ribosomal proteins), forms ribosomes.

ribosome   The site of protein synthesis in a cell. It is a multimolecular structure composed of proteins and ribosomal RNAs.

ribozyme   Enzyme made of RNA rather than protein.

RNA-dependent RNA polymerase   An enzyme used by both plus- and minus-strand RNA viruses to synthesize RNA, using RNA as a template.

RNA polymerase   The enzyme that unwinds the DNA double helix during transcription and assembles a chain of RNA nucleotides complementary to the exposed DNA sequence.

RNA splicing   The removal of introns from a molecule of mRNA, after transcription and before translation.

RNA virus   Virus utilizing RNA as its genetic material.

root nodule   A specialized structure found on the roots of some plants, which is colonized by nitrogen-fixing bacteria.

rough endoplasmic reticulum   Endoplasmic reticulum with ribosomes on its cytoplasmic surface.

salt   A compound held together by ionic bonds between the atoms.

saprophyte   An organism that obtains nutrients from dead organic material.

saturated fat   A type of fat in which all the carbon atoms in the fatty acids are linked together by single bonds.

scale-up   A sequence of stages in which production of a metabolite by a microorganism is transferred from small-scale to large-scale production.

science   A process of learning about nature through observation and experiment.

scientific method   A defined series of steps in which the scientist learns about nature. The scientific method requires the construction and testing of a proper hypothesis.

secondary consumer   An animal that feeds upon primary consumers (herbivores). A **carnivore**.

secondary immune response   An adaptive immune response that occurs upon a second exposure to a particular pathogen, and which is usually stronger and more rapidly effective than the primary response.

secondary metabolite   A metabolite that is only synthesized at maximum rate once the microorganism producing it has entered the stationary phase of growth.

selective toxicity   A property of some antimicrobial drugs, describing the ability to kill microorganisms without harming the host.

sepsis   An infection of a normally sterile area of the body such as the blood.

septum   In bacterial endospore formation, the membrane barrier that forms within a cell between the portion of the cell containing the original chromosome and the newly replicated chromosome destined to be enclosed within the endospore.

sex pilus   A long hollow fiber made of protein that is produced by a bacterium to connect it to another bacterium (plural: sex pili). Sex pili bring two bacterial cells of the same species together for conjugation and the exchange of genetic material.

sexual reproduction   A type of reproduction in which genes from two parent organisms are combined in the offspring.

single-cell protein   Bulk microorganisms that are used as a source of protein for humans or animals.

slime layer   A glycocalyx with an irregular and diffuse structure, produced by bacterial cells. It aids in their adherence and/or protection.

smooth endoplasmic reticulum   Endoplasmic reticulum lacking ribosomes.

solution   A mixture of two or more substances in the same phase (solid, liquid, or gas). In biological systems most solutions consist of water (the solvent) and hydrophilic material dissolved in the water (the solute).

species   The most fundamental level of classification within a hierarchical taxonomy.

sperm   The gamete produced by male animals.

spirillum   A type of bacterium characterized by a spiral shape (plural: spirilla).

spirochete   A type of bacterium characterized by a helical shape.

spleen   An organ in mammals in which adaptive immune responses to blood-borne pathogens are generated; it is also the organ in which worn-out red blood cells are destroyed.

spontaneous mutation   Any change in DNA sequence that arises in the absence of a mutagen, such as a random error in DNA replication.

staphylococcus   Common name for any member of the bacterial genus *Staphylococcus*, which are cocci that typically grow as grape-like clusters (plural: staphylococci).

stationary phase   The phase in microbial growth after the lag phase, during which the population is neither increasing nor decreasing.

sterilization   The process of completely eliminating all living things from an area or an object.

sterols   A class of lipid molecules, examples of which are cholesterol and certain hormones.

sticky ends   Short, single-stranded, overhanging ends on double-stranded DNA molecules that are formed when DNA is cut with certain restriction enzymes. Because the single-stranded portions are complementary in sequence to each other, two pieces of DNA cut by the same enzyme can be joined via pairing of these ends.

storage granule   A structure composed of carbohydrate or lipid and used as a reservoir of nutrients in prokaryotic cells.

strain improvement   A process of increasing the production of a metabolite in a strain of microorganisms beyond what a naturally occurring strain would synthesize.

streptobacillus   A general description of any rod-shaped bacterium that forms long chains of cells (plural: streptobacilli).

streptococcus   Common name for any member of the bacterial genus *Streptococcus*, which are cocci that typically grow linked into long chains (plural: streptococci).

strict anaerobes   Microorganisms that rely exclusively on anaerobic respiration and for which oxygen is toxic.

stroma   In chloroplasts, the compartment enclosed by the outer membranes and in which the reactions of carbon fixation take place.

stromatolite   A layered structure occasionally formed by mats of cyanobacteria and other microorganisms in warm shallow waters. Fossil stromatolites contain what are believed to be the earliest microbial fossils.

subclinical infection   An infection in which no symptoms are apparent.

substrate   The starting substance in an enzyme-catalyzed reaction, and which is converted into product in the reaction.

subunit vaccine   A nonliving vaccine consisting only of specific antigens from a pathogen.

sulfanilamide   The active, antibacterial agent in sulfa drugs used to treat bacterial infections.

superinfection   A second infection by a different microorganism that is resistant to the antibiotic used against the first infection.

synchronous growth   A growth pattern in which all the cells in a population divide at the same time.

synthesis stage   The portion of the viral replicative cycle in which viral proteins are synthesized and the viral genome is replicated.

T cell   The class of lymphocytes that carry out the cell-mediated part of an adaptive immune response. T cells develop in the thymus.

T-cell receptor (TCR)   Receptor protein present on the surface of T cells, which recognizes a specific antigen bound to an MHC molecule on the surface of an antigen-presenting cell.

taxonomy   The science of biological classification.

termination sequence   A nucleotide sequence found at the end of a gene that indicates where transcription will end.

theory   A hypothesis that has been repeatedly and rigorously retested and cannot be disproved. It represents the best available explanation for a scientific phenomenon.

thermocline   A sharp demarcation in temperature between the top layer of warmer water and the colder, deeper water in deep lakes in temperate regions during the spring and summer.

thermophile   A microorganism that grows best at high temperatures.

thermotaxis   Movement toward a favorable temperature (positive thermotaxis) or away from an unfavorable temperature (negative thermotaxis).

thylakoid   Membrane-bound vesicle found within chloroplasts of photosynthetic plant cells. The site of the light reactions in photosynthesis.

thymine dimer   A covalent linkage of two adjacent thymine bases in a molecule of DNA.

thymus   A lymphoid organ in mammals in which T cells mature.

time characteristics   A description of when individuals are becoming ill with a particular medical condition.

tissue   An assemblage of a characteristic set of cell types that carries out a particular physiological function in multicellular eukaryotes.

Toll-like receptor (TLR)   Any one of a class of receptors found mainly on cells of innate immunity such as macrophages and dendritic cells. Toll-like receptors recognize molecular features of microbes that are not present in animal cells.

toxins   A substance with poisonous properties. In microbiology, often refers to microbial molecules that interfere with normal host cell functions and therefore contribute to the pathogenicity of the microbe.

transcription   The synthesis of RNA by using a strand of DNA as a sequence template. RNA is complementary in sequence to its template DNA.

transcription factor   A protein that binds to promoter sequences in eukaryotic genes. After binding of the promoter by transcription factors, RNA polymerase then binds the transcription factors to begin transcription.

transduction   The transfer of bacterial DNA from one bacterium to another by a bacteriophage.

transfer RNA (tRNA)   A class of small RNA molecule that transports amino acids to the ribosome and matches them to the correct codon on mRNA during translation.

transformation   The process in which DNA released by a dead bacterium is absorbed by a living bacterium and incorporated into its chromosome.

transgenic organism   An organism into whose genome DNA from another species has been deliberately introduced.

translation   The process of assembling amino acids into a specific protein, guided by the genetic information encoded by a molecule of the corresponding mRNA.

transmembrane protein   A membrane protein that extends across the membrane, projecting into both the cytoplasmic and extracellular environments.

transport vesicle   A small sac of membrane in which material is transported between the components of the endomembrane system within a eukaryotic cell.

transposon   A short DNA sequence that can move from one site to another within the genome.

trophic level   The position of an organism in a food chain.

trophozoite   A general term describing the stage in a protozoan life cycle during which feeding and reproduction take place.

uncoating   The release of the nucleic acid of a virus into the cytoplasm of the host cell.

unicellular   Consisting of a single cell.

unsaturated fat   A type of fat in which there is at least one double bond between carbon atoms in the fatty acids. Because of the double bonds, unsaturated fats have fewer carbon–hydrogen bonds than saturated fats.

variolation   An early means of vaccination against smallpox, in which material from an infected individual was used to inoculate potentially susceptible individuals.

vasodilation   Increase in the diameter of a blood vessel.

vector   (1) An invertebrate that carries a pathogen from one host to the next. (2) An entity such as a virus or a plasmid used in genetic engineering to transfer genes into a cell.

vector-borne disease   A disease transmitted by an arthropod or other invertebrate.

vertical transmission   Pathogen transmission from mother to child, either across the placenta or in breast milk.

viral envelope   A phospholipid bilayer of host origin that is acquired by some types of viruses as they leave a host cell.

viral glycoprotein   A glycoprotein embedded in the envelope of enveloped viruses, which has an important role in host cell recognition and entry.

viral specificity   Describes the particular set of cell types that a given virus can infect.

virion   An individual complete viral particle.

virulence   The capacity of an organism to cause disease.

virulence factor   A characteristic that increases the disease-causing capacity of a microorganism.

virus   An acellular life form, which is an obligate parasite and can only replicate itself inside a living cell. Viruses are the cause of many diseases.

whey   The liquid portion of milk remaining after the separation of milk solids during the production of dairy products such as cheese.

wort   The liquid portion containing sugars, recovered after malting of barley.

zoonosis   An animal disease that is capable of infecting and causing disease in humans.

zooplankton   Consumer organisms in the plankton.

INDEX

The index covers the main text but not the Concept Questions or Glossary. The suffixes F and T indicate that relevant material occurs on that page only in a Figure (usually in the caption) or Table.

A

A subunit, cholera toxin 267
abiotic conditions
    and microbial growth 226
    modification by living things 222, 225
    synthesis of biomolecules 199–200
abiotic factors in ecology 222, 229, 231F
Acetobacter spp. 400
acetone from fermentation 120–1, 179
acid-fast cell walls 49, 53–4, 323
Acidithiobacillus spp. 244
    A. ferrooxidans 17, 190
acidity
    and basicity 26–7
    of common substances 28F
    gastric, antimicrobial action 268, 280–1
    toxin denaturing 36
acidophiles 77, 190, 193–5, 225, 420
acquired immunity 111
activation energy, enzymes 37
acute infections 100–2, 115–17, 271, 272F
adaptations
    pathogens to innate immune response 285
    V. cholerae 197
adaptive immune response
    antigen presenting cells 285–8, 289F, 290, 295, 301F
    components and process 284–96
    HIV and 301–2F
    humoral response 290–3, 294F, 295, 296F, 300–2
    lymphatic organs 286–7, 290, 295
    relationship to innate response 278, 282, 303
    and subsequent exposure 295–6
    and vaccination 298
adenine deamination 161
adenosine triphosphate *see* ATP
adhesion (to host) 47–8, 103, 158, 257, 264
ADP (adenosine diphosphate), ATP interconversion 167
adsorption (bacteriophage) 103
Aedes aegypti 355–6
aerobic respiration 176–7
    aeration at large scales 416
    ATP yield 176–7
    in composting 426
    growth rate and reproduction 177, 193
    obligate aerobes 187, 188F, 228, 400
aerosols *see* airborne transmission
aerotaxis 49

aerotolerant anaerobes 188
aflatoxin 163
Africa, pace of colonization 115, 116–17
agar
    adoption as a growth medium 127
    favoring Gram-positive bacteria 247
agglutination 291, 292F, 293
Agrobacterium tumefaciens 383
AIDS (acquired immune deficiency syndrome)
    beneficial effects of GB Virus C 2
    emergence 352, 358
    opportunistic infections and 300, 302, 303T, 359
    sexual transmission 352
    tuberculosis and 360
    worldwide distribution 299F
    see also HIV
air travel
    influenza infection case 278, 280
    measles infection case 337
    spread of emerging viruses 356–8
airborne transmission 111, 261–2, 263F, 273, 280, 361
alcohol
    as a biofuel 428–30
    disinfection 313–14
    incompatibilities 325
alcoholic beverages 73, 125–6, 389–90, 392, 396–9
algae
    as biofuels 429–30
    distinction from protozoa 78
    zooxanthellae 245–6
algal blooms 234, 244
alien (extraterrestrial) life 238
Aliivibrio fischeri 152
alleles 154, 158, 163
allergies and microbiota 132, 255, 256T
allergies to antibiotics 317, 321
alligators 334
amantadine 327–8
amebae
    as bacterial predators 248
    Entamoeba histolytica 80, 272–3, 281
    parasitic 262, 272–3
    phylogenetics 72F, 78F
    pithovirus susceptibility 93
    as protozoa 80
Americas, imported diseases 110, 114–16
amino acids
    commercial production 404

deamination 161F, 180
genetic code for 147F
hydrophilic and hydrophobic 34
protein assembly 145–9
as protein building blocks 34
ammonia 23–5, 183, 190, 198–9, 242F, 243
ammonification 242F, 243
Amphotericin B 325
ampicillin 29, 322, 330–2, 375F
anaerobes
    aerotolerant anaerobes 188
    ATP production 177
    bioremediation by 424
    on the early Earth 204
    electron acceptors 176
    facultative anaerobes 187–8, 193F, 194, 231, 246, 430
    obligate anaerobes 187, 188F, 203–4, 231–2, 246
    photosynthetic 429
    species surviving 237
    strict anaerobes 75
anatomical barriers to entry 281
animals
    antibiotics in feedstuffs 332–3, 334
    antimicrobials from 334
    domestic animals 116, 348, 429
    similarity of fungi to 83
    transgenic animals 372, 376, 379–80, 382–6
    see also reservoir species
Antarctica 67, 229, 230F, 238
anthrax 58–9, 123T, 127–8, 130, 297, 360
antibacterial drugs
    as bactericidal or bacteriostatic 52, 318
    see also antimicrobial
antibiotics
    broad-spectrum 319–22, 324
    cephalosporins 322, 417
    chloramphenicol 322–3F, 324
    choosing 317–21
    ciprofloxacin 324, 333–4
    in commercial production 417–18
    defined 317
    discovery of new forms 412
    gentamicin 322F, 324
    and human survival 315–16
    isoniazid 319T, 322F, 323–4
    limitations 82, 227
    mechanisms of action 321–4
    misuse and resistance 330–5
    narrow-spectrum 319–20, 322–3, 335
    peptidoglycan targeting 51
    ribosome targeting 57, 149, 323
    streptomycin 317, 319T, 320–4, 417
    tetracycline 63–4, 319–20, 323, 331, 417
    see also erythromycin; penicillins
antibiotic resistance
    antibiotic misuse and 330–5
    countermeasures 334–5
    gene cloning example 374–5
    genetic recombination and 158
    hospital environments 307
    natural selection and 212–13, 332
    reversal 213
    in S. aureus 159, 323, 330–1, 360
    superinfections 333
antibodies
    in the adaptive immune response 278, 283–5, 290–2
    antibody classes 292

in cross-reactivity 121
hantavirus 380
from plasma cells 83
see also immunoglobulins
anticodons 147–8
antigens, defined 285
antigen presenting cells 285–8, 289F, 290, 295, 301F
    see also dendritic cells; macrophages
antigenic drift/shift 347–8
antimicrobial drugs
    development 130–1
    health effects 315–17
    from insects and vertebrates 334–5
    targets 322F
    see also antibacterial
antiparallel strands 137
antisense molecules 329
antiseptics, distinguished from disinfectants 313
antiviral drugs 9, 100, 299, 326–30
antiviral state, induction 283–4
apicomplexans 72F, 78F, 80F, 81–2
Appert, Nicholas 308, 310
applied microbiology 7–10
Aquifex/Hydrogenobacter bacteria 73F, 74–5
archaea
    as ancestors of eukaryotes 206–7
    cell walls 54, 56, 77
    CRISPR/Cas system 377F
    growing within rocks 229
    as prokaryotes 4, 76–8
Archaea (domain) 72–3, 77
Aretino, Spinello 7, 9F
art, cleaning 7, 10
Arthrobacter simplex 418
Arthrobotrys spp. 248, 249F
artificial selection 212, 414
ascomycota 85F, 86–8
aseptic techniques 123T, 127, 315
asexual reproduction 153–5, 218–19
Ashbya gossypii 404, 414
Aspergillus spp. 401
    A. flavus 163
    A. niger 401, 413
assembly of virions 97
asthma 255, 256T
AT3 protein 372
Athens, plague of c430 BC 110–11
atoms, defined 18
atomic mass, number, weight 18–19
ATP (adenosine triphosphate)
    in cell respiration 170
    energy for synthesis 177
    energy supply as 167–8
ATP synthase 175, 177
attachment, virus replication 95–7
attenuation of vaccines 123T, 130, 297–8, 302, 381
autoclaves 308T, 309–10
autoimmune diseases 255, 256T
autoinducers 152, 271
autotrophs 83, 180–1, 183, 206, 239
avian influenza/bird flu 260, 349–50
azoles 325
Azospirillum spp. 243
Azotobacter spp. 243
AZT (zidovudine) 327F, 328–9
Aztec civilization 110, 114F, 115

B

B cells 290–3, 295–6, 300, 301F
B subunit, cholera toxin 267
bacilli, shape 45
Bacillus anthracis 58–9, 128, 360–1
Bacillus cereus 421F
Bacillus sphericus 58
Bacillus thuringiensis 384, 420
bacteria
    characteristics of earliest bacteria 74–5
    common shapes 45–6
    conjugation 48
    coordination of gene expression 151–2
    electron transport components in 175
    extracellular structures 46–9
    natural selection in 211F
    not culturable 370, 419
    origin of chloroplasts and mitochondria 65
    as prokaryotes 4
    rapid metabolism and reproduction 45
    wine contamination 126
Bacteria (domain) 72–4
bacterial fuel cells 430
bacterial lawns 11–13, 412
bactericidal and bacteriostatic drugs 52, 318
bacteriophages
    bacterial gene transfer 156–7, 370
    as cloning vectors 376–7
    defenses against 372
    discovery 335
    as earliest viruses 208
    genome sequencing 369
    infection and replication cycle 102–5
    phage lambda 104, 376
    phage therapy 335
    T4 bacteriophage 102, 103F, 104
bacteriostatic drugs
    contrasted with bactericidal 52, 318
    erythromycin and tetracyclines 323
Bacteroides gingivalis 75
bagels 406–7
barotolerance 238
barriers to entry 277–82, 294, 307
    anatomical, microbial and genetic 281
Bartonella henselae/B. quintana 273
base pairing in DNA 137
bases, DNA 38, 137, 143
basic and applied microbiology 7–10
basicity and acidity 26–7
basidiocarp 84F, 85
basidiomycota 85–6
Batrachochytrium dendrobatidis 82, 84–5
bats 2, 86–7, 358–9
Baytril® 320, 324, 334
Bdellovibrio spp. 248
beers 73, 389–90, 397–8
beneficial bacteria 29, 31
beneficial mutations 160
benthic zone 231–2, 234, 237–8
bilayers *see* lipid bilayers
binary fission 184, 192, 206
binomial nomenclature 68
bioconversions 418–19
biofilms 226–7, 229, 233, 400, 421–4
biofuels 428–30
biogeochemical cycles 238–9, 244, 249
biological vectors 263, 274
    see also mosquitoes; vector-borne diseases

biological warfare 113
biological weapons 360–1
bioluminescence 152
biomass
    defined 224
    methane producers and oxidizers 237
biomineralization 411
Biomphalaria spp. 70–1
biopesticides 420
bioprospecting 413–14, 425
bioremediation 158, 227, 370, 385, 421–6
biotechnology
    ethical and safety concerns 384–6, 415, 420
    industrial applications 384–5
    medical and veterinary applications 379–81
    techniques of 363–78
bioterrorism 59, 360–1
biotic interactions 222
bird flu/avian influenza 260, 349–50
birds
    milk contamination by 342–3
    vitamin B2 requirement 403–4
birth defects 79, 81F, 82, 216
the Black Death 109F, 113
blood vessels, permeability 283
Borrelia burgdorferi 263, 265, 352, 370
BOTOX® 268–9
botulism 33–4, 36, 230–2, 403
brain
    evolution and prion disease 217
    prion disease localization 209
Brazil 2, 354F, 355, 356F, 357
breadmaking 390, 392, 395–6, 406
Brevibacterium spp. 404
broad-spectrum antibiotics 319–22, 324, 333
broad-spectrum antivirals 330
BSE (bovine spongiform encephalopathy, mad cow disease) 7,
    105–6, 107F, 354F
Bt corn/cotton 382, 384, 420
bubonic plague 109F, 111, 113–14, 128
budding (protein export) 63, 285F
budding (virion release) 98, 101–2
buildings, preservation 411, 412F
Burnet, Macfarlane 251
butterflies 382, 384

C

calcium carbonate 239, 240F, 245, 411
calcium chloride 408
calories as potential energy 32
Calvin cycle 181, 183, 239
Camelidae 69F
Campylobacter spp. 252
    C. jejuni 342
cancers
    diagnosis 1
    Kaposi's sarcoma 303T, 359
    magnetic bacteria and 420–1
    melanoma and gene editing 381
    microbiota and 356
    mitomycin and 419T
    viruses and 124T, 271, 272F, 297
cancer chemotherapy 101, 254, 318
Candida albicans 6F, 87, 254–5, 302, 303T
cannibalism 105, 217
capsids 93–9, 104
    nucleocapsids 93–4, 96, 98–9

capsomeres 93, 95, 97
capsules, bacterial 47, 264
carbohydrates
    in cell respiration 170–1
    functions 29–31
    see also glucose; sugars
carbon
    fixation 183, 223, 239–40, 245
    release in the Krebs cycle 173–4
carbon cycle 239, 240F
carbon dioxide (CO_2)
    in photosynthesis 64, 181
    product of fermentations 396–8, 402
    product of oxidation reactions 168, 169F, 396
    reducing atmospheric 240
    structure 22–3
    use by autotrophs 166, 180
    use by methanogens 176
carbon fuels, oxidation 169
carbon-hydrogen bonds
    in hydrocarbons 24
    in lipids and fats 31–2, 33F, 180
    in oxidation reactions 168–71, 173–4, 177
carcinogens 424
carnivores, as secondary consumers 223
carrier proteins 55
Cas (Crispr associated protein) 377
case definitions 350–2
case–control studies 352–3
caspases 330
catalase 187–8
cats
    Bartonella and 272
    FLV, EBV and 94
    Toxoplasma and 79, 81–2
CD4 proteins 96, 300
cells
    discovery 41
    energy budgets 149–50
    minimum requirement of proteins 379
    origins 74, 201–2
    parent and daughter cells 153
    prokaryotic and eukaryotic 3F, 4, 42, 43F, 59
    sizes and efficiency 44–5, 59, 77, 208
cell division 184–5, 206
cell lysis 49, 97
cell-mediated response 289, 294–5, 296F, 300–2
cell membranes
    of archaea 77
    Chlamydia spp. 75
    of eukaryotes 60–1
    LPS outer membrane 51
    pathogen adherence to proteins 264
    and temperature preference 189–90
    see also plasma membranes
cell nucleus, eukaryotes 42
cell respiration
    process 170–5
    three stages 171
    without glucose 180
cell theory 3–4, 14, 41
cell-to-cell communication 4, 61
cell types
    in humans 59
    virus susceptibility 94–5
cell walls
    absence among animals 60

bacterial types 49
    Deinococcus radiodurans 75
    in fungi 83
    Mycobacterium tuberculosis 211, 212F
    in plants 60
    presence among groups of microorganisms 6F
    of prokaryotic cells 49–54
cellular transport, endomembrane system 62–3
cellulases 31, 384, 419
cellulose
    ethanol from 428
    plants cell walls 60
    as a polysaccharide 31
cephalosporins 322, 417
cheese and yogurt making 10, 179, 193–4, 391–2, 401–2, 408
chemical bond types 20–2
chemical control 313
chemically defined media 191
chemoautotrophs 181, 183, 239
chemotaxis 49
chemotherapeutic agents *see* antimicrobial drugs
chestnut blight 82, 86–7
chickenpox 272, 345F
chikungunya virus 354F, 357
childbed fever 123T, 126
China 122, 129, 325, 363, 378, 381
    novel influenza strains 279, 346, 349
Chipotle Mexican Grill 340–1
chitin 60, 83
Chlamydia spp. 319T
    C. psittaci 63–4
    C. trachomatis 76, 252
    as obligate intracellular parasites 76
Chlamydiae 72–3F, 75
chloramphenicol 322–3F, 324
chlorine disinfection 313
chlorine transporters 21–2
chlorophyll 182–3, 230F, 236
chloroplasts
    appearance 3F
    endosymbiotic origins 205–6, 379
    photosynthesis in 64–5, 182–3
chloroquine 325, 326F
cholera
    attempted phage therapy 335
    case example 266–8, 281
    fowl cholera 130
    Snow's epidemiological study 338, 339F, 341
    survival in water 259
    virulence variations 197, 216
cholera beds 267F
cholera toxin 105, 267
cholesterol 54, 60F, 61, 325, 378, 418–19
'Christmas trees' 142, 144, 145F
chromosomes
    DNA organization onto 137–8
    prokaryotic and eukaryotic 56–7, 138
chronic infections 100, 115, 271, 272F, 318, 343, 370
chronic wasting disease (CWD) 91, 105–16, 107F
Chytridiomycota 84–5
cilia 61, 80, 280
ciliates 72F, 78F, 80, 248
ciprofloxacin 324, 333–4
circular DNA
    in chloroplasts and mitochondria 65, 206
    plasmids 57
    in prokaryotes 138, 140

citric acid 401, 413
class, taxonomic 68
classification, biological *see* taxonomy
Claviceps purpurea 273, 419T
Cleopatra 399
climate, colder, and mutualism 86
climate change (global warming) 78, 85, 176, 237, 239–40, 246, 427
cloning genes 373
cloning hosts 376
cloning vectors 373–4, 376–7
Clostridium spp. 58, 243
    C. acetobutylicum 120
    C. bifermentans 425
    C. botulinum 34, 36, 58, 231–2, 268, 403
    C. difficile 333
    C. perfringens 58
    C. tetani 58, 259, 268, 273
clotting factors 364, 372, 379–80
clouds 238
co-infection 265, 348F
coagulase 211, 212F, 270
cocci, shapes 45–6
Coccidioides immitis 87, 258T, 259
coding and noncoding strands 144
codons 146–8, 159–61, 324
    start codons 148
    stop codons 147–8, 160F, 161
coevolution hypothesis, viral origins 209
coffee 403
colonial organisms 208F
commensalism 246
common source epidemics 340–2
communities, ecological 222
competition and growth rate 225
competitive exclusion 247–8
complex media 191
composting 426
condensation reactions 30, 32, 34, 35F
conjugation, bacterial 48, 157, 332–3
constitutive genes 150
contact transmission 261, 263F
contamination events (GM crops) 282, 284
contamination events (infection) 252, 261–2
control and experimental groups/variables 12–13
coral bleaching 246
coral reefs 245
cordite production 120–1
coronaviruses 358
cortisone 418
Corynebacterium spp. 246
    C. diphtheriae 191, 268
coupled reactions 166, 167F, 174
covalent bonds 22–5
    in archaea membranes 56
    polar and nonpolar 23–5
Crenarchaeota 72F, 77
Creutzfeldt–Jakob disease (CJD) 105–7, 209
    variant CJD 106, 107F
CRISPR/Cas system 124T, 132, 363, 363F, 377–8, 381
CRISPR (clustered regularly interspersed short palindromic
    repeats) 377
cross-reactivity 121
cross-species infections 116
Cryphonectria parasitica 86
Cryptosporidium spp. 303T
crystal violet 53
culture, large-scale 415

curd, in cheesemaking 402
curly toe, in birds 403–4
CWD (chronic wasting disease) 91, 105
cyanobacteria
    in biofilms 227F
    and the origin of chloroplasts 206, 379
    oxygenation event and 203
    photosynthesis and nitrogen fixation 76, 181–3, 239
    phylogenetic status 72–3F
    in polluted water 234
    as possible foods 394
    in soils and freshwater 229–30
cyclosporin 419T
cystic fibrosis 21–2
cysts 73, 79, 80, 81F, 82
    oocysts 81F, 82
cytochrome oxidase 175
cytokines 283, 285, 287, 288F, 289–90, 291F, 294
    see also interferons; interleukins
cytomegalovirus 300F, 303T, 329
cytopathic effects of viruses (CPE) 101–2, 271
cytoplasm 42
    as membrane bound 56
    organelles of 62
cytoskeleton 61–4
cytotoxic T cells (Tc cells) 289–90, 295, 300, 301F

D
dairy products *see* milk
Darwin, Charles 210
Darwinian medicine 216
Darwin's paradox 245
daughter and parental DNA 139
daughter cells 153
dead zones 236
deamination of amino acids 161F, 180
decline phase, growth curve 192, 195
decomposers 83–4, 86, 229–30, 239, 244, 249, 425
    ecological view 223F, 224
deer 91, 105
deer mice 258, 259F, 342–3
definitive hosts 81F, 82
Deinococcus radiodurans 75, 379, 425
DeJong, Randall 70–1
denaturing, proteins 35–6
dendritic cells 282, 284–7
dengue virus 329, 354F, 355, 357–8
dental plaque 46–7
deodorants 246
deoxyribonucleic acid *see* DNA
deoxyribose, ring numbering 137
'designer babies' 363, 386
designer drugs 124T, 328–9
detergents
    added proteases 419
    antibacterial action of soaps and 314–15
    pollution 233–4, 413
D'Hérelle, Felix 335
diabetes 255–6, 364, 378, 386
diarrhea
    in animals 29, 31, 403
    from *C. difficile* 333
    from *Campylobacter jejuni* 342, 343
    child deaths from 251
    in cholera 104, 197, 266–8, 274
    from *Cryptosporidium spp.* 303T
    Cyprian's plague 113

E. coli O157:H7 158, 340, 368
from *Entamoeba histolytica* 281
following antibiotic treatment 320
from *Giardia intestinalis* 80, 272
and seasonal canine illness 319
from *Shigella* 274
traveler's/hikers 42, 48, 80, 261
treatment 256
dinoflagellates 236
diphtheria 128, 191, 268, 297T, 354F, 360
direct contact transmission 261
disaccharides 30
disease control and reservoir type 260
disease development stages 252
disease eradication
hopes and expectations 251, 260
human reservoirs and 260
kuru 105
disease identification
emerging and re-emerging diseases 353-4
time, place and personal characteristics 351-2
disease incidence in epidemiology 338
disease outbreaks
emerging diseases 358
in the news 2, 251
disease resistance and likelihood of epidemics 343-5
disease thresholds and virulence 303
disinfectants, distinguished from antiseptics 313
Disneyland 337, 338F, 349
distillation 399
disulfide bonds 35F
DNA amplification 365-7
DNA (deoxyribonucleic acid)
adoption as genetic material 202, 364
as an information molecule 4, 135, 141
antibiotics targeting 324
of chloroplasts and mitochondria 65, 206
and evolutionary relatedness 70-1
importance of discovery 131-2
nuclear location 61
structure and function 38, 136-7
transforming factors 156
see also recombinant DNA
DNA ligases 375
DNA polymerases 139F, 140-1, 160F, 161-3, 327, 365-6, 419
Taq polymerase 366, 419
DNA repair mechanisms
Deinococcus radiodurans 75, 379, 425
DNA proofreading 162-3
mismatch repair enzymes 163
DNA replication 138-40, 324, 326-7
DNA sequencing 364-70, 371F
DNA viruses
assembly 97F
synthesis stage 98-9
Dobzhansky, Theodosius 209
dogs, genetically modified 363, 363F
Domagk, Gerhard 131, 316
domains, in taxonomy 72-3
domestic animals 116, 348, 429
'domesticated' microorganisms 415
double helix, DNA 38, 39F, 137
DRACO 329-30
drain cleaners 419
drug development *see* antimicrobial drugs
drug resistance
chloroquine resistance 325, 326F

re-emerging diseases 359
see also antibiotic resistance
dsRNA 329-30

E
E. coli (Escherichia coli)
026 strain 340-1, 368
bacteriophages infecting 102
as a cloning host 376
drug resistance 330
F$^+$ and F$^-$ cells 157
as a facultative anaerobe 187
gene editing 364
generation time 184-5
genetic differences between strains 135
as a model organism 8, 412
O157:H7 strain 158, 340-1, 368-9
opportunistic infections 254
as a proteobacterium 76
rapid metabolism and reproduction 45
virulence factors 48
Earth, conditions on early 74-5, 198
Ebola virus 2, 88, 251, 353, 354F, 358-9
eclipse phase 97
ecological zones
lakes 231-3
marine environment 234-8
ecology, defined 222
ecology, microbial 225
*Eco*RI enzyme 373
ecosystems
coral reefs 245
defined 222
dependence on microorganisms 222
EEE (eastern equine encaphalitis) 350
EEHV (elephant endotheliotropic herpesvirus) 91, 101
effector T cells 288F, 289, 291F, 295-6
egg and sperm, as gametes 153-4
Ehrlich, Paul 123T, 131, 316
El Niño 342-3
electron acceptors 170
final electron acceptors 175-6, 178-9, 181, 182F, 186-7, 430
see also FAD; NAD$^+$; oxygen
electron affinities 21, 23
electron shells 18F, 19, 20F, 22, 28
in covalent bonds 22-3
electron transfer in redox reactions 168
electron transport
membrane proton pumping 177
metals and 191
oxygen detoxification 204
in photosynthetic light reactions 182-3
third stage of cell respiration 171, 174-5
electrons, introduced 18-19
electrophoresis 366, 367F
elements, chemical 18
elephant endotheliotropic herpesvirus (EEHV) 91, 101
elephantiasis 128
emerging and re-emerging diseases 353-60
emus 350
encephalitis 350-3
endemic disease, defined 346
endergonic reactions 166-7, 172, 177, 181
endocytosis 63-4, 96, 267
endomembrane system 62-3
endospore formation 57-9, 75, 228, 309-11, 360-1
endosymbiosis 65, 204-8, 379

endotoxins
    Gram-negative membranes 269–70, 318
    LPS 52, 269
energy
    alternative 428–30
    flows in ecosystems 222–4
    lipid-rich foods 32
    production in chloroplasts and mitochondria 64–5
    sources for chemoautotrophs 183
energy budgets
    endergonic and exergonic reactions 166
    in metabolism 165
Entamoeba histolytica 80, 272–3, 281
Enterococcus faecalis 270F
environmental factors
    industrial replication 415–16
    influence on growth rate 186–92, 225–6, 308T
environmental reservoirs 259
environmental responsiveness
    as characteristic of living things 4, 5F
    endospore formation 57–9
environments
    direct DNA sampling 370
    DNA absorption from 156
    influence on phenotype 152–3
    microenvironments 225, 247
enzyme-substrate complexes 38F
enzymes
    as biological catalysts 30, 36–7
    commercially useful 419–20
    denaturing 189
    and DNA coding 141
    protein superiority to RNA 201–2
epidemic threshold 345
epidemics
    causative factors 342–3
    common source epidemics 340–2
    cycles 345
    defined 340
    disease resistance and likelihood 343–5
    host-to-host epidemics 341F, 342, 344
    London 1853 cholera epidemic 338, 339F
    plagues 109F, 110–14, 128
    Spanish flu, 1918 346, 348–9
epidemiology
    cholera investigation 338, 339F
    contribution of Florence Nightingale 127, 339
    emerging and re-emerging diseases 353–60
    of influenza 346–9
    scope 9, 337
Epstein-Barr virus (EBV) 94, 294
Epulopiscium fishelsoni 44
ER (endoplasmic reticulum), rough and smooth 62
ergosterol 61, 325
ergot 273, 419T
Erwinia dissolvens 403
erythromycin 146, 318, 319T, 322–3, 331, 417
    antiribosomal action 57, 149, 322F, 323
escaped-gene hypothesis, viral origins 208
Escherichia spp. *see E. coli*
estradiol 418–19
estrogen 418
ethanol *see* alcohol
Eukarya (domain) 72, 78
eukaryotes
    archaea as ancestors 206–7
    emergence 202–3, 204–7

eukaryotic and prokaryotic chromosomes 138
    microorganisms among 6, 73, 78
    multicellularity in 207–8
    regulation of gene expression 152
eukaryotic cells
    distinguished from prokaryotic 3F, 4, 42, 43F
    examples among microorganisms 6, 78
eukaryotic parasites 272–3
eukaryotic pathogens, selective toxicity 324–5
Euprymna scolopes 152F
Europa (satellite of Jupiter) 238
Euryarchaeota 72F, 77F, 78
eutrophs and eutrophication 233–4, 236, 244
evolution
    as characteristic of living things 4, 5F
    favoring reduced virulence 215
    as fundamental to biology 209–10
    natural selection in 210–12
    and the origin of living things 74–5, 198–203
    reflected in biological classification 68–9
    role of mutations 163
exergonic reactions 166–7, 168F, 172, 173F, 174, 177, 181
exobiology 238
exocytosis 63, 267
exons 145
exotoxins 267–9
experimental and control groups/variables 12–13
experimental design 10–13
explosives 120, 425
exponential growth 184, 185F
extinct bacilli 57
extraterrestrial life 238
extremophiles 76, 77F, 238
    Deinococcus radiodurans 75, 379, 425
    see also acidophiles; thermophiles
extremozymes 419–20

F
F⁺ and F⁻ cells 157, 158F
facultative anaerobes 187–8, 193F, 194, 231, 246, 430
FAD in the Krebs cycle 171F, 173–5
FADH₂ 171F, 173–7
families, in taxonomy 68
Faroe Islands 116
fastidious organisms 191, 378, 423
fatalities, from infections 294
fats
    in cell respiration 180
    and fatty acids 33F
    see also lipids
fatty acids
    and body odor 246
    and fats 33F
fecal transplantation 256
feces 254, 370
fermentation
    acetone production 120–1
    in cheese and yogurt making 193–4, 391–2
    fermented foods in history 391–2, 395–403
    fermented products as biofuels 428
    incomplete glucose oxidation 178–9, 395
    Pasteur's investigations 123T, 126
    see also alcoholic beverages
fermenters, laboratory, pilot plant and industrial 416–17
feta cheese 408
fetus, vertical transmission and 79, 81–2, 263–4, 355
feudal system 114

fever, in the innate immune response 283
filamentous fungi 83–4, 393
filtration to remove microbes 311
fimbriae 48, 158F, 254, 264, 314
final electron acceptors 175–6, 178–9, 181, 182F, 186–7, 430
Finlay, Carlos 119, 128
fish, deterioration if unfrozen 176
five-kingdoms classification system 72
flagellae 43F, 48, 49F, 61, 80
flagellates 72F, 78F, 80
    dinoflagellates 236
flavobacteria 75, 385
flavor enhancers 399–401, 404
fleas as vectors 113, 215, 258T, 273, 352
Fleming, Alexander 10–11, 123T, 131, 317
'flesh-eating bacteria' 270
fleshy fungi 84
flu *see* influenza
fluid mosaic membrane model 56
fluorescence 379F
FLV (feline leukemia virus) 94
folic acid 316–18
fomites 261
food, genetically-modified 363, 382, 384–5
food-borne illness 261–2, 315
    food and water-borne 262, 315, 341, 390
food chains 222–3, 224F, 234, 239, 241–2, 244, 249, 370
    pollutants in 422, 424
food microbiologists 9
food poisoning 192, 252, 266, 271, 340–3
food preservation 308, 312–13, 390–1, 399
food production, microbes in 390
food spoilage 390–1
food webs 223–4, 245
formivirsen 329
fossil fuel formation 239
fowl cholera 130
fowlpox 380–1
frameshift mutations 160F, 161–2
free radicals 162
freeze-drying 313
freshwater habitats 230–4
fuel cells, bacterial 430
fungi
    basic body plans 83–4
    in bioremediation 425–6
    ecological, medical and industrial importance 82–3
    edible mushrooms 393–4
    as eukaryotes 78
    filamentous fungi 83–4, 393
    five-kingdoms system 72
    fungal pathogens 273, 299, 325
    mutualism with termites 246
    number of species 83
    phylogeny 85F
fusion, enveloped viruses 96

G
Gajdusek, Carleton 105
Galápagos Islands 266
gametes 153–4
gamma radiation 311
Garfield, President James A. 122
GB Virus C 2
gel electrophoresis 366, 367F
GenBank database 367–9

genes
    alleles 154
    cloning 373
    clustering as operons 150
    copy numbers 154
    as DNA 38, 136
    encoding T cell receptors 288
    regulated and constitutive 150
    splicing toxin genes into crops 420
    for virulence 368–70
gene-editing 363–4, 377–8, 381, 386, 415
    and genetic diseases 381–2
gene expression
    chemical coordination 151–2
    as information flow 39F
    and phenotype 152–3, 159–60
    quorum sensing and 270–1
    regulation 149–50, 152
    as transcription and translation 142, 146
gene guns 384
gene sequencing *see* DNA sequencing
gene therapy 381–2
genera, in taxonomy 68
generation time 184
genetic barriers to entry 281
genetic code, amino acids in 147F
genetic diseases 381–2
genetic drift/genetic shift 348–9
genetic engineering 9, 156, 363–4, 371–4, 377–9, 384, 385F
    and metabolite yield 415
genetic information flow 39F
genetic material
    bacterial exchange 157, 363–4
    original form 200–2
    RNA as, in some viruses 93
genetic recombination 154, 158, 163, 231
    see also recombinant DNA technology
genetic shift 348–9
genetic variation
    horizontal gene transfer 154–6, 364
    mechanisms 153–8, 218
    mutations as source 159–61, 163
genetically modified organisms
    for bioremediation 423–4
    GM crops 382, 384–6, 420
genetics and biotechnology 378
genomes
    influenza virus 346–9
    and mitochondrial origins 206
    sizes 141, 378
genomics 368–70, 378
    metagenomics 370, 371F
genotype, distinguished from phenotype 152–3
gentamicin 322F, 324
germ theory of disease 14, 123T, 126–9, 315, 339
German measles (rubella) 263, 297T
Giardia intestinalis 73, 80, 258T, 264F, 272
global warming (climate change) 78, 85, 176, 237, 239–40, 246, 427
glucose
    energy conversion to ATP 177
    fermentation 178–9
    partial and complete digestion 171
    see also carbohydrates
glycocalyces 46F, 47, 226, 254, 264F
glycogen 31
glycolysis
    emergence 203

first stage of cell respiration 171–2
 respiration incomplete after 178
 in yogurt making 194
glycoproteins
 eukaryotic cell membranes 60–1
 in viral envelopes 94–6
Golden Age of microbiology 117, 123–4T
golden toads 82, 85
Golgi apparatus 62–3, 205, 285F, 380
gonorrhea 128, 264–6, 317F
Gorgas, William 119–20
gout 394
Gram stains 53, 54F
Gram-negative bacteria
 antibiotic choice 324
 genera 76
 growth inhibition by salt 399, 405
 membrane endotoxins 269–70, 318
 in the reproductive tract 253–4
 selective growth media for 247
 Vibrio cholerae as 104, 267
Gram-negative cell wall structure 49, 51–3
Gram-positive
 genera 76
 selective growth media for 247
 skin bacteria as predominantly 253
 vancomycin and 159
Gram-positive cell wall structure 49–51
green sulfur/nonsulfur bacteria 75
Greenland Viking settlement 114
Griffith, Frederick 155–6
growth curves 192–5
growth media, selective 247
growth rate
 antibiotic-fed animals 333
 in commercial applications 414–15
 and competition 225
 effects of acidity and salinity 190–1
 effects of oxygen 186–8, 194
 effects of temperature 188–90, 193–5
 of microbial populations 184–6
 and nutrient levels 191–2, 226
guano 243–4
guide RNA 378
gut bacteria 29, 31, 106, 246, 255–6, 333–4, 404
gut fermentation syndrome 178–9

H

Haemophilus influenzae 124, 369
Haiti 117
halophiles 78, 190–1, 235, 420
Haloquadratum walsbyi 77F
hand washing 123, 126–7, 262F, 307, 314
hantavirus 258, 259F, 342–3, 352–3, 358, 380
Harvoni® 329
Hawaii 81, 152, 399F
Hawaiian bobtail squid 152
heart attacks, streptokinase use 36–7
helical sheath, T4 phage 103–4
helical viruses 93, 98F
Helicobacter pylori 26–7, 190, 359
helminths 121–2
helper T cells (Th cells) 287–90, 295, 300–2
hemagglutinin 347–8
hemolysins 270
hepatitis A virus 262, 297T, 341
hepatitis B virus 94–5, 101, 124T, 265, 271, 281, 297–8, 381

hepatitis C virus 261, 271, 284, 329, 343
hepatitis D virus 265
herbivores, as primary consumers 223
herpesviruses
 acyclovir treatment 326–7, 328F
 as DNA viruses 99
 elephant endotheliotropic herpesvirus 91–2, 101
 human herpesviruses 92, 101, 303T, 359
 latent infections 102, 272F
Heterorhabditis marelatus/bacteriophora 221, 221F
heterotrophs 83, 180–1
hierarchy, in classification 68
hiker's disease 73, 80
Hippocrates 346
HIV (human immunodeficiency virus)
 attachment sites 96
 opportunistic infections 254, 299, 301, 303T
 treatment 328–9
 tuberculosis and 265, 303T, 360
 vertical transmission 263
 see also AIDS
H5N1 strain 2, 349, 354F
homeostasis 4, 5F
Hooke, Robert 41
horizontal gene transfer 154–6, 364
horses 29, 31, 330
hospital sanitation 126–7, 307, 315
hospital sanitation and hospital infections 126–7, 307–8, 315
hosts
 defined 214
 new, invasion by a known pathogen 358
 speciation and bacteria 218
host cells, mechanism of damage by viruses 101–2
host defenses *see* immune responses
host range, viruses 94
host-to-host epidemics 341F, 342, 344
human genome 142, 369, 378, 381
human growth hormone 376
human herpesvirus type 1 (HHV-1) 101
human herpesvirus type II 303, 359
human herpesviruses 101, 303, 359
Human Microbiome Project 370
human papillomavirus 99, 124T, 271, 272F, 297T
human survival 315F
human-to-human transmission 348–9, 358, 359F
humans
 number of cell types 59
 number of genes 378
humidity detection 420, 421F
humoral immune response 290–3, 294F, 295, 296F, 300–2
Humulin®, recombinant insulin 9, 124, 364, 379
humus 228
hyaluronidase 270
hydrocarbons
 bonding in 22
 example structures 29F
 hydrophobicity 29
hydrogen
 fuel from photosynthesis 429
 loss in oxidation reactions 169
hydrogen bonding
 in DNA 38, 39F, 137, 140, 162
 water and 24–5
hydrogen ions *see* protons
hydrogen peroxide 2, 163, 186–7, 188F, 429
hydrogen sulfide 176, 183, 200, 234, 421–2, 424
hydrolysis reactions 30, 32

hydrophilicity
  among amino acids 34
  defined 25
  Gram-negative outer membrane 51
  and hydroxyl groups 29
  of lipids 31
  peptidoglycan cell walls 50
hydrophobicity
  among amino acids 34
  defined 25
  of hydrocarbons 29
hydrothermal vents 200, 207F, 238
hydroxyl groups 29
hydroxyl ion 311
hyphae 83F, 84–7, 248–9, 393
hypotheses
  elevation to theories 13–14
  in experimental design 11–13

I

Ichthyophthirius multifiliis 80
icosahedral viruses 93–4, 102–3
ID$_{50}$ 264, 268
Ideonella sakaiensis 425
IL-1 etc. *see* interleukins
immune responses
  consequences of failures 294–5
  innate and adaptive distinguished 277–9
  pathology resulting 265–6, 294
  primary and secondary 295–6
  reponse to LPS 52, 269
immune systems
  acquired immunity 111
  development and microbiota 255–6
  influence of microorganisms 216
  influenza antigens 347
  suppression by HIV 300–2
immunity, population 343–5
  see also disease resistance
immunocompromised individuals 82, 254, 270, 302, 318, 320
immunodeficiency *see* AIDS; SCID
immunoglobulins
  IgA 292–3
  IgD 292
  IgE 292
  IgG 292–3, 296F
  IgM 292–3, 296F
immunological memory
  adaptive immunity and 278, 293, 295
  and antigenic drift 347
  and epidemic probability 343
  in prokaryotes 377
  vaccines 297–8, 302, 381
immunology, scope 277
immunosuppression 254, 359
inactivated viruses 297, 302
Inca Empire 114
inclusion bodies 57, 271
incubation period 115T, 264, 265F, 341
inducers 150F, 151, 414
  autoinducers 152, 271
  inducible operons 151
inductive period 293, 303F
industrial and commercial applications 411–12
industrial fermenters 416–17
infections
  acute, chronic and latent 100–1, 115, 271

barriers to entry 277, 278F, 279–82, 294, 307
  control in surgery 122–3
  defined 252, 266
  location of, and treatment choice 320–1
  opportunistic 254–5, 269–70, 294, 299–302, 303T, 359
  subclinical 296
  transmission characteristics 258T
inflammation, in innate immune response 283
inflammatory response 266
influenza
  aircraft infection case 278
  as an RNA virus 99, 346–9
  antigens H and N 286, 347–8
  avian (bird flu) 260, 349–50
  drug therapy 327–8
  epidemiology 346–9
  H5N1 strain 2, 349, 354F
  secondary bacterial infections 301
  Spanish flu epidemic, 1918 346, 348–9
  swine flu 346, 349
  vaccines 381
innate immune response 278, 282–5, 292
inorganic precursors of life 198–200
insects
  antimicrobials from 334
  predation on 221, 248–9
  see also mosquitoes; vector-borne diseases
insulin, recombinant 9, 124, 364, 379
intercalating agents 162
interferons 283–4, 329
interleukins 283, 289–90, 294, 364, 380
  IL-1 283
  IL-2 289
  IL-4 and IL-5 290
  IL-10 294
intermediate hosts 81F, 82
intertidal zone 234
intracellular pathogens *see* obligate intracellular parasites
introns 145, 376
iodine 53, 308T, 313
ion formation 20
ionic bonds 21–2
Ireland 114, 118–19
iron
  marine environment 236
  ocean fertilization 240–1
isomers of sugars 30
isoniazid 319T, 322F, 323–4
Israel 121
Ivan IV (the Terrible) 109
Ivanowski, Dmitri 123T, 129
Ixodes scapularis 263F

J

Japan 217, 343, 425
Jenner, Edward 123T, 129–30
'jumping genes' 157

K

Kaposi's sarcoma 303T, 359
kingdoms, in taxonomy 68, 72
'kissing disease' (mononucleosis) 94, 294
Klebsiella spp. 254
koala 334–5
Koch, Robert 123T, 127–8, 129F, 392
Koch's postulates 123T, 127–8, 129F, 130
Korarchaeota 72F, 77

Korea 358, 399F
Krebs cycle 171, 173-5, 177
kuru 105, 217

L
laboratory fermenters 417
lac operon 150-1
lactase 30
lactate, from fermentation 178F, 179
Lactobacillus spp. 179F, 191
    L. acidophilus 193, 194F, 195
    L. brevis 399
    L. bulgaricus 396, 402, 407
    L. plantarum 396, 399, 405F
    L. sanfrancisco 396
Lactococcus lactis 408
lactose 30, 35, 150-1, 408, 414
lag phase, growth curve 192
Lake Erie 233-4
Lake Vostok 67
lakes
    ecology 230-3
    seasonal changes 232
landfill 427, 429
latent infections 100F, 101-2, 271-2
LD$_{50}$ 264
legume boosters 241, 243
Leishmania spp. and leishmaniasis 6F, 80
leprosy 53, 189
lethal genes, in gene cloning 374-5
leucocytes
    in innate immunity 282
    neutrophils and macrophages as 282
Leuconostoc mesenteroides 399
lice, as vectors 117-18, 339
lichens 229, 230F
life *see* living things
life expectancy *see* human survival
light reactions 181-2
Linnaeus, Carolus 68
lipases 32, 180
lipids
    in archaea and bacteria 77
    function 31-2
    saturated and unsaturated fats 32
    and temperature preference 189-90
    see also fats; sterols
lipid A 51-2
lipid bilayers 55-6, 60, 94, 96, 201F, 313
liposomes 201-2
Lister, Joseph 123T, 127-8, 313
Listeria spp. 252
    L. monocytogenes 310
littoral zone 231, 234-6, 245
liver, hepatitis specificity 94-5
living things
    artificial cells 379
    characteristics of 3-4, 5F
    defining 7
    evolutionary timelines 74F, 198F, 202F, 207F
    extraterrestrial life 238
    origins 74-5, 198-203
    spontaneous generation 124, 125F, 198
    status of viruses 7, 93
    water as essential 25
locomotion in ciliates 80
logarithmic growth curves/phases 185, 186F, 192, 194

Lokiarchaeota 207
the Louisiana Purchase 117, 118F
LPS (lipopolysaccharides) 51-2, 269
luciferase operon 152, 153F
lumen, endoplasmic reticulum 62, 63F
Lyme disease 263, 265, 297, 352-3, 370
lymph nodes 128, 273, 286F, 287-90, 291F, 302, 330
lymphatic organs in adaptive immunity 286-7, 290, 295
lymphocytes, B cells and T cells as 287
lysogeny 104
lysosomes 64, 271, 285
lytic cycle 104, 105F

M
macrophages
    in innate immunity 282, 284-8, 291
    LPS receptors 269
    M. tuberculosis and 324
    phagocytosis by 64
mad cow disease (BSE) 7, 105-6, 107F, 354F
magnetic bacteria 420-1
major histocompatibility complex (MHC) 285, 287-90
malaria
    in Ancient Rome 112
    discovery of mosquito vector 123T, 128
    drug treatment 325, 326F
    and El Niño 342
    in the New World 114, 119-20
    resistance 281
    in syphilis treatment 189
    transmission characteristics 258T, 263
    see also Plasmodium
Malarone® 325
malting barley 397
mammals, phylogenetic tree 69
marine environment
    deep sea adaptations 237-8
    ecological zones 234-8
Mars 238, 420
Marshall, Barry 26
maternal immunity transmission 293
maternal pathogen transmission *see* vertical
matrix, mitochondrial 64
MDRTB (multi-drug-resistant tuberculosis) 331
measles 99, 111, 114-16, 258T, 297T, 337
meat eating
    contamination 262
    fermented meat products 403
    morning sickness 216-17
    tenderizers 419
mechanical vector transmission 263
media, chemically defined, minimal and complex 191
medical applications of biotechnology 379-80
membrane proteins and virus specificity 94
memory B cells 290, 291F, 293, 296
memory T cells 289-90, 293, 295-6
meningitis 310
mercury 425
MERS (Middle East respiratory syndrome) 358-9
mesophiles 188-9, 195, 229, 309-10, 408, 419
Mesophilic-A 408
metabolic efficiency 45, 184
metabolic pathways, evolutionary sequence 203-4
metabolism
    as characteristic of living things 4, 5F
    common features 165
    energy budgets and 165, 166-8

and growth rate 184
heterotrophs and autotrophs 180-1
stepwise oxidation in 169
metabolites
commercially useful 413-15, 418-20
primary and secondary 414
metagenomics 370, 371F
metal-plated microorganisms 420
metals, toxicity in 315
meteorites 200, 238
methane 165, 169, 176, 427, 429
methanogens 78, 176, 231, 237, 427, 429
metronidazole 320, 325
MHC (major histocompatibility complex) 285, 287-90
mice
deer mice 258, 259F, 342-3
microbiota and timidity 255, 256
microaerophiles 188
microbes *see* microorganisms
microbial barriers to entry 281
microbial chemistry, scope 18
microbiology
basic, applied and their subdisciplines 7-10
groups of living things included 4-6
and history 109
history of 122-32
microbiota
beneficial foods 404-5
Human Microbiome Project 370
immune system and obesity 255-6
normal 252-5, 257, 266, 314-15
and timidity in mice 255, 256
microcephaly 355, 356F
microenvironments/microhabitats 225, 247
microorganisms
classification 67
detection 124
ecosystem dependence on 222
energy sources 168
eukaryotic, examples 6, 73, 78
experimental usefulness 3, 8
and food 389
in industry and commerce 411-12
link to specific diseases 127
methods for controlling growth 308T
numbers and importance 1
reproduction and population growth 184-5
unicellular and multicellular 3
viruses as acellular 6, 92
Micropia Zoo, Amsterdam 1-2
microRNAs 152
microtubules 61
milk
cheese and yogurt making 10, 179, 193-4, 391-2, 401-2
immunoglobulins 292F, 293
pasteurization 310
spoiling 30, 35, 342-3
vertical transmission 263
milkmaids 129
Miller, Stanley 199
mining waste 17
mismatch repair enzymes 163
missense mutations 160-1
mites 319
mitochondria
electron transport chain 175

endosymbiotic origins 65, 205, 379
in energy production 64-5
mitomycin 419T
model organisms 8, 131
modes of transmission 261-4
molasses 401, 417
molds
Ashbya spp. 404, 414
Aspergillus spp. 162-3, 400-1, 413
in blue cheeses 10
characteristics 84-5
and miracles 186
as obligate anaerobes 228
Penicillium spp. 10-13, 131, 317, 402, 414
molecular clocks 71, 357
Monera 72
monoculture, dangers 118-19
mononucleosis 94, 294
monosaccharides 30
Montagu, Lady Mary 129
morning sickness 216-17
mosquitoes
and elephantiasis 128
and encephalitis 352
and malaria 112, 119, 123T, 128, 263
and myxoma virus 215
and other viruses 357
and West Nile virus 263, 350, 357
and yellow fever 123T, 128-9
and Zika virus 355
motility in prokaryotes 48-9
mRNA (messenger RNA) 39, 142, 145-6, 148F, 323F
MRSA (methicillin-resistant *Staphylococcus aureus*) 331
mucus/mucous membranes 21, 80, 105, 280-1, 293
multicellular organisms
common among eukaryotes 59-60
defined 3
emergence 207-8
multiple sclerosis 255, 256T
muscular dystrophy 378
must, grape 396
mutagens 161-2, 204, 211, 311, 414
mutations
and genetic variation 159-61
repair mechanisms 162-3
role in and evidence of evolution 70, 163, 211, 213
speed, RNA viruses 347
types of mutation 159-61
mutT gene 379
mutualistic relationships
corals and algae 245
host and pathogen 215
mycorrhizal associations 85-6, 229
nematodes and bacteria 221-2
plants and fungi 83
replication in the laboratory 415
termites, bacteria and fungi 246
mycelia 83F, 84, 86, 393
Mycobacterium spp. 53, 323, 419
M. avium 303T
M. leprae 53, 189
M. tuberculosis 135, 211, 212F, 252, 264, 365
see also tuberculosis
Mycococcus xanthus 411
Mycoplasma spp. and *M. pneumoniae* 54, 146, 149, 319, 323, 378-9
mycoremediation 425-6
mycorrhizal associations 85-6, 229

mycoses 87
Myotis lucifugus 87F
myxoma virus 214–15

N

NAD⁺ (nicotinamide adenine dinucleotide) 170, 171F, 172–5,
    178, 395
    in photosynthesis 181, 182F
NAG (*N*-acetylglucosamine) 50, 51F, 213F
NAM (*N*-acetylmuramic acid) 50, 51F, 213F
naming and taxonomy 68
Nanoarchaeota 72F, 77
Nanoarchaeum equitans 72F, 77–8
nanopore sequencing 369
nanotechnology 420
Napoleon (Bonaparte) 117–18, 308
narrow-spectrum antibiotics 319–20, 322–3, 335
Nasonia spp. 218
native conformation, proteins 34
natural selection
    as driving evolution 210–12
    limitations 211–12
    in microorganisms 212–13
necrotizing fasciitis 270
Negri bodies 102, 271F
Neisseria gonorrhoeae 264, 266
neuraminidase 347–8
neutral mutations 159–60
neutralization, infections 121, 291, 292F, 293
neutrons 18–19
neutrophils 282, 291
New World, imported diseases 110, 114–16
newborns, immunity 293
next-generation sequencing 369
Nightingale, Florence 123T, 127–8, 339
nitrification 242F, 243, 246
Nitrobacter 243, 246
nitrofurantoin 320
nitrogen
    fixation 76, 241–3
    and growth rate 191, 241
    in marine environments 235, 241
nitrogen cycle 242–3
Nitrosomonas 243, 246
nitrous acid 161
nomenclature *see* taxonomy
nonbiological precursors of life 198–200
nonenveloped viruses 94–6, 98, 313
nonfastidious organisms 191, 378, 423
nonpolar covalent bonds 23–5
nonsense mutations 160F, 161
normal microbiota 252–5, 257, 266, 314–15, 320, 403
norovirus 261–4, 271, 273
nosocomial infections 307
nuclear envelope 61–2
nucleic acids, functions 37–9
    see also DNA; RNA
nucleocapsids 93–4, 96, 98–9
nucleoid area 56
nucleotides
    intercalation between 162
    in structure of DNA 38, 135–6
nucleus, atomic 18, 23–4
nucleus, cell, characteristic of eukaryotes 42
nucleus, for condensation 238
nutrient flow, ecosystems 222–4

nutrient levels
    and growth rate 191–2, 226
    marine environment 235–7

O

O polysaccharides 51, 52F
obesity 256, 371F
obligate aerobes 187, 188F, 228
obligate anaerobes 187, 188F, 203–4, 231–2, 246
obligate intracellular parasites
    all viruses as 7F, 83, 92–3, 208–9
    Chlamydiae as 75–6
    Mycoplasma pneumoniae as 378
    Rickettsia prowazekii as 379
oceans
    ecological zones 234–8
    iron fertilization 240–1
    phosphorus accumulation 244
oenology 397
oil-eating bacteria 371
oil spills 32F, 423
oligotrophs 233, 236–7
omnivores 223
oncogenes 271
one health movement 353
oocysts 81F, 82
operator sequences 151
operons 150–3, 191
opportunistic infections
    after microbiota disturbance 255
    after tissue damage 254, 269–70, 315
    associated with AIDS 299–302, 303T, 359
    following respiratory problems 294
opsonins 291–3
orders, in taxonomy 68
organelles
    as characteristic of eukaryotic cells 3F, 4, 42, 59
    endomembrane system 62–3
    origins of 204–6
    variety of 61–2
organic molecules and life 28
organs, as groupings of tissues 60
origins of replication 140, 141F
oxidation
    carbon fuels 169
    as loss of electrons 168
oxidizing agents 187
oxygen
    absence from the early Earth 198
    and ATP synthesis 167
    in cell respiration 170
    dissolved, in lakes 232–4
    as final electron acceptor 176
    growth of atmospheric 203
    and microbial growth rate 186–8, 194
    release in some photosynthesis 183
oxygen detoxification 204
ozone layer 74, 198F, 199, 202F, 203–4, 205F, 207F

P

PABA (*p*-aminobenzoic acid) 316–17, 322F
paintings, cleaning 7, 10
Panama Canal 119–20
pandemics 114, 346–9
papillomaviruses 99, 124T, 271, 272F, 297T
Paramecium 80

parasites
    parasitic archaea 78
    parasitic fungi 82-3
    parasitic protozoa 79
    parasitism defined 121-2
    see also obligate intracellular
parent cells 153
parental and daughter DNA 139
parrot fever (psittacosis) 63, 285
Pasteur, Louis 123T, 124-7, 130, 198, 392
pasteurization
    alternatives 311
    discovery 123T, 126
    flash pasteurization 308T, 310, 396
    pasteurized milk 310, 407
    ultrapasteurization 310
pathogens
    defenses against the immune system 294-5
    defined 53, 214, 252, 266
    destroyed by pasteurization 310
    implication in a known disease 359
    mechanism of disease causation 266-73
    modes of transmission 261-4
    requirements for disease causation 256-7
    and sexual reproduction 218-19
pattern recognition 282, 283F
PCBs (polychlorinated biphenyls) 424
PCR (polymerase chain reaction) 365-7, 419
pectin 403
pelagic zone 234-6
penetration, viruses 96, 104
penicillin (penicillin G)
    discovery 10, 131, 317
    mass production 413-14
penicillinase 322, 331
penicillins
    activity spectrum 319T
    ampicillin 29, 322, 330-2, 375F
    peptidoglycan targeting 51, 321
    structural variants 322
Penicillium spp.
    P. camemberti 402
    P. chrysogenum 10, 131, 414
    P. citrinum 419T
    P. notatum (now *P. chrysogenum*) 10, 12F, 131, 414
    P. roqueforti 10, 11-12F, 13, 402
pentaglycine bridges 50-1, 321
pentamidine 300
peptic ulcers 26-7
peptide bonds 34, 35F, 37, 147-8, 323F, 324
peptidoglycans
    in bacterial cell walls 50-3, 58, 62, 74F, 212-13, 282, 321
    penicillin and 50-1, 321
    vancomycin and 212, 213F
peripheral proteins 56
peritonitis 269
Perkin, William 120
personal characteristics, case definition 351-2
Peru 365, 367
pH
    animal digestive systems 420
    effects on growth rate 190
pH scale 28
PHA (poly-β-hydroxyalkanoates) 427-8
phage lambda 104, 376
phage therapy 335
phages *see* bacteriophages

phagocytosis/phagocytic cells
    accumulation as pus 266
    adaptive immunity 285
    endosymbiosis and 207
    humoral immune response 291-3
    innate immune response 282-4
    of invading microorganisms 47, 64, 278
    resistance of acid-fast bacteria 53
    see also macrophages
'pharming' 380
phenol 313
phenotypes
    distinguished from genotypes 152-3
    mutation effects 159
phenylalanine deaminase 180
Phillips, Carol 342-3
phospholipids, plasma membrane 55
phosphorus
    and growth rate 191
    in household detergents 233-4
    phosphorus cycle 243-4
phosphorylation 327, 328F
photic zone 231, 234-5, 237
photoautotrophs 181
Photorhabdus spp. 221
photosynthesis
    and the carbon cycle 239
    in cyanobacteria 76
    described 64
    distribution across groups of microorganisms 6F
    emergence 203, 205
    hydrogen fuel from 429
    in some archaea 77
    in sulfur bacteria 75
phyla, in taxonomy 68, 71-2
phylogenetic trees 69-71, 72F, 73-6
phylogenetics and taxonomy 69
Phytophthora infestans 119
phytoplankton 230-1, 234, 236, 240-1, 245
pickling 391, 399, 405-6
pili 43F, 48
    sex pili 48, 157, 158F
pilot plant fermenters 417
pithovirus 93F
place characteristics, case definition 351
placenta 79, 81-2, 263-4, 293
plagues
    The Black Death 109F, 113
    Great Plague of Cyprian 112-13
    Plague of Athens, 110-11
    Plague of Justinian 113
    see also bubonic plague
plant tumors 383
plant viral infections 102
plasma cells 62, 290, 291F, 292, 295-6
plasma membranes
    archaea 56
    bacteria and eukaryotes 55
    common to all cells 42
    detergent effects 314
    electric charge 44
    electron transport chain 175
    lysis 49
    structure 56
plasmids
    as circular DNA 57, 138F
    as cloning vectors 373-6, 383

inserting animal DNA 364, 372F
transfer via conjugation 157, 158F
virus origins and 209
Plasmodium
    as an apicomplexan protozoon 80F, 81, 272
    responsibility for childhood mortality 60
    transmission by mosquitoes 119, 258T, 263
plastic waste 425
plastics, microbial 427–8
Pneumocystis jirovecii 299–300, 302, 303T
pneumonia
    secondary to HIV infection 299
    'walking pneumonia' 146, 323, 378
Poe, Edgar Allan 92
Poland 117, 121
polar covalent bonds 23–5, 29
poliomyelitis
    epidemic threshold 345
    eradication hopes 251, 260
    twentieth century epidemic 354, 355F
poliovirus
    as an RNA virus 100
    DRACO effectiveness 330
    Salk and Sabin vaccines 297T, 298
    size 94F
    target cells 96
    transmission 262
pollution
    coastal waters 236
    into the food chain 422, 424
    Lake Erie 233–4
    Pseudomonas stutzeri and 7
    Shamokin Creek, Pennsylvania 17
    see also bioremediation
polyene antimycotics 325
polymerase chain reaction (PCR) 365–7, 419
polysaccharides
    in biofilms 226
    examples 30–1
    in the glycocalyx 46F, 47
population density
    and acute disease persistence 116
    and disease susceptibility 110–11, 227
    quorum sensing 151–2, 227, 270
population growth, microbial
    environmental influences 186–92
    phases 192–5
    process 184–6
populations within an ecological community 222
porins 51–2
portals of entry 257F, 258T, 263–4
portals of exit 257F, 273
potatoes 33, 36, 102, 127
the potato famine 118–19
pravastatin 419T
prebiotics 405
predation
    distinguished from competition 247–8F
    by microorganisms 248–9
Preempt® 252–4, 404
Premarin® 418
primary consumers 223
primary immune response 295, 297
primer sequences 366
prions/prion diseases
    chronic wasting disease 91, 105–16, 107F
    known prion diseases 105–6, 107F

kuru 105, 217
    mad cow disease 7, 105, 354F
    possible origins 209
    scrapie 7F, 106
probiotics 404–5
producers, in food chains 223
production problems, biotechnology 415–18
profundal zone 231, 234, 237
progesterone 418
prokaryotes
    archaea and bacteria as 4, 76
    chromosome structure 56–7
    classification as Monera 72
    emergence 202–3
    extracellular structures 46–9
    immune systems 132, 377
    lacking cell walls 54–5
    in the marine environment 237
    predominantly unicellular 59
prokaryotic cells
    cell walls 49–54
    chromosomes 56–7, 138
    classification by shape 45
    distinguished from eukaryotic 3F, 4, 42, 43F, 138
    reproduction 184
    sizes 44
    transcription and translation 149F
promoter sequences 143–4, 145F, 150F, 151–2
prontosil 131, 316
prophage state 104
Propionibacterium 402, 404
proteases 37, 97, 180, 329, 402, 419
protease inhibitors 124T, 327F, 329
proteins
    denaturing 35, 309
    determining cell characteristics 140–2
    enzymes as 36, 201–2
    function 33–6
    minimum requirement 379
    misfolding 106, 107F
    plasma membrane 55–6
    see also enzymes; prions
protein synthesis
    assembly during translation 145–9
    nucleotide coding 136
    on ribosomes 57, 145–6
protein transport 62–3
proteobacteria
    Gram-negative species 76
    in human microbiota 371F
    phylogenetics 69T, 72–3F
Proteus spp. 243
    P. vulgaris OX19 121
proton concentrations, acidity and basicity 26–7
    see also pH
proton gradients and energy storage 177
protons
    in biological oxidations 169
    in the electron transport process 174–5, 183
    introduced 18–19
protozoa
    distinction from algae 78
    five-kingdoms system 72
    groupings 78–82
    protozoan parasites 272, 325
    trypanosomes as 227
    as unicellular eukaryotes 59, 78

PrP proteins 106, 107F, 209
Pseudogymnoascus destructans 2, 86–8
Pseudomonas spp. 175, 404, 423
    P. aeruginosa 270–1
    P. stutzeri 7, 9F, 10
psychrophiles 188–9, 194F, 237, 309, 420
psychrotolerance 189, 194F, 195
pure culture technique 123T, 127–8, 129F, 225, 392, 414
purines 38, 137
pyrimidines 38, 137, 143
pyruvate
    and fermentation 178, 395
    oxidation in Krebs cycle 173
    production in glycolysis 172

Q

quinine 325
quorum sensing 151–2, 227, 270

R

R groups
    amino acids 34
    penicillin analogs 322
rabbits and myxoma virus 214–15
rabies
    as an enveloped virus 94
    as an example virus 92
    as an RNA virus 99
    diagnosis 271
    reservoirs and transmission 258T
    symptoms 100, 102
    vaccine 123T, 126F, 130, 297T
radiation as a mutagen 163
radiation sterilization 311
radioactive waste 424–5
radioresistant bacteria 75
    Deinococcus radiodurans 75, 379, 425
re-emerging diseases 353, 354F, 359
reactions, endergonic and exergonic 166
reactions, oxidation/reduction 168–70, 174–5, 199
reactivation, latent infections 100F, 101
reactive oxygen compounds 187
reactivity and electronic structure 19–20
receptors
    cytoplasmic, for autoinducers 152
    phagocytic cells 269, 282, 283F, 285, 288, 290, 292F
    toxins 267–8
    virus attachment 96–7, 103, 264
recipes 405–9
recombinant DNA technology 124T, 132, 364, 371–3, 374F, 376F
    vaccines from 380–1
recovery phase/recovery period 265
Red Queen hypothesis 218–19
red tides 236
Redi, Francesco 124, 125F, 198, 298
redox reactions 168–70, 174–5, 199
reduction, as electron gain 168
refrigerators 310
regulated genes 150
regulatory T cells (T_R cells) 293
release of virions 97–8
rennin, in cheesemaking 402, 408
replication, in experimental design 13
replication, of viruses 95–9
replication forks 140, 141F, 162
repressor proteins/sequences 150F, 151, 191, 414
reproduction speed, and cell volume 45, 208

reservoirs
    and disease control 260
    environmental 259
reservoir species
    domestic and wild 116–17, 348
    E. coli pathogenic strains 341
    emerging virus diseases 357
    human reservoirs 258–60
    infectious diseases 116, 257–61
    influenza 346–7, 349
restriction enzymes 372–4, 375–6F, 383F
retroviruses 99, 100F, 101, 300, 366
reverse transcriptase 99, 329, 366
    RT-PCR (reverse transcriptase-PCR) 366
rheumatic fever 252
rhinoviruses 99
Rhizobium spp. 243F
Rhizopus stolonifer 85
*rh*izosphere 229
Rhodobacter sphaeroides 428F
Rhodococcus rhodochrous 88
ribavirin 329
riboflavin (vitamin B2) 403–4, 414
ribosomes
    antibiotics targeting 57, 149, 323
    of chloroplasts and mitochondria 206
    protein synthesis 57, 145–7
    see also rRNA
ribozymes 201
Rickettsia spp. 208, 320
    R. prowazekii 117, 121, 379
    R. typhi 339
rifampin 322F, 324, 417
rivers and streams, ecology 233
RNA polymerase 142–5, 150F, 151–2, 324, 329
    RNA-dependent 99
RNA (ribonucleic acid)
    double-stranded (dsRNA) 329–30
    genetic material in viruses 93, 99
    guide RNAs, gene editing 378
    microRNAs 152
    mRNA (messenger RNA) 39, 142, 145–6, 148F, 323F
    structure and function 39
    transfer RNA (tRNA) 147, 323
    viral, recognition by TLRs 282
    see also rRNA
RNA splicing 145
RNA viruses
    influenza as 346–9
    mutation speed 347
    replication 99, 100F
RNA world hypothesis 200–2
rock, microorganisms in 229, 237
Roman Empire, disease in 112–13, 308
Roosevelt, Theodore 119
root nodules 243
Ross, Ronald 119, 123T, 128
rRNA (ribosomal RNA) 147
rRNA genes 70–2, 74, 77, 83
RT-PCR (reverse transcriptase-PCR) 366
rubella (German measles) 263, 297T
rust (parasite) 86F

S

Saccharomyces spp. 389, 397
    S. carlsbergensis 73, 397
    S. cerevisiae 60, 73, 86, 179, 369, 394–6, 406

safety concerns, biotechnology 384-6, 415, 420
safranin 53
Salmonella spp.
    competitive exclusion 254-5
    fermentation waste 179F
    and food poisoning 252, 266, 271, 333
    in genetic engineering 384
    reservoir species 258-9
    S. typhi 258T, 259
    transmission 258, 262
salt
    cystic fibrosis 21
    effects on growth rate 190-1
    in sauerkraut 399, 405
    in seawater 234
    as sodium chloride 21-2
salvarsan 131, 316
sand flies 6F, 80
sanitation
    hospital sanitation 126-7, 307, 315
    and human survival 315-16
saprophytes 83-6, 248
SARS (severe acute respiratory syndrome) 353, 354F, 358
saturated and unsaturated fats 32, 33F
sauerkraut 399, 405
scaling-up 416-17
scarlet fever 111
Schistosoma mansoni 70
schistosomiasis 70-1, 122
SCI (seasonal canine illness) 319-20, 349
SCID (severe combined immunodeficiency) 381
scientific method 10-13
scrapie 7F, 106
sea lion feces 370
sea water
    marine environment 234-8
    salt content 234
    virus content 104
    see also oceans
seasonings 399-401
secondary bacterial infections 301
secondary consumers 223
secondary immune response 296
secondary metabolites 414
segmented genomes 346-7
selective toxicity 316, 321, 323-7
Semmelweis, Ignaz 123T, 126-8
sepsis 52
septum 58
sex pili 48, 157, 158F
sexual reproduction
    and genetic variation 153-4
    and predation 218-19
sexually transmitted infections 75-6, 261, 273, 352, 355
shellfish 236
Shewanella spp. 430
Shiga toxin 341, 368-9
Shigella spp. 274, 369-70
    S. dysenteriae 158
    S. enteritis 341
shingles 272, 297T, 303T
side effects 254, 283-4, 320-1, 324-5, 329-30, 418
silicon, unsuitability for biology 28
the Silk Road 391F
single-cell protein 394
skin
    as a barrier 50, 59, 277, 279-80F, 281
    typical microorganisms 50, 253, 315

skin cancer 359, 381
SLE (St Louis encephalitis) 350-3
sleeping sickness 117, 263, 272, 274F
    see also trypanosomes
slime layers 47
smallpox
    bioterrorism 361
    elimination 260
    historical impact 110-11, 114-16, 123T
    size and structure 92-4
    vaccination 129-30, 297T
    variolation 123T, 129
snails and schistosomiasis 70-1
Snow, John 338, 339F
soaps and detergents 314-15
social effects of the Black Death 114
sodium chloride (table salt) 21-2, 25
soil
    composting 426
    microbial communities 228-9, 412
    microenvironments 225
solar system formation 198
solubility, sucrose example 25
soy sauce 400-1
species, in taxonomy 68
specific toxicity 321
specificity, virus 94-5
sperm and egg, as gametes 153-4
spice trade 391
spirilla 45
spirits, distilled 398-9
spirochetes 46F, 72-3F, 75
Spirulina platensis 394-5
spleen 286F, 287, 288F, 290
spontaneous generation 124, 125F, 198
spontaneous mutations 161-2
spore formation
    anthrax 360-1
    biopesticidal 420
    C. difficile 333
    endospores 57-9, 75, 228, 309-11, 360-1
    fungi 34F, 36, 85-8, 229, 246, 248-9
    tetanus 259
squid 152
staining
    and cell wall types 44, 53-4
    for light microscopy 44
Staphylococcus spp.
    coagulase production 211, 212F, 270
    drug resistance in 159, 323, 330-1, 360
    as facultative anaerobes 187
    as Gram-positive example 50
    in penicillin discovery 11-12, 131
    S. aureus 50, 159, 163, 212, 323, 330-1, 360
    S. epidermidis 11, 13, 252, 334, 412
    shapes 46
starch 31, 427
stationary phase, growth curve 192, 414
sterilization 309
steroid hormone production 418-19
sterols 54, 61
    see also cholesterol; egosterol
Stewart, William 251
sticky ends 373-4, 375-6F, 383F
stone-washed denim 2, 384-5, 419
stop codons 147-8, 160F, 161
storage granules 57, 428

strain improvement 414, 416F
strep throat 42, 258, 270, 330
streptobacilli 46
Streptococcus spp. 46, 226F, 252, 270
    S. mitis 270F
    S. mutans 47
    S. pneumoniae 155-6, 285
    S. pyogenes 111, 126, 252, 258, 270, 330
    S. thermophilus 396, 401, 407
streptokinase 36-7, 270
Streptomyces spp. 417-18
    S. caespitosus 419T
    S. griseus 317
streptomycin 317, 319T, 320-4, 417
strict anaerobes 75
stroma, chloroplasts 64
stromatolites 202-3
subatomic particles 18-19
subclinical infections 296
substrates, enzyme 37
subunit vaccines 298, 381
sucrase 38
sucrose 25, 30
sugars
    mono-, di- and polysaccharides 30
    solubility 25, 29
    see also glucose
sulfa drugs 123T, 131, 316-18, 319T
sulfanilamide 131, 316
superinfections 333
superoxide ion 187, 204, 311
surface to volume ratio 45
surgery, infection control 122-3
swamp gas 165, 176
symbiosis
    endosymbiosis 65, 204-8, 379
    laboratory replication 415
    legume boosters 241, 243
    lichens 230F
    mycorrhizal associations 85-6, 229
    in squid bioluminescence 152
    vitamin B12 from 404
symptoms that assist disease transmission 274
synchronous growth 185
synthesis stage, viral replication 97-9
syphilis
    causative organism 75, 189
    in history 109, 116, 353
    transmission 258T, 263
    treatment 123T, 131, 189, 316

T

T cell receptors 288-90
T cells 287-90, 293, 295, 300, 302F
    cytotoxic T cells (Tc cells) 289, 290, 295, 300, 301F
    effector T cells 288F, 289, 291F, 295-6
    helper T cells (Th cells) 287-90, 295, 300-2
    memory T cells 289-90, 295-6
    regulatory T cells (T$_R$ cells) 293
T4 bacteriophage 102, 103F, 104
Taiwan 122, 231
Taq polymerase 366, 419
taxis, positive and negative 48-9
taxonomy 367
    of archaea 72-3, 77
    binomial nomenclature 68
    five-kingdoms system 72

    of fungi 84
    microorganisms 67
    reflecting relatedness 68-9
    use of rRNA genes 70-2, 74, 77, 83
    see also phylogenetics
temperature
    coral bleaching 246
    dry and moist heat 309
    growth rate effects 188-90, 193-5, 308-10, 414
    and the innate immune response 283
    PCR reactions 365-6, 419
    regulation of, as homeostasis 4
temperature tolerance *see* thermophiles
termination sequences/sites 142, 143F, 144
termites 246
tetanus 59, 259, 268, 273, 297T
tetracyclines 63-4, 319-20, 323, 331, 417
Texas red water fever 123T, 128
thalassemia 381
Theileria microti 265
theories, distinguished from hypotheses 13-14
Thermoactinomyces vulgaris 58
thermoclines 232
thermocyclers 366
thermophiles 74-7, 188-90, 193, 194F, 200, 366
thermotaxis 49
Thermus aquaticus 366, 419
Thiomargarita namibiensis 44-5
thrush 87, 254-5, 302, 303T
Thucydides 111
thylakoids 64, 65F, 182
thymidine kinase 327, 328F
thymine dimers 161F, 162, 311
thymus 286F, 287, 288F, 289-90
ticks, as vectors 123T, 128, 263, 265, 352
time characteristics, case definition 351
timelines, for microbiology 123-4T
tissues, as cell groupings 59
TNT (trinitrotoluene) 425
tobacco mosaic virus 102F, 129
Toll-like receptors (TLRs) 269, 282-3, 285
Tolypocladium inflatum 419T
toxins
    aflatoxin 162
    Bacillus thuringiensis 420
    bioremediation of environmental 423
    C. difficile 333
    cholera 104, 266-8
    Claviceps purpurea 273
    Clostridium tetani 59, 268
    Corynebacterium diphtheriae 191
    from dinoflagellates 236
    E. coli O157:H7 (Shiga toxin) 158, 341, 368-70
    exotoxins and endotoxins 267
    Gram-negative LPS as 52-3, 268-9, 318
    oxygen as 49
    from pathogens 266-73
    Pseudomonas aeruginosa 271
    see also botulism
Toxoplasma gondii 79, 81-2, 216
trace elements and growth rate 191
trachoma 76, 252F, 315
transcription/transcription factors
    in eukaryotes and bacteria 144-5
    in gene expression 142-5
transduction 157-8, 333, 341, 370, 377
transformation in bacteria 156, 333, 376

transforming factor 155-6
transgenic animals 372, 376, 379-80, 382-6
transgenic bacteria 376, 380
transgenic plants 382-3
translation
    in gene expression 142, 145-8
    inhibition by antibiotics 324
    inhibition by interferons 284
    protein assembly in 145-9
transmembrane proteins 56
transmission
    animal-to-human 116
    human-to-human 348-9, 358, 359F
    modes, for pathogens 261-4
    population immunity and 343-5
    see also vector-borne diseases
transport vesicles 62-4
transposons 157, 158F, 209
tree of life see phylogenetic trees
Treponema pallidum 75, 189, 258T, 316
Trichoderma reesi 428
trimethylamine oxide 176
tRNA (transfer RNA) 147, 323
trophic levels 223-4, 245
trophozoites 79, 81
trp operon 151
trypanosomes 60, 272
Trypanosoma brucei 59F
Trypanosoma cruzi 258T
tsetse fly 117, 263, 274F
tuberculosis
    and AIDS 360
    causative agent 53, 123T, 128
    as a chronic disease 115T
    drug-resistant 135, 153, 321, 324, 331
    HIV co-infection 265, 303T, 360
    inflammatory response 266
    in the New World 365, 367
    potential eradication 251
    rifampin and 324
    transmission characteristics 258T
    vaccine 297T
tulips 102
tumors see cancers
turkeys 334, 380
twin studies 256
Typhoid Mary (Mary Mallon) 259
typhoid virulence variations 216
typhus 117-18, 121, 339

U
UFO sightings 165
ultraviolet (UV) radiation
    DNA damage 163, 311, 312F, 414
    on the early Earth 74, 203-4
    in electrophoresis 367F
    phage reactivation 104
uncoating 97, 327F
unicellular organisms defined 3
unsaturated fats 32, 33F
uracil 38F, 39, 143
uranium 18, 424
urbanization and disease susceptibility 111, 117, 355

V
vaccination and epidemic probability 344-5
vaccines
    development 129-30

Ebola virus 386
HIV and AIDS 302
inactivated and attenuated viruses 123T, 130, 297-8, 302, 381
influenza 347
measles 337
from recombinant DNA technology 380-1
subunit vaccines 298, 381
valley fever 87, 258T, 259
Vampirococcus spp. 248
Van Leeuwenhoek, Anton 123T, 124
vancomycin 159, 161-3, 322-3, 331, 417
    resistance 212-13, 331
varicella zoster virus 271-2, 297T, 303T
vasodilation 283
vCJD 106, 107F
vector-borne diseases
    bubonic plague as 113
    defined 112
    as a transmission mode 263
vectors
    bats 358, 359F
    fleas 113, 215, 258T, 273, 352
    lice 117-18, 339
    mites 319
    myxoma virulence and 215
    parasitic protozoa 6F, 79, 80-1
    ticks 123T, 128, 263, 265, 352
    see also mosquitoes
Vegemite 393-4, 409
vertical gene transfer 154
vertical pathogen transmission 79, 81-2, 263-4, 355
veterinarians 353
Vibrio cholerae 105, 197, 259, 264, 266-8, 274
vinegar 399, 400F
viral envelopes 94-8, 264F, 286, 288, 292-3, 347
viral infections
    antibiotic misuse 332
    asymptomatic 100-1
    of plants 102
viral RNA, recognition by TLRs 282
virions 93, 95-9, 101F, 103-4, 300
virulence
    fimbriae and 158F
    genes for 368-70
    natural selection and 214-16
    quorum sensing and 152, 270-1
    vaccine reversion to 298
virulence factors 47-8, 57, 158F, 267, 271, 303
viruses
    as acellular obligate parasites 6, 83, 92
    basic structures 93-4
    host specificity 94-5
    origins 208-9
    replication stages 95-9
    sizes 92, 93F
    whether living 7, 93
    see also bacteriophages; vaccines
virus diseases
    cytopathic effects 101, 271
    Ebola 2, 88, 251, 353, 354F, 358-9
    myxoma and resistance 214-15
    newsworthiness 91
    yellow fever as 129
    Zika 2, 354-9
    see also AIDS; influenza
vitamins
    B vitamins 175, 394, 403-5, 414
    in protein synthesis 191

Volvox 208F
vomit 254, 261–2, 325
VRSA (vancomycin-resistant *Staphylococcus aureus*) 331
VSV (vesicular stomatitis virus) 99F

W
'walking pneumonia' 146, 323, 378
Warren, Robert 26
wasps 218
water
    characteristics of liquid 26
    as essential to life 25
water-borne diseases
    cholera identified as 197, 338, 339F
    Entamoeba histolytica 273
    food and water-borne 261–3, 315, 341, 390
    sanitation and virulence 216F, 315
watercourses, ecology of 233
waxes, acid-fast cell walls 53
Weil-Felix test 121F
Weizmann, Chaim 120–1
West Nile virus 263, 329, 350–3, 354F, 357–9
whey, in cheesemaking 402, 415
white nose syndrome 2, 86–7
white spot (fish) 80
whole genome sequencing 368–9
winemaking 123T, 125–7, 392, 396–7
witchcraft 273
Wolbachia spp. 218
World Health Organization 251, 260, 346, 348–9, 356, 359F

World War I 120, 346, 413
World War II 121, 317, 325, 343
wort 397

X
Xenopus laevis 364

Y
yeasts in breadmaking 390, 392, 395–6
    see also Saccharomyces
yellow fever
    in African colonization and in Haiti 117
    in Cuba 260
    mosquito vector 123T, 128–9
    and the Panama Canal 119, 260, 353
    transmission characteristics 258T
    vaccine 297T
Yersin, Alexander 123T, 128
Yersinia pestis 123T, 128, 258T
yogurt 193–5, 391, 401–2, 407–8

Z
zanamivir 327–8
zebra fish 379F
zidovudine (AZT) 327F, 328–9
Zika virus 2, 354–9
zoonoses 257–8, 273, 353, 359
zooplankton 230, 234
zooxanthellae 245, 246
zygomycota 85–6